Texts and Monographs in Physics

Springer
Berlin
Heidelberg
New York
Barcelona
Budapest
Hong Kong
London
Milan
Paris
Santa Clara
Singapore
Tokyo

Professor Dr. Rudolf Haag

Waldschmidtstrasse 4b, D-83727 Schliersee-Neuhaus, Germany

Editors

Library of Congress Cataloging-in-Publication Data. Haag, Rudolf, 1922– Local quantum physics : fields, particles, algebras / Rudolf Haag, – 2nd rev. and enl. ed. p. cm. – (Texts and monographs in physics) Includes bibliographical references and index. ISBN 3-540-61049-9 (Berlin : alk. paper) 1. Quantum theory. 2. Quantum field theory. I. Title. II. Series. QC174.12.H32 1996 530.1'2–dc20 96-18937

ISBN 978-3-540-61049-6 ISBN 978-3-642-61458-3 (eBook)
DOI 10.1007/978-3-642-61458-3

ISSN 0172-5998

ISBN 978-3-540-61049-6 2nd Edition (Softcover)
Springer-Verlag Berlin Heidelberg New York

ISBN 978-3-540-53610-9 1st Edition (Hardcover) Springer-Verlag Berlin Heidelberg New York

Springer-Verlag Berlin Heidelberg New York
a member of BertelsmannSpringer Science+Business Media GmbH

Typesetting: Data conversion by K. Mattes, Heidelberg
SPIN: 10841432 (soft) / 10542737 (hard) 55/3111-5 4 3 2 1 - Printed on acid-free paper

Dedicated to Eugene P. Wigner in deep gratitude
and to the memory of
Valja Bargmann and Res Jost

Preface to the Second Edition

The new edition provided the opportunity of adding a new chapter entitled "Principles and Lessons of Quantum Physics". It was a tempting challenge to try to sharpen the points at issue in the long lasting debate on the Copenhagen Spirit, to assess the significance of various arguments from our present vantage point, seventy years after the advent of quantum theory, where, after all, some problems appear in a different light. It includes a section on the assumptions leading to the specific mathematical formalism of quantum theory and a section entitled "The evolutionary picture" describing my personal conclusions. Altogether the discussion suggests that the conventional language is too narrow and that neither the mathematical nor the conceptual structure are built for eternity. Future theories will demand radical changes though not in the direction of a return to determinism. Essential lessons taught by Bohr will persist. This chapter is essentially self-contained.

Some new material has been added in the last chapter. It concerns the characterization of specific theories within the general frame and recent progress in quantum field theory on curved space-time manifolds. A few pages on renormalization have been added in Chapter II and some effort has been invested in the search for mistakes and unclear passages in the first edition.

The central objective of the book, expressed in the title "Local Quantum Physics", is the synthesis between special relativity and quantum theory together with a few other principles of general nature. The algebraic approach, that is the characterization of the theory by a net of algebras of local observables, provides a concise language for this and is an efficient tool for the study of the anatomy of the theory and of the relation of various parts to qualitative physical consequences. It is introduced in Chapter III. The first two chapters serve to place this into context and make the book reasonably self-contained. There is a rough temporal order. Thus Chapter I briefly describes the pillars of the theory existing before 1950. Chapter II deals with progress in understanding and techniques in quantum field theory, achieved for the most part in the 1950s and early 1960s. Most of the material in these chapters is probably standard knowledge for many readers. So I have limited the exposition of those parts which are extensively treated in easily available books to a minimum: in particular I did not include a discussion of the path integral and functional integration techniques. Instead I tried to picture the panorama, supply continuous lines of argument, and discuss concepts, questions, and conclusions so that, hopefully,

Chapters I–III by themselves may serve as a useful supplementary text for standard courses in quantum field theory. Chapters IV–VI address more advanced topics and describe the main results of the algebraic approach.

The remarks on style and intentions of the book are quoted below from the preface to the first edition.

April 1996 *Rudolf Haag*

From the Preface to the First Edition

... Physical theory has aspects of a jigsaw puzzle with pieces whose exact shape is not known. One rejoices if one sees that a large part of the pieces fits together naturally and beautifully into one coherent picture. But there remain pieces outside. For some we can believe that they will fit in, once their proper shape and place has been recognized. Others definitely cannot fit at all. The recognition of such gaps and misfits may be a ferment for progress. The spirit of tentative search pervades the whole book.

The book is not addressed to experts. I hope that at least the essential lines of argument and main conclusions are understandable to graduate students interested in the conceptual status of presently existing theory and sufficiently motivated to invest some time, thinking on their own and filling in gaps by consulting other literature.

Local Quantum Physics is a personal book, mirroring the perspective of the author and the questions he has seriously thought about. This, together with the wish to keep the volume within reasonable bounds implied that many topics could only be alluded to and that I could not attempt to produce a well balanced list of references.

Thanks are due to many friends for valuable advice and correction of errors, especially to Detlev Buchholz, Klaus Fredenhagen, Hans Joos, Daniel Kastler, Heide Narnhofer, Henning Rehren, John Roberts, Siegfried Schlieder, and Walter Thirring. I am grateful to Wolf Beiglböck for his unwavering support of the project at Springer-Verlag ..., to Michael Stiller, Marcus Speh, and Hermann Heßling for their perseverance in preparing files of the manuscript.

Last but not least I must thank Barbara, without whose care and constant encouragement this book could not have been written.

November 1991 *Rudolf Haag*

Organization of the Book

The book is divided into chapters and sections. Some sections, but not all, are subdivided. Thus III.2.1 denotes Subsection 1 of Section 2 in Chapter III. Equations are consecutively numbered in each section. For example (III.2.25) is equation number 25 in Section 2 of Chapter III. All types of statements (theorems, definitions, assumptions ...) are consecutively numbered in each *subsection* and the chapter number is suppressed; thus lemma 3.2.1 is the first statement in Subsection 3.2 of the running chapter. The index is divided into three parts:

1. Bibliography (listing books).

2. Author Index and References; it combines the listing of authors with the quotation of articles in journals and the indication of the place in the text (if any) where the respective article is referred to.

3. Subject Index. Only those page numbers in the text are given where the item is defined or first mentioned or where some significant further aspect of it appears.

References from the text to the bibliography are marked in italics with the full name of the author written out; references to journal articles are indicated by an abbreviated form of an author's name and the year of publication such as [Ara 61].

For the benefit of the reader who likes to start in the middle, here are a few remarks on notation and specific symbols. Vectors and matrix elements in Hilbert space are usually written in the Dirac notation: $|\Psi\rangle$, $\langle\Psi|A|\Phi\rangle$. In the case of vectors the bracket is sometimes omitted. Generically, the symbol $\mathcal{A}$ is used to denote a general *-algebra, whereas the symbol $\mathfrak{A}$ is used in the case of a C^*-algebra generated by observables and $\mathcal{R}$ is used for a von Neumann ring or W^*-algebra.

Contents

IV. Charges, Global Gauge Groups and Exchange Symmetry 149

V. Thermal States and Modular Automorphisms 199

I. Background

I.1 Quantum Mechanics

Quantum mechanics was born in two different guises, both inspired by the problem of explaining the existence of stable energy levels in an atom. Heisenberg had started from the idea that, in some sense, the position coordinates $x^{(i)}$ and momentum coordinates $p^{(i)}$ of electrons should be replaced by doubly indexed arrays $x^{(i)}_{nm}$, $p^{(i)}_{nm}$ where the indices n, m label the energy levels and the numbers $x^{(i)}_{nm}$, $p^{(i)}_{nm}$ are related to the intensity of spectral lines for a transition between levels n and m. It was quickly realized by Born and Jordan that, mathematically, the double indexed schemes should be regarded as infinite matrices and the essential relations could be simply expressed in terms of matrix multiplication. Schrödinger's starting idea is seen in the title of his first paper on the subject: "Quantization as an eigenvalue problem". Following up the suggestion by de Broglie he assumed that classical mechanics results from a more fundamental wave theory by the same approximation by which geometric optics results from wave optics. He replaced the Hamilton function of classical mechanics by a differential operator acting on a wave function. Its eigenfunctions should correspond to the stationary states, its eigenvalues to the allowed energy levels.

The equivalence of the two approaches and the underlying common abstract structure emerged soon by the work of Schrödinger, Dirac, Jordan, v.Neumann. Abstractly one considers a Hilbert space $\mathcal{H}$ and linear operators acting on it. To each measurable quantity of the classical theory there corresponds a self adjoint operator acting on $\mathcal{H}$. The spectral values of this operator are the possible values which may occur in a measurement of the quantity in question. It is irrelevant for the physical consequences how the Hilbert space is described and which basis one chooses in it. One only needs a characterization of the "observables" i.e. the operators corresponding to the measurable quantities. Two formulations using differently described Hilbert spaces $\mathcal{H}^1$, $\mathcal{H}^2$ are equivalent if there exists a unitary map U from $\mathcal{H}^1$ onto $\mathcal{H}^2$ such that the operators $A^{(1)}$, $A^{(2)}$ corresponding to the same observable are related by

$$A^{(2)} = UA^{(1)}U^{-1}. \qquad \text{(I.1.1)}$$

Some comments are warranted here. First, the characterization of a specific theory proceeds from the knowledge of a classical theory which tells what the relevant observables are and how they are related. Specifically one starts from the canonical formalism of classical mechanics. There is a configuration space whose points are described by coordinates q_k $(k = 1, \ldots n)$ and to each position coordinate there is a canonically conjugate momentum coordinate p_k. The "quantization" consists in replacing these variables by self adjoint operators satisfying *canonical commutation relations*

$$[q_k, q_l] = 0; \quad [p_k, p_l] = 0; \quad [p_k, q_l] = \frac{\hbar}{i}\delta_{kl}. \tag{I.1.2}$$

Next, it can be shown (v.Neumann, Rellich, Stone, H.Weyl) that under natural requirements (irreducibility, sufficient regularity) these relations fix the representation of the operators p_k, q_k in Hilbert space up to unitary equivalence provided that n is finite. This statement is mostly called von Neumann's uniqueness theorem. It does no longer apply if n, the number of degrees of freedom, becomes infinite, a situation prevailing in quantum field theories and in the thermodynamic limit of statistical mechanics. For such systems there exists a host of inequivalent, irreducible representations of the canonical commutation relations which defies a useful complete classification. This problem will have to be faced in later chapters.

Returning to quantum mechanics we must keep in mind that even there the described "quantization" process is far from unique. On the one hand there are interesting observables besides p_k, q_k. In the classical theory they are functions of the canonical variables. There is an ambiguity in defining the corresponding functions of the noncommuting operators (choice of the order of factors in a product). Furthermore the classical theory is invariant under canonical transformations, the quantum description under unitary transformations. The group of classical canonical transformations is not isomorphic to the unitary group of the Hilbert space in which the p_k, q_k are irreducibly represented. This means that the result of the quantization depends on the choice of the classical canonical system. Luckily in practice these ambiguities have not caused a serious problem because one has usually a natural, preferred set of basic variables and simplicity then turns out to be a good guide.

The next comment leads on to the interpretation, the philosophy and conceptual structure of quantum physics. Dirac introduced the term "observable" to indicate that these objects do not normally have any numerical value. A number is produced by the act of measurement which "forces the system to give a definite answer". A "measuring result" can not be interpreted as revealing a property of the system which existed (though unknown to us) prior to the act of measurement. Take the example of a position measurement on an electron. It would lead to a host of paradoxa if one wanted to assume that the electron has some position at a given time. "Position" is just not an attribute of an electron, it is an attribute of the "event" i.e. of the interaction process between the electron and an appropriately chosen measuring instrument (for instance a screen), not of the electron alone. The uncertainty about the position of the electron prior to the measurement is not due to our subjective ignorance.

It arises from improperly attributing the concept of position to the electron instead of reserving it for the event.

In the general abstract scheme this interpretation is generalized as follows: One considers a *physical system* whose behaviour we want to study and, on the other hand, an *observer* with his instruments capable of making *measurements* resulting in unambiguous phenomena (events). The need to subdivide the world into an observed system and an observer has been especially emphasized by Bohr and Heisenberg ("Heisenberg cut"). Bohr points out that because "we must be able to tell our friends what we have done and what we have learnt" it is necessary to describe everything on the observer side, the instruments and phenomena, in common language which may, for brevity and without harm, be supplemented by using concepts and laws from classical physics because of their unambiguous correspondence to common experience and the deterministic character of the laws. Heisenberg showed in examples that to some extent the cut between system and observer may be shifted. Part of the observer's equipment may be included in the definition of the system. But he stressed that always a cut must be made somewhere; a quantum theory of the universe without an outside observer is a contradiction in itself. This is one source of a basic uneasiness most drastically felt in quantum field theory where one would like to take the universe as the system under consideration. Does the terminology "observer", "measurement" mean that the cut is ultimately between the physical world and the mind to which physical laws are not applied? More concretely, does an event never become absolutely factual until the mind intercedes? Schrödinger tried to emphasize this dilemma by his paradox of the cat which is neither dead nor alive until the veterinarian looks at it. Most physicists seem to feel that "it has nothing to do with the mind".[1] Perhaps one has to realize that this problem is due to an overidealization. Replacing the word "measuring result" by "event" it is clear that with a high level of confidence a flash on a scintillation screen, the blackening of a grain on a photographic plate, the death of a cat can (for the purposes of physics) be considered as facts which happen irrespective of the presence of an observer. But can one isolate events with absolute precision? Can one describe the universe as a set of events increasing in number as time goes on? This would necessitate the introduction of irreversibility on a fundamental level and a revision of the concepts of space and time. In quantum mechanics this is circumvented by limiting the class of events considered to the interaction of the "microscopic system" with a macroscopic measuring device. Other idealizations could be considered, depending on the regime of physics under consideration.[2]

In quantum *mechanics* in contrast to quantum *field theory* the "system" is characterized by its material content i.e. typically it is a certain number of electrons, protons or other particles. One may define the system also by the set of *observables* which can be measured on it or by the set of its possible *states*. The term "state" suggests intuitively something like mode of existence for an individual system. However, since the theory is not deterministic and the optimal

---

[1] I am grateful to J.A. Wheeler for this clear statement (private conversation).

[2] We shall return to these questions in Chapter VII.

attainable knowledge of the state allows only the prediction of a probability for each possible result in a subsequent measurement we have to consider statistical ensembles of individual systems in order to test the predictions of the theory. Thus the word *state* really refers to the "source of the system" i.e. to the experimental arrangement by which the system is isolated and influenced prior to the intended measurements. Alternatively speaking, it refers to the statistical ensemble of individual systems prepared by this source. We may also say that the state subsumes our knowledge of (the relevant part of) the past history of the system.

In the mathematical formalism the vectors of unit length in Hilbert space represent states. In fact they represent "pure states" i.e. optimal attainable knowledge (finest preparation). General states, corresponding to less than optimal knowledge are called "mixtures"; they can be mathematically described by positive, self adjoint operators with unit trace ("statistical operator", "density operator"). The special case of a pure state results if this density operator degenerates to a projection operator on a 1-dimensional subspace. Then it is equally well characterized by a unit vector which defines this 1-dimensional subspace. Among the observables, mathematically represented by self adjoint operators on Hilbert space, the conceptually simplest ones are the *propositions* i.e. measurements which have only two alternative outcomes. We may call them yes and no and assign the measuring value 1 to the "yes"-answer and 0 to the "no"-answer. Then such a proposition is mathematically represented by an orthogonal projector, a self adjoint operator E with $E^2 = E$. The probability for a yes-answer in this measurement, given a state with density operator ϱ, is given by

$$p(E;\varrho) = \mathrm{tr}(\varrho E); \tag{I.1.3}$$

the right hand side denotes the trace of the product of the two operators ϱ and E. In the special case of a pure state (unit vector Ψ) this degenerates to

$$p(E) = \langle \Psi, E\Psi \rangle = \|E\Psi\|^2. \tag{I.1.4}$$

The discussion of general observables can be reduced to this situation. We describe it here only for observables with a discrete spectrum. The result of measurement of such an observable (represented by the operator A) can be any one of a discrete set of values a_k. To each outcome a_k we have a projector E_k corresponding to the proposition yes for this particular outcome. These projections are mutually orthogonal

$$E_i E_k = \delta_{ik} E_i \tag{I.1.5}$$

and sum up to the unit operator

$$\sum E_k = \mathbb{1}. \tag{I.1.6}$$

The observable A itself combines the specification of this orthogonal family of projectors $\{E_i\}$ with the assignment of a real number a_i (the measuring result) to each of them. The operator representing the observable is then

$$A = \sum a_i E_i. \tag{I.1.7}$$

Given this operator the projectors E_k and the values a_k are obtained by the spectral resolution of A. The probability of result a_k in a measurement on a state ϱ is given by $\operatorname{tr}(\varrho E_k)$. The mean value or *expectation value* of the observable A in the state ϱ is given by

$$\langle A \rangle_\varrho = \sum a_i p(E_i; \varrho) = \operatorname{tr} A\varrho. \tag{I.1.8}$$

It should be noted that due to (I.1.5), (I.1.6) and (I.1.7)

$$A^2 = \sum a_i^2 E_i, \tag{I.1.9}$$

or generally, if F is any real valued function of one real variable,

$$F(A) = \sum F(a_i) E_i. \tag{I.1.10}$$

Thus the change from the operator A to the operator $F(A)$ does not mean that we change the apparatus; it only labels the measuring results differently, assigning the value $F(a)$ to the event which was formerly labeled by the value a. Therefore it would be more appropriate to consider the Abelian algebra generated by A rather than A itself as the mathematical representor of the measuring apparatus. The choice of a particular operator in this algebra only fixes a particular labeling of the measuring values.

Extending this argument Segal [Seg 47] proposed to consider as the primary object of the mathematical formalism of quantum mechanics the algebra generated by the *bounded* observables, equipped with the norm topology. This leads to the notion of a C*-algebra which will be defined in Chapter III, section 2 and used extensively afterwards. In this view Hilbert space appears only in a secondary rôle, as the representation space of the algebra. In ordinary quantum mechanics where one deals with a unique equivalence class of representations, the C*-algebra formulation and the Hilbert space formulation are equivalent.

One of the most salient aspects of the quantum mechanical description is the so called *superposition principle*. Probabilities are calculated as the absolute squares of amplitudes. In the naive picture of matter waves the superposition of two waves, corresponding to the addition of wave functions, has a simple intuitive meaning. The intensity, being given by a quadratic expression in the wave function, shows the interference phenomena familiar from wave optics. In quantum mechanics this superposition, the fact that a linear combination of state vectors with complex coefficients gives again a state vector, is somewhat more mysterious. On the one hand, if the system consists of more than one particle, the state vector describes a wave in configuration space, not in ordinary space. Secondly, while in Maxwell's electromagnetic theory of light the components of the wave function are in principle measurable quantities, namely electric or magnetic field strengths, the value of the quantum mechanical wave function can not possibly be measured. In fact the state vectors Ψ and $\lambda\Psi$ where λ is a phase factor (complex number of absolute value 1) represent the same physical state. *Pure states correspond to rays rather than vectors* in Hilbert space.

This is physically extremely significant. Otherwise there could be no particles with half integer spin. Classical electrodynamics is not the quantum theory of a single photon but a correspondence limit of the quantum theory for infinitely many photons. We must recognize therefore that the quantum mechanical superposition principle does not have a simple operational meaning. Is it possible to understand by some deeper lying principle why the formalism is based on a linear space with complex numbers as coefficients? Is the linearity due to an approximation and is the field of complex numbers just the simplest possibility? We shall defer such questions to Chapter VII.

Let us first return to the correspondence between classical and quantum mechanics. It is a striking fact that notwithstanding the revolutionary change in our understanding of the laws of nature the formal correspondence between quantum mechanical relations and those of the canonical formalism in classical mechanics is very close. This hint was followed in the subsequent development of quantum electrodynamics and further in quantum field theories of elementary particles where one is way beyond the regime in which experimental information about an underlying classical theory exists. The idea that one must first invent a classical model and then apply to it a recipe called "quantization" has been of great heuristic value. In the past two decades the method of passing from a classical Lagrangean to a corresponding quantum theory has shifted more and more away from the canonical formalism to Feynman's path integral. This provides an alternative (equivalent?) recipe. There is, however, no fundamental reason why quantum theory should not stand on its own legs, why the theory could not be completely formulated without regard to an underlying deterministic picture. Sometimes Niels Bohr's epistemological arguments are interpreted as denying this possibility as a matter of principle. Bohr emphasizes the need for a common language in which phenomena can be objectively described. This is certainly a prerequisite for doing physics. But, while he used the term "classical" in this context this does not mean that one needs a full fledged deterministic mirror theory from which the true one can be obtained by "quantization". In fact, Bohr did not believe that a classical picture of half integer spin was useful.[3] A simple but impressive example of a self contained quantum theory is provided by Wigner's analysis sketched in section 3 of this chapter.

The quantum mechanical formalism is an extension of ordinary probability theory in the following sense. In standard probability theory one has an event space Σ which is a set of subsets of a (total) set Ω. The elements of Σ correspond to events or propositions. Σ forms a Boolean algebra with unit element Ω; the union of two subsets in Σ (event 1 *or* event 2) and the intersection of subsets (event 1 *and* event 2) are again elements of Σ. Ω itself corresponds to an event (the trivial, the certain one). On Σ one has a probability measure. This measure defines in a natural way an extension of the Boolean algebra to an Abelian W*-algebra **M** (this concept is defined in Chapter III). The measure may be regarded as a normalized, positive linear form on **M** (a "state"). The quantum mechanical generalization starts from the fact that one is dealing with

[3] Private conversation 1954.

a large family of event spaces simultanously (one for every maximal observable) and with a large family of states. Each state defines a probability measure on all event spaces. In this one assumes that one can measure any observable in any state.[4]

Within the set of states the process of mixing i.e. the throwing together of several ensembles with arbitrary weights is an operationally well defined procedure, inherent in the probability concept. With any pair of states, abstractly denoted here by ω_1, ω_2 one has the family of states

$$\omega_\lambda = \lambda\omega_1 + (1-\lambda)\omega_2; \quad 0 \leq \lambda \leq 1, \tag{I.1.11}$$

the mixtures of ω_1 and ω_2 with weights λ and $1-\lambda$ respectively. This implies that the set of states is a convex body $\mathcal{K}$. Linear combinations of states with positive coefficients adding up to 1 are again states and the linearity is respected in the probabilities for all subsequent measuring results. One may embed this convex body in a real linear space V by considering formal linear combinations of states with arbitrary real coefficients. The part $V^+ = \{a\mathcal{K} : a \geq 0\}$ is a convex cone in V. The normalization of states is provided by a distinguished linear form e on V so that $\mathcal{K} = \{x \in V^+ : e(x) = 1\}$. The dual space of V, denoted by V^*, is the set of all linear forms on V. It also has a positive cone $V^{*+} = \{A \in V^* : A(\omega) \geq 0 \text{ for all } \omega \in \mathcal{K}\}$. The elements of V^* correspond to observables and $A(\omega)$ to the expectation value of A in the state ω.

The general structure expresses essentially only the probabilistic setting and the assumption that an arbitrary observable can be measured in any state. The specific mathematical structure of Quantum Mechanics where V^+ is the set of positive operators of trace class on a Hilbert space and V^* the set of self adjoint operators is not so easily derivable from operational principles. Much effort has been devoted to this and we shall describe some of it in Chapter VII. Here it may suffice to indicate the most important deviation of quantum mechanical calculus from its classical counterpart. The convex set $\mathcal{K}$ is not a simplex. This means that the same mixed state can arise in many different ways as a convex combination (mixture) of pure states. It does not uniquely define the pure components contained in it.

I.2 The Principle of Locality in Classical Physics and the Relativity Theories

The German term "Nahwirkungsprinzip" is more impressive than the somewhat colourless word "locality". Certainly the idea behind these words, proposed

[4]Note that, to justify this, it is important to distinguish between observables and *observation procedures*. Many procedures yield the same observable. An observable is an equivalence class of procedures; for instance two different measuring devices operated at different times may correspond to the same observable due to the existence of a dynamical law. This seemingly evident distinction will turn out to be important also in several other contexts. Attention to this was drawn by Ekstein [Ek 69].

by Faraday around 1830, initiated the most significant conceptual advance in physics after Newton's Principia. It guided Maxwell in his formulation of the laws of electrodynamics, was sharpened by Einstein in the theory of special relativity and again it was the strict adherence to this idea which led Einstein ultimately to his theory of gravitation, the general theory of relativity.

In contrast to the picture underlying mechanics, whereby material bodies influence each other by forces acting at a distance, one assumes that all points of space participate in the physical processes. Effects propagate from point to neighbouring point. The vehicle by which this idea is put to work is the concept of fields. Each point in space is thought to be equipped with dynamical variables, the field quantities. In electromagnetic theory they are the vectors $\mathbf{E}$ and $\mathbf{B}$ of electric and magnetic field strength. The knowledge of these as functions of the position in space and time is the goal in the description of a particular physical situation. All laws governing their behaviour are strictly local i.e. they are a system of partial differential equations.[1] Initial conditions and boundary conditions may be freely varied thus allowing the great variety of possible physical situations. From the specific form of the field equations in electrodynamics, the Maxwell equations, two further properties can be abstracted which were raised to the status of general principles, obligatory for all theories, by Einstein in the development of the concepts of special relativity. The first is the existence of a limiting velocity for the propagation of effects (hyperbolic character of the field equation). The second is the invariance under the Lorentz group.

I.2.1 Special Relativity, Poincaré Group, Lorentz Group, Spinors, Conformal Group

The basic observation leading to special relativity is the following: the comparison of times at different places in space is not an objectively well defined procedure if all laws are local and the speed of signals is limited. It needs a convention for the synchronization of different clocks and there is a certain amount of arbitrariness in this convention. This prevents us from considering space and time as two distinct, objectively meaningful, concepts. Rather we must consider the 4-dimensional manifold of possible pointlike events, the 4-dimensional space-time manifold. This consequence of Einstein's argument about the relativity of the notion of equal time was first emphasized by Minkowski. Still the distinction of past and future of such a pointlike event remains objectively meaningful. Associated to a point x there is a division of space-time into three parts:[2] the forward cone V^+ containing all events which can be causally influenced from x, the backward cone V^- containing all events from which an influence on x can come and the complement S of these two cones, the "space-like region" to the point x. The latter contains all points which can have no causal relationship with x (Fig. I.2.1.). One may call V^+ the future, V^- the past and S the

[1] This last conclusion was not drawn by Faraday but is due to Maxwell.

[2] I apologize for having used the symbol V^+ in different context before.

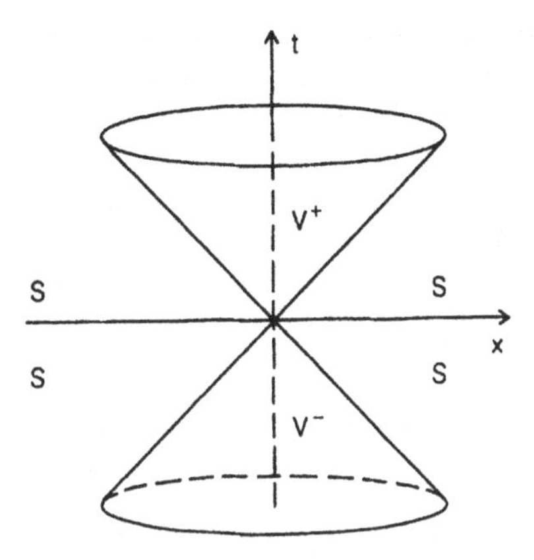

Fig. I.2.1.

present of the event x. The boundary of V^+ is formed by all possible events which can be reached by the fastest possible signals sent from x and these are identified with light signals in physics. The locality principle is thus sharpened to Einstein's causality principle: No physical effect can propagate faster than light.

In special relativity one assumes that the causal structure of space-time is a priori globally given. There is supposed to exist a preferred class of coordinate systems, the "inertial systems" in which the causal future of a point $x = (t, \mathbf{x})$ consists of all points $x' = (t', \mathbf{x}')$ with

$$t' - t \geq |\mathbf{x}' - \mathbf{x}|. \tag{I.2.1}$$

Notation. In the context of special relativity we shall always use inertial coordinates. As implied in (I.2.1) we choose units in which the velocity of light equals unity, i.e. times are measured in cm (1 s = 3×10^{10} cm). For a space-time point we denote the time coordinate by x^0, the three Cartesian space coordinates by x^i (i =1, 2, 3). A Euclidean 3-vector with components x^i is denoted by a boldface letter $\mathbf{x}$. Always Roman indices $i, k, \cdots$ will be employed in this context to label space components; Greek indices $\alpha, \beta, \ldots \mu, \nu, \ldots$ will denote both space and time components and run from 0 to 3. The use of upper indices for the coordinates is conventional; it is the customary usage in general relativity. We use the metric tensor

$$g_{\mu\nu} = \begin{pmatrix} 1 & 0 & 0 & 0 \\ 0 & -1 & 0 & 0 \\ 0 & 0 & -1 & 0 \\ 0 & 0 & 0 & -1 \end{pmatrix} \tag{I.2.2}$$

and the summation convention: if an index appears in an expression twice, once in an upper position and once in a lower position, a summation over the values of this index is automatically implied (contraction of indices). The "Lorentz distance" square of two space-time points, replacing the Euclidean distance square of two points in 3-dimensional space, is then given by

$$\Delta x^2 = g_{\mu\nu}\,(x'^{\mu} - x^{\mu})\,(x'^{\nu} - x^{\nu}). \tag{I.2.3}$$

It is positive for time-like separation, negative for space-like separation and zero if x and x' can be connected by a light signal. The 4-dimensional continuum equipped with the (indefinite) metric (I.2.2), (I.2.3) is called Minkowski space and will be denoted by $\mathcal{M}$.

Poincaré Group, Lorentz Group. The invariance group of Minkowski space is the Poincaré group, also called the "inhomogeneous Lorentz group". We may understand invariance in two different ways. Either we consider mappings from $\mathcal{M}$ to $\mathcal{M}$ described in a fixed coordinate system, where the point x is shifted to $x' = gx$. This is the active interpretation. g is then a diffeomorphism of $\mathcal{M}$. Or we may consider g as a coordinate transformation so that x and x' denote the same point in two different coordinate systems (passive interpretation). In the present context it does not matter which interpretation we choose. The invariance group consists of all maps g which do not change the Lorentz distance between any two points, or alternatively, in the passive interpretation, which keep the formal expression for the Lorentz distance unchanged by the transformation. This implies that[3]

$$x'^{\mu} = a^{\mu} + \Lambda^{\mu}_{\ \nu} x^{\nu}, \tag{I.2.4}$$

where a^{μ} and $\Lambda^{\mu}_{\ \nu}$ are constant and Λ satisfies

$$g_{\mu\nu}\Lambda^{\mu}_{\ \alpha}\Lambda^{\nu}_{\ \beta} = g_{\alpha\beta}. \tag{I.2.5}$$

If we regard $\Lambda^{\mu}_{\ \nu}$ and $g_{\mu\nu}$ as 4×4 matrices and Λ^T denotes the transposed matrix (I.2.5) can be written as

$$\Lambda^T \mathbf{g} \Lambda = \mathbf{g}. \tag{I.2.6}$$

The general element $g = (a, \Lambda)$ of the Poincaré group $\mathfrak{P}$ consists of a translation 4-vector a^{μ} and a Lorentz matrix Λ. The (full) Lorentz group may be described as the set of all real 4×4 matrices satisfying (I.2.6). This relation implies

$$(\det \Lambda)^2 = 1 \;\rightarrow \det \Lambda = \pm 1 \tag{I.2.7}$$

$$(\Lambda^0_{\ 0})^2 = 1 + \sum_i (\Lambda^i_{\ 0})^2; \quad \Lambda^0_{\ 0} \left\{ \begin{array}{l} \geq +1 \\ \leq -1 \end{array} \right. . \tag{I.2.8}$$

Thus the full Lorentz group consists of four disconnected pieces depending on the sign combinations of $\det \Lambda$ and $\Lambda^0_{\ 0}$. The branch which is connected to the identity, i.e. the branch with $\det \Lambda = 1$, $\Lambda^0_{\ 0} \geq 1$ will be denoted by $\mathfrak{L}$ and we

[3] See the subsequent discussion of the conformal group.

shall reserve in the sequel the term Lorentz group for this branch. The other branches are obtained from $\mathfrak{L}$ by reflections in time and (or) space. These will always be separately discussed. The question as to whether and in which way the reflections correspond to physical symmetries has been an exciting issue in fundamental physics during the past decades.

The use of both upper and lower indices and of the "metric tensor" (I.2.2) appears to be unduly cumbersome for computations. It is avoided mostly in the early literature by the Minkowski trick of using instead of the time coordinate x^0 the imaginary coordinate $x^4 = ix^0$. The (negative) metric tensor will then have the Euclidean form $\delta_{\alpha\beta}$ $(\alpha, \beta = 1, \cdots 4)$ and one may identify upper and lower indices. Conceptually, however, the notation is very appropriate and eases the transition to general relativity. We have to remember that in differential geometry a vector field $V^\mu(x)$ maps a manifold into the tangent bundle.[4] In other words, for a given point $x \in \mathcal{M}, V^\mu(x)$ is an element of the tangent space at x. On the other hand a "covariant vector field" $W_\mu(x)$, also called a 1-form, maps into the cotangent bundle. It gives a cotangent vector at x, a linear function on tangent space, assigning to the tangent vector V the number $W(V) = W_\mu V^\mu$. The metric provides a mapping between tangent and cotangent spaces. From the vector V^μ we may go over to the covector $V_\mu = g_{\mu\nu} V^\nu$ and, vice versa, we can define $W^\mu = g^{\mu\nu} W_\nu$ where $g^{\mu\nu}$ is the inverse matrix to $g_{\mu\nu}$. In special relativity $\mathbf{g}$ is the constant numerical matrix (I.2.2) which is its own inverse; so $g^{\mu\nu}$ is also given by (I.2.2).

The fields appearing in classical physics have simple transformation properties under the Poincaré group $\mathfrak{P}$. Their value at a point refers to the tangent space at this point in a way described by a certain number of upper and lower indices. For a scalar field $\Phi(x)$ (no indices) the transformed field under $g = (a, \Lambda)$ is

$$\Phi'(x) = \Phi(g^{-1}x). \tag{I.2.9}$$

For a vector field:

$$V'^\mu(x) = \Lambda^\mu_{\ \nu} V^\nu(g^{-1}x), \tag{I.2.10}$$

for a 1-form

$$W'_\mu = \tilde{\Lambda}_\mu^{\ \nu} W_\nu(g^{-1}x), \tag{I.2.11}$$

where the matrix $\tilde{\Lambda}$ is the "contragredient" to Λ i.e.

$$\tilde{\Lambda} = (\Lambda^T)^{-1}. \tag{I.2.12}$$

For tensor fields we have one factor Λ resp. $\tilde{\Lambda}$ for each index e.g.

$$T'^{\ \nu}_{\mu}(x) = \tilde{\Lambda}_\mu^{\ \alpha} \Lambda^\nu_{\ \beta} T_\alpha^{\ \beta}(g^{-1}x). \tag{I.2.13}$$

---

[4]The reader unfamiliar with these concepts may find an easily readable introduction in [*Warner 1971*] or [*Thirring 1981, Vol.1*].

Spinors. For many physical consequences, possibly even for the origin of Lorentz symmetry, it is of great importance that $\mathfrak{L}$ is isomorphic in the small to the group SL(2,$\mathbb{C}$), the group of complex 2×2-matrices with determinant 1. This connection can be most easily seen if one arranges the four components of a 4-vector V in the form of an Hermitean 2×2-matrix $\widehat{V}$:

$$\widehat{V} = \begin{pmatrix} V^0 + V^3 & V^1 - iV^2 \\ V^1 + iV^2 & V^0 - V^3 \end{pmatrix} \tag{I.2.14}$$

and notes that

$$\det(\widehat{V}) = g_{\mu\nu} V^\mu V^\nu. \tag{I.2.15}$$

A transformation

$$\widehat{V} \to \widehat{V}' = \alpha \widehat{V} \alpha^*, \tag{I.2.16}$$

where α is any complex 2×2-matrix with determinant 1 and α^* its Hermitean adjoint, leads to an Hermitean $\widehat{V'}$ with the same value of the determinant. Hence it defines a real linear transformation of the vector components

$$V'^\mu = \Lambda^\mu_{\ \nu} V^\nu \tag{I.2.17}$$

which conserves the Lorentz square of V. Thus Λ in (I.2.17) is a Lorentz matrix. To compute it we write (I.2.14) as

$$\widehat{V} = V^\mu \sigma_\mu \tag{I.2.18}$$

with

$$\sigma_0 = \begin{pmatrix} 1 & 0 \\ 0 & 1 \end{pmatrix}, \quad \sigma_1 = \begin{pmatrix} 0 & 1 \\ 1 & 0 \end{pmatrix},$$

$$\sigma_2 = \begin{pmatrix} 0 & -i \\ i & 0 \end{pmatrix}, \quad \sigma_3 = \begin{pmatrix} 1 & 0 \\ 0 & -1 \end{pmatrix}. \tag{I.2.19}$$

One has

$$V^\mu = \frac{1}{2} \operatorname{tr} \widehat{V} \sigma_\mu; \quad \Lambda^\mu_{\ \nu}(\alpha) = \frac{1}{2} \operatorname{tr} \alpha \sigma_\nu \alpha^* \sigma_\mu. \tag{I.2.20}$$

A closer study shows that as α runs through all SL(2,$\mathbb{C}$) the corresponding $\Lambda(\alpha)$ runs twice through $\mathfrak{L}$. Obviously α and $-\alpha$ give the same Λ. As a topological group $\mathfrak{L}$ is not simply connected. There are closed paths in $\mathfrak{L}$ which cannot be continuously contracted to a point. An example is a rotation by 2π which is an uncontractible path in $\mathfrak{L}$ from the identity to the identity transformation. SL(2,$\mathbb{C}$) on the other hand is simply connected. It is the covering group of $\mathfrak{L}$ and we shall denote it therefore by $\overline{\mathfrak{L}}$. An element of $\overline{\mathfrak{L}}$ may be regarded as an element of $\mathfrak{L}$ together with the homotopy class of a path connecting this element with the identity. There are two classes of paths. Correspondingly we denote by $\overline{\mathfrak{P}}$ the covering group of the Poincaré group. Its elements are (a, α) with $\alpha \in$ SL(2,$\mathbb{C}$) $= \overline{\mathfrak{L}}$. The consideration of $\overline{\mathfrak{L}}$, $\overline{\mathfrak{P}}$ instead of $\mathfrak{L}$, $\mathfrak{P}$ leads to the concept of spinors which play a prominent rôle in relativistic quantum theory.

The defining representation of SL(2,$\mathbb{C}$) is given by the matrices α described above. It acts in a complex 2-dimensional vector space. Elements of this space are called covariant spinors of rank 1 and the components will be denoted by Ψ_r ($r = 1, 2$). Intimately related to this representation there is

a) the complex conjugate representation $\alpha \to \overline{\alpha}$,

b) the "contragredient" representation $\alpha \to \beta = (\alpha^T)^{-1}$,

c) the representation $\alpha \to \overline{\beta}$.

Van der Waerden suggested that spinors transforming under a) should be denoted by a dotted index: $\Psi_{\dot{r}}$, spinors transforming under b) by an upper index: Ψ^r , those transforming under c) by an upper dotted index. This reflects the fact that the representation b) is equivalent to the defining representation, namely one has

$$\beta = \epsilon \alpha \epsilon^{-1}, \tag{I.2.21}$$

or, with fully written indices

$$\beta^r{}_s = \epsilon^{rr'} \alpha_{r'}^{s'} \epsilon_{s's}, \tag{I.2.22}$$

with

$$\epsilon_{rs} = -\epsilon^{rs} = \begin{pmatrix} 0 & 1 \\ -1 & 0 \end{pmatrix}. \tag{I.2.23}$$

So ϵ may be regarded as a metric tensor in spinor space. The representations α and $\overline{\alpha}$ are inequivalent but $\overline{\alpha}$ and $\overline{\beta}$ are again related by ϵ.

Taking tensor products of these representations one comes to spinors of higher rank, having several dotted and undotted, upper or lower indices. One finds that all equivalence classes of finite dimensional irreducible representations of SL(2,$\mathbb{C}$) are obtained by considering spinors with n dotted, m undotted lower indices which are totally symmetric with respect to permutations of each of the two types of indices. If $n + m$ is even such a spinor belongs to a true (single valued) representation of $\mathfrak{L}$, it corresponds to a Lorentz tensor of some rank. The simplest case may be read off from (I.2.14), (I.2.16). A 4-vector V^μ is written in spinorial notation as $V_{\alpha\dot{\beta}}$. The next simple cases, symmetric spinors with two undotted or two dotted indices correspond to (complex) self dual skew symmetric tensors $F_{\mu\nu} \pm i\epsilon_{\mu\nu\varrho\sigma}F^{\varrho\sigma}$ (for the notation see equation (I.2.64) below). For an excellent brief review of spinor calculus see Laporte and Uhlenbeck, [Lap 31] and [*van der Waerden 1932*].

Conformal Group.[5] The metric (I.2.2) combines two aspects. On the one hand it defines the causal structure (Fig. I.2.1.); in addition it defines distances in space-like and time-like directions. One may separate the two aspects and consider the causal structure as more fundamental. If the rest masses of all particles were zero only this aspect would remain. Such an idealization may be of interest as an asymptotic description of the high energy regime and possibly also if one wants to explain the origin of masses.

The group which conserves only the causal structure, not the metric, is the conformal group. We want to determine the transformation formulas between

[5]This will be used only in a few sections and can be skipped at a first reading.

two coordinate systems in which the light cones are characterized by the vanishing of the right hand side of (I.2.3) with $g_{\mu\nu}$ given by (I.2.2). Let us look at a small neighbourhood of a point x and an infinitesimal coordinate transformation.

$$x^\mu \to x^\mu + \varepsilon\, u^\mu(x), \quad \varepsilon \to 0.$$

The line element from x to $x + dx$

$$ds^2 = g_{\mu\nu} dx^\mu dx^\nu$$

changes by

$$\delta ds^2 = \varepsilon\, g_{\mu\nu} \left(\frac{\partial u^\mu}{\partial x^\varrho} dx^\varrho dx^\nu + \frac{\partial u^\nu}{\partial x^\varrho} dx^\varrho dx^\mu \right) = \varepsilon\, G_{\mu\nu} dx^\mu dx^\nu,$$

$$G_{\mu\nu} = \frac{\partial u_\mu}{\partial x^\nu} + \frac{\partial u_\nu}{\partial x^\mu}\,; \quad u_\mu = g_{\mu\nu} u^\nu. \tag{I.2.24}$$

We have used the fact that $g_{\mu\nu}$ is constant. If the form of the light cone equation shall not change by the coordinate transformation we must have

$$G_{\mu\nu}(x) dx^\mu dx^\nu = 0 \quad \text{whenever} \quad g_{\mu\nu} dx^\mu dx^\nu = 0. \tag{I.2.25}$$

Thus $G_{\mu\nu}$ must be a (possibly position dependent) metric tensor which has the same light like directions as $g_{\mu\nu}$. This means that

$$G_{\mu\nu}(x) = \lambda(x) g_{\mu\nu}. \tag{I.2.26}$$

We can eliminate λ by contracting with g

$$\lambda = \frac{1}{4} G_{\mu\nu} g^{\mu\nu}$$

and get the following condition for the displacement field u of the infinitesimal transformation:

$$\frac{1}{2}\left(\frac{\partial u_\mu}{\partial x^\nu} + \frac{\partial u_\nu}{\partial x^\mu} \right) - \frac{1}{4} g_{\mu\nu} \partial_\varrho u^\varrho = 0. \tag{I.2.27}$$

The general solution of this equation can, for smooth functions, be easily found. If one makes a Taylor expansion around the point $x = 0$

$$u_\mu(x) = a^{(1)}_\mu + a^{(2)}_{\mu\nu} x^\nu + a^{(3)}_{\mu\nu\varrho} x^\nu x^\varrho + \cdots \tag{I.2.28}$$

one notes that the homogeneous polynomials of different degree decouple and (I.2.27) gives separate conditions for each a

$$a^{(n)}_{\mu\nu\varrho\sigma\cdots} + a^{(n)}_{\nu\mu\varrho\sigma\cdots} = \frac{1}{2} g_{\mu\nu} a^{(n)\lambda}{}_{\lambda\varrho\sigma\cdots} \tag{I.2.29}$$

with $a^{(n)}_{\mu\nu\varrho\sigma\cdots}$ totally symmetric in the $n-1$ last indices. For $n = 1$ this gives no restriction for $a^{(1)}$. For $n = 2$ one gets

$$a^{(2)}_{\mu\nu} = \omega_{\mu\nu} + \lambda g_{\mu\nu} \quad \text{with} \quad \omega_{\mu\nu} = -\omega_{\nu\mu}. \tag{I.2.30}$$

For $n = 3$ one finds

$$a^{(3)}_{\mu\nu\varrho} = g_{\mu\nu}c_\varrho + g_{\mu\varrho}c_\nu - g_{\nu\varrho}c_\mu. \tag{I.2.31}$$

For $n \geq 3$ there is no solution (if the dimensionality of space-time is higher than 2). This follows from the permutation symmetry of the last indices of $a^{(n)}$.

The admissible form of u is then

$$u^\mu(x) = a^\mu + g^{\mu\lambda}\omega_{\lambda\nu}x^\nu + d\,x^\mu + 2x^\mu c_\lambda x^\lambda - x_\lambda x^\lambda c^\mu, \tag{I.2.32}$$

where a^μ, d, c^μ are arbitrary constants and ω is antisymmetric. The family of differential operators $u^\mu(x)\partial_\mu$ are the generators of a 15-parametric Lie group, the conformal group of 4-dimensional space-time with the causal structure (I.2.1).[6] The part with the parameters $a^\mu, \omega_{\mu\nu}$ generates the Poincaré transformations (I.2.4), the part with d generates dilations, the one with c_μ the "proper conformal transformations", also called "conformal translations"

$$x'^\mu = \frac{x^\mu - (x,x)c^\mu}{1 - 2(x,c) + (x,x)(c,c)}. \tag{I.2.33}$$

These transformations become singular on the submanifold where the denominator vanishes. Therefore they do not give global symmetries of $\mathcal{M}$ but can be defined physically only as local diffeomorphisms in suitable regions for a limited range of the parameters c. One can, however, compactify $\mathcal{M}$ so that the conformal transformations act as diffeomorphisms defined everywhere in the resulting space. The conformal group plays a rôle in the high energy asymptotics of quantum field theory.

Computations involving the conformal group are greatly simplified using a representation of the group by linear transformations. One possibility is to take a 6-dimensional real space (coordinates $\xi^\alpha, \alpha = 0, 1, \cdots 5$) equipped with a metric

$$g_{\alpha\beta} = \left\{ \begin{array}{c} +1 \\ -1 \end{array} \right. \delta_{\alpha\beta} \left\{ \begin{array}{ll} \text{for} & \alpha = 0, 5 \\ \text{for} & \alpha = 1, 2, 3, 4. \end{array} \right. \tag{I.2.34}$$

The group of pseudo-orthogonal transformations SO(4,2)

$$\xi'^\alpha = M^\alpha_\beta \xi^\beta, \tag{I.2.35}$$

with matrices M satisfying

$$g_{\alpha\beta}M^\alpha_\gamma M^\beta_\delta = g_{\gamma\delta}; \quad \det M = 1, \tag{I.2.36}$$

yields such a representation. Its relation to conformal transformations in Minkowski space is established as follows. $M \in$ SO(4,2) transforms the cone

[6] Note that in 2-dimensional space the argument that $a^{(n)} = 0$ for $n > 3$ is no longer true. In this case the conformal group has infinitely many parameters. It is the group of analytic functions of a complex variable with nonvanishing derivative in the domain of definition.

$$C^{(6)} = \{\xi : g_{\alpha\beta}\xi^\alpha\xi^\beta = 0\} \tag{I.2.37}$$

into itself. Defining

$$x^\mu = (\xi^4 + \xi^5)^{-1}\xi^\mu, \quad \mu = 0, 1, 2, 3 \tag{I.2.38}$$

the transformation (I.2.35) induces a transformation $x \to x'$ in $\mathbb{R}^4$ which turns out to be a conformal transformation. Introducing

$$\eta = \xi^4 + \xi^5, \quad \kappa = \xi^4 - \xi^5, \tag{I.2.39}$$

one gets from (I.2.37), (I.2.38)

$$(x, x) = g_{\mu\nu}x^\mu x^\nu = \eta^{-1}\kappa. \tag{I.2.40}$$

The various subgroups:

(i) The 6-parametric subgroup leaving ξ^4 and ξ^5 fixed

$$M = \begin{pmatrix} \Lambda^{\mu\nu} & & \\ & 1 & \\ & & 1 \end{pmatrix}, \quad \Lambda \in \mathfrak{L} \tag{I.2.41}$$

yields the homogeneous Lorentz transformations;

(ii) the 4-parametric subgroup

$$\xi'^\mu = \xi^\mu + \eta a^\mu; \quad \eta' = \eta; \quad \kappa' = \kappa + 2(\xi, a) + \eta(a, a) \tag{I.2.42}$$

yields the translations

$$x'^\mu = x^\mu + a^\mu.$$

(Note that the transformation law of κ follows from the others by (I.2.40));

(iii) the 4-parametric subgroup

$$\xi'^\mu = \xi^\mu - \kappa c^\mu; \quad \kappa' = \kappa; \quad \eta' = \eta - 2(\xi, c) + \kappa(c, c) \tag{I.2.43}$$

yields the "conformal translations" (I.2.33);

(iv) the dilations

$$x'^\mu = \lambda x^\mu$$

result from

$$\xi'^\mu = \xi^\mu; \quad \kappa' = \lambda\kappa; \quad \eta' = \lambda^{-1}\eta. \tag{I.2.44}$$

By a conformal transformation the "Lorentz distance square" between two points $x_1, x_2 \in \mathcal{M}$ is changed by a conformal factor

$$(x_1' - x_2')^2 = N(x_1)^{-1} N(x_2)^{-1} (x_1 - x_2)^2, \tag{I.2.45}$$

where, of course $N = 1$ for Poincaré transformations, $N = \lambda^{-1}$ for dilations. For the "conformal translations" (I.2.43) one finds that N is the denominator in (I.2.33)

$$N = \eta^{-1} \eta'. \tag{I.2.46}$$

Instead of realizing the conformal group by pseudo-orthogonal transformations in a 6-dimensional real space one may also use a realization by pseudo-unitary transformations in a 4-dimensional complex space. This is an extension of the spinor calculus for the Lorentz group to the case of the conformal group. The 2-component spinors are replaced by 4-component bispinors $\Psi_\alpha (\alpha = 1, \cdots 4)$, familiar from Dirac wave functions at a point. Following the suggestion by Penrose we shall call them *twistors*.[7] The conformal group is isomorphic in the small to the group SU(2,2), the group acting on twistor space which leaves the (indefinite) Hermitean form

$$(\psi, \psi) = \langle \psi | \gamma^0 | \psi \rangle \tag{I.2.47}$$

invariant. Here $\langle \psi | \psi \rangle$ denotes the (naive) positive definite Hermitean scalarproduct and γ^0 is an Hermitean matrix with two eigenvalues $+1$ and two eigenvalues -1. The Lie algebra of SU(2,2) is the algebra of the 15 Dirac matrices

$$\gamma^\mu; \quad \gamma^{\mu\nu} \equiv i\gamma^\mu \gamma^\nu (\mu \neq \nu); \quad \gamma^{5\mu} \equiv i\gamma^5 \gamma^\mu; \quad \gamma^5 \equiv \gamma^0 \gamma^1 \gamma^2 \gamma^3, \tag{I.2.48}$$

generated from the γ^μ defined in section 3.3 below. The correspondence to SO(4,2) is established by identifying the generators of
- rotations in the ξ^5-ξ^μ plane with γ^μ
- Lorentz transformations with $\gamma^{\mu\nu}$
- rotations in the ξ^4-ξ^μ plane with $\gamma^{5\mu}$
- dilations (rotations in the ξ^4-ξ^5 plane) with γ^5.

I.2.2 Maxwell Theory

"War es ein Gott, der diese Zeichen schrieb?" This quotation from Goethe's Faust was Boltzmann's reaction to Maxwell's "Treatise on Electricity and Magnetism". The enormous wealth of physical phenomena predictable from a few basic equations borders indeed on the miraculous. These equations (in pre-relativistic vector notation)

$$\frac{\partial \mathbf{B}}{\partial t} + \operatorname{curl} \mathbf{E} = \mathbf{0}; \quad \operatorname{div} \mathbf{B} = 0, \tag{I.2.49}$$

[7] See [Pen 67]. There is an essential difference in the transformation law between Dirac wave functions and twistors. In Dirac theory the translations do not change the index α; in the case of twistors the translations (and conformal transformations) act on the index α.

$$\operatorname{curl}\mathbf{H} - \frac{\partial \mathbf{D}}{\partial t} = 4\pi\,\mathbf{j}; \quad \operatorname{div}\mathbf{D} = 4\pi\varrho, \tag{I.2.50}$$

$$\mathbf{D} = \varepsilon\,\mathbf{E}; \quad \mathbf{B} = \mu\,\mathbf{H} \tag{I.2.51}$$

connect the electric resp. magnetic field strengh $\mathbf{E}, \mathbf{B}$ with the electric current density $\mathbf{j}$ and charge density ϱ. In the absence of polarizable media we may choose units so that $\mathbf{D}$ is identified with $\mathbf{E}$ and $\mathbf{H}$ with $\mathbf{B}$.

Neither Maxwell nor Boltzmann could know that these equations contain the germs for developments in theoretical physics far transcending electromagnetism. We have already mentioned Lorentz invariance and special relativity. It is miraculous how these equations simplify if one uses the 4-dimensional Minkowski space notation. The electric and magnetic field strengths combine to a skew symmetric tensor (2-form)

$$F_{\mu\nu} = \begin{pmatrix} 0 & -E_1 & -E_2 & -E_3 \\ E_1 & 0 & B_3 & -B_2 \\ E_2 & -B_3 & 0 & B_1 \\ E_3 & B_2 & -B_1 & 0 \end{pmatrix}, \tag{I.2.52}$$

the charge and current densities combine to a 4-vector

$$j^\mu = (\varrho, \mathbf{j}). \tag{I.2.53}$$

The homogeneous Maxwell equations (I.2.49) become

$$\partial_\mu F_{\varrho\sigma} + \partial_\varrho F_{\sigma\mu} + \partial_\sigma F_{\mu\varrho} = 0. \tag{I.2.54}$$

They can be formally solved by introducing a 1-form $A_\mu(x)$, the vector potential, putting

$$F_{\mu\nu} = \partial_\mu A_\nu - \partial_\nu A_\mu. \tag{I.2.55}$$

The inhomogeneous Maxwell equations (I.2.50) become

$$\partial_\mu F^{\mu\nu} = -4\pi j^\nu. \tag{I.2.56}$$

Charge conservation which is locally expressed by the continuity equation

$$\frac{\partial \varrho}{\partial t} = -\operatorname{div}\mathbf{j}$$

becomes

$$\partial_\mu j^\mu = 0 \tag{I.2.57}$$

and is read off immediately from (I.2.56) due to the antisymmetry of $F^{\mu\nu}$.

It is tempting at this stage to remark that the equations fit naturally into differential geometric concepts, specifically the theory of integration over submanifolds (of Minkowski space) and the associated de Rham cohomology. Considering a piece of a p-dimensional submanifold characterized in parametric form by $x = x(\lambda_1, \cdots \lambda_p)$ the oriented "surface element" [8]

[8] The determinant in (I.2.58) is the Jacobian of the p coordinates in question with respect to the parameters.

$$d\sigma^{\mu_1\cdots\mu_p} = \det\left|\frac{\partial x^{\mu_k}}{\partial\lambda_l}\right| d\lambda_1\cdots d\lambda_p, \tag{I.2.58}$$

is invariant under change of parametrization and it transforms like a contravariant tensor of rank p under coordinate transformation. It is totally skew symmetric in its indices. Therefore a totally skew symmetric covariant tensor field $X_{\mu_1\cdots\mu_p}$ assigns in an intrinsic (coordinate independent) way and without reference to the metric a measure on each p-dimensional submanifold of $\mathcal{M}$ by the integral $\int X_{\mu\nu\ldots}d\sigma^{\mu\nu\cdots}$. This is the reason why such a tensor is called a p-form. One finds that the curl-operation, the antisymmetric differentiation, leads (again in an intrinsic way) from a p-form to a $(p+1)$-form, denoted by dX

$$(dX)_{\mu\mu_1\cdots\mu_p} = \text{Antisymmetr. } \partial_\mu X_{\mu_1\cdots\mu_p} \tag{I.2.59}$$

Furthermore this operation, called the *coboundary operation* in mathematics, has the property

$$d^2 = 0. \tag{I.2.60}$$

One recognizes that the first Maxwell equation means that the 2-form F satisfies

$$dF = 0 \tag{I.2.61}$$

and that the Ansatz (I.2.55) which reads now

$$F = dA \tag{I.2.62}$$

solves (I.2.61) due to (I.2.60). A p-form X satisfying $dX = 0$ is called a *cocycle*. If $X = dY$ then X is called a *coboundary* which certainly implies that X is also a cocycle. The question as to whether each cocycle is a coboundary is the basic question of cohomology theory. For a contractible region, a region which can be shrunk to a point by continuous deformations, the answer is "yes" (Poincaré's lemma). Thus (I.2.55) gives a general solution of (I.2.54) as long as we do not have any holes in Minkowski space or boundary conditions at infinity to worry about. The potential A_μ is not uniquely determined by (I.2.62); there remains the freedom of adding an arbitrary cocycle which locally can be written as $d\Phi$ where Φ is an arbitrary scalar field. This innocent observation contains the germ of a further very important extrapolation from Maxwell theory, the principle of local gauge invariance.

Turning now to the second Maxwell equations we may note that it would be more natural to consider the current density as a 3-form because it is by integrals over 3-dimensional submanifolds that we obtain from it the physically interesting quantity: the electric charge. This can be done by defining

$$J_{\mu\nu\varrho}(x) = \epsilon_{\mu\nu\varrho\sigma} j^\sigma(x) \tag{I.2.63}$$

where ϵ is the Lorentz-invariant Levi-Civita numerical tensor

$$\epsilon_{\mu\nu\varrho\sigma} = \begin{cases} +1 & \text{if } \mu\nu\varrho\sigma \text{ is an even permutation of } 0123 \\ -1 & \text{if } \mu\nu\varrho\sigma \text{ is an odd permutation of } 0123 \\ 0 & \text{if any two indices are equal.} \end{cases} \tag{I.2.64}$$

One can use ϵ to define to each p-form X in 4-dimensional space-time a $(4-p)$-form called its Hodge dual and denoted by $*X$. For instance $J = *j$ if we consider j as the 1-form $j_\mu = g_{\mu\nu} j^\nu$, and

$$(*F)_{\mu\nu} = \frac{1}{2}\epsilon_{\mu\nu\varrho\sigma} F^{\varrho\sigma}. \tag{I.2.65}$$

The second Maxwell equation reads then

$$d * F = J \tag{I.2.66}$$

and the continuity equation

$$dJ = 0. \tag{I.2.67}$$

On a formal level one might consider the 2-form $G = *F$ as a potential for the current density J since J is the coboundary of G. This determines G from J up to a cocycle. Barring cohomological obstructions this arbitrariness can precisely be lifted by requiring

$$d * G \equiv dF = 0, \tag{I.2.68}$$

which is the first Maxwell equation.

On this level one could say that the Maxwell equations result from Poincaré invariance and charge conservation purely by a convenient definition of a "charge potential" G. However, the field F is not merely a computational device. It is an observable which can be measured by its effect on the motion of the charges. A point particle of charge q moving with velocity $\mathbf{v}$ in an electromagnetic field experiences the *Lorentz force*

$$\mathbf{K} = q(\mathbf{E} + \mathbf{v} \times \mathbf{B}). \tag{I.2.69}$$

In relativistic kinematics the trajectory of such a particle is a *world line* in Minkowski space which is conveniently described by giving the position x^μ as a function of the *proper time* τ (defined by $d\tau^2 = g_{\mu\nu} dx^\mu dx^\nu$). The 4-velocity

$$u^\mu = \frac{dx^\mu(\tau)}{d\tau} \tag{I.2.70}$$

is a 4-vector having the constant Lorentz square 1 along the trajectory. The relativistic adaptation of Newton's second law with the Lorentz force acting on the particle is

$$m\frac{d^2 x^\mu}{d\tau^2} = -\,qF^{\mu\nu}(x(\tau))u_\nu(\tau). \tag{I.2.71}$$

To obtain a complete theory one must add to the Maxwell equations a model for the current density and describe the effect of $F_{\mu\nu}$ on the development of j. If one assumes the current to be built up from point particles, numbered by an index i, then

$$j^{\mu}(x) = \sum_i q_i \int u_i^{\mu}(\tau_i)\delta^4(x - x_i(\tau_i))d\tau_i \tag{I.2.72}$$

and for each particle we have the equation of motion (I.2.71). But this does not give a well defined theory because the insertion of the current density (I.2.72) into the Maxwell equations gives a field which is singular on the trajectories of the particles. In (I.2.71) we get an infinite or undefined *self force* on each particle resulting from the electromagnetic field generated by it. We cannot just claim that this self force is zero because an accelerated charge radiates electromagnetic energy and the energy balance demands that there is a radiation reaction force on the particle. In a beautiful analysis Dirac has extracted the relevant finite part of the self force of a point particle [Dir 38]. This may be considered as a "renormalization" prodedure in the classical theory. It has, however, some defects. In the form given by Dirac the electromagnetic field is split into an "external field" and a "self field". The equation of motion of the particle is a third order differential equation in which the external field at the position of the particle enters. A proper formulation of the initial value problem replaces this differential equation by an integro-differential equation which determines the orbit from the incoming electromagnetic field and the initial (straight line) asymptote of the world line of the particle. This equation has solutions only if the incoming field is not too rapidly varying [Haag 55b], [Rohr 61]. For further discussion of these problems see for instance [*Rohrlich 1965*], [*Thirring 1981, Vol.2*]. From the present day point of view it appears that the point particle model for the current, while adequate for many problems, is not a good starting point for a fundamental theory. There are not only the mentioned defects but we have learned that electromagnetism is closely tied to the principle of local gauge invariance which will be discussed later. It demands that the charge should be carried by a complex field and that j is a quadratic expression in this matter field, as it is in Schrödinger's matter waves. But then we are leaving the province of classical theory.

I.2.3 General Relativity[9]

In special relativistic electrodynamics the principle of locality is still not fully implemented. There remains the rigid metric structure of Minkowski space assumed to be known a priori, fixing the causal order between far apart points irrespective of the physical situation in between. Also the theory of gravitational forces existing around 1910 did not fit naturally into the picture of a local field theory in Minkowski space. The clue to remove both defaults, leading to the development of the general theory of relativity, was recognized by Einstein to lie in the equivalence of inertial and gravitational mass. It suggests that the transition from one coordinate system to another one, arbitrarily accelerated

[9]Our purpose here is only to give an outline of basic concepts and ideas underlying general relativity. They will only be used in Chapter VIII of this book. For an easily readable introduction we refer to [*Bergmann 1976*], for a critical survey of the physical background, computational techniques and applications to astrophysics to [*Weinberg 1972*].

with respect to the first, cannot be physically distinguished from a change of the gravitational field. Therefore the existence of a preferred class of coordinate systems, the *inertial* ones, should not be assumed.

The mathematical structure needed to incorporate these physical ideas is Riemann's geometry. It refers the concept of metric to the infinitesimal neighborhood of each point, more precisely to the tangent spaces of the points of the space-time manifold. In a coordinate system, a tangent vector z at a point x may be characterized by the derivative of the coordinates of a parametrized curve $x(s)$ passing through the point:

$$z^{\mu} = \left. \frac{dx^{\mu}}{ds} \right|_{s=0} ; \quad x(0) = x. \tag{I.2.73}$$

The scalar product of two such tangent vectors at x is defined as

$$(z_1, z_2) = g_{\mu\nu}(x) z_1^{\mu} z_2^{\nu} \tag{I.2.74}$$

where the "metric tensor" $g_{\mu\nu}$ does not have the a priori given form (I.2.2) but depends on the point x. One has a *metric field* $g_{\mu\nu}(x)$. Only the signature of the metric is the same as that of (I.2.2) at every point. One has time-like vectors (positive length square). The orthogonal complement of a time-like vector is a 3-dimensional subspace on which the metric is negative definite; it contains only space-like vectors. Now the notions of tangent bundle, cotangent bundle become really relevant because we have no natural identification of tangent spaces at different points. Nor do we have a preferred class of coordinate systems related by linear transformations. Therefore all relations should be expressed in a way valid in any coordinate system and we have to handle general coordinate transformations, $x \to x'$

$$x'^{\mu} = x'^{\mu}(x^0, \cdots x^3) \tag{I.2.75}$$

where the four functions on the right hand side may be quite arbitrary apart from being smooth and such that (I.2.75) is invertible i.e. solvable for the x^{μ} in terms of x'^{μ} within the region to which the coordinate systems are applied.

It has been disputed whether the requirement of *general covariance of all laws* is a physical principle or merely an esthetic demand which can always be met irrespective of the physical content of the theory. It is empty indeed if nothing is said about the nature of the quantities appearing in the theory. But it is very restrictive if we know that the theory deals with a certain number of fields which are related in a specific way to the tangent spaces. Thus, a vector field $V^{\mu}(x)$ assigns to each point of the manifold a vector in the attached tangent space; it is a "section in the tangent bundle". Under the coordinate change (I.2.75) it transforms as

$$V'^{\mu}(x') = \frac{\partial x'^{\mu}}{\partial x^{\nu}} V^{\nu}(x). \tag{I.2.76}$$

A covector field $X_{\mu}(x)$, a section of the cotangent bundle, transforms as

$$X'_{\mu}(x') = \frac{\partial x^{\nu}}{\partial x'^{\mu}} X_{\nu}(x). \tag{I.2.77}$$

The matrices Λ and $\tilde{\Lambda}$ in (I.2.10), (I.2.11) are replaced respectively by the Jacobian matrix of the transformation (I.2.75) or its inverse. For tensor fields with several upper or lower indices the transformation law is as in (I.2.13) replacing the Lorentz matrices by the respective Jacobian matrices.

Since the laws shall be formulated as differential equations for fields and since the field components at a point have to be related to a basis in the tangent space of the point we must be able to compare vectors in the tangent spaces of infinitesimally close points even though there is no natural identification of the tangent spaces of different points in general. The necessary additional structure is the notion of *parallel transport* or *affine connection.* In a coordinate system it is described by *connection coefficients* $C^{\mu}_{\nu\varrho}$. The change of a component of a vector V at x under parallel transport to $x + \delta x$ is of the form

$$\delta V^{\mu} = -C^{\mu}_{\nu\varrho} V^{\nu} \delta x^{\varrho}, \tag{I.2.78}$$

(the minus sign is conventional). The splitting

$$C^{\mu}_{\nu\varrho} = \Gamma^{\mu}_{\nu\varrho} + T^{\mu}_{\nu\varrho} \tag{I.2.79}$$

where Γ is symmetric, T antisymmetric in the lower indices has an intrinsic meaning, it does not depend on the coordinate system. T is called *torsion* and transforms as a tensor under coordinate changes. Γ is not a tensor but obeys the inhomogeneous transformation law

$$\Gamma'^{\lambda}_{\mu\nu} = \frac{\partial x'^{\lambda}}{\partial x^{\varrho}} \frac{\partial x^{\sigma}}{\partial x'^{\mu}} \frac{\partial x^{\tau}}{\partial x'^{\nu}} \Gamma^{\varrho}_{\sigma\tau} + \frac{\partial x'^{\lambda}}{\partial x^{\varrho}} \frac{\partial^2 x^{\varrho}}{\partial x'^{\mu} x'^{\nu}}. \tag{I.2.80}$$

In the standard theory of general relativity the torsion T is assumed to vanish. Γ may be interpreted as the gravitational field. The motion of a point particle which is under the influence of no other than gravitational forces is assumed to be an *isoparallel* i.e. the orbit is generated by parallel transport of the velocity 4-vector $u^{\mu} = dx^{\mu}/d\tau$ where τ is the proper time ($d\tau^2 = g_{\mu\nu} dx^{\mu} dx^{\nu}$) :

$$\frac{du^{\lambda}(\tau)}{d\tau} = -\Gamma^{\lambda}_{\mu\nu} u^{\mu} u^{\nu}. \tag{I.2.81}$$

Because of the equivalence of inertial and gravitational mass the particle mass does not appear in this equation of motion and, moreover, one can always choose a coordinate system so that the inertial forces balance the gravitational forces at a chosen point which means that the apparent acceleration and hence Γ vanish at this point. On the side of the mathematical algorithm this possibility arises from the fact that Γ is not a tensor but has an inhomogeneous term in its transformation law.

It is clear that the connection coefficients must bear a relation to the metric field because parallel transport should respect the metric structure; the scalar product of two vectors should not change under parallel transport. It is, however, surprising that this requirement determines Γ completely in terms of the metric field. One finds

$$\Gamma^{\lambda}_{\mu\nu} = \frac{1}{2} g^{\lambda\varrho}(\partial_\nu g_{\varrho\mu} + \partial_\mu g_{\varrho\nu} - \partial_\varrho g_{\mu\nu}). \tag{I.2.82}$$

H. Weyl called this relation "the fundamental fact of differential (Riemannian) geometry".

The affine connection allows us to define in an intrinsic way the notion of differentiation of vector and tensor fields. We must, in a coordinate system, compare the field components at the point $x + \delta x$ with those resulting from parallel transport of the field at x to this neighboring point. The resulting concept of *covariant differentiation* gives for a vector field the definition

$$D_\varrho V^\sigma(x) = \partial_\varrho V^\sigma(x) + \Gamma^{\sigma}_{\nu\varrho}(x) V^\nu(x). \tag{I.2.83}$$

This allows a very simple transcription of laws from special to general relativity; for instance electrodynamics in the presence of a gravitational field results if we replace in Maxwell's equations the partial derivatives ∂_μ by the symbols of covariant differentiation D_μ. One may note here that in the curl-operation (the coboundary operation) the connection coefficients drop out. Thus, if the Maxwell system is written using only the d-symbol, then one does not need to know Γ for these relations. But not all important differential relations in physics can be written in terms of the curl-operation. An important example is energy-momentum conservation which, locally, is expressed as a continuity equation for the energy-momentum tensor $T_{\mu\nu}$. In general relativity this reads

$$D_\mu T^{\mu\nu} = 0. \tag{I.2.84}$$

The replacement of ordinary differentiation by covariant differentiation relates to the fact that $T_{\mu\nu}$ includes gravitational contributions.

A further essential concept tied to the existence of an affine connection is *curvature*. Transporting a vector V parallel around an infinitesimal loop may lead to a change of its direction by an infinitesimal Lorentz transformation:

$$\delta V^\mu(x) = R^{\mu}{}_{\nu\varrho\sigma}(x) V^\nu d\sigma^{\varrho\sigma} \tag{I.2.85}$$

where dσ is the surface element of the loop. The coefficients $R^{\mu}{}_{\nu\varrho\sigma}$ form a tensor, the *Riemann curvature tensor*. It is, by definition, antisymmetric in the indices $\varrho\sigma$ since dσ is antisymmetric and it is an infinitesimal Lorentz matrix in the first two indices which means that

$$R_{\lambda\nu\varrho\sigma} \equiv g_{\lambda\mu} R^{\mu}{}_{\nu\varrho\sigma} \tag{I.2.86}$$

is antisymmetric also in $\lambda\nu$. There are further general properties of $R_{\lambda\nu\varrho\sigma}$ which follow from its construction viz. the symmetry under the exchange of the first pair with the second pair of indices

$$R_{\lambda\nu\varrho\sigma} = R_{\varrho\sigma\lambda\nu}, \tag{I.2.87}$$

the cyclicity in the last three indices

$$R_{\lambda\nu\varrho\sigma} + R_{\lambda\sigma\nu\varrho} + R_{\lambda\varrho\sigma\nu} = 0, \tag{I.2.88}$$

and the Bianchi identities

$$D_\mu R_{\lambda\nu\varrho\sigma} + D_\sigma R_{\lambda\nu\mu\varrho} + D_\varrho R_{\lambda\nu\sigma\mu} = 0. \tag{I.2.89}$$

From the Riemann tensor one obtains by contraction the symmetric *Ricci tensor*

$$R_{\mu\nu} = g^{\lambda\varrho} R_{\lambda\mu\varrho\nu}, \tag{I.2.90}$$

and, by further contraction, the curvature scalar

$$R = g^{\mu\nu} R_{\mu\nu}. \tag{I.2.91}$$

The Bianchi identities yield then the crucial relation

$$D_\mu \left(R^{\mu\nu} - \frac{1}{2} g^{\mu\nu} R \right) = 0. \tag{I.2.92}$$

The remaining hard problem is to replace Newton's law of gravitation

$$\text{div}\, \mathbf{F} = 4\pi G \varrho \tag{I.2.93}$$

(ϱ is the mass density, $\mathbf{F}$ the gravitational field, G the gravitation constant). Einstein solved this, replacing the mass density by the energy-momentum tensor $T_{\mu\nu}$ (which is a function of the local fields, provided that one has a field theoretic model of matter) while the left hand side of (I.2.93) is replaced by a tensor obtained from the curvature. The choice of this is severely limited by the fact that due to (I.2.84) the covariant divergence of it has to vanish. One possible choice is suggested by (I.2.92) and this fits with the non relativistic approximation (I.2.93) when the gravitational force is related to the connection coefficients by comparing (I.2.81) with Newton's equations of motion. One obtains then the Einstein equations

$$R_{\mu\nu} - \frac{1}{2} g_{\mu\nu} R = -8\pi G T_{\mu\nu}. \tag{I.2.94}$$

Apart from an additional, so called "cosmological term" $\lambda g_{\mu\nu}$ the left hand side of (I.2.94) is the only tensor of this type with vanishing covariant divergence which can be formed from the metric tensor and its first and second derivatives without using higher powers of the curvature. In this the Einstein equations are the simplest, most natural possibility allowed in the setting.

I.3 Poincaré Invariant Quantum Theory

I.3.1 Geometric Symmetries in Quantum Physics. Projective Representations and the Covering Group

It was stressed in section I.1 that "states" and "observables" in quantum theory have their physical counterparts in instruments which serve as sources of an ensemble or detectors of an event. Within the regime of special relativity we may distinguish two different types of information in the description of an instrument. On the one hand there is the intrinsic structure of the apparatus, given essentially by a workshop drawing. On the other hand, we have to specify where and how this equipment is to be placed within some established space-time reference system in the laboratory. We can put its center of mass in different positions, orient its axes, trigger the apparatus at a time of our choice, let it rest in the laboratory or move it with a constant velocity. There are altogether 10 parameters which specify the "placement". They correspond to the 10 parameters of the Poincaré group. Poincaré invariance of the laws of nature means that the result of a complete experiment does not depend on the placement. Specifically, if both the source and the detector are subjected to the same shift of placement, then the counting rates will remain unaltered.

Suppose we have an ideal source which prepares an ensemble in a pure state. For a specified placement of this source in our space-time reference system the prepared state is mathematically described by a ray in a Hilbert space $\mathcal{H}$, say by $\widehat{\Psi} = \{\lambda\Psi\}$, where Ψ is a vector in $\mathcal{H}$ and λ an arbitrary complex number. Similarly, if we have an ideal detector, giving a yes-answer in a pure state described by the ray $\widehat{\Phi}$ and the answer no in the orthogonal complement to Φ, then the probability of detecting an event in this set up of source and detector is given by the *ray product*

$$[\widehat{\Phi}|\widehat{\Psi}] = \frac{|(\Phi,\Psi)|^2}{(\Phi,\Phi)(\Psi,\Psi)}, \tag{I.3.1}$$

where (Φ,Ψ) denotes the scalar product of Hilbert space vectors. Shifting the placement[1] of the source by a Poincaré transformation $g = (a, \Lambda)$ we denote the state prepared by this source by $\widehat{\Psi}_g$. Correspondingly, shifting the detector by the same element $g \in \mathfrak{P}$ leads from $\widehat{\Phi}$ to $\widehat{\Phi}_g$. Symmetry means that the probability of an event is unaffected if both source and detector are shifted by the same group element:

$$[\widehat{\Phi}_g|\widehat{\Psi}_g] = [\widehat{\Phi}|\widehat{\Psi}]. \tag{I.3.2}$$

Keeping g fixed and letting $\widehat{\Psi}$ run through all states (here assumed to correspond precisely to all rays of $\mathcal{H}$) we get a map $\widehat{T}_g$ of the rays corresponding to the shift $g \in \mathfrak{P}$

$$\widehat{T}_g\widehat{\Psi} = \widehat{\Psi}_g. \tag{I.3.3}$$

It must leave the *ray product* of any pair of rays invariant

$$[\widehat{T}_g\widehat{\Phi}|\widehat{T}_g\widehat{\Psi}] = [\widehat{\Phi}|\widehat{\Psi}] \tag{I.3.4}$$

and it must satisfy the composition law of the Poincaré group

$$\widehat{T}_g\widehat{T}_{g'} = \widehat{T}_{gg'}. \tag{I.3.5}$$

[1] We understand the Poincaré transformations here in the "active" sense i.e. not as a change of the space-time reference system but as a change of placement of hardware in the reference system.

In particular (I.3.5) implies that $\hat{T}_g$ has an inverse. So $\hat{T}_g$ maps the set of all rays onto itself.

A ray transformation may be replaced in many different ways by a transformation of the vectors of the Hilbert space. All we have to ask for is that the image vector shall lie on the image ray. However one has the following important fact, known as the *Wigner unitarity-antiunitarity theorem.*

Theorem 3.1.1
A ray transformation $\hat{T}$ which conserves the ray product between every pair of rays of a Hilbert space can be replaced by a vector transformation T which is additive and length preserving, i.e. which satisfies

$$T(\Psi_1 + \Psi_2) = T\Psi_1 + T\Psi_2, \quad \text{and} \quad \| T\Psi \|^2 = \| \Psi \|^2 . \tag{I.3.6}$$

This vector transformation is determined uniquely up to an arbitrary phase factor (complex number of modulus 1) by $\hat{T}$ and the requirements (I.3.6). It satisfies either

$$Ta\Psi = aT\Psi \tag{I.3.7}$$

or

$$Ta\Psi = \overline{a}T\Psi, \tag{I.3.8}$$

where a is a complex number, $\overline{a}$ its complex conjugate. In the first case T is a linear, unitary operator, in the second case it is called antilinear and antiunitary.

We shall not prove it here.[2] It is closely related to the "fundamental theorem of projective geometry". In the case of a continuous group the possibility of antiunitary operators is excluded because each group element is the square of another one and the product of two antiunitary operators is unitary. We conclude: To every Poincaré transformation $g = (a, \Lambda)$ there corresponds a unitary operator $U_g \equiv U(a, \Lambda)$ acting in the Hilbert space, determined up to a phase factor $e^{i\alpha}$. The multiplication law (I.3.5) becomes

$$U_{g_1} U_{g_2} = \eta U_{g_1 g_2}. \tag{I.3.9}$$

Here η is a phase factor which may depend on g_1 and g_2. Thus U is a *representation* up to a *factor*, a *projective representation* of the Poincaré group. Since there remains the freedom of changing every operator U_g by an arbitrary phase factor one may use this to simplify the phase function η. One finds [Wig 39], [Barg 54]:

Theorem 3.1.2
Any ray representation of the Poincaré group can, by a suitable choice of phases, be made into an ordinary representation of the covering group $\overline{\mathfrak{P}}$.

Thus the factor η can be eliminated in (I.3.9) if we understand by g_k elements of $\overline{\mathfrak{P}}$ rather than $\mathfrak{P}$ and it is reduced to a sign ± 1 if we consider the Poincaré group itself.

[2]For a proof see the book by Wigner, appendix to chapter 20 [*Wigner 1959*].

To summarize: Poincaré invariance in quantum theory means that we have a unitary representation of $\overline{\mathfrak{P}}$, describing the effect of Poincaré transformations on the state vectors of Hilbert space. The occurence of the covering group results from the fact that states correspond to rays rather than vectors of $\mathcal{H}$.

I.3.2 Wigner's Analysis of Irreducible Unitary Representations of the Poincaré Group

The building blocks of general unitary representations of $\overline{\mathfrak{P}}$ are the irreducible ones i.e. those which act in a Hilbert space not containing any (proper, closed) subspace which is transformed into itself. These representations should yield the state spaces of the simplest physical systems in relativistic quantum theory and it is a purely mathematical task to classify them all (up to unitary equivalence). In a fundamental paper E.P. Wigner recognized these facts and largely solved the classification problem [Wig 39]. It is curious that this paper remained practically unknown or unappreciated in the physics community for more than a decade. We sketch the essential line of argument and the results without aspiring to mathematical rigour. For more thorough discussions of this subject the reader is referred to the books [*Hamermesh 1964*], [*Ohnuki 1988*].

Let $U(a)$ denote the representor of a translation by the 4-vector a_μ and $U(\alpha)$ that of an element $\alpha \in \overline{\mathfrak{L}}$. Hopefully the distinction by the type of argument of U will be sufficient to avoid confusion. The Roman character stands for a translation 4-vector, the Greek α for a unimodular 2×2 matrix which determines a homogeneous Lorentz transformation $\Lambda(\alpha)$.

The translation subgroup is commutative. We can write

$$U(a) = e^{iP_\mu a^\mu} \tag{I.3.10}$$

where the infinitesimal generators P_μ are four commuting, self adjoint operators. The multiplication law in $\overline{\mathfrak{P}}$ gives the relation

$$U(\alpha)^{-1}\, P^\mu\, U(\alpha) = \Lambda^\mu_{\ \nu}(\alpha) P^\nu; \quad P^\mu \equiv g^{\mu\nu} P_\nu. \tag{I.3.11}$$

In words: the operators P^μ transform like the components of a 4-vector under the homogeneous Lorentz group. Since they commute they have a simultaneous spectral resolution. The spectral values are a subset of a 4-dimensional space (p-space) and we may represent a general vector $\Psi \in \mathcal{H}$ by the set of spectral components Ψ_p

$$\Psi = \{\Psi_p\}. \tag{I.3.12}$$

Ψ_p is a vector in a degeneracy space $\mathcal{H}_p$. If the simultaneous P-spectrum were discrete $\mathcal{H}_p$ would be a subspace of $\mathcal{H}$. In the case of continuous spectrum, which is the case of primary interest, we may consider $\mathcal{H}_p$ as an infinitesimal (improper) subspace of $\mathcal{H}$. The scalar product in $\mathcal{H}$ becomes

$$\langle \Psi' | \Psi \rangle = \int \langle \Psi'_p | \Psi_p \rangle d\mu(p), \tag{I.3.13}$$

where the bracket under the integral denotes the scalar product of the components in $\mathcal{H}_p$ and $d\mu$ is some positive measure in p-space. In other words, we represent $\mathcal{H}$ as the direct integral of the spaces $\mathcal{H}_p$ with respect to the measure $d\mu$.

According to (I.3.11) the operator $U(\alpha)$ maps $\mathcal{H}_p$ onto $\mathcal{H}_{p'}$ where $p' = \Lambda(\alpha)p$. Let us call the set of all points in p-space which arise from a single point p upon application of all homogeneous Lorentz transformations the "orbit of p". Clearly then all spaces $\mathcal{H}_p$ with p-vectors lying on one orbit will be mapped on each other by the action of the operators $U(\alpha)$ whereas spaces with p-vectors on different orbits will not be connected by any $U(a)$ or $U(\alpha)$. Therefore in an *irreducible* representation the P-spectrum has to be *concentrated on a single orbit.* This gives a first division of the irreducible representations into the following classes:

class	orbit			
m_+	Hyperboloid in forward cone;	$p^2 = m^2$	and	$p^0 > 0$.
0_+	Surface of forward cone;	$p^2 = 0$	and	$p^0 \geq 0$.
0_0	The single point $p^\mu = 0$.			
κ	Space-like hyperboloid;	$p^2 = -\kappa^2$	(κ real).	
m_-	Hyperboloid in backward cone;	$p^2 = m^2$	and	$p^0 < 0$.
0_-	Surface of backward cone;	$p^2 = 0$	and	$p^0 \leq 0$.

Of course $p^2 = g_{\mu\nu}p^\mu p^\nu = (p^0)^2 - \mathbf{p}^2$. We shall be concerned only with the classes m_+ and 0_+ for the following reason. The operator P^0, considered in its rôle as an observable, is interpreted as the total energy of the system and, correspondingly, P^i are the components of the linear momentum.[3] One of the most important principles of quantum field theory, ensuring the stability, demands that the energy should have a lower bound. This is not the case in the last three classes. In class 0_0 all states have zero energy-momentum, all states are translation invariant. If they are also Lorentz invariant we have the trivial representation appropriate to the vacuum state. Other possibilities in 0_0 which carry a non trivial unitary representation of the Lorentz group have only marginal physical interest, though, mathematically, the analysis of this class, carried through by Bargmann and by Naimark, is very interesting. See [Barg 47] and the books [*Naimark 1964*], [*Ohnuki 1988*].

Since p^μ is interpreted as energy-momentum the classification parameter m (resp. 0) of the first two classes is, of course, the rest mass. The measure $d\mu$ is fixed (up to an arbitrary normalization factor) by the requirement that it must be Lorentz invariant and concentrated on a single hyperboloid. In 4-dimensional notation it is[4]

[3] Here we go over to natural units, putting $\hbar = 1$. This will be done from now on throughout this book.

[4] The step function

$$\Theta(p^0) = \begin{cases} 1 \\ 0 \end{cases} \text{for} \begin{cases} p^0 > 0 \\ p^0 < 0 \end{cases}$$

combined with the δ-function on the hyperboloid is Lorentz invariant; it just singles out the positive energy mass hyperboloid.

$$d\mu = \delta(p^2 - m^2)\Theta(p^0)d^4p \tag{I.3.14}$$

or, if we use the 3-momentum $\mathbf{p}$ to coordinatize the hyperboloid

$$d\mu = \frac{d^3p}{2\varepsilon_{\mathbf{p}}}; \quad \varepsilon_{\mathbf{p}} = (\mathbf{p}^2 + m^2)^{1/2}. \tag{I.3.15}$$

To complete the analysis we pick a point $\overline{p}$ on the orbit and choose for every p on the orbit an element $\beta(p) \in \overline{\mathfrak{L}}$ so that

$$\Lambda(\beta(p))\overline{p} = p. \tag{I.3.16}$$

In the case m_+ we may take $\overline{p} = (m, 0, 0, 0)$. According to the definition we have

$$\widehat{\overline{p}} = m\mathbb{1}; \quad \widehat{p} = p^0\mathbb{1} + \mathbf{p}\boldsymbol{\sigma}. \tag{I.3.17}$$

Thus we can choose

$$\beta(p) = m^{-1/2}(\varepsilon_{\mathbf{p}} + \mathbf{p}\boldsymbol{\sigma})^{1/2} \tag{I.3.18}$$

which is well defined because $\widehat{p}$ is a positive matrix. We can use $U(\beta(p))$ to identify the degeneracy spaces $\mathcal{H}_p$ with $\mathcal{H}_{\overline{p}}$ so that the degeneracy spaces can all be considered as samples of one "little Hilbert space" $\mathfrak{h}$. Then we can write

$$\mathcal{H} = \mathfrak{h} \otimes \mathcal{H}_C \tag{I.3.19}$$

where $\mathcal{H}_C = \mathcal{L}^{(2)}(\mathbb{R}^3)$ with measure (I.3.15). This is most simply described in Dirac's notation. Choosing a basis in $\mathfrak{h}$ labelled by an index k we have an "improper basis" $|p; k\rangle$ in $\mathcal{H}$ and

$$|p; k\rangle = U(\beta(p))\,|\overline{p}; k\rangle. \tag{I.3.20}$$

Now, for any given p on the orbit and any given $\alpha \in \overline{\mathfrak{L}}$ we can make the decomposition

$$\alpha = \beta(p')\gamma(\alpha, p)\beta^{-1}(p); \quad p' = \Lambda(\alpha)p. \tag{I.3.21}$$

Then γ, defined by (I.3.21) is an element of the stability group of $\widehat{\overline{p}}$, i.e. $\Lambda(\gamma)$ leaves $\overline{p}$ invariant. With our choice of $\overline{p}$ the stability group is SU(2), the covering group of SO(3). To obtain an irreducible representation of $\widetilde{\mathfrak{P}}$ in $\mathcal{H}$ we need an irreducible representation of the "little group" SU(2) in $\mathfrak{h}$. These latter are finite dimensional. The representors are the well known $(2s+1)$-dimensional rotation matrices D^s to angular momentum s ($s = 0, 1/2, 1 \cdots$). In our chosen basis we can write the complete representation as

$$U(a)|p; k\rangle = e^{i(p^0a^0 - \mathbf{pa})}|p; k\rangle, \tag{I.3.22}$$

$$U(\alpha)|p; k\rangle = \sum_l D^s_{lk}(\gamma(\alpha, p))\,|p'; l\rangle; \quad p' = \Lambda(\alpha)p. \tag{I.3.23}$$

One might think that the case of zero mass can be dealt with by taking the limit $m \to 0$ in the preceding discussion. This is not exactly true, however.

We can pick now the momentum vector $\overline{p} = (1/2,\ 0,\ 0,\ 1/2)$ on the orbit. The corresponding matrix is

$$\hat{\overline{p}} = \begin{pmatrix} 1 & 0 \\ 0 & 0 \end{pmatrix} \tag{I.3.24}$$

and the little group (stability group) is generated by the unimodular matrices

$$\gamma_\varphi = \begin{pmatrix} e^{i\varphi} & 0 \\ 0 & e^{-i\varphi} \end{pmatrix} \quad \text{and} \quad \gamma_\eta = \begin{pmatrix} 1 & \eta \\ 0 & 1 \end{pmatrix}, \tag{I.3.25}$$

$\varphi \in \mathbb{R} \bmod 2\pi, \eta$ arbitrary complex. This 3-parameter group is isomorphic to the Euclidean group of the plane, the generator γ_φ corresponding to rotations, and γ_η to translations. The irreducible representations of this group can be found by applying the same technique as used for the Poincaré group, now with two dimensions less. If γ_η is non-trivially represented the representation space is infinite dimensional. These representations have so far found no application in physics. So we restrict attention to the representations with $U(\gamma_\eta) = 1$. They lead to a 1-dimensional "little Hilbert space" and

$$D(\gamma_\varphi) = e^{in\varphi}; \quad n \in \mathbb{Z}. \tag{I.3.26}$$

The restriction of n to (positive or negative) integers comes from the fact that we need a true representation of the matrix group γ_φ, not a representation of its covering group to get a true representation of $\overline{\mathfrak{P}}$. Noting that $\Lambda(\gamma_\varphi)$ is a rotation in the x_1-x_2-plane by an angle 2φ we can interpret

$$s = \frac{n}{2} \tag{I.3.27}$$

as the angular momentum along the axis of the linear momentum p. Instead of the spin (magnitude of the angular momentum in the rest frame) we have in the zero mass case the *helicity*, the component of the angular momentum in the direction of p.

I.3.3 Single Particle States. Spin

The irreducible representations of $\overline{\mathfrak{P}}$ of class $(m_+,\ s)$ yield the quantum theory of a single particle with rest mass m and spin s alone in the world. A general state vector is here characterized by a $(2s+1)$-component wave function $\widetilde{\psi}_k(p)$ in p-space, the probability amplitude in the basis (I.3.20). The scalar product is given by

$$\langle \Psi' | \Psi \rangle = \sum \int \overline{\widetilde{\psi}}'_k(\mathbf{p}) \widetilde{\psi}_k(\mathbf{p}) \, \frac{d^3p}{2\varepsilon_{\mathbf{p}}}. \tag{I.3.28}$$

We can, instead, use a wave function in x-space, defining

$$\psi_k(x) = (2\pi)^{-3/2} \int \widetilde{\psi}_k(p) e^{-ipx} d\mu(p)$$

$$= (2\pi)^{-3/2} \int \widetilde{\psi}_k(\varepsilon_{\mathbf{p}}, \mathbf{p}) e^{i(\mathbf{px} - \varepsilon_{\mathbf{p}} t)} \frac{d^3p}{2\varepsilon_{\mathbf{p}}}. \tag{I.3.29}$$

Due to the fact that p is restricted to the mass hyperboloid $\psi_k(x)$ satisfies the Klein-Gordon-equation

$$(\square + m^2)\psi_k(x) = 0; \quad \square = \frac{\partial^2}{\partial t^2} - \Delta. \tag{I.3.30}$$

One may note in passing that this apparently "dynamical" equation results in the present context from the irreducibility condition for the representation. The "Casimir operator" $P^\mu P_\mu$ commutes with all representors $U(a, \alpha)$ and must therefore, in an irreducible representation, be represented by a number. We must remember still that only the positive energy solutions of (I.3.30) are admitted. This implies that the initial values ψ and $\partial\psi/\partial t$ at a fixed time are not independent; in fact $\partial\psi/\partial t$ is determined by ψ. The differentiation $\partial/\partial t$ is described in momentum space by the multiplication operator $-i\varepsilon_{\mathbf{p}}$. This corresponds in position space at fixed time to a convolution with an integral kernel

$$\mathcal{E}(\mathbf{x} - \mathbf{x}') = -i(2\pi)^{-3/2} \int (\mathbf{p}^2 + m^2)^{1/2} e^{i\mathbf{p}(\mathbf{x}-\mathbf{x}')} d^3p, \tag{I.3.31}$$

so that

$$\frac{\partial\psi}{\partial t}(t, \mathbf{x}) = \int \mathcal{E}(\mathbf{x} - \mathbf{x}')\psi(t, \mathbf{x}')d^3x'. \tag{I.3.32}$$

In the case of spin zero the full description of the representation in terms of the wave function $\psi(x)$ is simple. The transformation law reads

$$(U(a)\psi)(x) = \psi(x - a); \quad (U(\Lambda)\psi)(x) = \psi(\Lambda^{-1}x) \tag{I.3.33}$$

and the scalar product can be written as

$$\langle \Psi' | \Psi \rangle = i \int_{x^0 = t} \left(\overline{\psi}' \frac{\partial\psi}{\partial x^0} - \frac{\partial\overline{\psi}'}{\partial x^0} \psi \right) d^3x. \tag{I.3.34}$$

The right hand side is independent of the choice of t (the space-like surface over which we integrate). One might be tempted to interpret the x-space wave function (I.3.29) in analogy to wave mechanics as the probability amplitude for finding the particle at time $t = x^0$ at the position $\mathbf{x}$. That this cannot be done is seen from the form of the normalization integral following from (I.3.34) and from the non local relation between $\partial\psi/\partial t$ and ψ at equal times. The question of how "position of the particle" should be defined in this context has been discussed by Newton and Wigner [New 49]. If one absorbs the factor $(2\varepsilon_{\mathbf{p}})^{-1}$ in (I.3.28), defining the "Newton-Wigner wave function" in momentum space as

$$\psi^{NW}(\mathbf{p}) = (2\varepsilon_{\mathbf{p}})^{-1/2}\psi(\mathbf{p}) \tag{I.3.35}$$

and, correspondingly, the Newton-Wigner wave function in x-space as its Fourier transform then the scalar product takes the same form as in the Schrödinger theory

$$\langle \Psi' | \Psi \rangle = \int_{x^0=t} \overline{\psi}'^{NW}(x) \psi^{NW}(x) d^3x. \qquad \text{(I.3.36)}$$

One may regard $\psi^{NW}(x)$ as the probability amplitude for finding the particle at time $t = x^0$ at the position $\mathbf{x}$.

Remark: This definition of localization is dependent on the Lorentz frame. A state Ψ_0, strictly localized at time $t = 0$ at the point $\mathbf{x} = 0$, given by

$$\psi_0^{NW}(0, \mathbf{x}) = \delta^3(\mathbf{x}),$$

is not strictly localized at $x^\mu = 0$ if viewed in a different Lorentz frame. However, for a massive particle the ambiguity in defining the localization is small, namely of the order of the Compton wave length. The Newton-Wigner wave function is related to the covariant wave function (I.3.29) by

$$\psi^{NW}(t, \mathbf{x}) = \int \mathbf{K}(\mathbf{x} - \mathbf{x}') \psi(t, \mathbf{x}') d^3x', \qquad \text{(I.3.37)}$$

and the integral kernel $\mathbf{K}$, being the Fourier transform of $(2\varepsilon_{\mathbf{p}})^{1/2}$, is a fast decreasing function of $m|\mathbf{x} - \mathbf{x}'|$. Thus, for massive particles, the covariant wave function tells us approximately where the particle is localized if no accuracy beyond the Compton wave length is needed. For massless particles the concept of localization is not appropriate.

Spin 1/2 and the Dirac Equation. For particles with spin the orthogonal basis in the little Hilbert space is not well suited for a description of the state vectors by wave functions in x-space. The transformation law (I.3.23) contains a matrix depending in a highly non trivial fashion on p. This means that one has a very complicated transformation law in x-space. One can give, however, a manifestly Lorentz invariant formulation. In the case of $s = 1/2$ this is achieved by going over to

$$\widetilde{\varphi}_r(p) = m^{1/2} \sum_k (\varepsilon_{\mathbf{p}} - \mathbf{p}\boldsymbol{\sigma})_{rk}^{-1/2} \widetilde{\psi}_k(p), \qquad \text{(I.3.38)}$$

and the corresponding x-space wave function $\varphi_r(x)$ obtained by a Fourier transform as in (I.3.29). These transform like spinorial wave functions namely

$$(U(\alpha)\varphi)_r(x) = \sum \alpha_{rs} \varphi_s(\Lambda^{-1}(\alpha)x). \qquad \text{(I.3.39)}$$

Each component φ_r satisfies the Klein-Gordon equation (I.3.30). The scalar product is

$$\langle \varphi' | \varphi \rangle = m^{-1} \int \overline{\widetilde{\varphi}}'(p)(\varepsilon_{\mathbf{p}} - \mathbf{p}\boldsymbol{\sigma}) \widetilde{\varphi}(p) d\mu(p). \qquad \text{(I.3.40)}$$

It is convenient to introduce the (conjugate contravariant) spinor χ by

$$(\varepsilon_{\mathbf{p}} - \mathbf{p}\boldsymbol{\sigma})\widetilde{\varphi} = m\widetilde{\chi} \qquad \text{(I.3.41)}$$

with the inversion

$$(\varepsilon_{\mathbf{p}} + \mathbf{p}\boldsymbol{\sigma})\tilde{\chi} = m\tilde{\varphi}, \tag{I.3.42}$$

and use the 4-component wave function

$$\tilde{\psi} = \begin{pmatrix} \tilde{\varphi} \\ \tilde{\chi} \end{pmatrix}. \tag{I.3.43}$$

This replaces in x-space the Klein-Gordon equation by a first order system, the Dirac equation

$$(i\gamma^{\mu}\partial_{\mu} - m)\psi = 0, \tag{I.3.44}$$

where the Dirac matrices are given by

$$\gamma^0 = \begin{pmatrix} 0 & \mathbb{1} \\ \mathbb{1} & 0 \end{pmatrix}; \quad \gamma^i = \begin{pmatrix} 0 & -\sigma_i \\ \sigma_i & 0 \end{pmatrix}. \tag{I.3.45}$$

The transformation law is

$$(U(\alpha)\psi)(x) = S(\alpha)\psi\left(\Lambda^{-1}(\alpha)x\right), \tag{I.3.46}$$

$$S(\alpha) = \begin{pmatrix} \alpha & 0 \\ 0 & \alpha^{*-1} \end{pmatrix}. \tag{I.3.47}$$

In order to conform with customary notation we denote the complex conjugate of the Dirac spinor by ψ^* and reserve the symbol $\overline{\psi}$ for $\psi^*\gamma^0$. The scalar product in x-space can then be written

$$\langle\psi|\psi\rangle = \int_{x^0=t} \psi^*(x)\psi(x)d^3x \tag{I.3.48}$$

independent of the choice of t.

To sum up: the state vectors of a massive spin 1/2 particle may be described as the positive energy solutions of the Dirac equation. This is a purely group theoretic result. The motion of an electron in an electromagnetic field and the prediction of antiparticles, tied to Dirac's original approach, is another matter. There further physical principles, in particular locality and gauge invariance play an essential rôle.

One comment should be added about the choice (I.3.45) for the Dirac matrices. One may subject ψ to an arbitrary position independent unitary transformation in the component space. Then the form (I.3.44) remains but the γ-matrices are changed by a similarity transformation. The algebraic relations

$$\gamma^{\mu}\gamma^{\nu} + \gamma^{\nu}\gamma^{\mu} = 2g^{\mu\nu} \tag{I.3.49}$$

and the property that γ^0 is Hermitean, γ^k skew Hermitean stays unaltered. The choice (I.3.45) is group theoretically distinguished because of the block diagonal form of the transformation matrix (I.3.47); the upper and lower components of ψ are then separately Lorentz covariant spinors.

Higher Spin. It is tempting to apply here the familiar phrase: "we leave this problem as an exercise to the reader". But let us add a few remarks. The simplest manifestly covariant generalization to the case of spin $s = n/2$ is to use a wave function $\tilde{\varphi}_{r_1 \cdots r_n}(p)$, $(r_k = 1, 2)$ which is a spinor of rank n, totally symmetric in its indices. This has indeed $(2s + 1)$ independent components. The scalarproduct is then

$$(\varphi, \varphi) = m^{-n} \int \overline{\tilde{\varphi}} \prod_1^n (\varepsilon_{\mathbf{p}} - \mathbf{p}\boldsymbol{\sigma}^k) \tilde{\varphi} d\mu(p) \qquad \text{(I.3.50)}$$

with $\boldsymbol{\sigma}^k$ the Pauli matrix acting on the k-th index. In x-space the product under the integral sign becomes a differential operator of order n. There are many equivalent descriptions if one introduces redundant components. Take for instance the case $s = 1$. A symmetric spinor of rank 2 corresponds to a self dual antisymmetric Minkowski tensor. Instead one might use a 4-vector φ_μ with the auxiliary condition $p^\mu \tilde{\varphi}_\mu = 0$ which again brings the number of independent components to 3. If the single particle wave equations are used heuristically to develop a quantum field theory with interactions then different starting points (equivalent for a single free particle) suggest different theories.

I.3.4 Many Particle States

Bose-Fermi Alternative. In quantum *mechanics* the Hilbert space of state vectors of an N-particle system is the N-fold tensor product of the single particle Hilbert spaces provided that the particles are not of the same species. For identical particles one has to take into account a further principle, the permutation symmetry: two state vectors which result from each other by a permutation of the indices distinguishing the particles must describe the same state. It may be reasoned that this principle is a consequence of the fact that in quantum theory we cannot keep track of individual particles without disturbing the system continuously by observations. We cannot know who is who. If a permutation of the particles is irrelevant for the state and if a state is represented by a ray of state vectors then such a permutation can change a state vector only by a numerical factor. These factors must give a 1-dimensional representation of the permutation group and there are only two such representations: the trivial one where each permutation is represented by the number 1 and the antisymmetric one where odd permutations are represented by -1 and even ones by $+1$. Which of the two possibilities is realized must depend on the particle species. One says that a particle "obeys Bose statistics", in short that the particle is a boson, if the trivial representation of the permutation group applies and that "it obeys Fermi statistics" (the Pauli principle), if the antisymmetric representation has to be used.

There has been an extended discussion as to whether the Bose-Fermi alternative really exhausts all possibilities. More general schemes called *para-statistics*, have been proposed following a suggestion by H.S.Green [Green 53]. The argument for the Bose-Fermi alternative sketched above looks convincing but one has to recognize that the starting assumption, namely that states of several indistinguishable particles may be described uniquely up to a factor by wave functions in configuration space, is not operationally justified. The position of an individually named particle is not an observable, only symmetric functions of the single particle observables are meaningful. Chapter IV of this book will be devoted to a derivation of the intrinsic significance of statistics as a consequence of the locality principle and to an analysis of the possibilities.

Le us denote by $(\mathcal{H}^{\otimes N})_S$ the symmetrized, by $(\mathcal{H}^{\otimes N})_A$ the antisymmetrized N-fold tensor product of the Hilbert space $\mathcal{H}$. These are the representation spaces for a system of N indistinguishable bosons (resp. fermions) and, if the particles do not interact, the representation of $\overline{\mathfrak{P}}$ in this space is determined from the irreducible representations in the single particle space:

$$U = \begin{cases} \left(U_{m,s}^{\otimes N}\right)_S & \text{for bosons,} \\ \left(U_{m,s}^{\otimes N}\right)_A & \text{for fermions.} \end{cases} \tag{I.3.51}$$

Fock Space. Creation Operators. In high energy particle physics processes in which particles are created or annihilated are essential. Then the particle number N becomes a dynamical variable and we need for a kinematical description of all relevant states the "Fock space " $\mathcal{H}_F$ which is the direct sum (denoted by $\sum^{\oplus}$) of all the N-particle spaces

$$\mathcal{H}_F = \sum_{N=0}^{\infty} {}^{\oplus} \left(\mathcal{H}_1^{\otimes N}\right)_{S,A}. \tag{I.3.52}$$

A state vector in $\mathcal{H}_F$ is an infinite hierarchy of symmetric (resp. antisymmetric) wave functions

$$\Psi = \begin{pmatrix} c \\ \psi_1(\xi) \\ \psi_2(\xi_1, \xi_2) \\ \dots \end{pmatrix} \tag{I.3.53}$$

where we have written the argument ξ to denote both position (or momentum) and spin component; ψ_N is the probability amplitude for finding just N particles and those in the specified configuration; $c \in \mathbb{C}$ is the probability amplitude to find the vacuum.

A more convenient way to describe vectors and operators in $\mathcal{H}_F$ is the following. We choose some complete and orthonormal basis of single particle state vectors, labeled by an index k ($k = 1, 2..$). Let

$$(n) = n_1, n_2, \cdots \tag{I.3.54}$$

denote an occupation number distribution; n_k is the number of particles in the k-th state. Note that (n) is an infinite sequence of occupation numbers since there are infinitely many orthogonal quantum states for one particle. But only such sequences are admitted for which the total particle number is finite, no matter how large:

$$\sum n_k < \infty. \tag{I.3.55}$$

In the case of fermions each n_k can, of course, only assume the values 0 or 1. Let $\Psi_{(n)}$ denote the normalized state vector corresponding to the occupation number distribution (n). These vectors form a complete orthonormal basis in $\mathcal{H}_F$. One may introduce then a system of annihilation and creation operators in this basis.

In the bosonic case the annihilation operator a_k for a particle in a state k is defined by

$$a_k \Psi_{(n)} = n_k^{1/2} \Psi_{(n-\delta_k)}. \tag{I.3.56}$$

Here $\delta_k = (0, 0 \cdots 1, 0 \cdots)$ with 1 in the k-th position. The creation operator (the adjoint of a_k) is

$$a_k^* \Psi_{(n)} = (n_k + 1)^{1/2} \Psi_{(n+\delta_k)}. \tag{I.3.57}$$

The factor $n_k^{1/2}$ in (I.3.56) is not a matter of convention but ensures the covariant transformation property of the annihilation operators under a change of the basis in single particle space. If one changes from the orthonormal system φ_k to φ_l' by a unitary matrix

$$\varphi_l' = \sum U_{lk} \varphi_k \tag{I.3.58}$$

then the creation operators in the primed basis are related to the original ones by

$$a_l'^* = \sum U_{lk} a_k^*. \tag{I.3.59}$$

This may be checked by direct calculation from the representation (I.3.53) of general state vectors in $\mathcal{H}_F$ but more simply by observing that (I.3.56), (I.3.57) imply the commutation relations

$$[a_k, a_l] = 0, \quad [a_k, a_l^*] = \delta_{kl}, \tag{I.3.60}$$

which generalize, for arbitrary single particle state vectors to

$$[a(\varphi_1), a(\varphi_2)] = 0, \quad [a(\varphi_1), a^*(\varphi_2)] = (\varphi_1, \varphi_2). \tag{I.3.61}$$

The creation operators depend linearly, the annihilation operators conjugate linearly on the wave function.

For fermions the definition (I.3.56) is replaced by

$$a_k \Psi_{(n)} = \begin{cases} 0 & \text{if} \quad n_k = 0 \\ (-1)^p \Psi_{(n-\delta_k)} & \text{if} \quad n_k = 1 \end{cases} \tag{I.3.62}$$

where $p = \sum_{i<k} n_i$ is the number of occupied levels preceding k. Instead of the commutation relations (I.3.60) one then has

$$[a_k, a_l]_+ = 0, \quad [a_k, a_l^*]_+ = \delta_{kl}, \tag{I.3.63}$$

where the symbol $[A, B]_+ = AB + BA$ denotes the anticommutator.

Empirically, particles with half integer spin are fermions, particles with integer spin are bosons. It has been one of the major achievements of general quantum field theory to show that this observed connection between spin and statistics follows from its general principles. This will be discussed in section 5 of Chapter II.

Separation of Center of Mass Motion. Let us consider a general (reducible) unitary representation of $\tilde{\mathfrak{P}}$ which contains no zero mass subrepresentation. Then we may split off the center of mass motion writing

$$\mathcal{H} = \mathfrak{h} \otimes \mathcal{H}_C, \tag{I.3.64}$$

where $\mathcal{H}_C = \mathcal{L}^{(2)}(\mathbb{R}^3)$ with measure $d\mu(p) = d^3p$. Vectors in $\mathfrak{h}$ may be viewed as corresponding to states in the center of mass system (total 3-momentum zero). Vectors in $\mathcal{H}_C$ describe the motion of the center of mass. A general state vector $\Psi \in \mathcal{H}$ is then given by a function

$$\mathbf{p} \in \mathbb{R}^3 \to \psi(\mathbf{p}) \in \mathfrak{h}, \tag{I.3.65}$$

with

$$\langle\Psi|\Psi\rangle = \int \langle\psi(\mathbf{p})|\psi(\mathbf{p})\rangle d^3p, \tag{I.3.66}$$

where the integrand on the right hand side is the scalar product in $\mathfrak{h}$.

In this description the generators of the Poincaré group are expressed in terms of operators P^k, X^k $(k = 1, 2, 3)$ acting in $\mathcal{H}_C$, a mass operator M and angular momentum operators l_k $(k = 1, 2, 3)$ acting in $\mathfrak{h}$; see Bakamjian and Thomas [Bakam 53]. The nonvanishing commutation relations are

$$[P^k, X^l] = -i\delta^{kl}, \tag{I.3.67}$$

$$[l_1, l_2] = il_3 \quad \text{and cyclic permutations.} \tag{I.3.68}$$

Of course the P^k, X^k commute with the l_k; M commutes with all the others. P^k are the generators of spatial translations. Time translations are generated by

$$P^0 = (\mathbf{P}^2 + M^2)^{1/2}. \tag{I.3.69}$$

The generators of spatial rotations are (in 3-vector notation)

$$\mathbf{L} = \mathbf{X} \times \mathbf{P} + \mathbf{l}. \tag{I.3.70}$$

The generators of boosts are[5]

$$\mathbf{K} = \frac{1}{2}(P^0\mathbf{X} + \mathbf{X}P^0) + (P^0 + M)^{-1}\mathbf{P} \times \mathbf{l}. \tag{I.3.71}$$

$\mathbf{X}$ may be (qualitatively) interpreted as the center of mass position at time zero. By (I.3.71) it is closely related to the generators of boosts. These formulas have served as a starting point for a Lorentz invariant formulation of phenomenological interparticle forces by writing $\mathbf{M}$ as a sum of single particle contributions plus an interaction potential (see in particular Coester [Coest 65]). It is a highly nontrivial task to ensure in this procedure even macroscopic locality if more than two particles are involved.

---

[5] These are the special Lorentz transformations changing the velocity; K^k generates pseudorotations in the x^0-x^k plane.

I.4 Action Principle

The belief that the actual world is the best of all possible worlds, or that God gave laws of nature optimally designed to achieve an end, has provided through centuries an inspiration to fundamental physics. It brought a teleological element which has been extremely fruitful but which, in spirit, appears to be quite opposed to the principle of locality. The simplest example illustrating the surprising reconciliation between the local and the teleological point of view is the fact that of all curves connecting two points the straight line is the one with the shortest length. Correspondingly, in Riemann's geometry an isoparallel is also a geodesic. The "straightness" (isoparallelism) is locally determined, the minimality of length (geodesic property) is an optimization of the path between a given beginning and end.

The field equations of classical physics mentioned in section 2 can all be regarded as the Euler equations of a variational principle, demanding that a certain quantity called *action* shall be stationary if the actual field distribution in space-time is infinitesimally varied. The *action* S is an integral of a local density[1] $\mathcal{L}$ called the *Lagrangean* which at a point x is a function of the fields and their derivatives

$$S = \int \mathcal{L}\, d^4x. \tag{I.4.1}$$

Some merits of the action principle are evident. It is much more economical to write down a Lagrangean than the full system of field equations resulting from it; it gives, at least if combined with requirements of simplicity, heuristic guidance of great value in the search for new theories. Closely tied to the action principle is the canonical formalism which historically played a central rôle in the evolution of quantum theory.

The single most salient feature which a classical theory formulated by an action principle has in common with quantum theory is the double rôle of each physical quantity. The quantity can be measured, yielding a number in any given physical situation. But it also defines an infinitesimal transformation of all other quantities, by the Poisson bracket in classical theory, by the commutator in quantum theory. The definition of Poisson brackets is usually given with the help of the machinery of the canonical formalism but, as Peierls has shown [Pei 52], they can be defined directly from the Lagrangean. This avoids the asymmetric treatment of the time coordinate and uses only relativistic causality instead. It assumes, however, that the quantity in question is an integral of a local function of the fields and their derivatives over a space-time region which is bounded at least in time-like extension. Given such a quantity X we consider the modified Lagrangean

$$\mathcal{L}_\varepsilon = \mathcal{L} + \varepsilon X, \quad \varepsilon \to 0 \quad \text{later.} \tag{I.4.2}$$

[1] Actually the integrand in (I.4.1) should be a 4-form. In special relativity we may regard $\mathcal{L}$ as a scalar (the Hodge dual of a 4-form). In general relativity this procedure replaces the integrand by $\mathcal{L}|g|^{1/2}$ where $|g|$ is the absolute value of the determinant of the metric tensor and $\mathcal{L}$ is a scalar.

The Euler equations of the modified Lagrangean agree outside of the support of X with those of $\mathcal{L}$. Let us generically denote all fields occurring in the theory by the single symbol Φ. Each solution $\Phi(x)$ of the Euler equations of the original Lagrangean determines a solution $\Phi^R_{\varepsilon X}$ of the Euler equations of $\mathcal{L}_\varepsilon$ which coincides with Φ in the remote past and another one, $\Phi^A_{\varepsilon X}$, which agrees with Φ in the remote future (the upper indices standing for retarded and advanced, respectively). Define

$$\delta^R_X\Phi = \frac{d}{d\varepsilon}(\Phi^R_{\varepsilon X} - \Phi)\bigg|_{\varepsilon=0}, \quad \delta^A_X\Phi = \frac{d}{d\varepsilon}(\Phi^A_{\varepsilon X} - \Phi)\bigg|_{\varepsilon=0}, \tag{I.4.3}$$

and

$$\delta_X\Phi = \delta^R_X\Phi - \delta^A_X\Phi. \tag{I.4.4}$$

One proves that $\Phi + \varepsilon\delta_X\Phi$ satisfies, to first order in ε, the Euler equations of the original Lagrangean. Thus δ_X gives an infinitesimal transformation in the solution space of the field equations.

We illustrate this fact in a simple example. It shows the essential elements for a proof in the general case. Suppose we have a single scalar field φ in Minkowski space and a Lagrangean density

$$\mathcal{L} = \frac{1}{2}\partial_\mu\varphi\partial^\mu\varphi - \frac{1}{4}g\varphi^4. \tag{I.4.5}$$

The wave equation is

$$\Box\varphi + g\varphi^3 = 0. \tag{I.4.6}$$

If φ is a solution of (I.4.6) the condition that $\varphi + \delta\varphi$ is again a solution reads, for infinitesimal $\delta\varphi$,

$$\Box\delta\varphi + 3g\varphi^2\delta\varphi = 0. \tag{I.4.7}$$

This is a linear equation for $\delta\varphi$ since we regard φ (the solution around which we vary) as given. On the other hand, taking the difference between the Euler equations for the modified and the original Lagrangean, one sees that $\delta^R_X\varphi$ and $\delta^A_X\Phi$ are solutions of an inhomogeneous equation

$$\Box\psi + 3g\varphi^2\psi = q_X(\varphi) \tag{I.4.8}$$

with the respective boundary conditions $\psi = 0$ for $t \to -\infty$ in the case $\psi = \delta^R_X\varphi$ and $\psi = 0$ for $t \to +\infty$ if $\psi = \delta^A_X\varphi$. The inhomogeneous term q_X drops out when we take the difference between the retarded and advanced solutions. Thus $\delta_X\varphi$ as defined in (I.4.4) satisfies (I.4.7). □

Now let Y be another quantity, expressable like X as an integral of a local function of the fields and their derivatives over a space-time region of bounded temporal extension. Then the action of δ_X on Y can be identified with the Poisson bracket $\{X, Y\}$ as originally defined within the canonical formalism. One has

$$\{X, Y\} = \delta_X Y = -\delta_Y X. \tag{I.4.9}$$

It suffices to verify this for the basic quantities (linear functionals of the fields) which is straightforward.

The double rôle of physical quantities is at the root of the very important relation between conservation laws and symmetries. A symmetry may be defined as a transformation $\Phi \to \Phi'$, $x \to x'$ with the two properties:

a) If Φ satisfies the field equations so does Φ'.

b) The transformation respects locality i.e. $\Phi'(x')$ depends only on $\Phi(x)$ (and possibly derivatives of Φ at this point).

Here, as above, Φ and Φ' are understood as denoting all fields which occur in the theory. We have seen that each physical quantity X defines an infinitesimal transformation δ_X satisfying a). If X is an integral of a local function of the fields over a 3-dimensional space-like surface Σ_0 then b) will be satisfied for those points which lie on this surface but not, in general, for later or earlier points. If, however, the integrand is a closed 3-form

$$X = \int_{\Sigma_0} J_{\alpha\beta\gamma} d\sigma^{\alpha\beta\gamma}, \tag{I.4.10}$$

$$dJ = 0 \tag{I.4.11}$$

then the integral (I.4.10) is independent of the choice of the surface; we may choose it to pass through any point. So b) will be satisfied and X defines an infinitesimal symmetry. In physics it is customary to write (I.4.10) as a surface integral of a conserved current; (I.4.11) is then the continuity equation:

$$X = \int_{\Sigma} j^\mu d\sigma_\mu, \tag{I.4.12}$$

$$\partial_\mu j^\mu = 0. \tag{I.4.13}$$

This results from (I.4.10), (I.4.11) by taking the Hodge duals (see section 2.3).

In special relativity the above reasoning is unaffected if j has additional tensor indices. For instance energy-momentum is the integral of the energy-momentum tensor $T^{\mu\nu}$ over a spacelike surface

$$P^\mu = \int_{\Sigma} T^{\mu\nu} d\sigma_\nu, \tag{I.4.14}$$

and the conservation law is locally expressed by the continuity equation

$$\partial_\mu T^{\mu\nu} = 0. \tag{I.4.15}$$

As a consequence P^μ, considered as defining an infinitesimal transformation by means of the Poisson bracket, generates a symmetry. It gives the infinitesimal translations in space-time. The same argument can be used for the angular momentum tensor $\Theta^{\mu\nu\varrho}$ in space-time (for its definition see e.g. [*Wentzel 1949*]). Its surface integral leads to the generators $M^{\mu\nu}$ of Lorentz transformations. Note incidentally that M^{0k} is not a constant of motion, corresponding to the fact that

Lorentz transformations do not commute with time translations. Therefore we have used the property b) to define the notion of symmetry which is more natural and more general than the frequently made statement that a symmetry should commute with time translations (compare the discussion in section 3.2 of chapter III). Correspondingly we regard the continuity equation as the expression of a conservation law rather than restricting attention to constants of motion.

The converse argument, leading from a continuous symmetry group to a conservation law, is known as *Noether's theorem*. Formulations and proofs of it at various levels of generality are found in many books.

In the past decades the direct use of a Lagrangean in the formulation of quantum theory has come more and more to the foreground. The basic idea, suggested by Feynman [Feyn 48], see also [*Feynman and Hibbs 1965*] is that in quantum mechanics every path leading from a point q_1 in configuration space at time t_1 to another point q_2, t_2 should be considered, not only the path following the classical equation of motion. The latter is the path giving an extremal value to the action. According to Feynman's proposal each path Γ contributes a probability amplitude exp $i/\hbar S(\Gamma)$ and the addition of all these amplitudes, the functional integration over all paths, gives the quantum mechanical transition amplitude from q_1, t_1 to q_2, t_2. The passage to the classical theory for $\hbar \to 0$ is visible then from the method of stationary phase in the evaluation of the functional integral. The path for which S is stationary with respect to small variations, i.e. the classical path, gives the main contribution.

The transfer of this idea to quantum field theory leads to the problem of integration over all sections of a field bundle. For its practical application in the computation of vacuum expectation values see e.g. [*Glimm and Jaffe 1987*] and the references given there.

I.5 Basic Quantum Field Theory

I.5.1 Canonical Quantization

The original objective of quantum field theory was to develop a quantum version of Maxwell's electrodynamics. This was to some extent achieved between 1927 and 1931 by applying to the Maxwell theory the same formal rules of "quantization" which had proved so successful in the transition from classical to quantum mechanics. For the free Maxwell equations (absence of charged matter) this approach was a full success. It led to a correct description of the properties of light quanta. The two free Maxwell equations taken together imply that each component of the field strength tensor $F_{\mu\nu}$ obeys the wave equation

$$\Box F_{\mu\nu} = 0; \quad \Box = g^{\mu\nu}\partial_\mu\partial_\nu. \tag{I.5.1}$$

In addition, the first order equations give constraints which couple the different components and yield the transversality of electromagnetic waves. Before discussing the complications due to the constraints let us treat as a preliminary exercise the case of a scalar field Φ obeying the Klein-Gordon equation, considered as a classical field equation

$$\Box\Phi + m^2\Phi = 0. \tag{I.5.2}$$

This example will already provide the essential lesson to be learnt from the canonical quantization of a linear field theory. The Lagrange density is

$$\mathcal{L} = \frac{1}{2} g^{\mu\nu}\partial_\mu\Phi\partial_\nu\Phi - \frac{1}{2}m^2\Phi^2. \tag{I.5.3}$$

To copy the quantization rules of section 1 one has to treat space and time in an asymmetric fashion, considering at some fixed time the spatial argument $\mathbf{x}$ of Φ as a generalization of the index i so that the field is regarded as a mechanical system with continuously many degrees of freedom. The partial derivatives $\partial/\partial q_i$ are replaced by the variational derivatives $\delta/\delta\Phi(\mathbf{x})$ and the Kronecker symbol δ_{ik} is replaced by the Dirac δ-function $\delta^3(\mathbf{x}-\mathbf{x}')$. The Lagrange function results from the Lagrange density (I.5.3) by integration over 3-dimensional space at a fixed time, say $t = 0$. The canonically conjugate momentum to Φ (at this time), denoted by $\pi(\mathbf{x})$, turns out to be equal to the time derivative of Φ and one imposes the canonical commutation relations for the initial values at $t = 0$

$$[\pi(\mathbf{x}), \Phi(\mathbf{x}')] = -i\delta^3(\mathbf{x}-\mathbf{x}'); \quad [\Phi(\mathbf{x}), \Phi(\mathbf{x}')] = [\pi(\mathbf{x}), \pi(\mathbf{x}')] = 0. \tag{I.5.4}$$

The Hamiltonian is

$$H = \frac{1}{2}\int \left(\pi^2 + (\mathrm{grad}\,\Phi)^2 + m^2\Phi^2\right) d^3x. \tag{I.5.5}$$

Since this procedure is described in all early presentations of quantum field theory, see e.g. [*Wentzel 1949*] there is no need to elaborate on it.

A more elegant method, avoiding the preference of a particular Lorentz system and fixed time is provided by Peierls' method (see section 4). One finds directly the commutation relations between fields at arbitrary (not necessarily equal) times[1]

$$[\Phi(x), \Phi(x')] = i\Delta(x-x') \tag{I.5.6}$$

where Δ is the difference between the retarded and the advanced Green's function of the Klein-Gordon equation (I.5.2). It is a singular integral kernel, solution of the Klein-Gordon equation with initial conditions

$$\Delta(z)\Big|_{z^0=0} = 0; \quad \frac{\partial\Delta}{\partial z^0}\Big|_{z^0=0} = -\delta^3(\mathbf{z}). \tag{I.5.7}$$

Two important remarks follow from (I.5.6) and (I.5.7):

[1]The commutation relation (I.5.6) was first proposed by Jordan and Pauli (1928) in the case $m = 0$.

a) The commutator of fields at different space-time points is a "c-number", (a multiple of the identity operator); it commutes with all observables.

b) The commutator vanishes if x and x' lie space-like to each other.

While property a) is special to free field theories (linear field equations) property b) has general significance and a simple physical reason: the measurement of the field in a space-time region $\mathcal{O}_1$ cannot perturb a measurement in a region $\mathcal{O}_2$ which is space-like to $\mathcal{O}_1$ because there can be no causal connection between events in the two regions. Therefore such observables are compatible (simultaneously measurable).

I.5.2 Fields and Particles

The most important consequence of the quantization is seen by Fourier transforming the field. Due to the field equations the support of the Fourier transform in p-space is concentrated on the two hyperboloids

$$p^0 = \pm(\mathbf{p}^2 + m^2)^{1/2} \tag{I.5.8}$$

and we can write

$$\Phi(t, \mathbf{x}) = (2\pi)^{-3/2} \int \left(a(\mathbf{p})e^{i(\mathbf{px}-\varepsilon_{\mathbf{p}}t)} + a^*(\mathbf{p})e^{-i(\mathbf{px}-\varepsilon_{\mathbf{p}}t)}\right) \frac{d^3p}{2\varepsilon_{\mathbf{p}}} \tag{I.5.9}$$

with

$$\varepsilon_{\mathbf{p}} = (\mathbf{p}^2 + m^2)^{1/2}. \tag{I.5.10}$$

If the classical field is real valued then the quantum field should be Hermitean which implies that $a^*(\mathbf{p})$ is the Hermitean adjoint of $a(\mathbf{p})$. The canonical commutation relations (I.5.4) give

$$[a(\mathbf{p}), a^*(\mathbf{p}')] = 2\varepsilon_{\mathbf{p}}\delta^3(\mathbf{p} - \mathbf{p}'); \quad [a(\mathbf{p}), a(\mathbf{p}')] = [a^*(\mathbf{p}), a^*(\mathbf{p}')] = 0. \tag{I.5.11}$$

Comparing with the discussion in Section 3, equation (I.3.61), we see that the canonical quantization of the free scalar field Φ leads to the quantum theory of a system of arbitrarily many identical, noninteracting, spinless particles obeying Bose statistics.

To be precise we have to supplement the relations (I.5.11) by the requirement that the representation space contains a vector Ψ_0, interpreted as the vacuum, which is annihilated by all the $a(\mathbf{p})$:

$$a(\mathbf{p})\Psi_0 = 0 \quad \text{for all} \quad \mathbf{p}. \tag{I.5.12}$$

Then the $a(\mathbf{p})$ can be interpreted as annihilation operators, the $a^*(\mathbf{p})$ as creation operators in the Fock space generated by repeated application of the creation operators on the vacuum. The requirement (I.5.12) has to be added to the algebraic relations (I.5.11) because the latter allow many inequivalent irreducible

Hilbert space representations of which (I.5.12) selects the one in Fock space. This is due to the fact (mentioned in Section 1) that in the case of infinitely many degrees of freedom one does not have von Neumann's uniqueness theorem for the representation of the canonical commutation relations.

There is another problem which has to be faced. The quantum field Φ at a point cannot be an honest observable. Physically this appears evident because a measurement at a point would necessitate infinite energy. The mathematical counterpart is that $\Phi(x)$ is not really an operator in $\mathcal{H}_F$. It is an "operator valued distribution" (see Chapter II, section 1.2) or, alternatively, it may be defined as a sesquilinear form on some dense domain in $\mathcal{H}_F$. This means that matrix elements $\langle\Psi_2|\Phi(x)|\Psi_1\rangle$ are finite if both vectors Ψ_1 and Ψ_2 belong to a dense domain $\mathcal{D} \subset \mathcal{H}_F$ characterized by the property that the probability amplitudes for particle configurations decrease fast with increasing momenta and increasing particle number. The problem is then that we cannot directly multiply "field operators" at the same point. This problem has plagued quantum field theory throughout its history. It is a fierce obstacle when one wants to define nonlinear local field equations. In the case of free fields to which our discussion so far was confined, there is, however, a simple solution. In this case (I.5.9) gives a clear separation of the annihilation part and the creation part of $\Phi(x)$. We may write

$$\Phi(x) = a(x) + a^*(x). \tag{I.5.13}$$

One observes [Wick 50] that any product of the form $a^*(x)^n a(x)^m$ is a well defined sesquilinear form on a dense domain or, alternatively speaking, it is an operator valued distribution. If A and B are two functionals of the field Φ one defines the *normal order product*, also called the *Wick product* of A and B by the prescription that each field operator occurring in A or B is expanded in the form (I.5.13) and in the resulting formal expression for the product AB the order of factors is changed so that all creation operators stand on the left, all annihilation operators on the right in each term. Denoting the normal order product by $:AB:$ we have for instance

$$:\Phi(x)^2: = a^*(x)a^*(x) + 2a^*(x)a(x) + a(x)a(x). \tag{I.5.14}$$

The Hamiltonian is obtained from the classical expression (I.5.5) by inserting for Φ the quantum field and taking the products in the normal order. One obtains

$$H = P^0 = \int \varepsilon_{\mathbf{p}} a^*(\mathbf{p})a(\mathbf{p})d\mu(p). \tag{I.5.15}$$

Similarly one gets the other generators of the Poincaré group from their classical counterparts.

The apparent miracle that canonical quantization of a free, scalar field leads to Fock space and an interpretation of states in terms of particle configurations has been seen as a manifestation of the wave-particle duality lying at the roots of quantum mechanics. The generalization of this observation to a field-particle duality has dominated thinking in quantum theory for decades and has been heuristically useful in the development of elementary particle theory. A prime example is Fermi's theory of the β-decay which associates a field with each of the relevant particle types: the proton, the neutron, the electron and the neutrino. From these fields taken at the same point in x-space one builds a Lagrangean density which has matrix elements for the transition n$\rightarrow p+e^-+\overline{\nu}_e$. Yet the belief in field-particle duality as a general principle, the idea that to each particle there is a corresponding field and to each field a corresponding particle has also been misleading and served to veil essential aspects. *The rôle of fields is to implement the principle of locality. The number and the nature of different basic fields needed in the theory is related to the charge structure, not to the empirical spectrum of particles. In the presently favoured gauge theories the basic fields are the carriers of charges called colour and flavour but are not directly associated to observed particles like protons.*

I.5.3 Free Fields

Equation (I.5.9) may be read either as giving the decomposition of a (scalar) quantum field into the set of creation-annihilation operators in the Fock space of a species of (spinless) particles or, conversely, as the construction of a covariant, local field from the representation $U_{m,s}$ of the Poincaré group (for $s=0$) via (I.3.51), (I.3.52), (I.3.61). The construction of a local, covariant field can also be carried through for the other relevant types of irreducible representations of $\mathfrak{P}$. The step from the single particle theory to the theory in Fock space is often called *second quantization* because, with a due amount of sloppiness, one regards the single particle theory to result by "quantization" from the classical mechanical model of a point particle and, interpreting the single particle theory as a classical field theory, the transition to Fock space is again a quantization, achieved by replacing the single particle wave functions by creation-annihilation operators with the "canonical commutation relations" (I.3.61).

For $s=1/2$ the construction leads to the Majorana field. If $\varphi_r(p)$ denotes a single particle wave function ($r=1,2$) as in equation (I.3.38) we may write the corresponding creation operator as[2]

$$a^*(\varphi)=\epsilon^{rs}\int \overline{a}_s(p)\varphi_r(p)d\mu(p) \tag{I.5.16}$$

which defines the (improper) creation operators $\overline{a}(p)$ in p-space.

To obtain a local quantum field $\Phi_r(x)$ we must combine the creation and annihilation operators as in (I.5.9)

$$\Phi_r(x)=a_r(x)+\overline{a}_r(x). \tag{I.5.17}$$

We assume Fermi statistics. The resulting commutation relations are then

$$[\Phi_r(x),\Phi_s(y)]_+=\epsilon_{sr}\Delta(x-y). \tag{I.5.18}$$

[2] ϵ^{rs} is the "metric tensor in spinor space" (see section 2).

Furthermore, since the Majorana field describes a single species of particles (no antiparticles) the Hermitean adjoint $\Phi^*_{\dot{r}}$ of Φ_r is related to Φ:

$$\Phi^*_{\dot{r}}(x) = \epsilon_{\dot{r}\dot{s}} m^{-1} \left(\frac{\partial}{\partial t} - \sigma \nabla \right)^{\dot{s}s'} \Phi_{s'}(x). \tag{I.5.19}$$

Thus $\Phi(x)$ anticommutes with $\Phi(y)$ and with $\Phi^*(y)$ if x and y lie space-like to each other. If we say that local observables are formed by taking the normal ordered product of an even number of factors Φ, Φ^* at the same point then we have commutativity of local observables at space-like separation as demanded by the locality principle. The transformation property of the field is

$$U(\alpha)\Phi_r(x)U^{-1}(\alpha) = (\alpha^{-1})_r^{r'}\Phi_{r'}(\Lambda x); \quad \Lambda = \Lambda(\alpha). \tag{I.5.20}$$

The analogous construction can be carried through for particles of higher spin if we describe the single particle wave functions without redundant components as symmetric spinors of higher rank (see section 3). One then arrives at local fields transforming like symmetrized spinor fields of higher rank. The case of rank 2 with mass zero gives the free Maxwell field. The symmetric spinors Φ_{rs} and their complex conjugates $\Phi^*_{\dot{r}\dot{s}}$ correspond to the self-dual tensors

$$\mathcal{F}^{\pm}_{\mu\nu} = \frac{1}{2}(F_{\mu\nu} \pm i * F_{\mu\nu}). \tag{I.5.21}$$

A plane wave solution of the Maxwell equations fixes them uniquely up to a normalization factor. They describe respectively right hand and left hand circularly polarized waves or, if one wishes, wave functions of a particle with sharp momentum, zero rest mass and helicity $+1$ and -1 respectively. We may introduce annihilation operators $a^{\pm}(p)$ for these modes and creation operators $a^{\pm *}(p)$. From these we can build up a local, Hermitean quantum field $F_{\mu\nu}(x)$ acting in the Fock space of photons. One cannot, however, find a covariant quantum field corresponding to the vector potential acting in this Fock space. The standard method to incorporate the vector potential is due to Gupta and Bleuler, [Gupt 50], [Bleu 50].[3] It uses a "Hilbert space with indefinite metric", a space equipped with a sesquilinear scalar product for which $\langle\Psi|\Psi\rangle$ is not positive definite. In this space one characterizes a certain subspace called the "physical state space", in which the metric is semidefinite. It still contains vectors of zero length. Then one forms equivalence classes; two vectors in the ("physical") space are called equivalent if their difference has zero length. Each such equivalence class correponds to a vector in the Fock space generated by the electromagnetic fields.

Note that the Fock space construction must always be modified if one uses single particle wave functions with redundant components, e.g. if one describes spin 1 particles by 4-vector wave functions. In that example we have a constraint

[3] For this and the remaining material in this section see any standard text on quantum field theory e.g. [*Jauch and Rohrlich 1976*] We restrict the exposition to a few remarks.

$p^\mu \varphi_\mu(p) = 0$ and the creation operators $a^*(\varphi)$ do not define creation operators $a^*_\mu(p)$ in p-space. If one wants to define such operators and, correspondingly, vector fields $\Phi_\mu(x)$ one needs a larger space than $\mathcal{H}_F$ and subsidiary conditions to single out the physical states.

Reviewing our construction of free fields from the irreducible representations of $\overline{\mathfrak{P}}$ we notice that the two most important fields, the Dirac field and the electromagnetic potential are not directly obtained. The Dirac field results if one starts from two types of particles (particle and antiparticle) whose description is interwoven so that it is not equivalent to two Majorana fields. We define the first two components of the field $\Phi_r(x)$ by combining the annihilation operators of the particle (as its negative energy part) with the creation operators of the antiparticle (as its positive energy part). Then Φ and Φ^* are independent; the Hermiticity condition (I.5.19) disappears. As in (I.3.41), (I.3.42) one goes over to the 4-component field $\psi = \begin{pmatrix} \Phi \\ \chi \end{pmatrix}$. The physically important point is seen if we introduce the *charge* as the difference between the number of particles and the number of antiparticles. All components of ψ lower the charge by one unit, the components of ψ^* raise it. This may serve as a reminder that the deeper significance of the fields is to effect a local change of charge, not of particle number. The observable fields are formed by multiplying components of ψ^* and ψ at the same point. The most important one is the current density which, in customary notation, is written as

$$j^\mu(x) = q : \overline{\psi}(x)\gamma^\mu\psi(x) : \tag{I.5.22}$$

with

$$\overline{\psi} = \psi^*\gamma^0. \tag{I.5.23}$$

Due to their construction the observables do not only commute with the charge but they remain unchanged if ψ is subjected to a *local gauge transformation*

$$\psi(x) \to e^{i\lambda(x)}\psi(x) \tag{I.5.24}$$

with an arbitrary function $\lambda(x)$.

I.5.4 The Maxwell-Dirac System

The Dirac equation, originally conceived as the wave equation for a particle with charge q in an electromagnetic field described by the vector potential A_μ, is

$$\{\gamma^\mu(i\partial_\mu + qA_\mu(x)) - m\}\,\psi(x) = 0. \tag{I.5.25}$$

It results from the free Dirac equation by the substitution

$$\partial_\mu \to D_\mu = \partial_\mu - iqA_\mu, \tag{I.5.26}$$

which corresponds to the old observation in the classical electron theory that the Lorentz force results in the Hamiltonian formalism by the simple replacement

$$p_\mu \rightarrow p_\mu + qA_\mu. \tag{I.5.27}$$

But the root of this rule is not visible in classical mechanics. It is local gauge invariance: if we subject ψ and A simultaneously to the gauge transformations[4]

$$\psi(x) \rightarrow \psi'(x) = e^{iq\lambda(x)}\psi(x); \quad A_\mu(x) \rightarrow A'_\mu(x) = A_\mu(x) + \partial_\mu \lambda(x) \tag{I.5.28}$$

then (I.5.25) remains valid for the transformed fields and A'_μ defines the same electromagnetic field strengths as A_μ. In differential geometric language the Dirac wave function is a section in a fiber bundle whose base space is Minkowski space and whose typical fibre is a complex 4-dimensional space with the structure group $U(1)$. The latter, because a point in the fiber (the value of the Dirac spinor ψ) is physically relevant only up to a phase factor which may be considered as an element of the group $U(1)$. The observable fields are of the form $\overline{\psi}\Gamma\psi$ with $\Gamma = \gamma^\mu, \gamma^\mu\gamma^\nu \cdots$. They remain all unchanged if ψ is multiplied by a phase factor i.e. if the section is subjected to a gauge transformation (I.5.28).

The vector potential provides a $U(1)$ connection in the bundle, analogous to the rôle of the affine connection, the parallel transport, in the case of the tangent bundle (see subsection 2.3). It allows the comparison of phases of ψ in infinitesimally neighbouring points. The electromagnetic field strength is the curvature of this connection. If it does not vanish then we have no natural comparison of phases at a distance. The operator D_μ of (I.5.26) is the covariant derivative in this bundle.

To complete the system of field equations we have to add to (I.5.25) the Maxwell equation (I.2.56) with the current density j^μ given by (I.5.22) and the expression (I.2.55) for the electromagnetic field in terms of the potential. These relations are obviously all invariant under the gauge transformations (I.5.28). The full system of equations follows from a variational principle with the Lagrange density

$$\mathcal{L} = \overline{\psi}(\gamma^\mu D_\mu - m)\psi - \frac{1}{4}F^{\mu\nu}F_{\mu\nu} \tag{I.5.29}$$

where it is understood that $F_{\mu\nu}$ is written in terms of A_μ and the basic variables which are varied independently in the variational principle are ψ, $\overline{\psi}$ and A. We can write

$$\mathcal{L} = \mathcal{L}_D + \mathcal{L}_M + \mathcal{L}_I \tag{I.5.30}$$

where $\mathcal{L}_D$ is the Lagrange density of the free Dirac theory

$$\mathcal{L}_D = \overline{\psi}(i\gamma^\mu\partial_\mu - m)\psi, \tag{I.5.31}$$

$\mathcal{L}_M$ the Lagrange density of the free Maxwell theory

$$\mathcal{L}_M = -\frac{1}{4}F^{\mu\nu}F_{\mu\nu} \tag{I.5.32}$$

[4]The second part of (I.5.28) is often called a gauge transformation of the second kind, while the first part, the transformation of the charged field, is called a gauge transformation of the first kind.

and $\mathcal{L}_I$ the Schwarzschild interaction Lagrangean coupling the Dirac and the Maxwell fields:

$$\mathcal{L}_I = j^\mu A_\mu. \tag{I.5.33}$$

Remarks: 1) As long as we consider $\psi_\alpha(x)$ and $A_\mu(x)$ as numerical valued functions we may omit the Wick ordering symbol in (I.5.22). The system of field equations (I.5.25), (I.2.56) with the definitions (I.2.55), (I.5.22) constitutes a mathematically well defined classical field theory which, moreover, is built from principles remarkably similar to those used in the construction of the general theory of relativity. But for physics we cannot interpret ψ as a classical field nor can we interpret it in general as a quantum mechanical wave function of a single particle because, if the vector potential is time dependent, one cannot separate the positive and negative energy parts of ψ. We must regard ψ and A as quantum fields. Then the definition of the products $\psi(x)A(x)$ occurring in the Dirac equation and $\overline{\psi}\psi$ occurring in (I.5.22) poses a problem which is not as easily dealt with as in the free field case where we can define a Wick product. In fact, this problem was considered as unsurmountable without a radical change of the basic concepts until the successful development of renormalized perturbation theory suggested that it could be overcome.

2) In classical Maxwell theory the vector potential is introduced as a convenient auxiliary tool. It serves to solve the homogeneous Maxwell equation (I.2.54) and it also allows to derive the Lorentz force on a charged particle from a variational principle by adding the Schwarzschild term (I.5.33) to the Lagrangean of the free motion. The interpretation of A_μ as the connection in a fiber bundle appears only when charged matter is described by a complex matter field, a feature which (at least in its physical interpretation) is tied to quantum physics. Otherwise there is nothing which A_μ connects. We may conclude therefore that in quantum physics the vector potential acquires a fundamental significance which it does not have in the classical theory. We encounter in the Maxwell-Dirac theory besides the pointlike observables like j^μ, $F_{\mu\nu}$ also "stringlike" observables associated to any path Γ from a point x to a point y, namely

$$A(\Gamma) = \overline{\psi}(x) \exp\left(-iq \int_\Gamma A_\mu(x')dx'^\mu\right) \psi(y). \tag{I.5.34}$$

This is brought into sharp focus by the observation of Aharonov and Bohm [Ahar 59]: If the accessible space available to an electron is not simply connected e.g. by excluding a toroidal tube, then we may produce a situation in which the electromagnetic field strengths vanish everywhere in the accessible region while the vector potential cannot vanish there if there is a magnetic flux inside the tube. The line integral of the vector potential around the tube must equal this flux. The time development of the electron wave function is influenced by the vector potential (i.e. by the total flux in the tube) even if the electron is confined entirely to the force-free region.

I.5.5 Processes

The quantum field theory based on the Maxwell-Dirac system is called Quantum Electrodynamics (QED). To study its physical consequences one started from the free field theory, resulting if $\mathcal{L}_I$ is neglected, as a zero order approximation. There we have the Fock space of noninteracting photons, electrons and positrons. $\mathcal{L}_I$ is considered as a perturbation producing transitions between particle configurations. The elementary process is the emission or annihilation of a photon combined with the corresponding change of momentum of an electron or positron or the creation or annihilation of an electron-positron pair. Originally Dirac's time dependent perturbation theory, giving a transition probability per unit time, ("the golden rule of quantum mechanics" as it was called by Fermi) was used to derive the probability for spontaneous decay of an excited state of an atom, the cross sections for the photo effect, Compton effect, electron-electron scattering, pair creation, Bremsstrahlung. An excellent presentation of this early stage of the theory is given in Heitler's book [*Heitler 1936*]. The results of the lowest order of perturbation theory were found to be in very good agreement with experiment. But there remained a dark cloud. One could not continue the approximation. The higher order terms lead to divergent integrals. The hope that QED contained any sensible physical information beyond the lowest order calculations dwindled during the thirties. Faith was restored in a spectacular way by joint progress in experimental precision and theoretical understanding between 1946 and 1949. Experiments established small deviation of the hydrogen spectrum from the theoretical predictions of wave mechanics ("Lamb shift") and a deviation of the gyromagnetic ratio of the free electron from the value 2 which it should have according to the 1-particle Dirac theory ("anomalous magnetic moment"). On the theoretical side it was the development of an elegant, manifestly covariant form of the perturbation expansion by Tomonaga, Schwinger, Feynman together with the idea of *renormalization*, suggested by H.A. Kramers and first applied to the Lamb shift problem by Bethe which allowed the computation of higher order corrections, and it turned out quickly that these corrections could account for the experimental findings. In subsequent years, as the experiments were refined and the theoretical computations were pushed to the 4-th order and ultimately to the 6-th order, the agreement between experiment and theory became truly spectacular.

We end this chapter with a quotation from the basic paper by Dyson [Dy 49]: "Starting from the methods of Tomonaga, Schwinger, Feynman and using no new ideas or techniques, one arrives at an S-matrix from which the well known divergences seem to have conspired to eliminate themselves. This automatic disappearance of divergences is an empirical fact, which must be given due weight in considering the future prospects of electrodynamics."

II. General Quantum Field Theory

II.1 Mathematical Considerations and General Postulates

The developments mentioned at the end of the last chapter had restored faith in quantum field theory. On the other hand it could not be overlooked that in spite of the great success of QED there remained ample reasons for dissatisfaction. It was not understood how the theory could be formulated without recourse to the perturbation expansion. The detailed renormalization prescriptions, needed to eliminate all infinities, had become quite complicated and not easily communicable to one who had not acquired familiarity with the procedure the hard way, namely by doing the computations and learning to avoid pitfalls. Apart from QED there existed models for a theory of weak interactions which could be compared in lowest order with experiment but which was not renormalizable, and there were the meson theories of strong interaction where perturbation expansions did not prove to be very useful. This mixture of positive and negative aspects of quantum field theory provided the motivation in the fifties to search for a deeper understanding of the underlying principles and for a more concise mathematical formulation. K.O. Friedrichs described his feelings about the literature on quantum field theory as akin to the challenge felt by an archeologist stumbling on records of a high civilization written in strange symbols. Clearly there were intelligent messages, but what did they want to say? (private conversation 1957). His answer to the challenge was his book on the mathematical aspects of quantum field theory [*Friedrichs 1953*] where, among other things, he pointed out the existence of inequivalent representations of the canonical commutation relations and discussed examples of these under the heading "myriotic fields".

II.1.1 The Representation Problem

The breakdown of von Neumann's uniqueness theorem (see I.1) in the case of infinitely many degrees of freedom had already been noted in 1938 by von Neumann himself but it took a long time until this fact became widely known and its significance for quantum field theory was recognized. The phenomenon is easily understood if we start from the relations in the form (I.3.60) which results from the Heisenberg relations (I.1.2) by substituting

$$a_k = 2^{-1/2}(p_k - iq_k); \quad a_k^* = 2^{-1/2}(p_k + iq_k) . \tag{II.1.1}$$

Using the notation of (I.3.54) we note that an occupation number n_k is an eigenvalue of the positive operator $N_k = a_k^* a_k$ and the commutation relations imply that n_k must be a non negative integer. An occupation number distribution (n) is an infinite sequence of such integers. We can divide the set of such sequences into classes, saying that $(n^{(1)})$ and $(n^{(2)})$ are in the same class if the sequences differ only in a finite number of places. Application of a creation or annihilation operator changes (n) only in one place. Therefore the basis vectors $\Psi_{(n)}$ introduced in subsection I.3.4, *with* (n) *restricted to one class*, already span a representation space of the a_k, a_k^* and it is evident that representations belonging to different classes cannot be unitarily equivalent. The set of representations obtained in this way does, however, not exhaust by far all possibilities. There are representations in which the $\Psi_{(n)}$ do not appear as normalizable vectors. Let us illustrate this in the case of the anticommutation relations (I.3.63) to which, of course, the same reasoning as above may be applied. There the numbers n_k are restricted to the value 0 or 1 and an occupation number distribution, being an infinite sequence of such numbers, is in one to one correspondence with a real number in the interval $[0, 1]$ because it may be regarded as defining an infinite binary fraction. One may then construct representations acting in a Hilbert space of functions over this interval, square integrable with respect to some measure.

A systematic study of the classification problem of irreducible representations of canonical commutation and anticommutation relations was undertaken by Gårding and Wightman [Gard 54a, 54b]. An exhaustive, practically usable list of representations appears unattainable. This follows from the study of the nature of the algebras, see [Mackey 57] and [Glimm 61].

The first examples in quantum field theory where "strange representations" appeared were given by Friedrichs loc. cit. and van Hove [Hov 52]. It soon became evident that the occurrence of such representations was not an exception but a generic feature [Haag 54, 55a]. Specifically: the standard quantization procedure of a classical field theory leads to field quantities which satisfy canonical commutation relations at a fixed time. Let us take the simplest example, a scalar field φ with Lagrangean (I.4.5). For $g = 0$ we have a free field. For $g \neq 0$ the field equations are non linear and one expects a theory with interacting particles. The equal time commutation relations between φ and its canonically conjugate momentum $\pi = \partial\varphi/\partial x^0$ are independent of g; they are given by (I.5.4) (replacing Φ by φ). We need a representation of the relations (I.5.4) in a Hilbert space $\mathcal{H}$, the space of physical states. In fact, we want an irreducible representation because the field at arbitrary time is determined by the canonical quantities due to the field equations and thus all observables can be expressed in terms of the $\varphi(\mathbf{x})$, $\pi(\mathbf{x})$.

We want to argue that the representations must be inequivalent for different values of g. If this were not so we could use one Hilbert space in which $\varphi(\mathbf{x})$ and

$\pi(\mathbf{x})$ are represented for different values of g by the same (improper) operators. For each g we expect to have one physically distinguished state, the physical vacuum. Its state vector Ω_g must depend on g because it is the lowest eigenstate of the Hamiltonian and the latter depends on g. On the other hand the operators $U(\mathbf{a})$, representing translations in 3-space, would not depend on g because, for any g, they give the same transformation of the canonical quantities

$$U(\mathbf{a})\varphi(\mathbf{x})U^{-1}(\mathbf{a}) = \varphi(\mathbf{x}+\mathbf{a}), \quad U(\mathbf{a})\pi(\mathbf{x})U^{-1}(\mathbf{a}) = \pi(\mathbf{x}+\mathbf{a}), \tag{II.1.2}$$

and, in an irreducible representation, they are determined (up to a numerical factor of modulus 1) by (II.1.2). The vacuum is invariant under space translations,

$$U(\mathbf{a})\Omega_g = \Omega_g. \tag{II.1.3}$$

For $g = 0$ we have a free field and the Fock representation as described earlier. There the vacuum state vector is the only invariant vector under the $U(\mathbf{a})$. If, for some other value of g, we also had the Fock representation then (II.1.3) would imply $\Omega_g = \Omega_0$. More generally, we shall see in III.3 that in an irreducible representation space of the field operators there can be at most one invariant vector under space translations. Thus if $\Omega_{g'} \neq \Omega_g$ then the representations must be inequivalent. We conclude therefore that the determination of the representation class of (I.5.4) is a dynamical problem. This conclusion is usually referred to as *Haag's theorem*, (compare also [Coest 60]).

Actually the above discussion touches only the tip of an iceberg. It appears that an infinite renormalization of the field is needed. In the renormalized perturbation expansion one relates formally the "true" field Φ to the canonical field φ which satisfies (I.5.4) by

$$\Phi = Z^{-1/2}\varphi \tag{II.1.4}$$

where Z is a constant (in fact zero). This means that the fields in an interacting theory are more singular objects than in the free theory and we do not have the canonical commutation relations (I.5.4).[1] While this decreases somewhat the interest in canonical commutation relations the representation problem remains significant whenever we have to deal with infinitely many degrees of freedom. We note that the algebraic structure, whatever it is, does not, in general, fix the Hilbert space representation. This phenomenon will concern us in chapter IV. For canonical commutation relations it plays an important rôle in the non relativistic many body problem, the statistical mechanics in the infinite volume limit. "Strange representations" appear naturally e.g. by Bogolubov transformations [Bog 58] and in the description of thermal states given by Araki and Woods [Ara 63a]. We shall come back to this in chapter V.

[1] In constructive field theory it was proved that field renormalization is finite in models based on 2-dimensional space-time. See [*Glimm and Jaffe 1987*]. So canonical commutation relations for interacting fields can be upheld in such cases. In Minkowski space theories an infinite renormalization seems to be needed and the relations (I.5.4) get lost.

II.1.2 Wightman Axioms

The need to base the discussion of quantum field theory on clearly stated postulates was felt acutely by several authors in the early fifties. Wightman and Gårding began in 1952 to isolate those features of quantum field theory which could be stated in mathematically precise terms and to extract general postulates which looked trustworthy in the light of the lessons learned from renormalization. This led to the "Wightman axioms".[2]

Let us state here the essential postulates. The problem of physical interpretation and description of particles will be postponed to sections 3 and 4.

A. Hilbert Space and Poincaré Group.

1. We deal with a Hilbert space $\mathcal{H}$ which carries a unitary representation of $\overline{\mathfrak{P}}$.

2. There is precisely one state (ray in $\mathcal{H}$), the physical vacuum, which is invariant under all $U(g)$, $g \in \mathfrak{P}$.

3. The spectrum of the energy-momentum operators P^μ is confined to the (closed) forward cone

$$p^2 \geq 0; \quad p^0 \geq 0. \tag{II.1.5}$$

B. Fields.

1. Fields are "operator valued distributions" over Minkowski space.[3]

   This needs an explanation. As already mentioned in subsection I.5.2 a quantum field $\Phi(x)$ at a point cannot be a proper observable. It may be regarded as a sesquilinear form on a dense domain $\mathcal{D} \subset \mathcal{H}$; this means that the matrix element $\langle \Psi_2 | \Phi(x) | \Psi_1 \rangle$ is a finite number when both Ψ_1 and Ψ_2 are in $\mathcal{D}$ and that it depends linearly on Ψ_1, conjugate linearly on Ψ_2. To obtain an operator defined on the vectors in $\mathcal{D}$ one has to average ("smear out") Φ with a smooth function f on Minkowski space i.e. take

$$\Phi(f) = \int \Phi(x) f(x) d^4x. \tag{II.1.6}$$

   If f belongs to the "test function space" then $\Phi(f)$ is an (unbounded) operator acting on $\mathcal{H}$, defined on some dense domain $\mathcal{D} \subset \mathcal{H}$. The test function space is usually taken as the space S of Laurent Schwartz, the space of infinitely often differentiable functions decreasing as well as their derivatives faster than any power as x moves to infinity in any direction. $\Phi(f)$

---

[2] The first published account is [Wight 57]. An extended version is given in [Wight 64]. Similarly motivated work [Haag 54, 55a], Lehmann, Symanzik and Zimmermann [Leh 55] placed less emphasis on mathematical precision and was more concerned with physical interpretation and particle aspects of the theory.

[3] Apart from the quoted work of Gårding and Wightman this has been recognized by Schmidt and Baumann [Schmidt 56].

depends linearly on f and its matrix elements are continuous functions of f with respect to the Laurent Schwartz topology of $\mathcal{S}$. If, as we shall always assume, the space of allowed test functions is $\mathcal{S}$, the distribution is called tempered. Its Fourier transform is again a tempered distribution. The mathematical theory of distributions is presented in [*Schwartz 1957*], [*Gelfand and Shilov 1964*].

2. The domain $\mathcal{D}$ should contain the vacuum state vector and be invariant under the application of the operators $U(a, \alpha)$ and $\Phi(f)$.

   Of course one will usually have to deal with several fields, each of which may have several tensor or spinor components. Correspondingly we must take test functions for each type (index i) and each component (index λ) and understand $\Phi(f)$ generically as

$$\Phi(f) = \sum_i \int \Phi^i_\lambda(x) f^{i\lambda}(x) d^4x. \tag{II.1.7}$$

Remarks: Renormalization theory suggests that it is essential to smear out Φ both in space and time, in contrast to the case of free fields, where an averaging over 3-dimensional space at a fixed time is sufficient. Due to the stronger singularities (see above) one cannot assume well defined commutation relations of fields at equal time.

C. Hermiticity.

The set of fields contains with each Φ also the Hermitean conjugate field Φ^*, defined as a sesquilinear form on $\mathcal{D}$ by

$$\langle \Psi_2 | \Phi^*(x) | \Psi_1 \rangle = \overline{\langle \Psi_1 | \Phi(x) | \Psi_2 \rangle}. \tag{II.1.8}$$

D. Transformation Properties.

The fields transform under $\overline{\mathfrak{P}}$ as

$$U(a, \alpha) \Phi^i_\lambda(x) U^{-1}(a, \alpha) = M^{(i)\varrho}_\lambda(\alpha^{-1}) \Phi^i_\varrho(\Lambda(\alpha)x + a), \tag{II.1.9}$$

here $M(\alpha)$ is a finite dimensional representation matrix of $\alpha \in \overline{\mathfrak{L}}$. Of course this should be properly written in terms of the smeared out fields but this rewriting is evident using (II.1.7).

E. Causality.

The fields shall satisfy causal commutation relations of either the bosonic or fermionic type. If the supports of the test functions f and h are space-like to each other then either

$$[\Phi^i(f), \Phi^j(h)] = 0 \tag{II.1.10}$$

or

$$[\Phi^i(f), \Phi^j(h)]_+ = 0. \tag{II.1.11}$$

F. Completeness.

By taking linear combinations of products of the operators $\Phi(f)$ one should be able to approximate any operator acting on $\mathcal{H}$. This may be expressed by saying that $\mathcal{D}$ contains no subspace which is invariant under all $\Phi(f)$ and whose closure is a proper subspace of $\mathcal{H}$. Alternatively one may say that there exists no bounded operator which commutes with all $\Phi(f)$ apart from the multiples of the identity operator (Schur's lemma).

G. "Time-slice Axiom". "Primitive Causality".

There should be a dynamical law which allows one to compute fields at an arbitrary time in terms of the fields in a small time slice

$$\mathcal{O}_{t,\varepsilon} = \{x : |x^0 - t| < \varepsilon\}. \tag{II.1.12}$$

Therefore the completeness, demanded under **F** should already apply when we restrict the support of the test functions f to $\mathcal{O}_{t,\varepsilon}$. This postulate was added somewhat later and in most of the work analyzing the consequences of the axioms it was not used. In chapter III we shall introduce a sharpened version of this postulate, taking the hyperbolic character of the dynamical laws in relativistic theories into account.

Remarks: The endeavour to study the consequences of the assumptions **A** to **G** (possibly supplemented by further assumptions concerning the particle spectrum and the physical interpretation) has, following Wightman, been commonly called *axiomatic quantum field theory*. This terminology is appropriate in the sense that the starting point, the total input, is clearly stated. On the other hand the word "axiom" suggests something fixed, unchangeable. This is certainly not intended here. Indeed, some of the assumptions are rather technical and should be replaced by more natural ones as deeper insight is gained. We are concerned with a developing area of physics which is far from closed and should keep an open mind for modifications of the assumptions, additional structural principles as well as information singling out a specific theory within the general frame. We shall endeavour to do this in later chapters and, following R. Jost, we prefer to use the term "general quantum field theory" rather than "axiomatic quantum field theory" because it allows more flexibility. We regard the postulates **A**–**G** as well as subsequent modifications and further structural assumptions as working hypotheses rather than as rigid axioms.

II.2 Hierarchies of Functions

The impact of covariant perturbation theory and the analysis of the consequences of the axioms lead in the fifties to a description of quantum field theory in terms of infinite hierarchies of functions.[1] Different choices of hierarchies were considered; the properties of the functions and their interrelation were studied, using as an input the general postulates (axioms supplemented by more specific information coming from experience with renormalized perturbation theory). The present section is intended as a brief description of the resulting picture.

II.2.1 Wightman Functions, Reconstruction Theorem, Analyticity in x-Space

We shall suppress the indices i, λ distinguishing different fields and components whenever they are not absolutely essential. Let us denote the state vector of the vacuum by Ω. The vacuum expectation values of products of fields

$$w^n(x_1, \cdots x_n) = \langle \Omega | \Phi(x_1) \cdots \Phi(x_n) | \Omega \rangle \tag{II.2.1}$$

are tempered distributions over $\mathbb{R}^{4n}$. This follows directly by the "nuclear theorem" of L. Schwartz from axiom **B**. The w^n are called Wightman distributions because, in a fundamental paper, Wightman pointed out the following facts [Wight 56]:

First, if the hierarchy of distributions w^n $(n = 0, 1, 2, \cdots)$ is known then the Hilbert space and the action of $\Phi(f)$ in it can be reconstructed. This *reconstruction theorem* is closely related to the Gelfand-Naimark-Segal construction which will be explained in the next chapter. The main point can be simply described. Due to the completeness axiom **F** the linear span of vectors

$$\Phi(f_1) \cdots \Phi(f_k) \Omega, \quad k = 0, 1, \cdots \infty \tag{II.2.2}$$

is dense in $\mathcal{H}$. If $\Phi(x)$ is contained in the set of fields so is $\Phi^*(x)$ (axiom **C**); hence the scalar product of two state vectors of the form (II.2.2) is known once we know all the w^n and the same is true for the matrix element of any $\Phi(f)$ between such state vectors.

Secondly, the w^n are boundary values of analytic functions of $4n$ complex variables. We sketch the argument using freely somewhat heuristic manipulations. A precise discussion is given in the books [*Streater and Wightman 1964*], [*Jost 1965*].

The vacuum is translation invariant

$$U(a)\Omega = \Omega. \tag{II.2.3}$$

Due to the translation covariance of the fields (axiom **D**)

$$\Phi(x) = U(x)\Phi(0)U^{-1}(x). \tag{II.2.4}$$

Expressing $U(x)$ in terms of the generators as in (I.3.10) we may write

[1] The term "function" is understood here in a broad sense. It includes generalized functions, distributions.

$$w^n(x_1,\cdots x_n) = \langle\Omega|\Phi(0)e^{-iP(x_1-x_2)}\Phi(0)e^{-iP(x_2-x_3)}\ldots\Phi(0)|\Omega\rangle$$

$$= W^n(\xi_1,\ldots\xi_{n-1}) \tag{II.2.5}$$

with

$$\xi_k = x_k - x_{k+1}. \tag{II.2.6}$$

By a Fourier transformation

$$W^n(\xi_1,\cdots\xi_{n-1}) = \int \widetilde{W}^n(q_1,\cdots q_{n-1})e^{-i\sum q_k\xi_k}d^4q_1\cdots d^4q_{n-1} \tag{II.2.7}$$

(II.2.5) becomes

$$\widetilde{W}^n(q_1,\cdots q_{n-1}) = (2\pi)^{-4(n-1)}\int W^n(\xi_1,\cdots\xi_{n-1})e^{i\sum q_k\xi_k}d^4\xi_1\cdots d^4\xi_{n-1}$$

$$= \langle\Omega|\Phi(0)\delta^4(P-q_1)\Phi(0)\delta^4(P-q_2)\cdots\delta^4(P-q_{n-1})\Phi(0)|\Omega\rangle. \tag{II.2.8}$$

Therefore $\widetilde{W}^n$ vanishes when any one of the q_k lies outside of the spectrum of P and thus, by axiom **A3**, at least when any one of the q_k lies outside the forward cone. If we now take complex 4-vectors

$$z_k = \xi_k - i\eta_k \tag{II.2.9}$$

instead of the real vectors ξ_k then the integrand on the right hand side of (II.2.7) becomes

$$\widetilde{W}^n e^{-i\sum q_k\xi_k}e^{-\sum q_k\eta_k}.$$

If all η_k are taken as positive time-like vectors then we get in (II.2.7) an exponential damping factor and the integral, as well as its derivatives with respect to any z_k, converges because $\widetilde{W}^n$ is a tempered distribution. Thus $W^n(z_1,\ldots z_{n-1})$ is an analytic function of all its arguments as long as the negative imaginary parts of all the z_k lie in the forward cone.

Actually the domain of analyticity of these functions is much larger than the *primitive domain*

$$\mathcal{T}^n\ :\ \ -\mathrm{Im}\, z_k \in V^+, k = 1,\cdots n-1 \tag{II.2.10}$$

found by the above argument. As a first step one shows that the domain of holomorphy includes all "points" $z = (z_1,\ldots z_{n-1})$ which arise from a point in the primitive domain $\mathcal{T}^n$ by a complex Lorentz transformation which is connected to the identity, i.e. by application of any complex matrix Λ with

$$\Lambda^T\mathbf{g}\Lambda = \mathbf{g};\quad \det\Lambda = \mathbf{1} \tag{II.2.11}$$

to all the z_k. By axiom **A2**

$$U(\alpha)\Omega = \Omega. \tag{II.2.12}$$

Therefore, according to **D** (Equ. (II.1.9)) a Wightman function with Lorentz transformed arguments is a linear combination of other Wightman functions with different components of the fields. The coefficients arise from products of representation matrices of $\overline{\mathfrak{L}}$. One notes that the number of spinorial fields in w^n must be even for non-vanishing w^n and therefore the product of representation matrices belongs to a true (tensorial) representation of $\overline{\mathfrak{L}}$. The coefficients are then single valued, analytic functions and the relation (II.1.9) may be extended to complex Lorentz transformations. The resulting domain reached by the application of complex Lorentz transformations to the primitive domain is called the *extended tube* $\mathcal{T}^{n'}$. It contains real points, the so called Jost points. These are the points where in (II.2.9) all ξ_k as well as their convex combinations are space-like and all $\eta_k = 0$:

$$\left\{z : \eta_k = 0; \quad \xi_k^2 < 0; \quad \left(\sum \lambda_k \xi_k\right)^2 < 0 \quad \text{for} \quad \lambda_k \geq 0, \quad \sum \lambda_k > 0\right\}. \tag{II.2.13}$$

A further extension of the domain of holomorphy results from local commutativity (axiom **E**). A permutation of the arguments of w^n again gives a Wightman function. Since the definition of the extended tube depends on the order of arguments in w^n, w^n will be an analytic function of the original arguments in a different domain. Due to **E** this function will coincide with the original one for space-like separation of the x_k and a careful study shows that there is a real open neighbourhood in the Jost regions of both extended tubes where the two functions agree. This is enough to conclude that the two functions are analytic continuations of each other. The domain of holomorphy of w^n is thus extended to the union of the "permuted extended tubes".

In the theory of functions of several complex variables a basic observation is that the domain of analyticity cannot have an arbitrary shape. If we know that a function is analytic in a certain region then it must be automatically analytic in a (usually larger) region, the so called *holomorphy envelope* of the given region. This leads to a further enlargement of the holomorphy domain of w^n. By these methods considerable information about the functions w^n for small n-values has been obtained.

Apart from the restrictions for the individual functions w^n of the hierarchy (analyticity, transformation properties) there is an infinite system of inequalities between the different w^n which must hold in order to make the metric in the Hilbert space, constructed from the w^n, positive definite. They are summarized in the demand

$$\langle \Omega | A^* A | \Omega \rangle \geq 0 \quad \text{for all} \quad A = \sum_n \int f^n(x_1, \cdots x_n) \Phi(x_1) \cdots \Phi(x_n) \prod d^4 x_k. \tag{II.2.14}$$

The analysis of these restrictions, directly in terms of the w^n, was sometimes called the "nonlinear programme".

II.2.2 Truncated Functions. Clustering

Consider a configuration $(x_1, \cdots x_n)$ consisting of several *clusters*. A cluster is a subset of points such that all the points in one cluster have a large space-like separation from all the points in any other cluster. Then we expect that

$$w^n(x_1, \cdots x_n) \sim \prod_r w^{n_r}(y_{r,1}, \cdots y_{r,n_r}), \tag{II.2.15}$$

the approximate equality becoming exact in the limit of infinite separation between the clusters. In (II.2.15) the index r labels the clusters, n_r is the number of points in the r-th cluster and $y_{r,k}$, $k = 1 \ldots n_r$, a subset of the x_k, denote the points belonging to this cluster. The reason for this asymptotic factorization is that the correlation between two compatible observables A, B in a state is measured by the difference

$$\langle AB \rangle - \langle A \rangle \langle B \rangle \tag{II.2.16}$$

where the bracket denotes the expectation value in the state. We expect that in the vacuum state the correlations of quantities relating to different regions decrease to zero as the separation of the regions increases to space-like infinity. This physically plausible guess is, in fact, not an independent assumption but follows from the axioms as will be discussed in detail in section 4.

It is therefore convenient to introduce another hierarchy of functions $\{w^n_T\}$ in terms of which the hierarchy $\{w^n\}$ is obtained as

$$w^n(x_1, \cdots x_n) = \sum_P \prod_r w^{n_r}_T(y_{r,k}). \tag{II.2.17}$$

Here P denotes a partition of the set of points $\{x_i\}$ into subsets (potential clusters), labeled by the index r. The sum extends over all possible partitions. We may note that $w^0_T = 0$, $w^0 = 1$. We call w^n_T the *truncated functions* or *correlation functions*. Given the hierarchy $\{w^n\}$ the truncated functions can be computed recursively:

$$\begin{aligned} &w^1_T(x) = w^1(x) \\ &w^2_T(x_1, x_2) = w^2(x_1, x_2) - w^1(x_1) w^1(x_2) \\ &\ldots \end{aligned} \tag{II.2.18}$$

The term w^n_T always occurs on the right hand side in (II.2.17) coming from the trivial partition where all the points are taken together. The other terms involve truncated functions with a smaller number of arguments. So (II.2.17) may be inverted, beginning with (II.2.18). The asymptotic property (II.2.15) of the w^n corresponds now to the simpler property

$$w^n_T(x_1, \cdots x_n) \rightarrow 0 \tag{II.2.19}$$

when any difference vector $x_i - x_j$ goes to space-like infinity. Actually the correlation functions in the vacuum tend also to zero with increasing time-like separation of the points. The rate of decrease in space-like and time-like directions depends on the details of the energy-momentum spectrum of the theory. We shall return to these questions in section 4 of this chapter.

For free fields the situation is particularly simple. All truncated functions with $n \neq 2$ vanish:

$$w^n_T = 0 \quad \text{for} \quad n \neq 2. \tag{II.2.20}$$

Thus, for a system of free fields it suffices to know the 2-point functions. The others are then given by (II.2.17) with the additional simplification that only such partitions contribute which are pairings of points.

Generating Functionals. Linked Cluster Theorem. Let us consider a hierarchy of functions $p^n(x_1, \ldots x_n)$, totally symmetric under permutations of their arguments.[2] This set of functions may be assembled into one generating functional $\mathcal{P}\{f\}$ of a function $f(x)$:

$$\mathcal{P}\{f\} = \sum_{n=0}^{\infty} \frac{1}{n!} \int p^n(x_1, \cdots x_n) f(x_1) \cdots f(x_n) dx_1 \cdots dx_n. \tag{II.2.21}$$

From it we can recover the p^n as the Taylor coefficients

$$p^n(x_1, \cdots x_n) = \frac{\delta^n}{\delta f(x_1) \cdots \delta f(x_n)} \mathcal{P}\{f\} \bigg|_{f=0}. \tag{II.2.22}$$

The relation between the hierarchy $\{p^n\}$ and the hierarchy of truncated functions $\{p^n_T\}$ is very simply expressed in terms of the respective generating functionals $\mathcal{P}$ and $\mathcal{P}_T$, namely

$$\mathcal{P}\{f\} = \exp \mathcal{P}_T\{f\}; \quad \mathcal{P}_T\{f\} = \ln \mathcal{P}\{f\}. \tag{II.2.23}$$

Note that we assume here $p^0 = 1$, $p^0_T = 0$. This relation, sometimes called the *linked cluster theorem*, combines in an elegant way the combinatorics involved in the definition (II.2.17) and its inversion (II.2.18).

We give an example of the application of (II.2.23) taken from the classical statistical mechanics of a 1−component real gas. It is the example from which the idea of introducing truncated functions originated (see Ursell [Urs 27], Mayer [May 37], Kahn and Uhlenbeck [Kahn 38]).

A grand canonical ensemble with inverse temperature β and chemical potential μ defines a hierarchy of functions p^n, the probability densities for finding n molecules with the configuration $\mathbf{x}_1, \cdots \mathbf{x}_n$ in 3-dimensional space. If the molecules interact by two-body forces, given by a potential V, the Gibbs prescription leads to

$$p^n(\mathbf{x}_1, \cdots \mathbf{x}_n) = v^{-n} e^{-\beta/2 \sum_{i \neq j} V_{ij}}. \tag{II.2.24}$$

Here $V_{ij} = V(\mathbf{x}_i - \mathbf{x}_j)$. The constant v has the dimension of a volume; it is proportional to $\beta^{3/2} e^{-\beta\mu}$ and is approximately equal to the specific volume per particle in the gas. The grand partition function

$$G = \sum \frac{1}{n!} \int p^n(\mathbf{x}_1, \cdots \mathbf{x}_n) d^3x_1 \cdots d^3x_n = \mathcal{P}\{1\} \tag{II.2.25}$$

is precisely the generating functional taken for the constant function $f = 1$. Thus, according to (II.2.23) , the "grand potential" Q, defined as the logarithm of G, is given by

[2] The Wightman functions of a single field are totally symmetric in their analyticity domain due to **E**. This will be used in subsection 2.7. They are of course, not symmetric for real points with time-like separation because a permutation changes the direction from which the boundary value should be approached. Examples of symmetric functions of real points in Minkowski space are the time ordered functions discussed below. The generalization to several fields is trivial. We just have to replace f by the multicomponent function $f^{i\lambda}$ as in (II.1.7).

$$Q = \ln G = \mathcal{P}_T\{1\} = \sum \frac{1}{n!} \int p_T^n(\mathbf{x}_1, \cdots \mathbf{x}_n) d^3x_1 \cdots d^3x_n. \tag{II.2.26}$$

The essential point is now that the truncated functions are practically zero unless the points in their argument form a *linked cluster* i.e. unless they can be connected by a bridge which at no step is wider than the range of the potential. Therefore the sum in (II.2.26) is essentially an expansion in powers of a^3/v where a is the range of the potential. In the case of low density this converges fast and only a few terms are needed in the sum over n to obtain the grand potential, from which all thermodynamic information can then be obtained by differentiating with respect to β and μ. The first truncated functions are easily computed from (II.2.18), beginning

$$p_T^1 = v^{-1}; \quad p_T^2 = v^{-2} \left(e^{-\beta/2V(\mathbf{x}_1 - \mathbf{x}_2)} - 1 \right); \quad \cdots \tag{II.2.27}$$

II.2.3 Time Ordered Functions

This is the most widely used hierarchy of functions in quantum field theory. It is computed in the Feynman-Dyson form of perturbation theory and it is closely related to the scattering matrix by the LSZ-formalism. One introduces a time-ordering operation T which, applied to a product of field operators $\Phi(x)$, shall mean that the actual order of the factors is taken to be the chronological one, beginning with the earliest time on the right, irrespective of the order in which the factors are written down. In the case of Fermi fields a sign factor -1 has to be added if the permutation needed to come to from the original order to the chronological order of the Fermi fields is odd. For example, for Bose fields one defines[3]

$$T\Phi(x_1)\Phi(x_2) = \begin{cases} \Phi(x_1)\Phi(x_2) & \text{if} \quad t_1 > t_2 \\ \Phi(x_2)\Phi(x_1) & \text{if} \quad t_1 < t_2. \end{cases} \tag{II.2.28}$$

If both fields are Fermi fields then the second line will be $-\Phi(x_2)\Phi(x_1)$. The vacuum expectation values of time ordered products will be called τ-functions. They are often also refered to as Green's functions or Feynman amplitudes:

$$\tau^n(x_1, \cdots x_n) = \langle \Omega | T\Phi(x_1) \cdots \Phi(x_n) | \Omega \rangle. \tag{II.2.29}$$

Of course the cluster decomposition leading to the hierarchy of truncated functions proceeds in exactly the same way as for the Wightman functions. Moreover, since the τ-functions are symmetric under permutation of their arguments, they can be subsumed by a generating functional and the linked cluster theorem can be applied.

---

[3]For $t_1 = t_2$ the upper and the lower line on the right hand side of (II.2.28) agree according to axiom **E** as long as the points do not coincide. For coincident points there is an ambiguity. The T-product is not defined as a distribution by (II.2.28). This ambiguity has been used by Epstein and Glaser in their treatment of renormalized perturbation theory [Ep 71]. Apart from the ambiguity for coincident points the definition of the T-product is Lorentz invariant because the chronological order depends on the Lorentz frame only when points lie space-like to each other and then the reordering is irrelevant.

II.2.4 Covariant Perturbation Theory. Feynman Diagrams. Renormalization

This is the backbone of the study of experimental consequences of QED and has remained one of the essential sources of guidance in the subsequent development of quantum field theory. Although it is described in many books we give here a brief sketch based on the "magic formula" of Gell-Mann and Low [Gell 51] using the terminology described above. We assume that the theory is obtained from a classical field theory with a Lagrange density

$$\mathcal{L} = \mathcal{L}_0 + \mathcal{L}_I. \tag{II.2.30}$$

$\mathcal{L}_0$ is the Lagrange density of a free field theory (compare (I.5.30)). For simplicity we assume that no derivatives occur in $\mathcal{L}_I$. We shall denote the interacting fields generically by $\Phi(x)$ and the corresponding free fields, which result if we drop the interaction Lagrangean $\mathcal{L}_I$, by $\Phi^0(x)$. For the generating functional of the τ-hierarchy of the interacting fields we shall write $\mathcal{T}\{f\}$. Gell-Mann and Low loc. cit. have written down an explicit expression for this functional in terms of quantities defined in the free field theory. This "magic formula of covariant perturbation theory" (still without renormalization) is

$$\mathcal{T}\{f\} = \frac{N\{f\}}{N\{0\}}, \tag{II.2.31}$$

$$N\{f\} = \langle \Omega_0 | T\, e^{i\Phi^0(f) + i\int \mathcal{L}_I^0(x) d^4x} | \Omega_0 \rangle. \tag{II.2.32}$$

Here Ω_0 is the vacuum state vector of the free theory, $\mathcal{L}_I^0$ means that the free fields $\Phi^0(x)$ instead of the interacting fields $\Phi(x)$ are inserted into the expression for the interaction Lagrangean density. [4] The formula (II.2.32) is symbolic in the sense that the exponential should be expanded into a power series and then in each term the chronological ordering should be applied. The term with n factors Φ^0 and p factors $\mathcal{L}_I^0$ gives the contribution to τ^n in the p-th order of perturbation theory. It is

$$\tau^n(x_1, \cdots x_n)^{(p)}$$
$$= \frac{i^p}{p! N\{0\}} \int \langle \Omega_0 | T \Phi^0(x_1) \cdots \Phi^0(x_n) \mathcal{L}_I^0(y_1) \cdots \mathcal{L}_I^0(y_p) | \Omega_0 \rangle d^4y_1 \cdots d^4y_p. \tag{II.2.33}$$

The integrand on the right hand side is a τ-function of free fields and can be evaluated easily using the cluster expansion, the analogue of (II.2.17), remembering that for free fields all truncated functions except the 2-point functions vanish. The integrand of (II.2.33) reduces therefore to a sum of products of 2-point τ-functions (Feynman propagators). Each term in this sum corresponds

[4] $\mathcal{L}_I^0(x)$ is a polynomial of fields at the same point. Since the fields entering here are free fields a prescription defining the product exists, namely the Wick ordering. We shall understand $\mathcal{L}_I^0$ as the Wick ordered product of its factors.

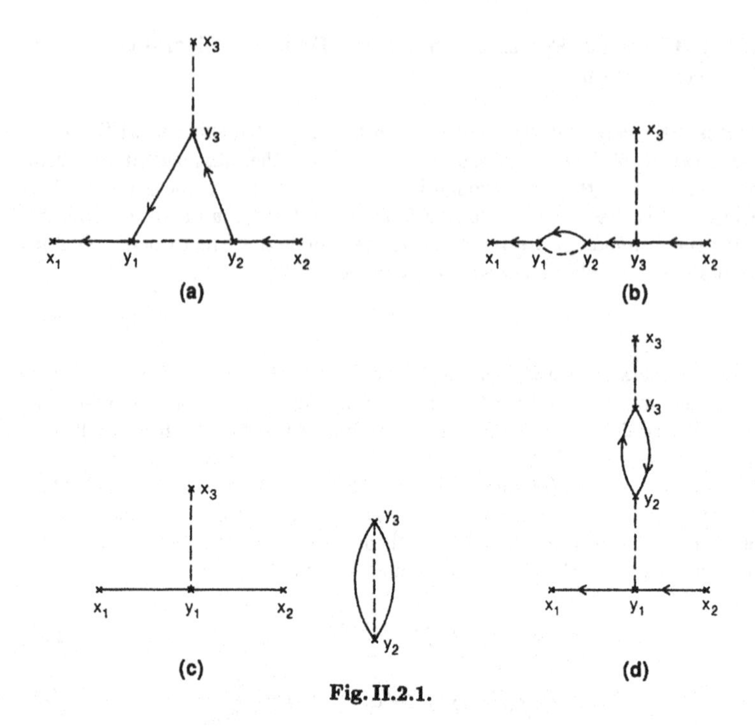

Fig. II.2.1.

to a complete pairing of the occurring fields and can be characterized by a Feynman diagram: the arguments x_i, y_j are represented by points, a *contraction* (a pairing of two fields) by a line connecting the respective points.

Let us illustrate this in QED where

$$\mathcal{L}_I^0 = e : \overline{\psi}^0 \gamma^\mu \psi^0 A_\mu^0 : . \tag{II.2.34}$$

Then each of the points y_i (*vertices*) is the source of three lines, one corresponding to ψ, one to $\overline{\psi}$ and one to A. The only nonvanishing propagators are the contractions of ψ with $\overline{\psi}$ and of A with A. This is incorporated in the graphical representation by using for ψ a solid line with an arrow pointing towards its source, for $\overline{\psi}$ a solid line with an arrow pointing away from its source and for A a dashed line. Thus, for instance, for the 3-rd-order contribution to $\langle \Omega | T \psi_\alpha(x_1) \overline{\psi}_\beta(x_2) A_\nu(x_3) | \Omega \rangle$ i.e. for the case $n = 3$, $p = 3$, one has the following four possible diagrams[5] (apart from a permutation of the y_j).

If we go over to the truncated τ-functions the situation simplifies still further. The contribution of a disconnected diagram (like the one in Fig. II.2.1.c) is just the product of the contributions of its connected parts. This is eliminated in the truncation process. So we may write

[5] Usually one does not mark the points x_i in a Feynman diagram. The lines emanating from them are called *external lines*.

$$\mathcal{T}_T\{f\} = \sum \text{conn. Feynman diagrams,} \tag{II.2.35}$$

which is a short hand notation for saying that a function τ_T^n results from summation over all the connected Feynman diagrams with n end points (external lines) in the expansion of $N\{f\}$. In particular, the normalization factor $N\{0\}$ is then eliminated. It corresponds to the sum over all Feynman graphs without external lines.

The connected graphs still lead to divergent integrals if they contain loops of certain types. Renormalization theory gives rules for removing these infinities. To convey some flavour of the computations we illustrate them in the example of diagram II.2.1.a. Its contribution is[6]

$$\int \{S_F(x_1 - y_1)\gamma^\varrho S_F(y_1 - y_3)\gamma^\mu S_F(y_3 - y_2)\gamma^\sigma S_F(y_2 - x_2)\}_{\alpha\beta}$$

$$\times \; g_{\varrho\sigma} D_F(y_3 - x_3) g_{\mu\nu} D_F(y_1 - y_2) d^4y_1 d^4y_2 d^4y_3. \tag{II.2.37}$$

By Fourier transformation this becomes

$$\int \exp\left(-i\left(p_1(x_1 - x_3) + p_2(x_3 - x_2)\right)\right) \widetilde{D}_F(p_1 - p_2)$$

$$\times \; \widetilde{S}_F(p_1)\Gamma_\nu(p_1, p_2)\widetilde{S}_F(p_2) d^4p_1 d^4p_2 \tag{II.2.38}$$

with

$$\Gamma_\nu(p_1, p_2) = \gamma_\varrho \int \widetilde{S}_F(q)\gamma_\nu \widetilde{S}_F(q + p_2 - p_1)\widetilde{D}_F(p_1 - q) d^4q \, \gamma^\varrho. \tag{II.2.39}$$

Inserting the expressions (II.2.36) for the propagators one sees that there are two powers of the components of q in the numerator, six in the denominator of the 4-dimensional integration. The integral diverges logarithmically at large values of q. However, the difference of Γ_ν at two different values of the arguments p_i gives a convergent integral and one may write formally

$$\Gamma_\nu = c\gamma_\nu + \Gamma_\nu^{\text{ren}}(p_1, p_2) \tag{II.2.40}$$

where c is an infinite constant and Γ^{ren} is a finite function. The divergent constant part of Γ_ν corresponds to the replacement of the triangle in the diagram by a single point with the local interaction $c\overline{\psi}(y)\gamma^\nu\psi(y)A_\nu(y)$ which is of the same form as the basic interaction Lagrangean and may be regarded as an (infinite) change of the parameter e ("charge renormalization") or cancelled by an (infinite) local counter term in the Lagrangean.

To show that a finite number of local counter terms in the Lagrangean suffices to subtract the infinities from all diagrams in a consistent manner is

[6] S_F is the contraction between ψ and $\overline{\psi}$, $g_{\mu\nu}D_F$ the contraction between A_μ and A_ν. The Fourier transforms of these Feynman propagators are

$$\widetilde{S}_F(q) = i(q^\mu\gamma_\mu - m)(q^2 - m^2 + i\varepsilon)^{-1}; \;\; \widetilde{D}_F(q) = (q^2 + i\varepsilon)^{-1}. \tag{II.2.36}$$

a rather involved combinatorial problem. It is achieved by a structure analysis of Feynman graphs, started by Dyson [Dy 49] and carried further by many authors.

We have presented the magic formula (II.2.31), (II.2.32) without giving any reasons. Let us try to justify it. For this purpose the simplest starting point is the so called "interaction representation" developed mainly by Tomonaga and Schwinger. To avoid confusion with other uses of the term "representation" we shall call it "interaction picture" or "Dirac picture". In the previous sections we used the "Heisenberg picture" for the description of the dynamics and we shall adhere to this with few exceptions throughout the book. In this picture a state is described by a vector (or density matrix) which does not change in the time interval between the completion of the preparation of the state and the subsequent measurement of an observable. The operator A representing an observable depends on the time of measurement. It changes with time according to the Heisenberg equation of motion

$$\frac{\partial A}{\partial t} = i[H, A]. \tag{II.2.41}$$

In non relativistic Quantum Mechanics the most commonly used description is the "Schrödinger picture" where the state vector changes according to the Schrödinger equation

$$\frac{\partial \Psi}{\partial t} = -iH\Psi,$$

whereas the operator A describing an observable depends only on the placement in 3-space, not in the time of measurement. Dirac's time dependent perturbation theory suggests a third picture which provided the intuitive basis for the development of the covariant perturbation expansion in Quantum Field Theory. If the Hamiltonian is split in the form

$$H = H_0 + H_I,$$

where H_0 is the Hamiltonian of a free field theory serving as the zero order approximation and H_I is regarded as the interaction, then one may divide the time dependence between observables and states, letting the observables change as in the free field theory and expressing the rest of the dynamical law as a change of the state vectors. We distinguish the quantities in the three pictures by an upper index H, S, D, respectively. In relating the mathematical expressions for the observables in the three pictures it must be understood that we talk about observables which are expressed in terms of a basic set of quantities referring to a sharp time, for instance the basic fields and their canonically conjugate momenta at the respective time. Note that in the Heisenberg picture H_0 and H_I are not fixed operators in $\mathcal{H}$ and that in the interaction picture the same holds for H and H_I. So to avoid confusion we should regard these symbols as denoting fixed functionals of the basic fields and momenta at a specified time and we shall express this by writing for instance $H_I\{\Phi^{(D)}(t)\}$ for the interaction Hamiltonian at time t in the Dirac picture. Since we assumed for simplicity that the interaction Lagrangean does not contain any derivatives of fields the form of H_I is

$$H_I\{\Phi(t)\} = -\int \mathcal{L}_I(\Phi(x)d^3x \quad \text{with} \quad x^0 = t. \tag{II.2.42}$$

$A^{(D)}(t)$ and $A^{(H)}(t)$ are related by a unitary operatur $U(t)$,

$$A^{(D)}(t) = U(t)A^{(H)}(t)U(t)^{-1}.$$

The time dependence of U is obtained by comparing (II.2.41) with the corresponding relation where $A^{(H)}$ is replaced by $A^{(D)}$ and H by H_0. We have

$$\frac{\partial A^{(D)}(t)}{\partial t} \equiv i\left[H_0\{\Phi^{(D)}(t)\}, A^{(D)}(t)\right]$$

$$= \frac{\partial U}{\partial t}A^{(H)}(t)U^{-1} + iU\left[H\{\Phi^{(H)}(t)\}, A^{(H)}(t)\right]U^{-1} - UA^{(H)}(t)U^{-1}\frac{\partial U}{\partial t}U^{-1}$$

$$= \frac{\partial U}{\partial t}U^{-1}A^{(D)}(t) + i\left[H\{\Phi^{(D)}(t)\}, A^{(D)}(t)\right] - A^{(D)}(t)\frac{\partial U}{\partial t}U^{-1}.$$

Since $H - H_0 = H_I$ and $A(t)$ is an arbitrary functional of the basic quantities at time t we have

$$\frac{\partial U(t)}{\partial t} = -iH_I\{\Phi^{(D)}(t)\}U(t). \tag{II.2.43}$$

Defining

$$U(t_2, t_1) = U(t_2)U(t_1)^{-1}, \tag{II.2.44}$$

we get an operator which establishes the relation between $A^{(D)}$ and $A^{(H)}$ at time t_2 if t_1 is chosen as the reference time at which $A^{(D)} = A^{(H)}$. The information from the differential equation (II.2.43) and the initial condition $U(t_1, t_1) = \mathbb{1}$ are combined in the integral equation

$$U(t_2, t_1) = \mathbb{1} - i\int_{t_1}^{t_2} H_I\{\Phi^{(D)}(t')U(t', t_1)dt'. \tag{II.2.45}$$

It can be solved by iteration, yielding the perturbation expansion (for $t_2 > t_1$)

$$\begin{aligned} U &= \mathbb{1} + \sum_{n=1}^{\infty} U^{(n)}, \\ U^{(n)}(t_2, t_1) &= (-i)^n \int_{t_2 > t'_n > \ldots t'_1 > t_1} H_I(t'_n)\ldots H_I(t'_1)dt'_1 \ldots dt'_n. \end{aligned} \tag{II.2.46}$$

Using the time ordering symbol T and inserting (II.2.42) we get

$$U^{(n)}(t_2, t_1) = \frac{i^n}{n!}\int_{t_2 > x_j^0 > t_1} T\mathcal{L}_I(x_1)\ldots\mathcal{L}_I(x_n)d^4x_1\ldots d^4x_n, \tag{II.2.47a}$$

or, symbolically

$$U(t_2, t_1) = T\exp i\int_{t_2 > x^0 > t_1} \mathcal{L}(x)d^4x. \tag{II.2.47b}$$

We have omitted the index D since all fields are now understood in the Dirac picture i.e. they are free fields. From the definition (II.2.44) follows the composition law

$$U(t_3, t_2)U(t_2, t_1) = U(t_3, t_1). \tag{II.2.48}$$

For Heisenberg fields coinciding with $\Phi^{(D)}$ at a time t_0 we get (for $x_2^0 > x_1^0$)

$$\Phi^{(H)}(x_2)\Phi^{(H)}(x_1) = U(x_2^0, t_0)^{-1}\Phi^{(D)}(x_2)U(x_2^0, x_1^0)\Phi^{(D)}(x_1)U(x_1^0, t_0).$$

Choosing some arbitrary final time t_f larger than x_2^0 we may write the first factor as $U(t_f, t_0)^{-1}U(t_f, x_2^0)$ and write symbolically in the case of n factors

$$\begin{aligned} &T\Phi^{(H)}(x_1)\ldots\Phi^{(H)}(x_n) \\ &= U(t_f, t_0)^{-1}T\Phi^{(D)}(x_1)\ldots\Phi^{(D)}(x_n)\exp i\int_{t_f>x^0>t_0}\mathcal{L}_I(x)d^4x \end{aligned} \tag{II.2.49a}$$

which is generalized for the generating functional of T-products of Heisenberg fields to

$$T\exp i\int\Phi^{(H)}(x)h(x)d^4x = U(t_f, t_0)^{-1}T\exp i\int_{t_f>x^0>t_0}\Big(\mathcal{L}_I(x) + \Phi(x)h(x)\Big)d^4x. \tag{II.2.49b}$$

If all the steps in the above argument were well defined then this remarkable formula would give us operators $\Phi^{(H)}(x)$ in Fock space, obeying the Heisenberg equations of motion in the time interval between t_0 and t_f, explicitly expressed in terms of free fields. Furthermore these fields would transform covariantly under the representation of the Poincaré group in the Fock space of free particles as long as the shifted arguments of the fields Φ stay in the time interval between t_0 and t_f. This latter circumstance would imply that in the limit $t_0 \to -\infty$ $t_f \to +\infty$ the physical vacuum should be represented by the Fock vacuum since this is the only Poincaré invariant vector in Fock space. Thus we would obtain the formula by Gell-Mann and Low.

The problems with the argument come from three sources. The first concerns the ultraviolet divergencies. $\mathcal{L}_I(x)$ is a Wick product of free fields. We shall call any Wick polynomial of the free fields and their derivatives at a point a "generalized field" or "composite field" (compare the end of section II.5.5). Any such field is an operator valued distribution on a dense domain of $\mathcal{H}$ (see section II.1.2). Ordinary products $\mathcal{L}_1(x_1)\ldots\mathcal{L}_k(x_k)$ of such composite fields are also well defined distributions for test functions of class $\mathcal{S}^{4k}$ i.e. smooth functions $f(x_1, \ldots x_k)$ of fast decrease. The time ordered product, however, differs from this by commutators of the $\mathcal{L}_j$ multiplied by step functions in the time differences. Their discontinuity introduces an ambiguity. Since the commutators of fields vanish for space-like separation of points the discontinuity of the step functions can only introduce a possible ambiguity when two or more points in the product of the $\mathcal{L}_j$ coincide. The ambiguities will therefore be of the form of some other composite field multiplied by an undefined (formally infinite) constant. The amazing fact is now that there exist finite sets of composite fields $\mathcal{L}_j$ ($j = 1\ldots N$) such that the ambiguous terms arising in the T-products between any of them lead only to composite fields within the same set. If the original Lagrangean belongs to such a set then the theory can be "renormalized" by redefining the T-products, replacing inductively in each subsequent order of the perturbation expansion the formally infinite coefficients of the troublesome terms by unknown finite constants. Then one obtains in each order a well defined expression involving a certain number (maximally N) unknown constants which cannot be determined from the theory but have to be fixed empirically.

The existence of renormalizable theories results from the fact that one can order the composite fields by their canonical dimension. A naive power counting argument indicates that in the T-product of two fields with dimensions d_1, d_2 the composite fields with ambiguous coefficients have dimensions $d_1 + d_2 - k$ with $k \geq 4$. So the theory has a chance of being renormalizable if the dimension of $\mathcal{L}_I$ is less or equal to 4, which means that the coupling constant should be dimensionless (or have the dimension of a positive power of energy). A precise proof of renormalizability is a highly nontrivial endeavor.

The second source of difficulties concerns the integration over infinite space-time. On the level of the operator formulas (II.2.49) this may be remedied by replacing the integral over $\mathcal{L}_I$ be $\int \mathcal{L}_I(x)g(x)d^4x$ where g is a smooth test function which is constant in a large region and vanishes at infinity. Then the formulas give operators which satisfy the Heisenberg equations of motion in a large region. The Poincaré covariance is modified and there is the problem of defining the physical vacuum. We shall not discuss this further. In some models of field theories this problem may present a serious obstacle as indicated by the slogan of "infrared slavery" in Quantum Chromodynamics.

The third question is the convergence of the perturbation series. For this we have to refer to work in Constructive Quantum Field Theory. For realistic theories the answer is presumably negative.

There has been a great effort to divorce the formalism from the canonical quantization of a classical field theory, and to derive from general principles the structural relations between amplitudes with different number of arguments, implied in the diagrammatic description. If one writes the τ-functions as a power series with respect to some parameter, starting with the τ-functions of a system of uncoupled free fields as the zero order terms, then the general postulates **A-E**, when supplemented by conditions limiting uniformly in all orders the growth (of the Fourier transforms) for large momenta, lead to the renormalized perturbation series with only a few possible choices of the interaction terms.[7]

II.2.5 Vertex Functions and Structure Analysis

Feynman diagrams suggest that τ-functions may be constructed from more primitive building blocks. A first step was the truncation process, reducing the problem to the consideration of connected diagrams only.

[7]Important contributions to this program are due to Bogolubov and Parasiuk, Zimmermann, Symanzik, Hepp, Epstein and Glaser, Steinmann. For detailed accounts see [Bog 57], [*Bogolubov and Shirkov 1959*], [*Hepp 1969*], [Zi 70], [Sym 70, 71], [Ep 71], [*Steinmann 1971*].

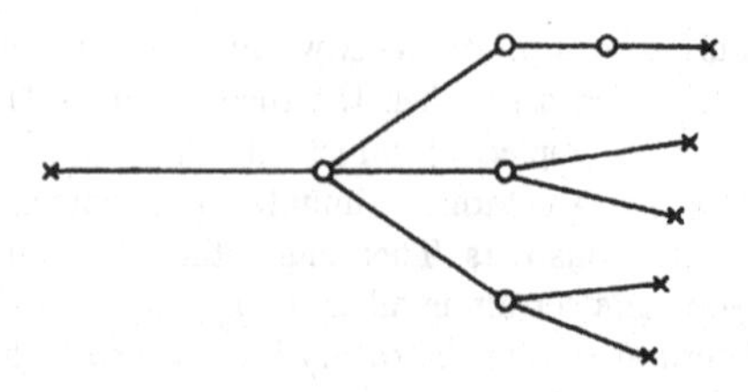

Fig. II.2.2.

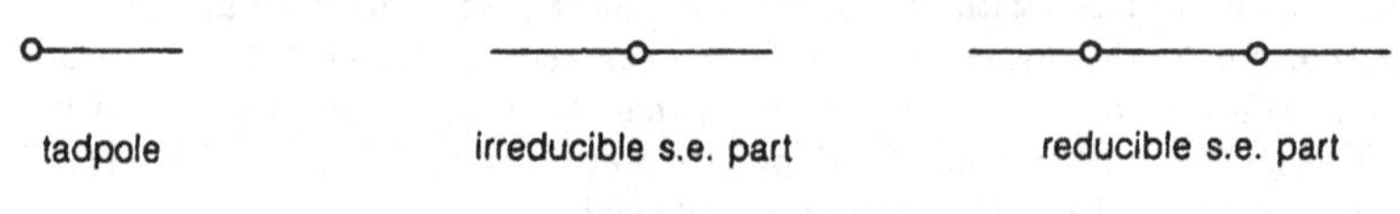

Fig. II.2.3.

To carry this one step further call two points of a diagram *strongly connected* if, by cutting a single line, the graph cannot be disconnected so that the two points lie in different pieces. One sees that the strong connection property is transitive. If x is strongly connected to y and y to z then x is strongly connected to z. So one can divide the points of a diagram uniquely into subsets of strongly connected points. If we picture each such subset by a circle and omit all lines connecting points within such a subset then the original diagram reduces to a *tree graph* i.e. a diagram without closed loops (Fig. II.2.2.). Let us call such a circle an f-vertex if f lines emanate from it. The cases $f = 1$ and $f = 2$ are special. A 1-vertex ("tadpole") occurs only as an end point of a diagram. A 2-vertex ("irreducible self energy part") followed by other 2-vertices still gives a subgraph with only two free lines. If we add the contributions from all possible diagrams to a truncated n-point function then the result can be expressed as a sum of contributions of tree graphs in which internally only f-vertices with $f \geq 3$ occur and where the connecting lines correspond to the full 2-point τ-function instead of the free Feynman propagator. An f-vertex corresponds analytically to a *vertex function* $\Sigma_f(x_1, \cdots x_f)$. It is the sum over all *1-particle irreducible diagrams*[8] which allow one external line to be attached to each of the points x_k $(k = 1, \cdots f)$. These attached lines are all considered as cut (*amputated*). The conclusion of this analysis is that the τ-functions can be constructed from the hierarchy of vertex functions Σ_f with $f \geq 3$ and the full 2-point function. The relation between the τ-hierarchy and the Σ_f-hierarchy can also be expressed in terms of their generating functionals [Sym 67]. See also [*Itzykson and Zuber 1980*].

[8]In our context we should better call it *1-line irreducible.* It means that we cannot disconnect the subdiagram by cutting a single line. In the intuitive picture a line is thought to correspond to a virtual particle. Symanzik carried the structure analysis further by considering *2-particle irreducibility* next.

Apart from simplifying the book-keeping and the discussion of renormalization in perturbation theory the vertex functions have the advantage of faster decrease for large separation of their points (also in time-like directions) as compared to the τ-functions. The lines in a diagram may be considered as ties between the points they connect. The more ties there are between two points the tighter they are pulled together.

II.2.6 Retarded Functions and Analyticity in p-Space

Another hierarchy, used especially in the study of *dispersion relations* are the *r-functions*, the vacuum expectation values of retarded, iterated commutators (resp. anticommutators); see Lehmann, Symanzik, Zimmermann [Leh 57], Nishijima [Nish 57], Glaser, Lehmann, Zimmermann [Glas 57]. They are defined as

$$\begin{aligned} & r^{(n)}(x; x_1, \cdots x_{n-1}) \\ &= (-i)^{n-1} \sum_{\text{Perm}} \Theta(x - x_1) \cdots \Theta(x_{n-2} - x_{n-1}) \\ &\times \langle \Omega | [\cdots [[\Phi(x), \Phi(x_1)], \Phi(x_2)] \cdots \Phi(x_{n-1})] | \Omega \rangle \end{aligned} \tag{II.2.50}$$

Note that the first argument is distinguished. The symmetrization (sum over permutations) refers to the remaining arguments $x_1, \cdots x_{n-1}$. Θ is the step function in time:

$$\Theta(x) = \begin{cases} 1 & \text{if} \quad x^0 > 0 \\ 0 & \text{if} \quad x^0 < 0. \end{cases} \tag{II.2.51}$$

Using space-like commutativity (axiom **E**) and the Jacobi identity for iterated commutators one shows that $r^{(n)}$ vanishes unless all the difference vectors

$$\xi_k = x - x_k \tag{II.2.52}$$

lie in the forward cone and that $r^{(n)}$ is a Lorentz covariant function of the ξ_k (apart from the ambiguity for coincident points which r-functions share with τ-functions). The support of the r-functions in x-space is analogous to that of the w-functions in momentum space. It is confined to the forward cone in all the ξ_k. Thus the Fourier transforms of the $r^{(n)}$ with respect to the ξ_k, the r-functions in momentum space, are analytic in an "extended tube".

II.2.7 Schwinger Functions and Osterwalder-Schrader Theorem

While the r-functions have gradually drifted into oblivion the "Euclidean Green's functions" or "Schwinger functions" have moved to the center of the stage. The Minkowski trick of replacing the time coordinate x^0 by

$$x^4 = ix^0 \tag{II.2.53}$$

changes the Lorentzian metric to the Euclidean metric

$$(x,x) = -\sum(x^\lambda)^2, \quad \lambda = 1,2,3,4. \tag{II.2.54}$$

Configurations of points x_k with real coordinates x_k^λ ($\lambda = 1, \cdots 4$) lie in the "permuted extended tube", the analyticity domain of the Wightman functions described in subsection 2.1. We denote the Wightman functions for configurations of points x_k in real Euclidean space (real coordinates x_k^λ, $\lambda = 1, \cdots 4$) by

$$S^n(x_1, \cdots x_n) = w^n(-ix_1^4, \mathbf{x}_1, \cdots - ix_n^4, \mathbf{x}_n). \tag{II.2.55}$$

These functions have some remarkable properties.

(i) If we exclude the coincidence of points i.e. remain outside of the submanifold

$$\Delta = \{x_1, \cdots x_n : x_k = x_j \quad \text{for some} \quad k \neq j\} \tag{II.2.56}$$

then the S^n are analytic functions. We may regard S^n as a distribution over $\mathbb{R}^{4n} - \Delta$. The appropriate space of test functions is $\mathcal{S}(\mathbb{R}^{4n} - \Delta)$, the smooth functions decreasing fast for large x and vanishing with all their derivatives on Δ.

(ii) S^n is Euclidean invariant.

$$S^n(gx_1, \cdots gx_n) = S^n(x_1, \cdots x_n), \tag{II.2.57}$$

where $gx = Rx + a$; $R \in \mathrm{SO}(4)$.
(transcription of axiom **D**).

(iii) S^n is symmetric under permutations.

$$S^n(x_{P1}, \cdots x_{Pn}) = S^n(x_1, \cdots x_n) \quad \text{for any permutation} \quad \begin{pmatrix} 1 & \cdots & n \\ P1 & \cdots & Pn \end{pmatrix}, \tag{II.2.58}$$

(transcription of axiom **E**).

Thus S^n for a general configuration of points is immediately given if it is known for (Euclidean) time ordered configurations in the positive half space

$$0 < x_n^4 < x_{n-1}^4 < \cdots x_1^4. \tag{II.2.59}$$

We denote the test functions with support in (II.2.59) by $\mathcal{S}_+(\mathbb{R}^{4n})$. So S^n may be regarded as a version of τ^n with complex arguments and this was the original point of view of Schwinger [Schwing 59] in considering this hierarchy. See also Ruelle [Rue 61], Symanzik [Sym 66].

Due to the permutation symmetry (iii) one can combine the hierarchy $\{S^n\}$ into one generating functional $\mathfrak{S}\{f\}$ and one has the linked cluster theorem. There is an analogue of the magic formula of Gell-Mann and Low, expressing $\mathfrak{S}\{f\}$ formally explicitly in terms of the interaction Lagrangean. This is the Feynman-Schwinger functional integral in the Euclidean domain

$$\mathfrak{S}\{f\} = \int e^{-\Phi(f) - \int \mathcal{L}_I(x) d^4x} d\mu(\Phi) \tag{II.2.60}$$

where $d\mu(\Phi)$ is the Gaussian measure over classical field distributions corresponding to free field theory. For a precise discussion and applications see [*Glimm and Jaffe 1987*].

There remains the question of whether and how one can return from the Schwinger functions to the physically interesting functions in Minkowski space. The conditions on the hierarchy $\{S^n\}$ which guarantee that their analytic continuation leads to distributions in Minkowski space satisfying the Wightman axioms have been found by Osterwalder and Schrader [Ost 73, 75]. This important work shows that besides the conditions (i) to (iii) the essential requirement is *reflection positivity*. Let $f_n \in \mathcal{S}_+(\mathbb{R}^{4n})$ and define an involution

$$(\Theta f_n)(x_1, \cdots x_n) = \overline{f}_n(\vartheta x_n, \cdots \vartheta x_1), \tag{II.2.61}$$

where $\vartheta x = (-x^4, \mathbf{x})$ gives the Euclidean time reflection. Let the symbol $\otimes$ denote the tensor product in the tensor algebra of test functions

$$\mathcal{T} = \sum_n{}^{\oplus} \mathcal{S}(\mathbb{R}^{4n}), \tag{II.2.62}$$

$$(f_n \otimes g_m)(x_1, \cdots x_n, y_1, \cdots y_m) = f_n(x_1, \cdots x_n) g_m(y_1, \cdots y_m). \tag{II.2.63}$$

Then the reflection positivity demands

$$\sum_{n,m} S_{n+m}(\Theta f_n \otimes f_m) \geq 0, \tag{II.2.64}$$

for all elements $f \in \mathcal{T}$ with $f_n \in \mathcal{S}_+(\mathbb{R}^{4n})$.

This is the counterpart of the positivity condition (II.2.14) in the Wightman hierarchy. To check it directly would again be a formidable task but the Feynman-Schwinger functional integral reduces this problem to a simpler one which can be checked in concrete models for $\mathcal{L}_I$. Thus one may regard (II.2.60) as a representation theorem guaranteeing that the hierarchy $\{S^n\}$ has the needed properties if $\mathcal{L}_I$ is suitably chosen.

The Osterwalder-Schrader theorem states that properties (i), (ii), (iii) together with (II.2.64) and some limitation of the growth of S^n with n are equivalent to the Wightman axioms. Therefore these conditions have been called the axioms of Euclidean quantum field theory. Moreover, given the Schwinger functions, one has a reconstruction theorem which gives the physical Hilbert space $\mathcal{H}$ with the action of the physical space-time translations on it. To obtain the action of the local field operators in $\mathcal{H}$ one needs, however, an analytic continuation.

II.3 Physical Interpretation in Terms of Particles

II.3.1 The Particle Picture

Suppose we have a theory satisfying the general postulates of section 1 and suppose even that we have "solved it", for instance by having computed the

w-functions or τ-functions. How does it relate to physical phenomena? Experiments in high energy physics are described in terms of collision processes of particles. Therefore we must know which state vectors in $\mathcal{H}$ correspond to specific configurations of particles before and after collision. In the present section we give a heuristic discussion of this question; it will be followed by a more thorough and precise analysis in section 4. In both sections the complications caused by the presence of zero mass particles and long range forces will be ignored. These will be treated in Chapter VI.

The first task is to characterize the *single particle space*, the subspace $\mathcal{H}^{(1)} \subset \mathcal{H}$ containing all state vectors which describe one particle, alone in the world. The answer is suggested by the analysis of Wigner, discussed in I.3, together with axiom **A**. In the reduction of the representation $U(a, \alpha)$ of $\mathfrak{P}$ some irreducible representations U_{m_i,s_i} should occur with a discrete weight. More simply, the spectrum of the operator $M^2 = P^\mu P_\mu$ should have a discrete part. $\mathcal{H}^{(1)}$ is the subspace belonging to the discrete part of the spectrum. The masses of particles appearing in the theory are the discrete eigenvalues of M. Their spins follow from a complete reduction of the representation $U(a, \alpha)$ in the subspace $\mathcal{H}^{(1)}$. The postulates of section 1 do not guarantee that there is any discrete part of the spectrum of the mass operator. For the moment we shall just add the assumption that $\mathcal{H}^{(1)}$ is not empty and postpone a deeper discussion of the question "when does a field theory describe particles" to later chapters.

The continuous part of the mass spectrum corresponds to the states of more than one particle. The disentangling of this part in physical terms is no longer a purely group theoretical reduction problem. We take as a guide the general quantum theory of collision processes as described by Brenig and Haag [Bren 59]. Let us take a complete orthonormal basis $\Psi_{i,\alpha}$ in $\mathcal{H}^{(1)}$. The index i shall denote the species of particle, thus

$$\mathcal{H}^{(1)} = \sum^{\oplus} \mathcal{H}_i^{(1)}. \tag{II.3.1}$$

The symbol $\sum^{\oplus}$ denotes the direct sum of Hilbert spaces. The index α labels basis vectors in $\mathcal{H}_i^{(1)}$. Our description is in the Heisenberg picture. So $\Psi_{i\alpha}$ describes the state "sub specie aeternitatis"; we may assign to it, as in (I.3.29), a wave function in space-time obeying the Klein-Gordon equation. This gives a rough picture of where one can find the particle at any given time.[1] Since $\Psi_{i,\alpha}$ is a normalized state the wave function is localized essentially in some finite region of space at time $t = 0$. For large positive or negative times it will move to infinity and spread out. We expect now, on physical grounds, that an arbitrary

[1]That this wave function cannot be precisely interpreted as a probability amplitude for finding the particle at time x^0 at the position $\mathbf{x}$ has already been pointed out in section I.3. We might use the Newton-Wigner wave function but this is related to the wave function (I.3.29) by convolution with an integral kernel $K(\mathbf{x} - \mathbf{x}')$ which is practically zero as soon as $|\mathbf{x} - \mathbf{x}'|$ exceeds a few times the Compton wave length of the particle. If the particle is not of zero mass this convolution does not change the qualitative picture when we deal with the large distance behaviour. So, for our purpose, we may ignore this problem.

state, if viewed at a sufficiently early or late time, can be described in terms of configurations of far separated particles. To express this mathematically let us first introduce some qualitative concepts which will be made precise at a later stage. We say that a state is localized at time t in a certain space volume $\mathcal{V}$ if it is orthogonal to the vacuum but "*looks like the vacuum*" *for observations at this time* in the spatial complement of $\mathcal{V}$. Example: a single particle state whose wave function at the time in question has its support (essentially) in $\mathcal{V}$. Suppose the state vectors Ψ_1, Ψ_2 describe states which at some particular time t are localized in separated space regions $\mathcal{V}_1$, $\mathcal{V}_2$. Then it should be meaningful to define a "product state vector"

$$\Psi = \Psi_1 \otimes^t \Psi_2 \tag{II.3.2}$$

which has two localization centers at time t, corresponding to Ψ_1 and Ψ_2 respectively. Note that, although in general no tensor product between vectors of $\mathcal{H}$ with values in $\mathcal{H}$ is defined, such a product becomes meaningful between states which are localized far apart at a particular time. The product refers to and depends on this time. We use it to define asymptotic products of single particle states, where *asymptotic* refers to times $t \to \pm\infty$. Let us write a single index λ instead of the double index i, α to label the basis in $\mathcal{H}^{(1)}$. We claim that

$$\Psi^{\text{in}}_{\lambda_1,\cdots\lambda_n} = \lim_{t\to-\infty} \Psi_{\lambda_1} \otimes^t \Psi_{\lambda_2} \otimes^t \cdots \Psi_{\lambda_n} \equiv \lim_{t\to-\infty} \Psi_{\lambda_1,\cdots\lambda_n}(t) \tag{II.3.3}$$

and

$$\Psi^{\text{out}}_{\lambda_1,\cdots\lambda_n} = \lim_{t\to+\infty} \Psi_{\lambda_1} \otimes^t \Psi_{\lambda_2} \otimes^t \cdots \Psi_{\lambda_n} \equiv \lim_{t\to+\infty} \Psi_{\lambda_1,\cdots\lambda_n}(t) \tag{II.3.4}$$

are well defined (Heisenberg) state vectors. We shall write in Dirac notation for (II.3.3) simply $|\lambda_1, \cdots \lambda_n\rangle^{\text{in}}$. A formal proof of this claim will be given in the next section. For the moment we should be content with a qualitative argument. If Ψ_λ and $\Psi_{\lambda'}$ are single particle states whose momentum space wave functions have disjoint supports, then at large positive or negative times the essential supports of their position space wave functions will be far apart. Therefore the product $\otimes^t$ will be defined between them for any large $|t|$. If t_1 and t_2 are large and have the same sign then the change in the meaning of the product at the two times is due to the interaction of the particles in the time interval $[t_1, t_2]$. It is the difference between composing the wave packets $|\lambda\rangle$ and $|\lambda'\rangle$ to a two particle state at t_1, letting it develop then under the full dynamics to time t_2, and the composition of these single particle wave packets (which develop without knowledge of the presence of the other particle) to a two particle state directly at t_2. We expect that the interaction decreases to zero as the separation of the localization centers increases to infinity. Therefore the sequence of state vectors on the right hand side of (II.3.3) or (II.3.4) will converge strongly to a limit vector as $t \to \pm\infty$.[2] If the supports of the momentum space wave functions are not disjoint the argument can still be carried through due to the spreading of

[2]This means that the sequence of Hilbert space vectors $\Psi(t)$ forms a Cauchy sequence in the norm topology of vectors, so $\| \Psi - \Psi(t) \| \to 0$.

the wave functions in position space. If one divides space into large cubes of fixed edge length R then in a single particle state the probability $w(t)$ of finding the particle in a particular cube at time t decreases proportional to t^{-3}; the number $N(t)$ of cubes in which one may find the particle increases proportional to t^3. The probability of finding two particles in the same cube or in adjoining cubes decreases like $N(t) \cdot w(t)^2 \sim t^{-3}$.

The decrease of the interaction with increasing separation can be related to the decrease of the correlation functions w_T for large space-like distances and this, in turn, can be derived from the general postulates if the energy-momentum spectrum of the theory has a gap between the vacuum and the lowest-lying single particle state (see section 4). All the above arguments refer, of course, to particles with non-vanishing mass. For the collision theory in the case of a theory without mass gap see Chapter VI. One shows now that the states $|\lambda_1, \cdots \lambda_n\rangle^{\text{in}}$ have the same orthogonality relations as the multiparticle states of a free theory. Specifically, the product $\Psi_\lambda \otimes^t \Psi_{\lambda'}$ is commutative if λ and λ' refer to two bosons or to one boson and one fermion, it is anticommutative if both indices refer to fermions. Thus we get for instance

$$^{\text{in}}\langle\lambda_1\lambda_2|\lambda_1'\lambda_2'\rangle^{\text{in}} = \langle\lambda_1|\lambda_1'\rangle\langle\lambda_2|\lambda_2'\rangle \pm \langle\lambda_1|\lambda_2'\rangle\langle\lambda_2|\lambda_1'\rangle. \tag{II.3.5}$$

Furthermore, these states transform under $U(a,\alpha)$ as if the theory were free i.e.

$$U(a,\alpha)|\lambda_1, \cdots \lambda_n\rangle^{\text{in}} = |\lambda_1', \cdots \lambda_n'\rangle^{\text{in}} \quad \text{with} \quad |\lambda_j'\rangle = U(a,\alpha)|\lambda_j\rangle. \tag{II.3.6}$$

Thus these states span a Fock space as in (I.3.51), (I.3.52) with the tensor product $\otimes^{\text{in}} = \lim_{t\to-\infty} \otimes^t$ and the action (II.3.6) of the Poincaré group on it. The equivalence class of the representation of $\tilde{\mathfrak{P}}$ in this Fock space is, as in the case of free fields given by

$$\begin{aligned} U &= \prod_i{}^{\otimes} U_{F,i}, \\ U_{F,i} &= \sum_N{}^{\otimes} \left(U_{m_i,s_i}^{\otimes N}\right)_{S,A}. \end{aligned} \tag{II.3.7}$$

The interaction can not be seen from the representation class of $U(a,\alpha)$.

The following picture of a 2-particle scattering process in classical mechanics is very instructive to illustrate the existence of two basis systems in $\mathcal{H}$ in which the representation $U(a,\alpha)$ looks like the one for free field theory. The full lines describe the actual trajectories of the two particles in space (we have indicated times t_i by markings on the trajectories). Together the two full lines correspond to a Heisenberg state of the 2-particle system; they give the complete history. Each trajectory has two asymptotes, drawn as dotted lines. They represent straight line motions with constant velocities and refer to the asymptotic times $t \to \pm\infty$. Each asymptote corresponds to a Heisenberg state of a single particle alone in the world. The Heisenberg state of the 2-particle system can be characterized by giving its asymptotes at $t \to -\infty$ (i.e. in classical mechanics by giving the asymptotic initial velocities and impact parameters).

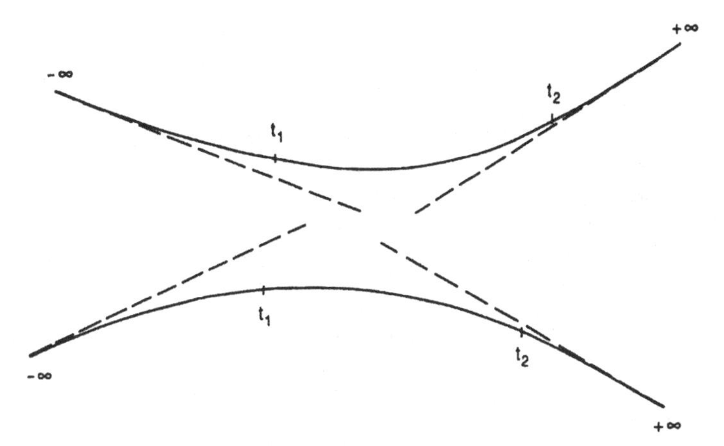

Fig. II.3.1.

Equally well, the state is characterized by giving the asymptotes at $t \to +\infty$. A Poincaré transformation shifts the trajectories in space-time but each new asymptote is just the shifted old asymptote, unaffected by the presence of a second particle.

From this picture we can also read off the time dependence of the product composition symbol $\otimes^t$

$$e^{iH\tau}\left(\Psi_1 \otimes^t \Psi_2\right) = \left(e^{iH\tau}\Psi_1\right) \otimes^{t+\tau} \left(e^{iH\tau}\Psi_2\right) \tag{II.3.8}$$

or

$$\Psi_1 \otimes^{t+\tau} \Psi_2 = e^{iH\tau}\left(e^{-iH\tau}\Psi_1\right) \otimes^t \left(e^{-iH\tau}\Psi_2\right). \tag{II.3.9}$$

The basic idea of this approach to collision theory is due to Ekstein, [Ek 56a and b]. It was expanded by Brenig and Haag [Bren 59]. Let us compare it with the more familiar description of single channel scattering theory in quantum mechanics. There the motion of particles without interaction is described by a "free Hamiltonian" H_0. Fixing t once and for all, say as $t = 0$, and choosing the Ψ_i as single particle states we may write (II.3.9) as[3]

$$\Psi_1 \otimes^\tau \Psi_2 = e^{iH\tau}e^{-iH_0\tau}\left(\Psi_1 \otimes^0 \Psi_2\right) \tag{II.3.10}$$

since the inner bracket in (II.3.9) describes the motion of each particle in the absence of the other. Thus

[3]This requires the definition of the product also for states which are not localized far apart at the time $t = 0$, but this is no problem; it means only that a certain freedom remains in the choice of the definition of this product. In wave mechanics one takes the tensor product of the wave functions at the time $t = 0$.

$$\Psi_1 \otimes^{\text{in}} \Psi_2 = \lim_{\tau \to -\infty} \Psi_1 \otimes^\tau \Psi_2 = \Omega^- \left(\Psi_1 \otimes^0 \Psi_2 \right) \tag{II.3.11}$$

where Ω^- is the *Møller operator*, the limit of the operator $e^{iHt}e^{-iH_0t}$ for $t \to -\infty$. The advantage of (II.3.3) over (II.3.11) is, that we do not have to refer to a splitting of H into a free part and an interaction term. This is already useful in nonrelativistic quantum mechanics if one has to deal with "bound states" ("composite particles") because then one has to take a different splitting of H for each channel. This cannot be done without abandoning the permutation symmetry, so one has to be careful to take the Pauli principle correctly into account if one uses the splitting method. In the field theoretic case, even in situations without "composite particles", the separation of "persistent interactions" from interactions which vanish in states where particles are far apart is not directly visible in the form of the Hamiltonian. This leads to the need for renormalization of parameters like mass and charge in (II.2.32) (see the remarks at the end of subsection II.2.4).

To sum up: given the existence of single particle states and the definition of the product $\otimes^t$ we can construct in $\mathcal{H}$ the state vectors of specific configurations of incoming or of outgoing particles. The above qualitative discussion indicates that this construction needs only the postulates of section 1 and the assumption of a mass gap in the P^μ-spectrum. This will be corroborated in the next section.

We shall assume in addition that all states can be interpreted in terms of incoming particle configurations, in other words, that the $|\lambda_1, \cdots\rangle^{\text{in}}$ span the whole of $\mathcal{H}$, that they provide a complete system of basis vectors. The same should be true, of course, for the vectors $|\lambda, \cdots\rangle^{\text{out}}$ too. This assumption, called *asymptotic completeness* is not a consequence of the postulates of section 1. In chapter VI we shall discuss natural postulates needed for this as well as for a deeper understanding of the notion of particles and collision processes in the case where the continuum of the energy spectrum begins at zero.

II.3.2 Asymptotic Fields. S-Matrix

We assume asymptotic completeness. Then $\mathcal{H}$ is isomorphic to the Fock space generated by means of the tensor product $\otimes^{\text{in}}$ from $\mathcal{H}^{(1)}$. We may therefore introduce annihilation and creation operators a^{in}_λ, $a^{*\text{in}}_\lambda$ as in section 3 of chapter I. The representation of $\mathfrak{P}$ respects this tensor structure i.e.

$$U(a, \alpha)(\Psi_1 \otimes^{\text{in}} \Psi_2) = U(a, \alpha)\Psi_1 \otimes^{\text{in}} U(a, \alpha)\Psi_2, \tag{II.3.12}$$

and in $\mathcal{H}^{(1)}_i$ it is given by U_{m_i, s_i}. We may therefore combine the creation and annihilation operators of incoming particles for each species of particle to a covariant free field $\Phi^{\text{in}}_i(x)$ as described in the subsection I.5.3. For spin zero we have a scalar field, for spin 1/2 a Majorana or Dirac field (depending on the existence of an antiparticle). For higher spin there is some arbitrariness in the choice as can be seen from the discussion in I.3. and I.5.

The same construction can, of course, be made with respect to the product $\otimes^{\text{out}}$, leading to fields $\Phi^{\text{out}}_i(x)$.

Since the bases $|\lambda_1, \cdots \lambda_n\rangle^{\text{in}}$ and $|\lambda_1, \cdots \lambda_n\rangle^{\text{out}}$ have the same orthogonality relations they are related by a unitary operator S:

$$|\lambda_1, \cdots \lambda_n\rangle^{\text{in}} = S|\lambda_1, \cdots \lambda_n\rangle^{\text{out}}, \tag{II.3.13}$$

or, in terms of the incoming and outgoing fields,

$$\Phi_i^{\text{in}}(x) = S\Phi_i^{\text{out}}(x)S^{-1}. \tag{II.3.14}$$

The operator S commutes with the Poincaré operators because Φ^{in} and Φ^{out} transform in the same way,

$$U(a, \alpha)SU(a, \alpha)^{-1} = S. \tag{II.3.15}$$

The quantities of experimental interest are the matrix elements of S in the basis of outgoing particle configurations. They give the probability amplitude for finding a specific outgoing configuration when the incoming configuration is known. Due to the unitarity of S these matrix elements are also equal to the matrix elements in the basis of incoming configurations

$$\begin{aligned} {}^{\text{out}}\langle\lambda_1', \cdots \lambda_m'|\lambda_1, \cdots \lambda_n\rangle^{\text{in}} &= {}^{\text{out}}\langle\lambda_1', \cdots \lambda_m'|S|\lambda_1, \cdots \lambda_n\rangle^{\text{out}} \\ &= {}^{\text{in}}\langle\lambda_1', \cdots \lambda_m'|S|\lambda_1, \cdots \lambda_n\rangle^{\text{in}}. \end{aligned} \tag{II.3.16}$$

Therefore we shall omit the labels *in* or *out* in writing the S-matrix elements. In almost all applications one will have only two particles in the initial state. One uses the improper basis of momentum eigenstates in $H^{(1)}$, so the translation covariance of S is simply expressed by writing

$$\langle p_1', \cdots p_m'|S - \mathbb{1}|p_1, p_2\rangle = \delta^{(4)}(p_1 + p_2 - \sum p_i')\langle p_1', \cdots p_m'|T|p_1, p_2\rangle. \tag{II.3.17}$$

We have suppressed the spin indices. The "T-matrix-element" is a Lorentz invariant function of the momenta and the spin indices. Its absolute square multiplied with a kinematical factor, depending on the relative velocities of the incoming particles and the choice of the independent variables describing the final state, gives the differential cross section for the process

$$p_1, p_2 \rightarrow p_1', \cdots p_m'.$$

II.3.3 LSZ-Formalism

Returning now to the question raised at the beginning of this section one answer suggests itself in analogy to Fig. II.3.1.. The basic local fields $\Phi(x)$ in terms of which the theory is formulated and to which the postulates of section 1 are supposed to apply, become asymptotically close to incoming fields when their time argument tends to $-\infty$ and to outgoing fields as $x^0 \rightarrow +\infty$. In other words, the fields serve to interpolate between Φ^{in} and Φ^{out} and this establishes the connection with the physical interpretation in terms of particles and collision cross sections. Of course this interpretation can be fully adequate only if we have a one to one correspondence between basic fields and species of stable particles. In the general case we need an interpretation which does not rely on field-particle dualism. This will be given in section 4. The purpose of the remainder of this section is to give a precise meaning to the intuitive assumptions

$$\Phi^{\text{in}}(x) \underset{t\to-\infty}{\longleftarrow} \Phi(x) \underset{t\to+\infty}{\longrightarrow} \Phi^{\text{out}}(x) \tag{II.3.18}$$

in the case of a one to one correspondence of fields and particles and to derive from it an algorithm relating S-matrix elements to τ-functions.

In order not to burden the discussion with too many indices we shall consider in the following the simplest case, where we have to deal with only one species of particle which has zero spin and nonvanishing mass and where we have one basic (scalar, Hermitean) field Φ in the theory. The generalization to particles with spin 1/2 and Dirac fields is obvious. For more general cases we refer to section 4.

Let φ denote a single particle state, $\varphi(x)$ its wave function which is a positive energy solution of the Klein-Gordon equation. The corresponding creation operator for incoming configurations is then given by (see (I.5.9))

$$a^{*\text{in}}(\varphi) = i\int_{x^0=t}\left\{\Phi^{\text{in}}(x)\frac{\partial}{\partial x^0}\varphi(x) - \left(\frac{\partial}{\partial x^0}\Phi^{\text{in}}(x)\right)\varphi(x)\right\}d^3x, \tag{II.3.19}$$

where the time t can be arbitrarily chosen. Consider the analogous expression

$$\Phi(\varphi;t) = i\int_{x^0=t}\left\{\Phi(x)\frac{\partial}{\partial x^0}\varphi(x) - \left(\frac{\partial}{\partial x^0}\Phi(x)\right)\varphi(x)\right\}d^3x. \tag{II.3.20}$$

Since the actual field does not obey the Klein-Gordon equation if there is any interaction (though φ does) the operator (II.3.20) will depend on the time of integration.[4] In fact one has

$$\frac{\partial\Phi(\varphi;t)}{\partial t} = -i\int_{x^0=t}(K\Phi)\varphi d^3x \tag{II.3.21}$$

where we have used the abbreviation K for the Klein-Gordon operator

$$K = \Box + m^2. \tag{II.3.22}$$

Lehmann, Symanzik, Zimmermann [Leh 55] proposed as the precise form of the *asymptotic condition*, replacing (II.3.18)

$$\langle\Psi_1|a^{*\text{in}}(\varphi)|\Psi_2\rangle \underset{t\to-\infty}{\longleftarrow} \langle\Psi_1|\Phi(\varphi;t)|\Psi_2\rangle \underset{t\to+\infty}{\longrightarrow} \langle\Psi_1|a^{*\text{out}}(\varphi)|\Psi_2\rangle \tag{II.3.23}$$

[4]We mentioned that Φ is not an operator valued distribution at a sharp time but needs smearing out with a smooth function also in time. Therefore the expression (II.3.20) is not an operator but only a sesquiliner form. But this could easily be remedied by an additional averaging over a small time interval and moreover we shall only use matrix elements of $\Phi(\varphi;t)$ in the following. So we can forget this complication.

for all state vectors Ψ_1, Ψ_2 belonging to a dense domain which contains the vectors generated from the vacuum by repeated application of the operators $a^{*\mathrm{in}}$ and $a^{*\mathrm{out}}$ with smooth wave functions. We may also express this by saying that $\Phi(\varphi;t)$ converges weakly on the mentioned domain to $a^{*\mathrm{in}}(\varphi)$, respectively $a^{*\mathrm{out}}(\varphi)$ as $t \to \mp\infty$.

In (II.3.23) the actual field is normalized so that it has the same matrix elements between the vacuum and the single particle states as the incoming field. From the general postulates of section 1 we have, with this normalzation

$$w^2(x_1,x_2) = i\Delta_+(x_1-x_2;m) + i\int \Delta_+(x_1-x_2;\kappa)d\mu(\kappa) \qquad \text{(II.3.24)}$$

where $d\mu$ is a positive measure with support in the interval $[2m,\infty]$. (see Lehmann [Leh 54]). The vacuum expectation value of the equal time commutator of $\pi(x)=\partial_0\Phi(x)$ with Φ is

$$[\pi(x_1),\Phi(x_2)] = -i\delta^{(3)}(\mathbf{x}_1-\mathbf{x}_2)(1+\int d\mu(\kappa))$$

which differs from the canonical commutation relation by the factor in the bracket, which is the (divergent) factor Z^{-1} mentioned in section 2.4.

From the asymptotic relations (II.3.23) one deduces the very useful *reduction formulas*

$$^{\mathrm{out}}\langle\varphi_1',\cdots\varphi_k'|T\Phi(x_1)\cdots\Phi(x_n)|\varphi_1,\cdots\varphi_l\rangle^{\mathrm{in}}$$
$$= i\int\left\{K_y\,{}^{\mathrm{out}}\langle\varphi_1',\cdots\varphi_k'|T\Phi(x_1)\cdots\Phi(x_n)\Phi(y)|\varphi_1,\cdots\varphi_{l-1}\rangle^{\mathrm{in}}\right\}\varphi_l(y)d^4y \qquad \text{(II.3.25)}$$
$$= i\int\left\{K_y\,{}^{\mathrm{out}}\langle\varphi_1',\cdots\varphi_{k-1}'|T\Phi(x_1)\cdots\Phi(x_n)\Phi(y)|\varphi_1,\cdots\varphi_l\rangle^{\mathrm{in}}\right\}\overline{\varphi}_k'(y)d^4y \qquad \text{(II.3.26)}$$

Here $K_y = \Box_y + m^2$ is the Klein-Gordon operator with respect to y and we have assumed that an orthogonal system of single particle states is used and that the incoming and outgoing configurations contain no common state.

The *reduction formulas* allow one in a very elegant manner to reduce the number of particles in the incoming or outgoing configuration by adding one more field under the T-product. By repeated application we thus convert an S-matrix element to an expression in terms of τ-functions. Using the momentum space basis we get, if the final momenta are all different from the initial momenta

$$\langle p_1',\cdots p_k'|S|p_1,\cdots p_l\rangle$$
$$= (-i)^{k+l}\prod(p_j'^2-m^2)(p_i^2-m^2)\tilde{\tau}(p_1',\cdots p_k',-p_1,\cdots -p_l) \qquad \text{(II.3.27)}$$

where $\tilde{\tau}$ is the Fourier transform of τ. Looking at the Feynman rules for computing the functions $\tilde{\tau}$ (see section 2.4) we see that these functions have poles at $q^2=m^2$ coming from the propagators of external lines. The factors (p^2-m^2) in (II.3.27) cancel these. The S-matrix elements are the τ-functions with amputated external lines, taken for momenta which lie all on the *mass shell* (the hyperboloid $p^2=m^2$).

II.4 General Collision Theory

We now want to divorce the treatment of collision processes from the assumption of a field-particle duality. Specifically we want to cover situations where particles occur which are not related to one of the basic fields via an asymptotic condition (II.3.23) ("composite particles") or where basic fields occur which have no counterparts among the species of observable particles (e.g. quark fields). The physical interpretation of the quantum fields will not be primarily attached to particles but to local operations. Specifically an operator $\Phi(f)$ (a basic field smeared out with a test function f) represents a physical operation performed on the system within the space-time region given by the support of f.[1] Naively speaking, the argument x of a basic field has direct physical significance. It marks the point where $\Phi(x)$ applied to a state produces a change. This sounds obvious and in fact one may regard it as the reason for axiom **E**. However one should remember that the argument x in $\Phi^{\text{in}}(x)$ does not have this direct physical meaning though Φ^{in}, being formally a free covariant field, also satisfies axiom **E**. Also it is not obvious at this stage that such a purely space-time-geometric interpretation of the basic fields will suffice to analyze the phenomenological consequences of the theory since it introduces no distinction between the different types of fields occurring. The suggested interpretation suffices indeed to construct the states $|\lambda_1, \cdots\rangle^{\text{in}}$ and $|\lambda_1, \cdots\rangle^{\text{out}}$. This will be demonstrated under some simplifying assumptions in this section and discussed more fully in Chapter VI.

II.4.1 Polynomial Algebras of Fields. Almost Local Operators

An operator

$$Q = \sum_n \int f^n(x_1, \cdots x_n)\Phi(x_1) \cdots \Phi(x_n) \prod d^4x_i \qquad \text{(II.4.1)}$$

will be called local (with localization region $\mathcal{O}$) if all f^n have their support in $\mathcal{O}^{\times n}$ (i.e. vanish whenever one of the points x_i lies outside of $\mathcal{O}$). The set of operators with localization region $\mathcal{O}$ generates an algebra, the polynomial algebra $\mathcal{A}(\mathcal{O})$ of the region $\mathcal{O}$. Our interpretation of the theory will be based on the assertion that the elements of $\mathcal{A}(\mathcal{O})$ correspond to physical operations performed in $\mathcal{O}$ [Haag 57].

The operator (II.4.1) will be called *almost local* if all f^n belong to $\mathcal{S}^{4n}$ (the relevant property in this context being that they decrease faster than any power if any point x_i moves to infinity). We denote by $Q(x)$ the translated operator

$$Q(x) = U(x)QU^{-1}(x). \qquad \text{(II.4.2)}$$

Of course if Q is almost local so is $Q(x)$ but if we consider very large translations then $Q(x)$ may be regarded as representing an operation centered around the

[1] Of course Φ and f are to be understood generically in the sense of (II.1.7) We do not explicitly write the indices distinguishing the types and components of the fields.

point x. The extension of the support of f^n in (II.4.1) becomes then irrelevant. In particular one has

Lemma 4.1.1

Equal time commutators[2] of almost local operators tend fast to zero with increasing separation. In application on the vacuum we have

$$\| \, [Q_1(t, \mathbf{x}_1), Q_2(t, \mathbf{x}_2)] |\Omega\rangle \, \| \; < \; c_N |\mathbf{x}_1 - \mathbf{x}_2|^{-N}, \tag{II.4.3}$$

for any positive N.

To justify (II.4.3) note that if the Q are localized in some finite region then axiom **E** implies that the commutator is zero as soon as $|\mathbf{x}_1 - \mathbf{x}_2|$ exceeds a certain finite distance. Almost local operators can be approximated by local ones with support in a 4-dimensional cube of edge length d so that the mistake on the left hand side of (II.4.3) decreases faster than any power of d as $d \to \infty$. For this latter claim it suffices to assume that the Wightman distributions are tempered.

The assumptions we want to add now to the axioms of section 1 are

α) The mass operator $M = (P_\mu P^\mu)^{1/2}$ has some discrete eigenvalues. In other words, the single particle subspace $\mathcal{H}^{(1)}$ is not empty.

β) In the subspace orthogonal to the vacuum the spectrum of M has a lower bound $m_0 > 0$.

γ) For any single particle state Ψ with smooth momentum space wave function[3] $\varphi(p)$ there exists an almost local operator Q generating it from the vacuum

$$\Psi = Q\Omega. \tag{II.4.4}$$

It should be understood that none of the assumptions α) – γ) nor all of the axioms of section 1 are of fundamental significance.[4]

In the following let Q denote any almost local operator. The basic tools in the subsequent analysis are

Lemma 4.1.2

The improper operator

$$\tilde{Q}(p) = \int Q(x) e^{-ipx} d^4x \tag{II.4.5}$$

changes the momentum of a state precisely by p. Equivalently, with

[2]respectively anticommutators in the case of fermionic operators.

[3]We suppressed the indices indicating the type of particle and the spin component. $\varphi(p)$ may be considered as a function of the 3-momentum since the energy is then determined by the mass. Smooth shall mean that φ is infinitely often differentiable with respect to the components of $\mathbf{p}$. This implies that the state is, roughly speaking, localized at finite times in a finite part of space. More precisely, that the position space wave function at finite times decreases faster than any power as $\mathbf{x}$ goes to infinity. Actually γ) is not really needed (see [Ara 67]). We assumed it here only for convenience.

[4]We ignore the infrared problems which invalidate assumptions β) and γ), we ignore aspects of local gauge theories where some of the axioms have to be modified.

$$Q(f) = \int Q(x)f(x)d^4x$$

we have

$$\langle\Psi_2|Q(f)|\Psi_1\rangle = 0 \quad \text{if} \quad (E_1 + \Delta) \cap E_2 \quad \text{is} \quad \text{empty} \tag{II.4.6}$$

where the momentum space regions E_i are the spectral supports of Ψ_i (in the spectral decomposition with respect to P^μ) and Δ is the support of

$$\tilde{f}(p) = \int e^{ipx} f(x)d^4x.$$

The proof of this lemma follows exactly the line of argument leading from (II.2.4) to (II.2.8).

We consider now the vacuum expectation value of products of shifted almost local operators, expressions of the form $\langle\Omega|Q_1(x_1)\cdots Q_n(x_n)|\Omega\rangle$. The corresponding truncated expectation values will be denoted by $\langle\cdots\rangle_T$. They are defined as in section 2, taking the Q_i instead of the basic fields, ignoring the internal structure of each Q_i. Then one has

Lemma 4.1.3 [Ruelle 62]
Let $x_k = (t, \mathbf{x_k})$ be a configuration of points at equal time and define $R = \max_{i,k}|\mathbf{x_i} - \mathbf{x_k}|$. Then, for any fixed set of almost local operators $\{Q_k\}$ and for any positive N there is a constant A_N such that

$$\langle Q_1(x_1)\cdots Q_n(x_n)\rangle_T \;<\; A_N R^{-N}. \tag{II.4.7}$$

This gives a uniform bound for the truncated functions in terms of the radius R of the configuration.

The proof given by Ruelle for this crucial lemma is rather complicated. We shall describe here the essential ideas used in it and refer the reader for a watertight proof to Ruelle's original paper.

First observation: Due to the assumption of a mass gap (assumption β)) we may, with the help of lemma 4.1.2, decompose an almost local operator into three parts

$$Q = Q^- + Q^+ + Q^0 \tag{II.4.8}$$

where

$$Q^-\Omega = 0, \quad Q^{+*}\Omega = 0, \quad Q^0\Omega = \langle\Omega|Q|\Omega\rangle\Omega, \tag{II.4.9}$$

and each part Q^i ($i = +, -, 0$) is still almost local. To see this let us take for Q the n-th term in (II.4.1) and define

$$\tilde{Q}(p) = (2\pi)^{-4} \int f^n(x_1 - y, \ldots x_n - y)\Phi(x_1)\ldots\Phi(x_n)e^{-ipy}d^4y \prod d^4x_i. \tag{II.4.10}$$

Then

$$Q = \int \tilde{Q}(p) d^4p = \int (F_+(p) + F_-(p) + F_0(p)) \tilde{Q}(p) d^4p \tag{II.4.11}$$

where we have made a smooth partition of unity

$$\begin{aligned} & F_+ + F_- + F_0 = 1, \\ & F_+(p) = 0 \quad \text{for} \quad p^0 < a, \\ & F_-(p) = 0 \quad \text{for} \quad p^0 > -a, \\ & F_0(p) = 0 \quad \text{for} \quad |p^0| > b, \\ & 0 < a < b < m_0. \end{aligned} \tag{II.4.12}$$

The three parts of Q arising from this partition satisfy (II.4.9) according to lemma 4.1.2 because of the spectral assumption β). Furthermore each Q_i ($i = +, -, 0$) is still almost local: the functions

$$f_i(x_1, \cdots x_n) = \int f(x_1 - y, \cdots x_n - y) e^{-ipy} F_i(p) d^4y d^4p$$

decrease faster than any power for large arguments because the $F_i(p)$ are infinitely often differentiable. Thus the decomposition (II.4.8) leads to almost local operators of which the first and the adjoint of the second annihilate the vacuum whilst the third reproduces the vacuum up to a factor.

Second observation: For the case of two factors i.e. for $n = 2$ in (II.4.7) lemma 4.1.3 follows directly from the above observations. We have

$$\begin{aligned} & \langle \Omega | Q_1(x_1) Q_2(x_2) | \Omega \rangle \\ & = \langle \Omega | \left(Q_1^-(x_1) + Q_1^0(x_1) \right) \left(Q_2^+(x_2) + Q_2^0(x_2) \right) | \Omega \rangle . \end{aligned}$$

For large $|\mathbf{x_1} - \mathbf{x_2}|$ we can commute Q_2^+ to the left and Q_1^- to the right where they give zero. The remaining term is

$$\langle \Omega | Q_1^0(x_1) Q_2^0(x_2) | \Omega \rangle = \langle \Omega | Q_1(x_1) | \Omega \rangle \langle \Omega | Q_2(x_2) | \Omega \rangle .$$

The neglected terms from the commutators decrease faster than any power with $|\mathbf{x_1} - \mathbf{x_2}|$ due to lemma 4.1.1.

For $n \geq 2$ one observes that for a configuration of "diameter" R as defined in lemma 4.1.3. there is a subdivision of the points into two subsets so that the spatial distance of any point in one subset from any point in the other subset is at least R/n. One may then apply the technique described above recursively.

II.4.2 Construction of Asymptotic Particle States

We choose now in $\mathcal{H}^{(1)}$ for each species of particle (index i) and each spin orientation (index α) a reference state vector $\widehat{\Psi}_{i,\alpha}$. Its momentum space wave function $\widehat{\varphi}_{i,\alpha}(p)$ shall be smooth and nowhere vanishing on the mass hyperboloid. According to assumption γ) we can pick an almost local operator $q_{i,\alpha}$ so that

$$q_{i,\alpha}\Omega = (2\pi)^{-3/2}\Psi_{i,\alpha}. \tag{II.4.13}$$

Let

$$h_i(x) = (2\pi)^{-3/2}\int \widetilde{h}_i(p)e^{-ipx}d\mu_i(p) \tag{II.4.14}$$

be a positive energy solution of the Klein-Gordon equation to mass m_i and define

$$q_{i,\alpha}(h_i;t) = i\int_{x^0=t}\left\{q_{i,\alpha}(x)\frac{\partial}{\partial x^0}h_i(x) - \left(\frac{\partial}{\partial x^0}q_{i,\alpha}(x)\right)h_i(x)\right\}d^3x. \tag{II.4.15}$$

Then $q_{i,\alpha}(h_i;t)\Omega$ is a single particle state (of the same species and spin orientation) with momentum space wave function

$$\varphi(p) = \widehat{\varphi}_{i,\alpha}(p)\widetilde{h}_i(p). \tag{II.4.16}$$

This state is independent of the choice of t.

We shall now drop the indices i, α and write for instance q_1 instead of q_{i_1,α_1}. Under the assumptions $\alpha) - \gamma)$ (and the standard axioms) one arrives at the following conclusion which we state as a theorem. Part a) gives the construction of states corresponding to specific configurations of incoming or outgoing particles. Part b) gives the orthogonality relations, part c) the transformation properties of these states.

Theorem 4.2.1 [Haag 58]
Let $q_k(h_k;t)$ be defined as in (II.4.15) with $\widetilde{h}_k(p)$ smooth. Then

a) The sequence of state vectors

$$\Psi(t) = q_1(h_1;t)\ldots q_n(h_n;t)\Omega \tag{II.4.17}$$

converges strongly for $t \to \pm\infty$. The limit states have the physical interpretation

$$\Psi^{\text{in}} \equiv \lim_{t\to-\infty}\Psi(t) = \Psi_1 \otimes^{\text{in}} \cdots \otimes^{\text{in}} \Psi_n \equiv |\varphi_1,\cdots\varphi_n\rangle^{\text{in}} \tag{II.4.18}$$

$$\Psi^{\text{out}} \equiv \lim_{t\to+\infty}\Psi(t) = \Psi_1 \otimes^{\text{out}} \cdots \otimes^{\text{out}} \Psi_n \equiv |\varphi_1,\cdots\varphi_n\rangle^{\text{out}} \tag{II.4.19}$$

(see section 3). Here the Ψ_k are single particle states whose type and spin orientation follows from the choice of q_k and whose momentum space wave function is

$$\varphi_k(p) = \widehat{\varphi}_k(p)\widetilde{h}_k(p). \tag{II.4.20}$$

b) The scalar product of vectors Ψ^{in} is

$$^{\text{in}}\langle\varphi_1',\cdots\varphi_n'|\varphi_1,\cdots\varphi_n\rangle^{\text{in}} = \sum_P(-1)^s\prod_k\langle\varphi'_{Pk}|\varphi_k\rangle \tag{II.4.21}$$

where P runs through all permutations of the indices $(1, \cdots n)$ and s is even or odd depending on the signature of the permutation of the Fermi factors. The scalar product between states with different numbers of incoming particles vanishes.

The same relation holds if the symbol *in* is replaced by *out*.

c) The Poincaré operators $U(a, \alpha)$ respect the product structure introduced by $\otimes^{\text{in}}$ and $\otimes^{\text{out}}$ i.e.

$$U(a, \alpha)\{\Psi_1 \otimes^{\text{in}} \Psi_2 \cdots\} = U(a, \alpha)\Psi_1 \otimes^{\text{in}} U(a, \alpha)\Psi_2 \cdots \tag{II.4.22}$$

with the same relations if $\otimes^{\text{in}}$ is replaced by $\otimes^{\text{out}}$.

Proof. The proof follows the line of argument sketched in section 3 with $\Psi_1 \otimes^t \Psi_2$ realized by $q_1(h_1; t)q_2(h_2; t)\Omega$. The decrease of interaction at large distances is replaced by the decrease of the truncated vacuum expectation values of products of almost local operators which is assured by lemma 4.1.3.

We first need the asymptotic form of single particle wave functions at large times. Let

$$f(t, \mathbf{x}) = \int \tilde{f}(\mathbf{p}) \exp\, i(\mathbf{p}\mathbf{x} - \varepsilon_{\mathbf{p}} t) \frac{d^3p}{2\varepsilon_{\mathbf{p}}}\,; \quad \varepsilon_{\mathbf{p}} = (\mathbf{p}^2 + m^2)^{1/2}, \tag{II.4.23}$$

with $\tilde{f}$ a smooth function, decreasing fast for $|\mathbf{p}| \to \infty$. The asymptotic form for large $|t|$ can be obtained by applying the method of stationary phase to the integral (II.4.23). Since the phase $\mathbf{p}\mathbf{x} - \varepsilon_{\mathbf{p}} t$ is rapidly varying as a function of $\mathbf{p}$ when t (and possibly also $\mathbf{x}$) is large, the dominant contribution to the integral comes from the point

$$\mathbf{p} = m\mathbf{v}(1 - \mathbf{v}^2)^{-1/2}; \quad \mathbf{v} = \mathbf{x}/t \tag{II.4.24}$$

where the phase is stationary. One finds the asymptotic expansion

$$f(t, \mathbf{x}) = \text{const.}\ t^{-3/2} \exp(-im\gamma^{-1}t) \left(\gamma^{3/2} \tilde{f}(m\gamma\mathbf{v}) + O(t^{-1})\right). \tag{II.4.25}$$

$$\gamma = (1 - \mathbf{v}^2)^{-1/2}; \quad \mathbf{v} = \mathbf{x}/t.$$

Whilst (II.4.25) describes the general asymptotic behavior of smooth wave functions a much stronger statement can be made for those directions in velocity space which lie outside the support of $\tilde{f}$.

Lemma 4.2.2 [Ruelle 62]
Let $f(x)$ be as in (II.4.23), Σ its support in velocity space i.e.
$\Sigma = \{\mathbf{v} = \mathbf{p}/p^0 : \mathbf{p} \in \operatorname{supp}\tilde{f}\}$ and $\mathcal{U}$ any open set containing Σ. Then

$$\text{(i) for} \quad \mathbf{v} \notin \mathcal{U} \quad \text{one has} \quad |f(t, \mathbf{v}t)| < C_N(1 + |\mathbf{v}|)^{-N}|t|^{-N} \tag{II.4.26}$$

for arbitrary positive N.

$$\text{(ii) for} \quad \mathbf{v} \in \mathcal{U} \quad \text{one has} \quad |f(t, \mathbf{v}t)| < C|t|^{-3/2}. \tag{II.4.27}$$

Next we consider scalar products between vectors obtained by applying several operators of the form

$$Q_f(t) = \int_{x^0=t} Q(x) f(x) d^3x \tag{II.4.28}$$

to the vacuum. Q shall be almost local with $\langle \Omega | Q | \Omega \rangle = 0$, f a smooth single particle wave function and all times are taken equal.

Lemma 4.2.3
Under the conditions stated immediately above one has

$$\langle \Omega | Q_{f'_1}(t)^* \cdots Q_{f'_m}(t)^* Q_{f_n}(t) \cdots Q_{f_1}(t) | \Omega \rangle$$

$$\rightarrow \delta_{mn} \sum_{\text{pairings}} (-1)^s \prod_{\text{pairs}} \langle \Omega | Q_{f'_k}(t)^* Q_{f_j} | \Omega \rangle. \tag{II.4.29}$$

The difference between the right and the left hand sides of (II.4.29) decreases in general like $|t|^{-3/2}$. If the supports of the functions f_k in velocity space as well as those of the functions f'_k are disjoint the decrease of the difference is faster than any inverse power of $|t|$.

In (II.4.29) the symbol Q is meant generically. The almost local operators may be different in each factor. The proof of this lemma is a rather straightforward application of lemmas 4.1.3 and 4.2.2 or equation (II.4.25). We make a cluster decomposition of $\langle \Omega | Q^*(y_1) \cdots Q^*(y_m) Q(x_1) \cdots Q(x_n) | \Omega \rangle$ and insert the estimate (II.4.25) for the wave functions. Numbering the clusters by an index r ($r = 1, \ldots N$) we denote the first point in the cluster r by $\mathbf{X}_r$ use for the other points in this cluster the relative coordinates $\boldsymbol{\xi}_i = \mathbf{x}_i - \mathbf{X}_r$. The truncated expectation values are, by lemma 4.1.3, fast decreasing functions of the ξ. They are independent of the $\mathbf{X}_r$ and of t. Therefore the integration over the ξ can be performed. It gives for cluster r with n_r factors Q^*, m_r factors Q a contribution of the form

$$C|t|^{-3/2(n_r+m_r)} \tag{II.4.30}$$

Changing the remaining integration variables from $\mathbf{X}_r$ to $\mathbf{V}_r = \mathbf{X}_r/t$ we obtain finally as the contribution to (II.4.29) from a clustering with N clusters

$$C'|t|^{3N-3/2(n+m)}. \tag{II.4.31}$$

Note that the $\mathbf{V}_r$ are restricted to the balls $|\mathbf{V}_r| < 1$, the boundary corresponding to the momenta $\mathbf{p} \rightarrow \infty$ in $\tilde{f}$. Therefore the integrations over $\mathbf{V}_r$ only produce a finite factor independent of t.

Since $\langle\Omega|Q|\Omega\rangle = 0$ each cluster must contain at least two points. For all clusterings which contain a cluster with more than 2 points $N < 1/2(n+m)$ and therefore (II.4.31) decreases at least as $|t|^{-3/2}$. Thus we can get a nonvanishing contribution for $|t| \to \infty$ only in the case of a complete pairing. Furthermore we get a nonvanishing contribution only if each cluster pairs a factor Q^* with a factor Q. A cluster with two factors Q contributes an expression

$$\int \tilde{w}(\mathbf{p})\tilde{f}_i(\mathbf{p})\tilde{f}_k(-\mathbf{p})e^{-i(\varepsilon_1+\varepsilon_2)t}d^3p$$

where $\tilde{w}$ is the 3-dimensional Fourier transform of

$$w(\mathbf{x}_1 - \mathbf{x}_2) = \langle\Omega|Q_1(t,\mathbf{x}_1)Q_2(t,\mathbf{x}_2|\Omega\rangle,$$

and therefore a smooth function of $\mathbf{p}$ due to the fast vanishing of w for large $|\mathbf{x}_1 - \mathbf{x}_2|$. The momentum space wave functions in the integral are smooth by assumption and the exponential factor is a fast oscillating function of $\mathbf{p}$ when t is large. Therefore the integral decreases faster than any power of $|t|$ for $|t| \to \infty$.

A complete pairing between factors Q^* and Q is possible only for $n = m$ and yields then the right hand side of (II.4.29). We may note that for $n = m$ the difference between the left and the right hand side of (II.4.29) decreases at least as $|t|^{-3}$, coming from a clustering where at least two clusters contain three factors. In the case of disjoint supports of the momentum space wave functions any clustering which is not contained on the right hand side of (II.4.29) gives a contribution vanishing faster than any power due to part (i) of lemma 4.2.2.

Corollary 4.2.4

Let $Q_f(t)$ be as in lemma 4.2.3, $k = 1, \cdots n$ and P a permutation sending k to Pk. Consider the vectors

$$\Psi = \prod Q_{f_k}(t)\Omega \quad \text{and} \quad \Psi' = (-1)^s \prod Q_{f_{Pk}}(t)\Omega$$

where s is the number of transpositions of Fermi factors arising from the permutation. Then as $|t| \to \infty$

$$\| \Psi - \Psi' \| < c|t|^{-3/2}. \tag{II.4.32}$$

To see this note that

$$\| \Psi - \Psi' \|^2 = \langle\Psi|\Psi\rangle - \langle\Psi|\Psi'\rangle - \langle\Psi'|\Psi\rangle + \langle\Psi'|\Psi'\rangle.$$

Each of the terms is of the form (II.4.29) with $m = n$ and they all give the same dominant contribution (the right hand side of (II.4.29)) because the pairings are independent of the order of factors. Therefore the dominant contributions cancel and the remainder decreases like $|t|^{-3}$.

After these preliminaries the proof of the theorem is quick.
We first show that $\| \partial\Psi(t)/\partial t \| < c|t|^{-3/2}$ for large $|t|$. This implies the strong convergence of $\Psi(t)$ with

$$\| \Psi^{\mathrm{ex}} - \Psi(t) \| < 2\,c\,|t|^{-1/2} \quad \text{for} \quad |t| \to \infty; \quad \mathrm{ex} = \begin{cases} \text{in} \\ \text{out} \end{cases} \quad \text{for} \quad t \to \begin{cases} -\infty \\ +\infty \end{cases}. \tag{II.4.33}$$

Why does $\| \partial\Psi(t)/\partial t \|$ decrease for large $|t|$? Differentiating $\Psi(t)$ we get a sum of terms of the same form as (II.4.17) with one of the $q(h;t)$ replaced by $\partial/\partial t\, q(h;t)$. We commute $\partial q/\partial t$ to the right till it stands directly in front of Ω where it gives zero because $q(h;t)\Omega$, as a single particle state, is independent of t. According to the above corollary this change of the order of factors introduces only a difference of order $|t|^{-3/2}$ in the norm.

After the convergence of $\Psi(t)$ is established part b) of the theorem follows directly from (II.4.29).

Coming to part c) of the theorem we remark that

$$U(a;\alpha)q_1 \ldots q_n\Omega = q_1' \ldots q_n'\Omega, \tag{II.4.34}$$

$$q_k' = U(a;\alpha)q_k U(a;\alpha)^{-1}. \tag{II.4.35}$$

Now

$$q_k'\Omega = \Psi_k' = U(a;\alpha)\Psi_k \tag{II.4.36}$$

is the transformed single particle state. For translations and rotations q_k' is again of the form (II.4.15) and the claim c) follows directly from part a) of the theorem. In the case of proper Lorentz transformations the verification of (II.4.22) needs a little more care since (II.4.35) will then involve almost local operators at different times. But the relevant configurations of points will still have large space-like separation and lemma 4.1.3 is valid in any Lorentz frame.

This concludes the proof of the theorem apart from the claim that the limit states (II.4.18), (II.4.19) must be interpreted as states with specified configurations of incoming or outgoing particles. Although this is rather evident from the qualitative discussion in section 3 we shall illustrate it by simulating the experimental procedure of particle detection more directly in the mathematical formalism.

II.4.3 Coincidence Arrangements of Detectors

Let us simulate a detector, centered near the point $x = 0$, by a positive operator C with $C\Omega = 0$, localized (essentially) in some finite region around the origin.[5] A coincidence arrangement of such detectors, centered at the points $x_1, \cdots x_m$ which lie far space-like separated from each other will then be represented by the product $C(x_1) \cdots C(x_m)$. We take the times equal and put $x_i = (t, \mathbf{v}_i t), \mathbf{v}_i \neq \mathbf{v}_j$. We are interested in the expectation value of this product in the state (II.4.17).

[5]Due to the Reeh-Schlieder theorem (see next section) C cannot be strictly localized in a finite region if we demand $C\Omega = 0$. We must either take it as almost local or allow a small response rate in the vacuum.

For simplicity we choose states which are composed of single particle states whose momentum space wave functions have disjoint support. We thus choose the h_i in (II.4.17) so that

$$\operatorname{supp} \tilde{h}_i \cap \operatorname{supp} \tilde{h}_j = \emptyset \quad \text{for} \quad i \neq j. \tag{II.4.37}$$

This is no serious restriction for most purposes. It just excludes thresholds where two or more particles with the same momentum occur in a configuration. It brings the advantage that the convergence of $\Psi(t)$ to Ψ^{ex} is much faster than in (II.4.32), namely

$$\| \Psi^{\mathrm{ex}} - \Psi(t) \| < C_N |t|^{-N} \tag{II.4.38}$$

for arbitrary positive N. Applying the familiar method we find that the contribution of a cluster decreases faster than any power of $|t|$ in any one of the following circumstances:

(i) the cluster contains more than one factor C (because $\mathbf{v}_i \neq \mathbf{v}_j$),

(ii) it contains more than one factor q or more than one factor q^* or q and q^* in one cluster are not associated with the same wave function (because of (II.4.37),

(iii) C is not accompanied by a factor q on the right and q^* on the left. (Since $C\Omega = 0$ and $C = C^*$ we get zero if C stands immediately adjoining the vacuum).

Thus one gets for any positive N

$$|t|^N \langle \Psi(t) | C(x_1) \cdots C(x_m) | \Psi(t) \rangle \to 0 \quad \text{if} \quad m > n \tag{II.4.39}$$

and unless each $\mathbf{v}_i$ is in the velocity support of one of the h_k.

If $\mathbf{v}_i \in \Sigma(h_i)$ for $i = 1, \cdots m$ and $m \leq n$ we get

$$\langle \Psi(t) | C(x_1) \cdots C(x_m) | \Psi(t) \rangle \to \prod_1^m \langle \Psi_i | C(x_i) | \Psi_i \rangle \prod_{m+1}^n \langle \Psi_k | \Psi_k \rangle . \tag{II.4.40}$$

A factor $\langle \Psi_i | C(x) | \Psi_i \rangle$ gives the response of a detector centered near the point x in the single particle state Ψ_i. Of course each such factor decreases like t^{-3} because the spreading of the wave function makes the probability of meeting the particle in some space region of fixed extension decrease. So (II.4.40) can be used only for large finite t, not for infinite times; this corresponds to the experimental situation of detector arrangements.

Remark: The preceding argument corroborates the physical interpretation of the states Ψ^{ex}. At the same time (II.4.39), (II.4.40) give the most natural definition of what is meant by particles. A state with n incoming particles is defined as a state which, at all times earlier than some time t, gives no signal in any coincidence arrangement of more than n far separated detectors (mathematically expressed in (II.4.39)), but does give a signal in some n-fold coincidence arrangements (Equ. (II.4.40)). This criterion is applicable even if one has to abandon the association of discrete eigenvalues of the mass operator with single particle states as one is forced to do in a serious treatment of infrared problems caused by photons because a charged particle alone in the world cannot be sharply distinguished from one where it is accompanied by some soft photons.

II.4.4 Generalized LSZ-Formalism

To close this section we want to establish the bridge to the LSZ-formalism and its generalizations. We compute matrix elements between states $\Psi(t)$, $\Psi'(t)$ of the form (II.4.17) for an almost local operator $A(x)$ centered around the point $x = (t, \mathbf{v}t)$ where, unlike the case of detectors, neither A nor A^* annihilates the vacuum but only the vacuum expectation value of A is required to vanish and, unlike (II.4.13) $A\Omega$ is not a single particle state. Let $\Psi(t)$ contain n factors, $\Psi'(t)$ m factors and assume that in each of these states the wave functions h_i have disjoint velocity support. With the technique, boringly familiar by now, we find that the only terms which do not decrease faster than any power for large $|t|$ come from clusterings where A is accompanied by one factor q or q^* or by one q^* *and* one q. This leads to

$$\begin{aligned}\langle\Psi'(t)|A(x)|\Psi(t)\rangle \underset{t\to-\infty}{\longrightarrow} &\sum\langle\Omega|A(x)|\varphi_i\rangle\langle\Psi'^{\text{in}}|a^{\text{in}}(\varphi_i)\Psi^{\text{in}}\rangle\\ &+\sum_j\langle\varphi'_j|A(x)|\Omega\rangle\langle a^{\text{in}}(\varphi'_j)\Psi'^{\text{in}}|\Psi^{\text{in}}\rangle\\ &+\sum_{i,j}\langle\varphi'_j|A(x)|\varphi_i\rangle\langle a^{\text{in}}(\varphi'_j)\Psi'^{\text{in}}|a^{\text{in}}(\varphi_i)\Psi^{\text{in}}\rangle,\end{aligned} \tag{II.4.41}$$

where

$$\Psi^{\text{in}} = |\varphi_1,\cdots\varphi_n\rangle^{\text{in}};\quad \Psi'^{\text{in}} = |\varphi'_1,\cdots\varphi'_m\rangle^{\text{in}}.$$

The first two terms on the right hand side decrease like $|t|^{-3/2}$; they are respectively positive energy and negative energy solutions of the Klein-Gordon equation to the mass associated with the wave function φ_i resp. φ'_j. The last term decreases like $|t|^{-3}$. To pick out matrix elements of $a^{\text{in}*}$ for a certain particle type, say of mass m, we choose an A which has nonvanishing matrix elements between the vacuum and single particle states of this type. If f is a positive energy solution of the Klein-Gordon equation to mass m and (compare (II.3.20))

$$A(f;t) = i\int\left\{A(x)\frac{\partial}{\partial x^0}f(x) - \left(\frac{\partial}{\partial x^0}A(x)\right)f(x)\right\}d^3x,\quad x^0 = t$$

then we get from (II.4.41)

$$\langle\Psi'(t)|A(f;t)|\Psi(t)\rangle \to \sum\langle\varphi'_j|A(f;t)\Omega\rangle\langle\Psi'^{\text{in}}|a^{\text{in}*}(\varphi'_j)|\Psi^{\text{in}}\rangle \tag{II.4.42}$$

because the first term in (II.4.41) is cancelled by the integration (since f and $\langle\Omega|A(x)|\varphi_i\rangle$ are wave functions with the same sign of the energy), the second is the term written down; it is independent of t. The last term goes to zero in the limit. We may omit writing the vectors Ψ and Ψ' in (II.4.42) if we use a complete orthonormal basis of single particle states of mass m, for instance $|p, s\rangle$ where p is the momentum, s the spin orientation.[6] Then (II.4.42) can be written as

$$A(f;t) \to a^{\text{in}*}(\varphi) \quad \text{as} \quad t \to -\infty$$
$$\text{with} \qquad \qquad (\text{II.4.43})$$
$$\tilde{\varphi}_s(p) = \langle p, s|A|\Omega\rangle \tilde{f}(p),$$

where the arrow is understood to mean weak convergence (convergence of matrix elements) on a certain domain.

In the LSZ-formalism, described in section 3, we were dealing with a spinless particle and took for A the basic (scalar) field operator at the point $x = 0$. The matrix element $\langle p|A|\Omega\rangle$ is then a constant. So $\tilde{\varphi}$ and $\tilde{f}$ can be identified if the normalization of the field is appropriately chosen. It is now evident how the "weak convergence method" (the generalized LSZ-formalism) can be extended to cover situations with *composite particles*. By these we mean here that the matrix elements of all basic fields from the vacuum to the single particle states in question vanish. One simply must look for some almost local operator A with nonvanishing matrix elements between the vacuum and these single particle states. Such operators are usually easy to find among the polynomials of the basic fields. This "weak convergence method" applied to the bound state problem was developed by Nishijima [Nish 58] and Zimmermann [Zi 58].

Remarks: In the proof of the main theorem 4.2.1 given in [Haag 58] the fast decrease of truncated functions in space-like direction was introduced as an additional assumption, made plausible on physical grounds, with the hope that this assumption could be shown to be a consequence of the axioms. Several authors obtained partial results in this direction. The situation was completely clarified by the work of Ruelle. He showed that if there is a lowest mass then the fast decrease of truncated functions follows (lemma 4.1.3). Ruelle also made all the estimates involved in the proof watertight. In particular he avoided the use of the asymptotic expansion (II.4.25) and replaced it by lemma 4.2.2. The truncated Wightman functions or, alternatively, the left hand side of (II.4.7) in the case when the Q_k are local instead of almost local, decrease proportional to e^{-mR}. This was proved by Araki, Hepp and Ruelle [Ara 62b]. An estimate independent of the type of the support regions was given by Fredenhagen [Fred 85b]. He obtained the estimate

$$|\langle\Omega|AB|\Omega\rangle - \langle\Omega|A|\Omega\rangle\langle\Omega|B\Omega\rangle|$$

[6] If there are several types of particles with the same mass then s must also include the index distinguishing these types.

$$\leq e^{-mR}\{\| A^*\Omega \| \| A\Omega \| \| B^*\Omega \| \| B\Omega \|\}^{1/2}, \qquad \text{(II.4.44)}$$

where m is the smallest mass, R the largest distance so that $e^{iHt}Ae^{-iHt}$ still commutes with B.

The discussion of the detector arrangements is taken from [Ara 67] and will be resumed in chapter VI. The relation to the generalized LSZ-formalism is discussed in [Hepp 65] and [Ara 67].

II.5 Some Consequences of the Postulates

II.5.1 CPT-Operator. Spin-Statistics-Theorem

In discussing the analyticity of Wightman functions $w^n(z_1, \cdots z_n)$ two important facts were mentioned in section 2. First, if $z_1, \cdots z_n$ is in the analyticity domain and Λ any complex matrix satisfying (II.2.11) then $z_1', \cdots z_n'$ with $z_k' = \Lambda z_k$ is in the analyticity domain and $w^n(z_1', \cdots z_n')$ may be found by applying (an extension of) the transformation law (II.1.9). Secondly in the complex Lorentz group the space-time reflection $x^\mu \to -x^\mu$ can be connected by a continuous path to the identity.

Let us look at the extension of (II.1.9) to the complex Lorentz group. The relation between $\mathfrak{L}$ and SL(2,$\mathbb{C}$) was established by writing a real 4-vector as the matrix (I.2.14) and its Lorentz transform by (I.2.16). If we keep (I.2.14) for complex 4-vectors and allow complex Lorentz transformation then (I.2.16) is generalized to

$$\widehat{V}' = \alpha \widehat{V} \beta^T \qquad \text{(II.5.1)}$$

where α and β are independent matrices from SL(2,$\mathbb{C}$). The special case of real Lorentz transformations is $\beta = \overline{\alpha}$. The covering group of the (connected part of the) complex Lorentz group is thus given by SL(2,$\mathbb{C}$)$\times$ SL(2,$\mathbb{C}$) i.e its elements are pairs (α, β). The complex Lorentz matrix Λ determines the pair up to an overall sign. Considering now transformations of spinors we see that the generalization (II.5.1) just means that a spinor φ_r (with undotted index) is transformed with the matrix α while a "conjugate spinor" $\varphi_{\dot{r}}$ (dotted index) is transformed by β instead of $\overline{\alpha}$. From this the transformation properties of spinors of higher rank with several undotted and dotted indices follow. The space-time reflection is represented by the pair $(\alpha, \beta) = (1, -1)$ or by $(-1, 1)$. To connect it to the identity we may, choosing the first alternative, pick a continuous path in SL(2,$\mathbb{C}$) leading from $\beta = 1$ to $\beta = -1$ (e.g. a rotation by 360 degrees) while keeping α fixed at 1. This will leave a spinor with undotted index unchanged and give a factor -1 for a spinor with dotted index. For a spinor of higher rank with n undotted and m dotted indices we will pick up a factor $(-1)^m$. It suffices to consider such "pure" spinorial fields since all covariant fields are (direct) sums of such.

This transformation law gives the following relation between Wightman functions

$$w^n(z_1, \cdots z_n) = (-1)^M w^n(-z_1, \cdots - z_n) \qquad \text{(II.5.2)}$$

where M is the total number of dotted indices of all fields occurring in w^n. Each z_k is attached to a specific component of a specific basic field. If we would let the imaginary parts of the z_k go to zero in (II.5.2) we would not get a relation between the Wightman distributions with real arguments because the imaginary parts of the difference vectors of successive points have opposite signs on the two sides. The relation holds, however for Jost points and there, due to the local commutativity axiom **E** we may change the order and write ([Jost 57])

$$w^n(z_1, \ldots z_n) = (-1)^P (-1)^M w^n(-z_n, \ldots - z_1). \qquad \text{(II.5.3)}$$

The first sign comes from the interchanges of anticommuting fields needed in the reversal of the order. Now, if $z_1, \ldots z_n$ is in the primitive domain so is $-z_n, \ldots - z_1$ and therefore we may let the imaginary parts go to zero and retain a relation between Wightman distributions valid for all real points. Specializing (II.5.3) to $n = 2$ with $\Phi_1 = \Phi$, $\Phi_2 = \Phi^*$ we get

$$\langle \Omega | \Phi(x) \Phi^*(y) | \Omega \rangle = (-1)^{P+M} \langle \Omega | \Phi^*(-y) \Phi(-x) | \Omega \rangle, \qquad \text{(II.5.4)}$$

and after integration with test functions $f(x) \cdot \overline{f}(y)$

$$\| \Phi^*(f)\Omega \|^2 = (-1)^{P+M} \| \Phi(\widehat{f})\Omega \|^2$$

with $\widehat{f}(x) = f(-x)$. So $P + M$ must be even.[1] In the case considered M is the sum of the numbers of dotted and undotted indices of Φ. It is even for true representations and odd for double valued representations. P is even if Φ commutes with Φ^*, odd if it anticommutes (at space-like distances). This suggests that we can make the Bose-Fermi alternative in axiom **E** more precise.

Definition 5.1.1
Φ is called a Bose field if

$$[\Phi(x), \Phi^*(y)] = 0 \quad \text{for} \quad (x - y)^2 < 0, \qquad \text{(II.5.5)}$$

and it is called a Fermi field if

$$[\Phi(x), \Phi^*(y)]_+ = 0 \quad \text{for} \quad (x - y)^2 < 0. \qquad \text{(II.5.6)}$$

We can then express the conclusion from (II.5.4) as the

"Spin-Statistics" Theorem 5.1.2
Fields belonging to integer spin representations are Bose fields, fields belonging to half odd integer spin representations are Fermi fields.

---

[1]The other possibility would be the vanishing of $\Phi\Omega$ and $\Phi^*\Omega$. But this can be excluded because it would lead to $\Phi = 0$ due to theorem 5.3.2 below.

To evaluate the factor $(-1)^P$ in (II.5.3) for general configurations of fields we need to know the (space-like) commutation relations between different fields. One has the following general restriction:

Lemma 5.1.3 [Dell' Ant 61]

$$\text{If} \quad [\Phi_1(x), \Phi_2^*(y)]_\pm = 0 \quad \text{for} \quad (x-y)^2 < 0$$

$$\text{then also} \quad [\Phi_1(x), \Phi_2(y)]_\pm = 0 \quad \text{for} \quad (x-y)^2 < 0.$$

Proof.

$$\langle \Omega | \Phi_1^*(f) \Phi_2^*(g) \Phi_2(g) \Phi_1(f) | \Omega \rangle = \| \Phi_2(g) \Phi_1(f) | \Omega \rangle \|^2 > 0.$$

If the supports of f and g lie space-like with a minimal separation R then the above expression equals

$$\begin{aligned} &\sigma \langle \Omega | \Phi_1^*(f) \Phi_1(f) \Phi_2^*(g) \Phi_2(g) | \Omega \rangle \\ \underset{R\to\infty}{\longrightarrow} \; &\sigma \langle \Omega | \Phi_1^*(f) \Phi_1(f) | \Omega \rangle \langle \Omega | \Phi_2^*(g) \Phi_2(g) | \Omega \rangle \\ &= \sigma \| \Phi_1(f) \Omega \|^2 \| \Phi_2(g) \Omega \|^2 . \end{aligned}$$

Here $\sigma = \pm 1$ if $[\Phi_2^*(g)\Phi_2(g), \Phi_1(f)]_\mp = 0$. The positivity of the norms squared implies $\sigma = +1$ which is the claim of the lemma. □

This suggests the following convention.

Normal Commutation Relations. At space-like distances any two Bose fields commute, any Bose field commutes with any Fermi field and any two Fermi fields anticommute.

Comment. We called this a convention because it can be shown [Ara 61] that although different possibilities exist they can always be transformed to the "normal" case by a redefinition of the fields in a way which does not change the physical consequences. This is related to *superselection rules* which will be discussed in a more general context in detail in chapter IV.

Assuming now normal commutation relations we have in (II.5.3)

$$P = \frac{1}{2} F(F-1)$$

where F is the number of Fermi fields in the brackets. For non vanishing Wightman functions F must be even, so

$$(-1)^P = (-1)^{F/2} = i^F .$$

Then we get from (II.5.3) the theorem

CPT-theorem 5.1.4 [Jost 57]

a) There exists an antiunitary operator Θ uniquely defined by

$$\Theta\Phi(x)\Theta^{-1} \equiv \Phi^C(x) = (-1)^m(-i)^F\Phi^*(-x), \tag{II.5.7}$$

$$\Theta\Omega = \Omega. \tag{II.5.8}$$

$$F = \begin{cases} 0 & \text{if} \quad \Phi \quad \text{is a Bose field} \\ 1 & \text{if} \quad \Phi \quad \text{is a Fermi field,} \end{cases}$$

m is the number of dotted spinor indices of Φ.

b)

$$\Theta U(a;\alpha) = U(-a;\alpha)\Theta. \tag{II.5.9}$$

c) Θ^2 commutes with all Bose fields and anticommutes with all Fermi fields. Alternatively stated

$$\Theta^2 = \begin{cases} +1 & \text{on the subspace of integer spin states} \\ -1 & \text{on the subspace of half odd integer spin states.} \end{cases}$$

d) Θ transforms a single particle state of species i into a single particle state of species $\bar{i}$, the *antiparticle* of i. It is not excluded that particle and antiparticle are identical. If they are different then we may use bases $|p\lambda i\rangle$ in $\mathcal{H}_{1,i}$ and $|p\dot{\lambda}\bar{i}\rangle$ in $\mathcal{H}_{1,\bar{i}}$ and where λ denotes an array of undotted and dotted spinor indices $\lambda = r_1, r_2 \cdots, \dot{s}_1, \dot{s}_2 \cdots$ $(r_i = 1,2; s_i = 1,2)$; $\dot{\lambda} = \dot{r}_1, \dot{r}_2 \cdots, s_1, s_2 \cdots$ so that

$$\Theta|p\lambda i\rangle = (-1)^m|p\dot{\lambda}\bar{i}\rangle \tag{II.5.10}$$

with m the number of dotted indices in λ.[2]

e)

$$\Theta|\varphi_1, \cdots \varphi_n\rangle^{\text{out}} = |\varphi_1^C, \cdots \varphi_n^C\rangle^{\text{in}} \tag{II.5.11}$$

with

$$|\varphi_k^C\rangle = \Theta|\varphi_k\rangle.$$

The wave function of the charge conjugate single particle state φ^C is related to that of φ by

$$\varphi^C_{\lambda i}(p) = (-1)^m\overline{\varphi}_{\dot{\lambda}\bar{i}}(p). \tag{II.5.12}$$

For the S-matrix elements this gives

[2] If particle and antiparticle are identical then we may still keep (II.5.10) but then the two bases refer to the same Hilbert space $\mathcal{H}_{1,i}$ and we have to give the relation between them. In the example of a Majorana particle

$$|pr\rangle = p_{r\dot{s}'}\varepsilon^{\dot{s}'\dot{s}}|p\dot{s}\rangle$$

where $p_{r\dot{s}}$ is the spinorial form of the 4-vector p i.e. the matrix $\widehat{p} = (p_0 + \mathbf{p}\boldsymbol{\sigma})_{rs}$.

$$^{\text{out}}\langle\varphi_1',\cdots\varphi_m'|\varphi_1,\cdots\varphi_n\rangle^{\text{in}} = {}^{\text{out}}\langle\varphi_1^C,\cdots\varphi_n^C|\varphi_1'^C,\cdots\varphi_m'^C\rangle^{\text{in}} \qquad \text{(II.5.13)}$$

and for the S-operator

$$\Theta S = S^{-1}\Theta. \qquad \text{(II.5.14)}$$

Proof. Equation (II.5.3) remains valid for real points and can then be written, using the knowledge of the sign factor $(-1)^P$ and inserting the definition of Φ^C from (II.5.7) as

$$\langle\Omega|\Phi_1(x_1)\cdots\Phi_n(x_n)|\Omega\rangle = \langle\Omega|\Phi_n^{C*}(x_n)\cdots\Phi_1^{C*}(x_1)|\Omega\rangle.$$

The antilinear operator Θ defined by

$$\Theta\Phi_1(f_1)\cdots\Phi_n(f_n)\Omega = \Phi_1^C(\overline{f}_1)\cdots\Phi_n^C(\overline{f}_n)\Omega \qquad \text{(II.5.15)}$$

is therefore antiunitary i.e. it maps $\mathcal{H}$ onto itself and satisfies

$$\langle\Theta\Psi_2|\Theta\Psi_1\rangle = \langle\Psi_1|\Psi_2\rangle.$$

The definition (II.5.15) for Θ coincides with the definitions (II.5.7), (II.5.8).

Concerning b) one verifies directly that the unitary operator $\Theta U(a;\alpha)\Theta^{-1}$ transforms each field in the same way as the operator $U(-a;\alpha)$. Because of the completeness axiom **F** the two operators can differ only by a numerical factor. But they have the same action on Ω, so they coincide. Part c) follows simply from the definition (II.5.7).

From b) it follows that Θ transforms a state with momentum p and spin λ to one with the same momentum but spin $\dot{\lambda}$. For single particle states, associated with an irreducible representation of $\overline{\mathfrak{P}}$, this fixes the image vector up to a phase which depends on the convention adopted for the matrix elements of the fields between the vacuum and the single particle state vectors. If particle and antiparticle are different then the only restriction is provided by c) and (II.5.10) is a consistent choice.

Part e) follows from a), b) together with (II.4.17), (II.4.18), (II.4.19).

II.5.2 Analyticity of the S-Matrix

The wish for better understanding of the mathematical problems and conceptual structure was one of the motivating forces in the development of general quantum field theory. Another, more pragmatic, motive was provided by the status of the theory of strong interactions. It had become clear that a perturbation treatment of the Yukawa theory of nucleon and pion fields could not lead to a quantitative comparison between theory and experiment. Therefore it became of interest to establish relations between different experimentally amenable quantities which did not depend on a detailed model but tested only general basic assumptions of the theory. A prime example was afforded by the dispersion relations for the forward scattering amplitude. It was recognized that the axioms of section 1 together with the reduction formulas of section 3 suffice to show that the forward scattering amplitude for pion-nucleon scattering is an analytic function of the energy in the cut complex plane. One may then represent this amplitude as a Cauchy integral in terms of the jump across the cut. The latter, called the "absorptive part" is related to the total cross section. One so obtains a *dispersion relation* which gives the forward scattering amplitude at energy E as an integral involving the total cross section at other energies.

The derivation of these relations from general principles and their experimental verification with significant accuracy has been hailed as one of the triumphs of general quantum field theory. It put a definite end to the conjecture about the existence of a fundamental length of 10^{-13} cm to 10^{-14} cm, below which the causal structure of space-time might become meaningless.[3] It stimulated much effort to derive further analytic properties of the S-matrix from first principles. These efforts met with some but not with spectacular success. There is no need to treat this topic here since there exist many extensive and easily available discussions in the literature.[4]

Another strategy, pursued and propagated in particular by G. Chew in the sixties, was to assume that S-matrix elements in momentum space are as analytic as they possibly can be, irrespective of what one can prove from quantum field theory. From a pragmatic point of view this appears eminently reasonable. From a fundamental point of view the renunciation of microscopic causality in space-time as envisaged earlier by Heisenberg's idea of a pure S-matrix theory seems to be premature. The subsequent development of elementary particle physics indicates that the special relativistic space-time concept is viable up to the highest energies available in present day accelerators.

II.5.3 The Theorem of Reeh and Schlieder

By the same argument which was used in section 2 to derive the analyticity of Wightman functions one finds that matrix elements

$$F(x_1, \ldots x_n) = \langle \Psi | \Phi(x_1) \cdots \Phi(x_n) | \Psi_1 \rangle \tag{II.5.16}$$

are boundary values of analytic functions in x-space for any $\Psi \in \mathcal{D}$ if the spectral support (in energy) of the vector Ψ_1 is bounded. The analyticity domain is the tube

$$z_1 = x_1 + i\eta_1, \quad z_2 = (x_1 - x_2) - i\eta_2, \cdots \quad z_n = (x_{n-1} - x_n) - i\eta_n \tag{II.5.17}$$

where all imaginary parts η_k lie in the forward cone, the real parts being arbitrary. It follows then from the "edge of the wedge theorem" that if F vanishes

[3]It was this conjecture which, in train of the frustration with quantum field theory in the late thirties, had led Heisenberg to propose a pure S-matrix theory from which concepts like fields, space and time are eliminated at the fundamental level. Heisenberg did, however, not adhere to this point of view for long. I remember him saying in a discussion in 1956: "The S-matrix is the roof of the theory, not its foundation."

[4]See e.g. [*Bjorken and Drell 1965*] and references quoted there. The dispersion relations do not imply the existence of point-like fields. Compare [Ep 69], [Buch 85].

within some open region of real configuration space then F vanishes for all configurations of points in Minkowski space. This has the startling consequence

Theorem 5.3.1 (Reeh and Schlieder, [Reeh 61])
The set of vectors $\mathcal{A}(\mathcal{O})\Omega$, generated from the vacuum by the polynomial algebra of any open region, is dense in $\mathcal{H}$.

To see this take $\Psi_1 = \Omega$ and Ψ in the orthogonal complement of $\mathcal{A}(\mathcal{O})|\Omega\rangle$. Then $F = 0$ as long as all $x_i \in \mathcal{O}$. Hence $F = 0$ identically. Hence $\Psi = 0$.

Remarks: (i) Intuitively one might have thought that with $Q \in \mathcal{A}(\mathcal{O})$ the vector $Q\Omega$ could be interpreted as representing a state *localized in* $\mathcal{O}$, i.e. a state looking practically like the vacuum with respect to measurements in the causal complement of $\mathcal{O}$. While, due to the cluster property of Wightman functions, this is qualitatively true if Q is picked at random in $\mathcal{A}(\mathcal{O})$ and measurements at a sufficiently large space-like distance from $\mathcal{O}$ are considered, the theorem tells us that for any chosen state vector Ψ one can always find an operator $Q \in \mathcal{A}(\mathcal{O})$ which, applied to the vacuum, produces a state vector arbitrarily close to Ψ. To achieve this the operator must judiciously exploit the small but nonvanishing long distance correlations which exist in the vacuum as a consequence of the spectral restrictions for energy-momentum in the theory. The theorem shows that the concept of *localized states*, if used in a more than qualitative sense, must be handled with care.
(ii) Obviously in the theorem the vacuum may be replaced by any vector with bounded energy.

Closely related to theorem 5.3.1 is

Theorem 5.3.2
If $\mathcal{O}$ has non void causal complement then $\mathcal{A}(\mathcal{O})$ does not contain any operator which annihilates the vacuum (or any other state vector with bounded energy).

II.5.4 Additivity of the Energy-Momentum Spectrum

The simultaneous spectrum of the energy-momentum operators P^μ is a closed subset of 4-dimensional p-space. We denote it by Spect P. If Δ is an open region of p-space which intersects Spect P then there exists an almost local operator Q with $\| Q\Omega \| \neq 0$ which, in the notation of lemma 4.1.2 , has supp $\tilde{Q}(p) \subset \Delta$. The state vector $\Psi = Q\Omega$ has spectral support in Δ. The translated vector $U(a)\Psi$ has the same spectral support.

Theorem 5.4.1
If p_1 and p_2 are in Spect P then $p_1 + p_2$ is in Spect P.

Proof. Take neighborhoods Δ_1, Δ_2 of the points p_1, p_2 respectively and almost local operators Q_1, Q_2 as described above. Then

$$\Psi_{21}(a) \equiv Q_2(a)Q_1\Omega \tag{II.5.18}$$

(where a is an arbitrary translation 4-vector) will have its spectral support in $\Delta_1 + \Delta_2$. We have to show that $\| \Psi_{21}(a) \|$ does not vanish for every a. The essential property needed to show this is the general cluster property. For $|a| \to \infty$ we get

$$\lim \| \Psi_{21}(a) \|^2 = \langle\Omega|Q_2^*(a)Q_2(a)|\Omega\rangle\langle\Omega|Q_1^*Q_1|\Omega\rangle = \| \Psi_2 \|^2 \| \Psi_1 \|^2 . \tag{II.5.19}$$

So $\Psi_{21}(a)$ cannot vanish for all a.

Remark: In Ruelle's lemma 4.1.3 the cluster property was proved using some assumptions about the spectrum, namely the positivity of the energy and the existence of a mass gap. Under these assumptions it was shown that the rate of decrease of truncated functions is faster than any inverse power of R. For the proof of theorem 5.4.1 we do not need to know the rate of decrease. The general cluster property, demanding only that the truncated functions vanish as the distances go to infinity, follows without any assumptions about the spectrum from the commutativity of space-like separated observables, and the existence of a normalizable, pure, translation invariant state (the vacuum) (see chapters III and V).

If Spect P is Lorentz invariant (which follows for instance if we accept axiom **A**1) then theorem 5.4.1 leaves only the possibilities

a) Spect $P \subset V^+$.
   Then Spect $P \supset \{p : p^0 \geq 0, p^2 \geq M^2 \text{ for some } M\}$, (II.5.20)

b) As a) with the sign of the energy reversed,

c) Spect $P = \mathbb{R}^4$.

To see this let $p \in V^+$ with $p^2 = m^2$. Varying the vectors p_1, p_2 on the hyperboloid with mass m the sum $p_1 + p_2$ will have a mass ranging continuously from $M = 2m$ to infinity. This demonstrates the second part of a). If $p_1 \in V^+$ and $p_2 \in V^-$ are in the spectrum then, using the second part of a) (resp. b)), we have vectors with equal mass but opposite energy in the spectrum. By addition of these we can reach a space-like p with arbitrary negative p^2. Finally, if a space-like p is in the spectrum, say with $p^2 = -\kappa^2$, then the addition of $p_\pm = (\sqrt{\mathbf{p}^2 - \kappa^2}, \pm\mathbf{p})$ will lead to vectors in V^+ with arbitrary mass as $|\mathbf{p}|$ varies from κ to infinity. Choosing the zero components of $p_\pm$ negative we can reach all mass values in V^-.

In conclusion we note that the spectral properties and the locality principle are interwoven. The alternatives a) and b) are physically equivalent. The one results from the other by changing the convention in the choice of the sign of the energy. Thus, if any open set in p-space is outside the spectrum, then Spect P has the "physical shape" demanded by axiom **A**3 and further restricted by (II.5.20). That the locality principle does not rule out the alternative c) has been demonstated in a model by Doplicher, Regge and Singer [Dopl 68 b].

II.5.5 Borchers Classes

Let Φ be a field system satisfying the axioms **A–F** and let Φ' be another field system acting in the same Hilbert space and satisfying **A–D** with respect to the same representation $U(a,\alpha)$ of $\overline{\mathfrak{P}}$. Then Φ' is called *local relative to* Φ if

$$[\Phi'(x), \Phi(y)] = 0 \quad \text{for space-like} \quad x - y.$$

One has

(i) if Φ' is local relative to Φ then Φ' satisfies axiom **E**, i.e. it is itself a local field system.

(ii) Let Φ, Φ', Φ'' be complete, local field systems in $\mathcal{H}$ for given $\{U(a,\alpha)\}$. If Φ' is local relative to Φ and Φ'' local relative to Φ' then Φ'' is local relative to Φ.

Thus relative locality for complete, local field systems is transitive and can serve to define an equivalence relation. We call two such field systems equivalent if they are local relative to each other. An equivalence class of complete, local field systems is called a *Borchers class* because Borchers introduced the notion of relative locality and derived the essential results in [Borch 60]. From the physical point of view the terminology "equivalence" is justified because one finds

(iii) The S-matrix depends only on the Borchers class, not on a specific field system chosen in this class.

A simple illustration of a Borchers class is available in the case of a free field φ. In this case the Wick ordered powers $:\varphi(x)^n:$ are fields in the Borchers class of φ and the Wick polynomials of φ and its derivatives at the same point exhaust the class [Ep 63], [Schroer, unpublished preprint 63]. Any one of these fields leads to $S = \mathbb{1}$.

III. Algebras of Local Observables and Fields

III.1 Review of the Perspective

In quantum physics just as in classical physics the concept of "field" serves to implement the principle of locality. In particular, a "quantum field" should not be regarded as being more or less synonymous with a "species of particles". While it is true that with each type of particle we may associate an "incoming field" and an "outgoing field", these free fields are just convenient artifacts and the discussion of collision theory in Chapter II, sections 3 and 4 should make it clear that the primary physical interpretation of the theory is given in terms of local operations, not in terms of particles. Specifically, we have used the basic fields to associate to each open region $\mathcal{O}$ in space-time an algebra $\mathcal{A}(\mathcal{O})$ of operators on Hilbert space, the algebra generated by all $\Phi(f)$, the fields "smeared out" with test functions f having their support in the region $\mathcal{O}$. We have interpreted the elements of $\mathcal{A}(\mathcal{O})$ as representing physical operations performable within $\mathcal{O}$ and we have seen that this interpretation tells us how to compute collision cross sections once the correspondence

$$\mathcal{O} \rightarrow \mathcal{A}(\mathcal{O}) \tag{III.1.1}$$

is known.

This suggests that the *net of algebras* $\mathcal{A}$, i.e. the correspondence (III.1.1), constitutes the intrinsic mathematical description of the theory. The mentioned physical interpretation establishes the tie between space-time and events. The rôle of "fields" is only to provide a coordinatization of this net of algebras. This point of view is supported by the observation of Borchers concerning equivalence classes of fields: different choices of a system of fields may yield the same S-matrix. Extrapolating from that we expect that no physical consequence of the theory depends on the choice of a specific field system within a Borchers class or, alternatively speaking, on the way how the net is coordinatized.

Bounded Operators. The algebras constructed from fields in the way described above are called *polynomial algebras* because their elements are obtained as sums of products $\Phi(f_1)\Phi(f_2)\ldots$ of smeared out fields. In the frame of section II.1 these elements are unbounded operators. Mathematically the polynomial algebras are somewhat unwieldy. Since one is dealing with unbounded operators the questions concerning their domains have to be considered carefully. Since one is dealing with operator valued distributions the topology in such an algebra is injected by the Laurent Schwartz topology of the test function spaces which is an extremely fine topology, too fine to be of intrinsic significance for the physical content of the theory. Thus, polynomial algebras of different field systems in the same Borchers class will not coincide in general. Remembering the discussion in I.1, in particular the approach by Segal [Seg 47], it appears that, without loss of generality, we may go over to bounded operators. Given an observable, mathematically represented by a self-adjoint (possibly unbounded) operator, we may consider instead its family of spectral projectors or the bounded functions of it. In the case of more general operations, represented by (closeable) unbounded operators, we can make a polar decomposition yielding an isometric operator and a positive self-adjoint one which may be spectrally decomposed.

For algebras of bounded operators on a Hilbert space there are several mathematically natural topologies and it will become apparent that they also have physical significance. They lead to the concepts of *von Neumann algebras* and *concrete C^*-algebras*. We shall outline some of the basic mathematical theory of such algebras in the next section. In most of the subsequent discussions we shall take the algebras $\mathcal{A}(\mathcal{O})$ in the correspondence (III.1.1) as algebras of bounded operators.

The question as to whether the Wightman axioms are equivalent to a theory formulated in terms af a net of algebras of bounded operators has been the subject of extensive discussions. One may ask whether an observable field, smeared out with a test function with compact support and defined on the Wightman domain, has an extension to a self adjoint operator and whether this extension is unique (Borchers and Zimmermann, [Borch 64]). Or one may ask under what conditions the construction of von Neumann algebras by polar decomposition of operators from polynomial algebras leads to a net respecting the causal structure. Such questions have been studied by Driessler, Summers and Wichmann [Driess 86] and by Borchers and Yngvason [Borch 90]. Conversely one may ask whether, given a net of algebras of bounded operators, one can define fields by a limiting process, shrinking the regions to points and whether this yields all field systems in a Borchers class (see [Fred 81b], [Summ 87]). We shall not enter here into a discussion of these questions but note that the Wightman axioms alone are not sufficient to guarantee the existence of a net of local algebras of bounded operators and that, conversely, such a net does not guarantee the existence of a field system satisfying the Wightman axioms. For most purposes, however, the difficulties of passing from one frame to the other may be ignored.

To what extent can the axioms of section II.1 be translated into strucure properties of a net of local algebras?

Axiom **A**, dealing with the representation of the Poincaré group, is unaffected. Besides the algebras $\mathcal{A}(\mathcal{O})$ there are the representors $U(a, \alpha)$ of the symmetry transformations and we shall keep the assumptions pertaining to them, most importantly the existence of a vacuum state and the positivity of the energy. We shall, however, try to relate these later to algebraic properties of the net.

Axiom **B**, introducing the notion of fields, suggests the *additivity property*

$$\mathcal{A}(\mathcal{O}_1 \cup \mathcal{O}_2) = \mathcal{A}(\mathcal{O}_1) \vee \mathcal{A}(\mathcal{O}_2) \tag{III.1.2}$$

where the symbol $\vee$ on the right hand side shall denote the operator algebra generated by the two algebras $\mathcal{A}(\mathcal{O}_i), i = 1, 2$. For a precise formulation see section 4.

Axiom **C** (Hermiticity) means that $\mathcal{A}(\mathcal{O})$ is an involutive algebra, a *-algebra: apart from the algebraic operations of forming products and linear combinations of elements one has within each $\mathcal{A}(\mathcal{O})$ the *involution* $A \to A^*$ which assigns to every element A its Hermitean adjoint A^*.

Axiom **D**, concerning the transformation properties of fields, becomes

$$U(a,\alpha)\mathcal{A}(\mathcal{O})U^{-1}(a,\alpha) = \mathcal{A}(\Lambda(\alpha)\mathcal{O} + a). \tag{III.1.3}$$

The geometric symmetry operations map the algebra of one region onto the algebra of the transformed region. The detailed transformation laws of tensorial or spinorial fields from which the algebra may be generated are lost.

Axioms **F** (completeness) and **G** ("primitive causality") are transcribed in an evident way.

Axiom **E** combines several features abstracted from conventional field theoretic models. The main principle expressed by it is the causal structure of events. Two observables associated with space-like separated regions are compatible. The measurement of one does not disturb the measurement of the other. The operators representing these observables must commute.

To avoid possible confusion it must be stressed that this has nothing to do with the discussion around the Einstein-Podolsky-Rosen paradox and Bell's inequality. There one is dealing with the joint probability distribution of measurements on two far separated particles coming from a common root e.g. as decay products of an unstable particle. If a neutral particle decays into two oppositely charged ones it will surprise nobody that a charge measurement on one of the decay products suffices to tell us the charge of the other one, no matter how far away it is. This is due to the correlation resulting from charge conservation, not to a causal influence between the charge measurements in space-like separated regions. The total experiment includes, of course, the preparation of the state of the unstable particle, by which the charge (resp. spin) are fixed. If, instead of the charge, we consider the angular momentum, the situation becomes indeed more curious. Instinctively one would like to associate with each of the particles (once they are sufficiently separated) an "objective", "real" state which determines the probability of finding a specific result in the subsequent measurement of the angular momentum component in any chosen direction. As Bell has shown

[Bell 64] this picture, together with angular momentum conservation, demands that a certain inequality must hold for the joint probability distributions for such measurements on the two particles. This inequality is not satisfied in the quantum mechanical description. Very fine experiments have been performed to check this inequality. They appear to speak for quantum mechanics and against the inequality. What is the message of this? It does not relate to a physical influence propagating faster than light but it illustrates in a particularly drastic way that the concept of a materially defined "physical system" has to be handled with extreme care. This latter is a mental construct whose correspondence to "reality" is (sometimes) questionable (compare [*d'Espagnat 1979*]). We shall give a thorough discussion of this problem in Chapter VII. Here we note only that the existence of correlations between far space-like separated events does not contradict the limitation of causal influences to time-like directions as demanded by axiom **E**.

Unobservable Fields. Returning after this aside to axiom **E** we may note that in quantum field theory there occur observable and unobservable fields. The former generate a net of algebras $\mathcal{A}_{\text{obs}}(\mathcal{O})$ in which the causality principle is expressed by

$$[A_1, A_2] = 0 \quad \text{if} \quad A_i \in \mathcal{A}_{\text{obs}}(\mathcal{O}_i),\ i = 1,\ 2 \quad \text{and} \quad \mathcal{O}_1 \text{ is space} - \text{like to } \ \mathcal{O}_2. \tag{III.1.4}$$

In the language of II.1 the observable fields are Bose fields though not all Bose fields need to be observable. But space-like commutativity is not postulated with a view of introducing Bose statistics but is a mirror of the causal structure of space-time. Why does one need unobservable fields at all and what are their commutation properties? A pragmatic answer to the first question comes from the existence of *superselection rules*. It has been recognized very early by Wigner that one cannot have an unrestricted superposition principle among the pure physical states. If Ψ_1 and Ψ_2 are state vectors carrying integer spin and half odd integer spin, respectively, then the relative phase between Ψ_1 and Ψ_2 can have no observable significance because a 360°-rotation (which is the identity transformation) changes the relative sign between these state vectors. Similarly one expects that the relative phase between vectors describing states with different electric charge should be meaningless. This phase is changed by a gauge transformation which has no effect on the observables. The situation is described by Wick, Wightman and Wigner [Wick 52] as follows: The Hilbert space $\mathcal{H}$ to which axiom **A** refers is a direct sum of subspaces $\mathcal{H}_k$ which we may call *coherent subspaces* or *superselection sectors*. Within each $\mathcal{H}_k$ one has the unrestricted superposition principle whereas phase relations between state vectors belonging to different sectors are meaningless.[1]

[1] Alternatively speaking, a linear combination of vectors from different sectors does not represent a pure state but a mixture and only the absolute square of the coefficients are relevant as the weights of the components.

Within our present context we may say that the observable algebras $\mathcal{A}_{obs}(\mathcal{O})$ transform each sector into itself; they do not connect states in different sectors. The rôle of the unobservable fields is to lead from one sector to a different one. Unobservable fields transfer some "superselection quantum number", some "generalized charge". They are "charge carrying fields". Still we may insist that the theory must be completely fixed by the net $\mathcal{A}_{obs}$, by the observable algebras alone. Since these do not connect different sectors we have in each sector a net of operator algebras $\mathcal{A}_{obs}|_k$, the restriction of $\mathcal{A}_{obs}$ to $\mathcal{H}_k$. These nets are not unitarily equivalent but algebraically isomorphic. Moreover any single one of them must contain all the physically relevant information since different sectors are distinguished by a global property (total charges). We may change the charge of a state by adding a charge arbitrarily far away and this will change the physical situation in any finite part of space-time arbitrarily little. The natural explanation of the situation is then the following: The intrinsic structure of the theory is fully characterized by *the algebraic relations* in the net of *observable algebras.* In other words, the basic object is a net of *abstract algebras* (as opposed to their representative operator algebras on a Hilbert space). We choose them as abstract C*-algebras, a concept which will be discussed in section 2.2. We shall henceforth denote the (abstract) observable algebra of the region $\mathcal{O}$ by $\mathfrak{A}(\mathcal{O})$. To emphasize the local point of view we may regard the algebras $\mathfrak{A}(\mathcal{O})$ to be defined only for finite regions.[2] From these we may define the algebra $\mathfrak{A}_{loc}$ of "all local observables" as

$$\mathfrak{A}_{loc} = \cup \mathfrak{A}(\mathcal{O}) \tag{III.1.5}$$

where $\cup$ denotes the set theoretic union taken over all finite regions[3] and the C*-algebra

$$\mathfrak{A} = \overline{\mathfrak{A}_{loc}}, \tag{III.1.6}$$

the completion of $\mathfrak{A}_{loc}$ in the norm topology. In physical terms $\mathfrak{A}$ may be called the algebra of all quasilocal observables; in mathematical terms it is the "C*-inductive limit" of the system $\{\mathfrak{A}(\mathcal{O})\}$.

Superselection rules arise if the net $\mathfrak{A}$ possesses several inequivalent representations by operator algebras acting on a Hilbert space. The divorce of the basic description of the theory from Hilbert space brings a tremendous additional freedom. It allows us, for instance, to consider thermodynamic equilibrium states corresponding to Gibbs' grand canonical ensembles in infinite space or, more generally, any kind of distribution of matter and energy extending to infinity. We shall indeed discuss such states in chapter V but of more direct interest in elementary particle physics are the charge superselection sectors. They arise if there remain inequivalent representation of $\mathfrak{A}$ even when we restrict the attention to states which "look like the vacuum" at space-like infinity. We treat them in chapter IV.

[2]By a "finite region" we always understand an open subset of Minkowski space with compact closure.

[3]$\mathfrak{A}_{loc}$ is an algebra because, according to the interpretation, the net must have the isotony property $\mathfrak{A}(\mathcal{O}_2) \supset \mathfrak{A}(\mathcal{O}_1)$ if $\mathcal{O}_2 \supset \mathcal{O}_1$ and for any two finite regions there exists another finite region containing them both.

Once we base the theory on abstract algebras we must reconsider the definition of Poincaré symmetry and axiom **A**. Poincaré invariance means now that to a transformation $g \in \mathfrak{P}$ there corresponds an automorphism α_g of the abstract net with the property

$$\alpha_g \mathfrak{A}(\mathcal{O}) = \mathfrak{A}(g\mathcal{O}). \tag{III.1.7}$$

In words: α_g maps the elements of $\mathfrak{A}(\mathcal{O})$ onto the elements of the algebra of the transformed region $g\mathcal{O}$ in such a way that all algebraic relations are conserved.

A representation π of $\mathfrak{A}$ is a homomorphism from the net $\mathfrak{A}$ to a net of operator algebras $\pi(\mathfrak{A})$ i.e. π assigns to each algebraic element A its "representor" $\pi(A)$, an operator acting in a Hilbert space, in such a way that the mapping conserves the algebraic relations. Given a representation the automorphism α_g may or may not be *implementable*. α_g is called implementable in the representation π if there exists a unitary operator $U(g)$ acting in the representation space such that

$$U(g)\pi(A)U^{-1}(g) = \pi(\alpha_g A). \tag{III.1.8}$$

Axiom **A** can now be replaced by the requirement that the abstract net should possess an irreducible representation π_0 in which α_g is implementable and which is furthermore distinguished by the feature that $U(g)$ satisfies **A2** and **A3**. We shall call this representation the *vacuum sector* of the theory (leaving aside for the moment the possibility of vacuum degeneracy). The implementability of α_g in a representation depends on global features of the class of states described by vectors in the representation space because $g \in \mathfrak{P}$ acts on all regions, no matter how far away. Thus axiom **A** brings in a global aspect whose relation to local properties of the theory is not well understood. One can, however, formulate the contents of axiom **A** in purely algebraic terms and we shall do this in section 3.

The other structural assumptions are simple and natural in the algebraic language. We collect them here once more:

a) The theory is characterized by a net of abstract C*-algebras

$$\mathcal{O} \rightarrow \mathfrak{A}(\mathcal{O}) \tag{III.1.9}$$

where $\mathcal{O}$ denotes an open, finitely extended region of Minkowski space. The self-adjoint elements of $\mathfrak{A}(\mathcal{O})$ are interpreted as observables which can be measured in the region $\mathcal{O}$.

b) The Poincaré group is realized by a group of automorphisms of the net ($g \in \mathfrak{P} \rightarrow \alpha_g$) with the geometric significance (III.1.7).

c) The causality structure is expressed by

(i) $\mathfrak{A}(\mathcal{O}_1)$ commutes with $\mathfrak{A}(\mathcal{O}_2)$ when the two regions lie space-like,

(ii) Let $\hat{\mathcal{O}}$ denote the causal completion of $\mathcal{O}$ (see section 4) then

$$\mathfrak{A}(\hat{\mathcal{O}}) = \mathfrak{A}(\mathcal{O}). \tag{III.1.10}$$

This stipulates that there is a dynamical law respecting the causal structure. It corresponds to the hyperbolic propagation character of fields (see, however, the remarks at the end of section 3).

Remark. The prototype of a causally complete region is a *diamond* or *double cone.* An open diamond is the intersection of the interior of the forward light cone from a point x_1 with the interior of the backward light cone from a point x_2 which lies in the future of x_1. The standard diamond, denoted by $\mathcal{O}_R$, is obtained by taking $x_1 = (-R, \mathbf{0})$, $x_2 = (R, \mathbf{0})$. It is the causal completion of the ball $| \mathbf{x} |< R$ in the hyperplane $x^0 = 0$.

The above frame opens the possibility of discussing the intrinsic significance of Fermi statistics and possible generalizations since it does not inject the Bose-Fermi-alternative as a basic assumption, it does not tie it to commutation relations between unobservable fields. Rather we have now the task of analyzing the superselection structure, the composition law of charges, the effect of a permutation of identical charges, the construction of unobservable fields. This will be the topic of chapter IV. The frame also provides a natural approach to the description of thermal equilibrium states and more general hydrodynamic states without the need to define a material system enclosed in a box. This will be discussed in chapter V.

Remarks and references. The proposal to base the theory on a net of local algebras corresponding to space-time regions originated in [Haag 57]. There the algebras were thought of to be generated by both observable and unobservable fields and taken as operator algebras acting on Hilbert space. In the first applications to physical and structural questions polynomial algebras or von Neumann rings were used according to convenience and the step from the former to the latter was considered as unproblematic. The main reward was the deeper understanding of collision theory as described in II.4 which eliminated the distinction between elementary and composite particles and the assumption of a correspondence between elementary particles and basic fields. A survey of the postulates from this point of view was given by Haag and Schroer [Haag 62a]. The development of the theory in terms of nets of von Neumann algebras of observables is largely due to Araki. He built a fairly self-contained frame on solid mathematical ground [Ara 62a]. The idea that one should take an abstract C*-algebra as the basic object and that the representation problem (see II.1) is irrelevant for physics, was advocated by Segal [Seg 47], [Seg 57]. An adequate physical interpretation was lacking, however. His proposal that the S-matrix should be considered as an automorphism of the algebra was not acceptable. It took some years till it was realized that a *net* of (abstract) C*-algebras of observables provided the natural setting for understanding superselection rules and the rôle of unobservable fields (Haag and Kastler, [Haag 64]).

III.2 Von Neumann Algebras. C*-Algebras. W*-Algebras

III.2.1 Algebras of Bounded Operators

The purpose of this section is to provide a brief glossary of mathematical terms, concepts and theorems which will be subsequently used. Proofs are only sketched or omitted altogether and the reader is referred to the bibliography at the end of the section to supplement this minimal exposition.

The set of all bounded, linear operators acting in a Hilbert space $\mathcal{H}$ is denoted by $\mathfrak{B}(\mathcal{H})$. A subset $S \subset \mathfrak{B}(\mathcal{H})$ is called an algebra if, with $A, B \in S, \alpha, \beta$ complex numbers also $\alpha A + \beta B \in S$ and $AB \in S$. It is called a *-algebra if furthermore with $A \in S$ also the adjoint A^* belongs to S. We consider *-subalgebras of $\mathfrak{B}(\mathcal{H})$.

For all processes of analysis one needs a topology i.e. a definition of what we mean by a neighborhood of an element. The most evident topology in $\mathfrak{B}(\mathcal{H})$ is provided by the norm of operators. The statement that $A \in \mathfrak{B}(\mathcal{H})$ is bounded means that

$$\| A \| \equiv \sup_{\Psi \in \mathcal{H}} \| A\Psi \| \cdot \| \Psi \|^{-1} \qquad \text{(III.2.1)}$$

is finite. On the right hand side $\| \ \|$ denotes the norm (length) of vectors in the Hilbert space, the left hand side is the norm of the operator A. One checks that

$$\| \alpha A \| = | \alpha | \, \| A \|, \qquad \text{(III.2.2)}$$
$$\| A + B \| \leq \| A \| + \| B \|, \qquad \text{(III.2.3)}$$
$$\| AB \| \leq \| A \| \cdot \| B \|, \qquad \text{(III.2.4)}$$
$$\| A^*A \| = \| A \|^2 . \qquad \text{(III.2.5)}$$

The first two relations say that $\mathfrak{B}(\mathcal{H})$ is a normed linear space.

An ε-neighborhood of A is the set of operators B with $\| A - B \| \leq \varepsilon$. The topology based on this concept of neighborhoods is called the norm topology or *uniform topology* in $\mathfrak{B}(\mathcal{H})$. Closure of a set S in this topology means that we add to S all elements which are limits of uniformly converging Cauchy sequences in S.

Definition 2.1.1
A *-subalgebra of $\mathfrak{B}(\mathcal{H})$ which is uniformly closed is called a (concrete) C*-algebra.

Other important topologies in $\mathfrak{B}(\mathcal{H})$ are defined by means of seminorms. Thus we may pick an arbitrary vector Ψ in $\mathcal{H}$ and define the "Ψ-seminorm" of the operator A as the vector norm $\| A\Psi \|$. Using these seminorms for all choices of Ψ one obtains the so called *strong operator topology* in $\mathfrak{B}(\mathcal{H})$. Thus a sequence of operators A_n is strongly convergent if for every $\Psi \in \mathcal{H}$ the sequence of vectors $\Psi_n = A_n\Psi$ is strongly convergent i.e. if $\| \Psi_n - \Psi_m \| \rightarrow 0$ for $n, m \rightarrow \infty$.

For topologies which are defined by a system of seminorms it is not enough to consider only the convergence of sequences. The closure of a set is obtained by adding the limit points of "generalized sequences" (nets). Still, as a first orientation, it is useful to look at ordinary sequences.

The *weak operator topology* is obtained if we use the absolute values of matrix elements $| \langle \Phi | A | \Psi \rangle |$ between arbitrary state vectors as a system of seminorms. Thus a sequence of operators converges weakly if all matrix elements converge. We have encountered weak and strong convergence before in the context of collision theory.

Still another system of seminorms is provided by $| \operatorname{tr} \varrho A |$ where ϱ is an arbitrary density matrix. The resulting topology was originally called "ultraweak". This is, apart from the norm topology, the most natural one. It is the *weak *-topology* induced by the set of "normal states" (see subsection 2.2).

All the mentioned topologies are different. The following examples may serve as an illustration. Let Ψ_n be a complete, orthonormal basis in $\mathcal{H}$, E_n the projector on the subspace spanned by the first n basis vectors and T_n the operator defined by $T_n \Psi_k = \Psi_{n+k}$. Then the sequence of operators E_n converges strongly to the unit operator for $n \to \infty$ because the component of each fixed vector Ψ in the orthogonal complement of $E_n \mathcal{H}$ tends to zero. This sequence does, however, not converge uniformly since $\| E_n - E_m \| = 1$ for any $n \neq m$. The sequence T_n converges weakly to zero but not strongly.

It is evident that the uniform topology is the strongest, the weak operator topology the weakest. The weak *- and the strong operator topology cannot be compared.

Definition 2.1.2

A weakly closed *-subalgebra of $\mathfrak{B}(\mathcal{H})$ which contains the unit operator will be called a *von Neumann algebra* or (synonymously) a *von Neumann ring.*

Remark. The requirement that the unit operator shall be contained in it is not always considered as part of the definition of a von Neumann ring. It is added here to simplify the formulation of some later theorems.

Quite generally, if a subset is closed in one topology it is automatically closed in every stronger topology because, with decreasing strength of the topology, one gets more and more limit points. Thus a von Neumann algebra is also a concrete C*-algebra whereas the converse is not true. However von Neumann has shown that for a *-subalgebra of $\mathfrak{B}(\mathcal{H})$ the closures in the three other mentioned topologies (strong operator, weak *, weak operator) all coincide. Therefore we meet only two kinds of topological *-subalgebras of $\mathfrak{B}(\mathcal{H})$ namely concrete C*-algebras and von Neumann algebras.

Definition 2.1.3

The *commutant* of an arbitrary subset $S \subset \mathfrak{B}(\mathcal{H})$ is the set of all bounded operators which commute with every element of S. It is denoted by S'.

Theorem 2.1.4
Let $S \subset \mathfrak{B}(\mathcal{H})$ be a self adjoint set (i.e. S shall contain with each element also its adjoint). Then

(i) S' is a von Neumann ring.
(ii) $S'' \equiv (S')'$ is the smallest von Neumann ring containing S.
(iii) $S''' = S'$.

Comment. If A and B commute with X and X^* then obviously also linear combinations of A and B, the product AB and the adjoint A^* will commute with X. The unit operator certainly commutes also with X. Thus S' is a *-algebra with unit. To obtain (i) one must show that S' is closed (strongly or weakly). If A_n is a sequence of bounded operators converging strongly to A and satisfying $[A_n, X] = 0$ then one sees easily that also $[A, X] = 0$. One shows (not quite trivially) that it suffices to add the limits of strongly convergent *sequences* to obtain the strong closure of a *-algebra in $\mathfrak{B}(\mathcal{H})$. Statement (ii) is the famous von Neumann double commutant theorem. We shall not sketch its proof here. Statement (iii) follows from

$$S''' \equiv \left((S')'\right)' = \left(S''\right)' = (S')''; \quad S \subset S''; \quad S' \subset (S')'',$$

noting that $S \subset T$ implies $T' \subset S'$.

We shall from now on use the symbol $\mathcal{R}$ to denote von Neumann algebras. From the theorem it follows that a von Neumann algebra may be characterized as a self-adjoint subset of $\mathfrak{B}(\mathcal{H})$. satisfying

$$\mathcal{R}'' = \mathcal{R}. \tag{III.2.6}$$

Denoting the smallest von Neumann algebra containing $\mathcal{R}_1$ and $\mathcal{R}_2$ by $\mathcal{R}_1 \vee \mathcal{R}_2$, the largest von Neumann algebra contained in $\mathcal{R}_1$ and in $\mathcal{R}_2$ by $\mathcal{R}_1 \wedge \mathcal{R}_2$ one has

$$\mathcal{R}_1 \vee \mathcal{R}_2 \equiv (\mathcal{R}_1 \cup \mathcal{R}_2)'' \tag{III.2.7}$$

and

$$(\mathcal{R}_1 \vee \mathcal{R}_2)' = \mathcal{R}_1' \wedge \mathcal{R}_2' = \mathcal{R}_1' \cap \mathcal{R}_2'. \tag{III.2.8}$$

The set of von Neumann algebras on a Hilbert space forms an orthocomplemented lattice with respect to the operations $\wedge$, $\vee$, and $'$.

Isomorphies. Let $\mathcal{A}_1$, $\mathcal{A}_2$ be two *-algebras of bounded operators acting on the Hilbert spaces $\mathcal{H}_1$, $\mathcal{H}_2$, respectively. An *isomorphism* between $\mathcal{A}_1$ and $\mathcal{A}_2$ means that there exists a bijective map from $\mathcal{A}_1$ to $\mathcal{A}_2$ (a one to one correspondence between elements) which conserves the relevant structure. In our case three possibilities are of interest:

1) *Algebraic isomorphism.* We require only that the algebraic structure (linear combinations, products, adjoints) is conserved.

2) "*Complete isomorphism*". The algebraic and topological structures are conserved.

3) "*Spatial isomorphism*" or *unitary equivalence*

$$\mathcal{A}_2 = V \mathcal{A}_1 V^{-1} \tag{III.2.9}$$

where V is a unitary map from $\mathcal{H}_1$ onto $\mathcal{H}_2$.

It is an interesting fact that in the cases of C*-algebras and von Neumann algebras algebraic isomorphism entails complete isomorphism. For two algebraically isomorphic C*-algebras also the norms of corresponding elements agree and for two algebraically isomorphic von Neumann algebras the norms and the weak *-topologies agree. Two algebraically isomorphic von Neumann algebras are also called *quasiequivalent*. Spatial isomorphism is more restrictive.

Reduction. Suppose that $\mathcal{R}$, acting in $\mathcal{H}$, transforms some proper subspace $\mathcal{H}_1$ into itself. Then the projector P_1 on this invariant subspace belongs to $\mathcal{R}'$ and the restriction of $\mathcal{R}$ to $\mathcal{H}_1$ is

$$\mathcal{R}_1 \equiv P_1 \mathcal{R} P_1 = P_1 \mathcal{R} = \mathcal{R} P_1 . \tag{III.2.10}$$

It is a von Neumann algebra in $\mathcal{H}_1$. The map

$$A \in \mathcal{R} \to P_1 A \tag{III.2.11}$$

is a homomorphism conserving the algebraic and topological structure. One says that the decomposition of $\mathcal{H}$ into $\mathcal{H}_1$ and its orthogonal complement gives a reduction of $\mathcal{R}$, or that $\mathcal{R}_1$ is a subrepresentation of $\mathcal{R}$. $\mathcal{R}_1$ will be quasi-equivalent to $\mathcal{R}$ if (III.2.11) maps no non-zero element of $\mathcal{R}$ to zero. Let $\mathfrak{J}$ be the set of elements of $\mathcal{R}$ mapped to zero by (III.2.11) i.e. $\mathfrak{J} = \{A \in \mathcal{R} : P_1 A = 0\}$. Obviously $\mathfrak{J}^* = \mathfrak{J}$ and $\mathfrak{J}\mathcal{R} = \mathcal{R}\mathfrak{J} = \mathfrak{J}$. Thus $\mathfrak{J}$ is a 2-sided *-ideal in $\mathcal{R}$ (a *-subalgebra which is stable under multiplication with any element of $\mathcal{R}$ from left or right) and $\mathfrak{J}$ is weakly closed. It is not difficult to show (see e.g. [*Naimark 1972*])

Theorem 2.1.5
A weakly closed 2-sided *-ideal of a von Neumann ring $\mathcal{R}$ is always of the form

$$\mathfrak{J} = \mathcal{R} E_0 \tag{III.2.12}$$

where E_0 is a projection operator belonging to the *center* $\mathfrak{Z} = \mathcal{R} \cap \mathcal{R}'$ of $\mathcal{R}$. E_0 is called the *principal unit of* $\mathfrak{J}$. It satisfies

$$E_0 X = X E_0 = X \quad \text{for} \quad X \in \mathfrak{J}. \tag{III.2.13}$$

The subspace $(\mathbb{1} - E_0)\mathcal{H}$ consists of all vectors which are annihilated by every element of $\mathfrak{J}$.

Definition 2.1.6
The von Neumann ring $\mathcal{R}$ is called a *factor* if its center is trivial (consists only of the multiples of the unit element):

$$\mathfrak{Z} = \mathcal{R} \cap \mathcal{R}' = \{\lambda \mathbb{1}\}, \tag{III.2.14}$$

or, equivalently,

$$\mathcal{R} \vee \mathcal{R}' = \mathfrak{B}(\mathcal{H}). \tag{III.2.15}$$

(III.2.15) is called a *factorization* of $\mathfrak{B}(\mathcal{H})$.

From the preceding discussion we conclude

Theorem 2.1.7
a) A self adjoint set $S \subset \mathfrak{B}(\mathcal{H})$ is *irreducible* (transforms no proper subspace into itself) iff $S' = \{\lambda \mathbb{1}\}$ or, equivalently, $S'' = \mathfrak{B}(\mathcal{H})$ ("Schur's lemma").
b) If $\mathcal{R}$ is a factor then its restriction (III.2.10) to any invariant subspace is quasi-equivalent to $\mathcal{R}$.

Thus a factor contains only one quasi-equivalence class of subrepresentations. The reduction of a von Neumann algebra into factors is unique and results from the "diagonalization" of its center. The further reduction into irreducibles is not unique; it is afforded by the choice of some maximally Abelian subalgebra[1] in the commutant and its diagonalization.

Classification of Factors. Let $\mathcal{R}$ be a factor. The basic classification theory of Murray and von Neumann starts from a comparison and ordering of the projectors contained in $\mathcal{R}$.

Definition 2.1.8
Two projectors $P_i \in \mathcal{R}$ $(i = 1, 2)$ are called equivalent with respect to $\mathcal{R}$, in symbols $P_1 \sim P_2$, if there exists an operator $V \in \mathcal{R}$ with

$$P_1 = V^*V; \quad P_2 = VV^*. \tag{III.2.16}$$

These relations say that V maps the subspace $\mathcal{H}_1 = P_1\mathcal{H}$ isometrically on the subspace $\mathcal{H}_2 = P_2\mathcal{H}$ and annihilates all vectors of the orthogonal complement of $\mathcal{H}_1$. V is therefore called a *partial isometry from* $\mathcal{H}_1$ *to* $\mathcal{H}_2$. We write $P_1 > P_2$ or $P_2 < P_1$ if the P_i are not equivalent but there exists a subspace of $\mathcal{H}_1$ whose projector $P_{11} \sim P_2$ (of course this requires also $P_{11} \in \mathcal{R}$). One has

Theorem 2.1.9
Let $\mathcal{R}$ be a factor, P_i projectors belonging to $\mathcal{R}$. Then precisely one of the following relations holds

$$P_1 > P_2, \quad P_1 \sim P_2, \quad P_1 < P_2. \tag{III.2.17}$$

[1] An algebra $\mathcal{R}_1 \subset \mathcal{R}'$ satisfying $\mathcal{R}_1' = \mathcal{R}_1''$.

The proof of this theorem uses the following basic lemma which is of interest for its own sake.

Lemma 2.1.10
If $\mathcal{R}$ is a factor, $A \in \mathcal{R}$, $B \in \mathcal{R}'$ then $AB = 0$ implies that either $A = 0$ or $B = 0$.

The ordering of projectors in a factor allows the introduction of a *relative dimension* for them. $\text{Dim}\, P$ is a positive number (possibly ∞) with the properties

(i)

$$\text{Dim}\, P_1 \begin{matrix} > \\ = \\ < \end{matrix} \text{Dim}\, P_2 \quad \text{if} \quad P_1 \begin{matrix} > \\ \sim \\ < \end{matrix} P_2. \tag{III.2.18}$$

(ii) If P_1 is orthogonal to P_2 (i.e. $P_1 P_2 = 0$) then

$$\text{Dim}\,(P_1 + P_2) = \text{Dim}\, P_1 + \text{Dim}\, P_2 \tag{III.2.19}$$

(iii)

$$\text{Dim}\, 0 = 0. \tag{III.2.20}$$

These three properties determine a dimension function on the set of projectors in a factor uniquely up to normalization. The following alternatives exist:

Type I. $\mathcal{R}$ contains minimal projectors. P is called minimal if it is not zero but $\mathcal{R}$ contains no nonzero projector $P_1 < P$. Normalizing the dimension function so that it takes the value 1 on minimal projectors, $\text{Dim}\, P$ ranges through the positive integers up to a maximal value n or up to ∞. If n is finite $\mathcal{R}$ is called of type I_n, otherwise of type I_∞. The factor I_n is quasi-equivalent to the full matrix ring of complex $n \times n$ matrices. I_∞ is quasi-equivalent to $\mathfrak{B}(\mathcal{H})$; in fact one can write $\mathcal{H}$ as a tensor product $\mathcal{H}_{\text{irr}} \otimes \mathcal{H}_{\text{deg}}$ so that $\mathcal{R} = \mathfrak{B}(\mathcal{H}_{\text{irr}}) \otimes \mathbb{1}_{\text{deg}}$. Thus a factor of type I differs from an irreducible algebra only by an additional multiplicity resulting from the tensoring with some degeneracy space.

Type II. $\text{Dim}\, P$ ranges through a continuum of values which may either have an upper bound, in which case one can normalize the values to fill the closed interval $[0, 1]$ and $\mathcal{R}$ is called a factor of type II_1; or it may be the whole positive real line, including 0 and ∞, and $\mathcal{R}$ is said to be of type II_∞.

Type III. $\text{Dim}\, P$ takes only the values 0 and ∞. All proper projectors in $\mathcal{R}$ have infinite dimension and, if $\mathcal{H}$ is separable (which is the only case with which we shall be concerned), they are all equivalent.

The Relative Trace. A trace on a von Neumann algebra $\mathcal{R}$ assigns to each positive operator in $\mathcal{R}$ a positive number, possibly ∞, so that

$$\operatorname{tr} \lambda A = \lambda \operatorname{tr} A; \quad (\lambda > 0),$$
$$\operatorname{tr}(A + B) = \operatorname{tr} A + \operatorname{tr} B$$

and such that elements which are conjugate by a unitary operator from $\mathcal{R}$ have the same trace:

$$\operatorname{tr} U A U^{-1} = \operatorname{tr} A \quad \text{for all} \quad U \in \mathcal{R}. \tag{III.2.21}$$

Again, if $\mathcal{R}$ is a factor, the trace is uniquely determined by these properties up to normalization. In the type III case all non zero elements have infinite trace. In the type I case we may choose the normalization so that minimal projectors have trace 1. Then we get the customary definition of the trace as the sum of the diagonal elements of the matrix representing the operator in an orthonormal basis. In type I_∞ as well as in II_∞ the trace is finite for some operators, infinite for others. For factors of type II_1 the trace is finite for all elements and we may normalize it by setting $\operatorname{tr} \mathbb{1} = 1$.

Remarks. A finer subdivision of the types II and III has been attained in the past decades. This development has been an example of a very fruitful interaction between theoretical physics and mathematics. We shall come to it in chapter V but mention already here that the local von Neumann algebras relevant in relativistic quantum physics are of type III.

III.2.2 Abstract Algebras and Their Representations

An abstract algebra, like an abstract group, is just a set in which certain relations are defined between its elements. In our case the relations are formed by linear combinations with complex coefficients, an associative (in general non-commutative) product and an involution (*-operation). These operations obey the familiar rules, specifically the distributive law for products of sums and, with A, B elements of the algebra and $\alpha \in \mathbb{C}$

$$A^{**} = A; \quad (\alpha A)^* = \overline{\alpha} A^*; \quad (A + B)^* = A^* + B^*; \quad (AB)^* = B^* A^*. \tag{III.2.22}$$

If the algebra is equipped with a norm $\| \ \|$ then the norm should satisfy at least (III.2.2), (III.2.3) and (III.2.4) in order to define a reasonable topology. Furthermore the elements with zero norm should be identified with the zero element of the algebra

$$\| A \| = 0 \quad \text{implies} \quad A = 0. \tag{III.2.23}$$

If (III.2.23) is not adopted then $\| \ \|$ is called a seminorm. A normed *-algebra $\mathcal{A}$ which is complete in the norm topology (which contains all the limit points of Cauchy sequences) is called a *Banach *-algebra* if the norm satifies (III.2.2) - (III.2.4) and (III.2.23). If, in addition, the norm satisfies also (III.2.5) it is called a C*-norm and $\mathcal{A}$ an (abstract) *C*-algebra.* The requirement (III.2.5) looks at first sight somewhat ad hoc in the case of abstract algebras. The fact that the operator norm in $\mathfrak{B}(\mathcal{H})$ satisfies it implies, of course, that if $\mathcal{A}$ has any faithful representation by operators in $\mathfrak{B}(\mathcal{H})$ then we can equip $\mathcal{A}$ with a C*-norm. Disregarding the question of Hilbert space representations we shall see that the C*-norm is distinguished and uniquely determined by the algebraic structure through its relation with the spectrum of elements.

Definition 2.2.1
Let $\mathcal{A}$ be an algebra with unit element (not necessarily a Banach algebra). The *spectrum* of an element $A \in \mathcal{A}$, denoted by Spect (A), is defined as the set of complex numbers λ for which $(\lambda \mathbb{1} - A)^{-1}$ does not exist in $\mathcal{A}$. The spectral radius of A is defined as

$$\varrho(A) = \sup |\lambda|, \quad \lambda \in \text{Spect}\,(A). \tag{III.2.24}$$

From this definition one obtains by purely algebraic arguments, not involving any limiting processes

Theorem 2.2.2
Let $\mathcal{A}$ be an algebra with unit (not necessarily a Banach algebra) and $A,\ B \in \mathcal{A}$.

(i) If $F(x)$ is a polynomial in the variable x and
$\lambda \in \text{Spect}\,(A)$ then $F(\lambda) \in \text{Spect}\, F(A)$. (III.2.25)

(ii) If A^{-1} exists and $\lambda \in \text{Spect}\,(A)$ then $\lambda^{-1} \in \text{Spect}\,(A^{-1})$.

(iii) If $\lambda \in \text{Spect}\,(AB)$ and $\lambda \neq 0$ then $\lambda \in \text{Spect}\,(BA)$.

(iv) If $\mathcal{A}$ is a * – algebra and
$\lambda \in \text{Spect}\,(A)$ then $\overline{\lambda} \in \text{Spect}\,(A^*)$.

The demonstration of these basic properties of the spectrum is elementary (see e.g. [*Bratteli and Robinson 1979, section 2.2*]).

The spectrum depends only on the set $\mathcal{A}$ and the algebraic relations within it, not on the topology of $\mathcal{A}$. However, if $\mathcal{A}$ is a Banach algebra then

$$\varrho(A) \leq \| A \|\,. \tag{III.2.26}$$

This follows from the power series expansion

$$(\lambda \mathbb{1} - A)^{-1} = \lambda^{-1}\left(\mathbb{1} + \frac{A}{\lambda} + \frac{A^2}{\lambda^2} + \dots\right) \tag{III.2.27}$$

which is convergent in the norm topology if $\| A \| < |\lambda|$.

One can obtain a precise relation between the spectral radius of A and the norms of powers of A. The convergence of (III.2.27) is determined not by $\| A \|$ but by the behaviour of $\| A^n \|$ for large n. Let

$$r_n = \| A^n \|^{1/n}\,. \tag{III.2.28}$$

One shows that

$$r = \lim_{n\to\infty} r_n \tag{III.2.29}$$

exists and satisfies

$$0 \leq r \leq \| A \| . \tag{III.2.30}$$

The sequence (III.2.27) converges in norm if $|\lambda| > r$. On the other hand one shows that if $(\lambda \mathbb{1} - A)^{-1}$ would also exist for all $| \lambda |= r$ then this would require $\| A^n \| / r^n \to 0$ in contradiction to the definition of r.[2] Thus one has

Theorem 2.2.3
In a Banach algebra with unit the spectral radius of an element A is

$$\varrho(A) = r(A) \equiv \lim_{n\to\infty} \| A^n \|^{1/n} . \tag{III.2.31}$$

The theorems suggest that, given a Banach *-algebra one may equip it with another, more natural, norm using the spectral radius of elements. On the one hand one sees from (III.2.31) that if A is self adjoint and the norm is a C*-norm (i.e. satisfies (III.2.5)) then $\| A \|$ and $\| A^n \|^{1/n}$ do not differ, hence the norm must be equal to the spectral radius. On the other hand one sees from theorem 2.2.2. that in a Banach *-algebra generated by a single self adjoint element A_0 (an algebra in which the polynomials of A_0 with complex coefficients form a dense set) the spectral radius of elements satisfies the conditions (III.2.2) - (III.2.5). The spectral radius provides a C*-seminorm on such an algebra. It may happen that $\varrho(A) = 0$ for some $A \neq 0$. This will be the case if

$$\lim \| A^n \|^{1/n} = 0. \tag{III.2.32}$$

We may call such elements *generalized zero elements.* They constitute a closed ideal $\mathfrak{J} \subset \mathcal{A}$. If we throw them out (identify them with the zero element) the quotient algebra $\mathcal{A}/\mathfrak{J}$ (whose elements are classes of elements in $\mathcal{A}$ modulo $\mathfrak{J}$) can be completed to a C*-algebra with respect to the norm $\| A \| = \varrho(A)$. It turns out that the elements of this C*-algebra are in one to one correspondence with the complex valued, continuous functions on the spectrum of A_0. A spectral value of A_0 characterizes a maximal ideal in this algebra namely the set of all functions which vanish at this point.

The last remark is relevant for the adaptation of the discussion to the case of general *commutative* Banach *-algebras which leads to Gelfand's theory of commutative C*-algebras. A closed maximal ideal in such an algebra $\mathcal{A}$ is called an element of the *spectrum of the algebra.* It corresponds to a set of simultaneous spectral values of the elements of $\mathcal{A}$. We denote the spectrum of $\mathcal{A}$ (set of maximal ideals) by $\mathcal{K}$. Gelfand has shown that a natural topology can be introduced on $\mathcal{K}$ and that $\mathcal{K}$ is then a compact Hausdorff space. One has

[2]See [*Bratteli and Robinson 1979, proposition 2.2.2*].

Theorem 2.2.4 (Gelfand)
A commutative C*-algebra $\mathcal{A}$ is isomorphic to the C*-algebra of complex valued, continuous functions on a compact Hausdorff space $\mathcal{K}$, the spectrum of $\mathcal{A}$. The norm of an element is given by the supremum of the absolute value of the corresponding function.

For a non-commutative Banach *-algebra one has

Theorem 2.2.5
If a Banach *-algebra admits a C*-norm then this norm is uniquely determined as

$$\| A \| = \varrho(A) \quad \text{if } A \text{ and } A^* \text{ commute,}$$
$$\| A \| = \{\varrho(A^* A)\}^{1/2} \quad \text{in general.}$$

The spectrum of an element A in a C*-algebra is already determined by its spectrum in the smallest C*-algebra containing A.

For the proof see Bratteli and Robinson loc. cit. section 2.2.

The theorems show that the C*-condition (III.2.5) for the norm is necessary to allow a viable spectral theory. In particular in a C*-algebra the spectrum of an element A is the same in any C*-subalgebra containing A. This also allows a simple characterization of positive elements in terms of the *-operation.

Definition 2.2.6
An element $A \in \mathcal{A}$ is called *positive* if $\operatorname{Spect} A \subset \mathbb{R}^+ \cup \{0\}$ (the non negative reals). The set of positive elements of $\mathcal{A}$ is denoted by $\mathcal{A}^+$.

Theorem 2.2.7
Let $\mathcal{A}$ be a C*-algebra. Then

(i) $\mathcal{A}^+$ is a convex cone i.e.
with $A,\ B \in \mathcal{A}^+$and $a,\ b \in \mathbb{R}^+$one has $aA + bB \in \mathcal{A}^+$.
(ii) $A \in \mathcal{A}^+$ if and only if $A = B^*B$ with $B \in \mathcal{A}, B \neq 0$.

For the proof see [*Sakai 1971*].
Positive Linear Forms and States.

Definition 2.2.8
(i) A function φ from an algebra $\mathcal{A}$ to the complex numbers is called a *linear form over* $\mathcal{A}$ if

$$\varphi(\alpha A + \beta B) = \alpha\varphi(A) + \beta\varphi(B); \quad A, B \in \mathcal{A}, \quad \alpha, \beta \in \mathbb{C}.$$

(ii) If $\mathcal{A}$ is a Banach algebra we call a linear form φ bounded if

$$| \varphi(A) | \leq c \| A \| .$$

The lowest choice for the bound c is called the norm of φ,

$$\| \varphi \| = \sup_{A \in \mathcal{A}} | \varphi(A) | \| A \|^{-1} . \qquad \text{(III.2.33)}$$

(iii) Let $\mathcal{A}$ be a *-algebra with unit, φ a linear form. φ is called

real	if	$\varphi(A^*) = \overline{\varphi(A)}$,
positive	if	$\varphi(A^* A) \geq 0$,
normalized	if	$\| \varphi \| = 1$.

A normalized positive linear form is called a *state*.

Theorem 2.2.9
A positive linear form over a Banach *-algebra with unit is bounded and

$$\| \varphi \| = \varphi(\mathbb{1}). \qquad \text{(III.2.34)}$$

It satisfies the Schwarz inequality

$$| \varphi(AB) |^2 \leq \varphi(AA^*) \varphi(B^* B). \qquad \text{(III.2.35)}$$

Remark. Note that a positive linear form over a *-algebra with unit is, a forteriori, real. The term *state* for a normalized positive linear form is, of course, borrowed from physics and indicates the physical relevance of the concept. If we assume that the self-adjoint elements of $\mathcal{A}$ correspond to observables and the unit element to the trivial observable having the value 1 in any physical state then a normalized positive linear form ω may be interpreted as an expectation functional over the observables i.e. the mathematical notion of state corresponds to the physical one.

The Gelfand-Naimark-Segal (GNS)-Construction. The following observation, elaborated by I.M. Gelfand and M.A. Naimark and by I.E. Segal is of fundamental importance. Each positive linear form ω over a *-algebra $\mathcal{A}$ defines a Hilbert space $\mathcal{H}_\omega$ and a representation π_ω of $\mathcal{A}$ by linear operators acting in $\mathcal{H}_\omega$.

One first notes that $\mathcal{A}$ itself is a linear space over the field $\mathbb{C}$ and that ω defines an Hermitean scalar product on $\mathcal{A}$ by

$$\langle A | B \rangle = \omega(A^* B), \quad A, \ B \in \mathcal{A}. \qquad \text{(III.2.36)}$$

Due to the positivity of ω this scalar product is semi-definite i.e.

$$\langle A | A \rangle \geq 0$$

and by (III.2.35)

$$|\langle B|A\rangle|^2 \leq \langle B|B\rangle\langle A|A\rangle.$$

Therefore the set $\mathfrak{J} \subset \mathcal{A}$ consisting of all elements $X \in \mathcal{A}$ with $\omega(X^*X) = 0$ is a *left ideal* i.e. a linear subspace of $\mathcal{A}$ which is stable under multiplication by an arbitrary element of $\mathcal{A}$ from the left:

$$X \in \mathfrak{J}, \quad A \in \mathcal{A} \quad \text{implies} \quad AX \in \mathfrak{J}. \tag{III.2.37}$$

We shall call $\mathfrak{J}$ the *Gelfand ideal of the state.* The set of classes $\mathcal{A}/\mathfrak{J}$ is thus a pre-Hilbert space, a linear space equipped with an Hermitean, positive definite scalar product. A vector Ψ in this space corresponds to an equivalence class modulo $\mathfrak{J}$ of algebraic elements

$$\Psi = \{A + \mathfrak{J}\}, \quad A \in \mathcal{A}.$$

We denote the class of the element A by [A]. The scalar product (III.2.36) does not depend on the choice of the algebraic element within one class and, if A is not in the class of the zero element

$$\langle\Psi|\Psi\rangle \equiv \|\Psi\|^2 > 0. \tag{III.2.38}$$

$\mathcal{H}_\omega$ is obtained by the completion of $\mathcal{A}/\mathfrak{J}$ in the norm topology given by (III.2.38). The product in $\mathcal{A}$ defines an action of $\mathcal{A}$ on the vectors in $\mathcal{A}/\mathfrak{J}$; it associates to each $A \in \mathcal{A}$ a linear operator $\pi_\omega(A)$, defined on the dense domain $\mathcal{A}/\mathfrak{J} \subset \mathcal{H}_\omega$ by

$$\pi_\omega(A)\Psi = [AB] \quad \text{if} \quad \Psi = [B]. \tag{III.2.39}$$

If $\mathcal{A}$ is a Banach *-algebra then the domain of definition of $\pi_\omega(\mathcal{A})$ can be extended to the whole of $\mathcal{H}_\omega$ by continuity.

Theorem 2.2.10
If $\mathcal{A}$ is a C*-algebra with unit and ω any state such that π_ω is a faithful representation (i.e. $\pi_\omega(A) \neq 0$ for $A \neq 0$) then the operator norm of $\pi_\omega(A)$ equals the C*-norm of A. (Uniqueness of the C*-norm).

A representation π is called *cyclic* if there exists a vector Ω in the representation space $\mathcal{H}$ such that $\pi(\mathcal{A})\Omega$ is dense in $\mathcal{H}$. Ω is called a *cyclic vector.* It is clear that, if $\mathcal{A}$ has a unit, the GNS-construction provides a cyclic representation with the cyclic vector Ω corresponding to the class of the unit element

$$\Omega = [\mathbb{1}].$$

Moreover

$$\omega(A) = \langle\Omega|\pi_\omega(A)|\Omega\rangle. \tag{III.2.40}$$

So the expectation value is expressed in a form familiar from quantum mechanics. We may say that the state ω is represented by the state vector Ω in $\mathcal{H}_\omega$. Any vector $\Psi \in \mathcal{H}_\omega$ defines a state

$$\omega_\Psi(A) = \langle \Psi | \pi_\omega(A) | \Psi \rangle, \qquad \text{(III.2.41)}$$

which may be approximated by $\omega(B^* AB)$ with $B \in \mathcal{A}$ since Ψ can be approximated by $\pi_\omega(B)\Omega$.

The GNS-construction shows that states over an abstract *-algebra come in families. One state ω determines a family of states, the states of form (III.2.41) with $\Psi \in \mathcal{H}_\omega$. We call these the *vector states of the representation* π_ω. More generally we may consider the states

$$\omega_\varrho(A) = \mathrm{tr}\, \varrho \pi_\omega(A) \qquad \text{(III.2.42)}$$

with ϱ a positive trace class operator in $\mathfrak{B}(\mathcal{H}_\omega)$. The set of all the states (III.2.42) we call the *folium of the representation* π_ω or the set of *normal states of the von Neumann algebra* $\pi_\omega(\mathcal{A})''$. It is, of course, determined by ω. Clearly one may now take a state ω' in this folium and make again the GNS-construction starting from ω'. Then ω' which was represented as a density matrix (positive trace class operator) in π_ω is now represented as a vector in $\mathcal{H}_{\omega'}$. To understand the relation between these different ways of representing the algebra and the states we have to take a closer look at the *dual space of* $\mathcal{A}$.

We begin with some generalities. Let V be a Banach space over $\mathbb{C}$. The dual space, denoted by V^*, is the set of all bounded, complex valued linear forms on V. V^* is again a Banach space with respect to the natural norm

$$\| \varphi \| = \sup_{X \in V} | \varphi(X) | \, \| X \|^{-1}; \quad \varphi \in V^*. \qquad \text{(III.2.43)}$$

Besides the norm topology there is another important topology on V^*. Choosing any finite set $X_1, \dots X_n$ of elements of V one can define a seminorm on V^* by

$$\wr\wr \varphi \wr\wr_{X_1,\dots X_n} = \sup_{k=1\dots n} | \varphi(X_k) |, \qquad \text{(III.2.44)}$$

and a corresponding neighborhood $\mathcal{N}(\varepsilon; X_1, \dots X_n)$ of the origin in V^* by

$$\mathcal{N}(\varepsilon; X_1, \dots X_n) = \{ \varphi \in V^* : \wr\wr \varphi \wr\wr_{X_1,\dots X_n} < \varepsilon \}$$
$$= \{ \varphi \in V^* : | \varphi(X_k) | < \varepsilon \quad \text{for} \quad k = 1, \dots n \}.$$

Such neighborhoods for arbitrary finite n and arbitrary choice of the elements X_k define the *weak topology* in V^* induced by V (also called the weak *-topology). In particular, a generalized sequence φ_α converges weakly to φ if

$$| \varphi_\alpha(X) - \varphi(X) | \to 0 \quad \text{for every} \quad X \in V. \qquad \text{(III.2.45)}$$

The following theorem will be used on several occasions.

Theorem 2.2.11 (Alaoglu)
The closed unit ball in V^* (i.e. the set $\{ \varphi \in V^* : \| \varphi \| \leq 1 \}$ is compact in the weak topology induced by V.
See [*Dunford and Schwartz 1958*].

We are interested first in the case where V is a C*-algebra $\mathcal{A}$. The dual space $\mathcal{A}^*$ [3] has a distinguished convex cone $\mathcal{A}^{*+}$, the set of positive linear forms on $\mathcal{A}$. The states over $\mathcal{A}$ are the elements of $\mathcal{A}^{*+}$ with norm 1. One has (see [*Sakai 1971*])

Theorem 2.2.12
The folium of a representation and the set of vector states of a representation are norm closed subsets of $\mathcal{A}^{*+}$.

On the other hand one has

Theorem 2.2.13 [Fell 60]
The folium of a faithful representation of a C*-algebra is weakly dense in the set of all states.

Remark. Since in physics we can never perform infinitely many experiments and since each experiment has a limited accuracy we can, by monitoring a state, never determine more than a weak neighborhood in $\mathcal{A}^{*+}$. By Fell's theorem this means that we cannot find out in which folium the state lies. Still there are reasons for considering more than one equivalence class of representations of $\mathcal{A}$. They may come from idealizations of the problem (injecting an infinite amount of information which we do not really have in order to simplify the problem). Examples will be seen in chapters IV and V. And there may also be reasons for excluding some folia as "physically not realizable" (see the remarks at the end of this section and the assumption of local definiteness in section 3).

Definition 2.2.14
Given two representations π_1, π_2 acting in the Hilbert spaces $\mathcal{H}_1$, $\mathcal{H}_2$ respectively. A bounded linear operator R from $\mathcal{H}_1$ to $\mathcal{H}_2$ is called an *intertwiner* (from π_1 to π_2) if

$$R\,\pi_1(A) = \pi_2(A)\,R \quad \text{for all} \quad A \in \mathcal{A}. \tag{III.2.46}$$

Its adjoint R^* is uniquely defined as a bounded linear operator from $\mathcal{H}_2$ to $\mathcal{H}_1$ by

$$\langle \Psi_1 | R^* | \Psi_2 \rangle = \langle R\psi_1 | \Psi_2 \rangle, \quad \Psi_k \in \mathcal{H}_k, \quad k = 1,\ 2. \tag{III.2.47}$$

It is an intertwiner in the opposite direction

$$R^* \pi_2(A) = \pi_1(A) R^*. \tag{III.2.48}$$

An intertwiner decomposes each of the two representation spaces into two invariant orthogonal subpaces $\mathcal{H}_k = \mathcal{H}_k' \oplus \mathcal{H}_k''$, $k = 1,\ 2$. $\mathcal{H}_1''$ is the kernel of R,

[3] Some confusion may arise because the symbol $\mathfrak{S}^*$ (with $\mathfrak{S} \subset \mathcal{A}$) is sometimes used to denote the set of all the adjoints of the elements of $\mathfrak{S}$. Hopefully the different meaning of the star will always be evident from the context.

the set of vectors in $\mathcal{H}_1$ which is annihilated by R; $\mathcal{H}_2'$ is the closure of the range of R (set of image vectors). The restriction of R to $\mathcal{H}_1'$ is a non-singular linear operator from $\mathcal{H}_1'$ onto $\mathcal{H}_2'$ which intertwines from the restriction of π_1 to $\mathcal{H}_1'$ to the restriction of π_2 to $\mathcal{H}_2'$. This means that these two subrepresentations are unitarily equivalent because one can make a polar decomposition

$$R = V(R^*R)^{1/2};$$

V is a unitary map from $\mathcal{H}_1'$ onto $\mathcal{H}_2'$ which also intertwines the subrepresentations since R^*R is in the commutant of π_1.

Definition 2.2.15
Two representations are called *disjoint* if they contain no subrepresentations which are unitarily equivalent i.e. if the set of (non zero) intertwiners is empty. Two representations are called *quasiequivalent* if every subrepresentation of the first contains a representation which is unitarily equivalent to a subrepresentation of the second. A representation is called *primary* if it is quasi-equivalent to each of its subrepresentations. A state is called primary if it leads to a primary representation by the GNS-construction.

Note that the set of intertwiners of a representation π with itself is a von Neumann algebra namely the commutant $\pi(\mathcal{A})'$. The notion of quasi-equivalence defined here agrees with the one given earlier for von Neumann algebras. A primary representation is one in which the center $\pi(\mathcal{A})' \cap \pi(\mathcal{A})''$ is trivial i.e. one in which $\pi(\mathcal{A})''$ is a factor.

We mention still

Theorem 2.2.16
The norm distance between any two states belonging to disjoint representations is 2 (the same as between two density matrices with supports in two orthogonal subspaces of a Hilbert space).

Purification. For two states ω_1, ω_2 and $0 < \lambda < 1$

$$\omega = \lambda\,\omega_1 + (1-\lambda)\,\omega_2 \tag{III.2.49}$$

is again a state, the mixture of ω_1 and ω_2 with weights λ and $1-\lambda$. If ω can be written in the form (III.2.49) then we say that ω dominates ω_1 (and ω_2). Excluding the trivial case $\omega_1 = \omega_2$ we also call ω_1 a *purification* of ω One sees readily that if ω dominates ω_1 then the GNS-representation π_{ω_1} is unitarily equivalent to a subrepresentation of π_ω. The intertwiner is given by

$$R\,\pi_\omega(A)\,\Omega = \pi_{\omega_1}(A)\,\Omega_1 = \pi_{\omega_1}(A)R\,\Omega. \tag{III.2.50}$$

This is consistent because $\omega(A^*A) = 0$ implies $\omega_1(A^*A) = 0$ and it suffices to define R because Ω is a cyclic vector. This leads to

Theorem 2.2.17
(i) The folium of a representation uniquely determines its quasi-equivalence class. If ω dominates ω_1 then π_{ω_1} is contained in π_ω.
(ii) The GNS-representation arising from a pure state is irreducible and the vector states in an irreducible representation are pure.
(iii) Every state in the folium of a primary representation is primary.

The theorem shows that the distinction between vector states and density matrices has an intrinsic significance only in the case of irreducible representations where it corresponds to the distinction between pure and impure states. There remains the question of the existence of "sufficiently many" pure states. This is answered by

Theorem 2.2.18
For a C*-algebra with unit the set of its states is weakly compact. It has extremal points, the pure states, and the whole set of states is generated from the pure states by convex combinations and weak closure.

Remark. The compactness follows from theorem 2.2.11 noting that $\varphi(\mathbb{1}) = 1$. If $\mathcal{A}$ has no unit it may happen that sequences of states converge weakly to zero. A typical example, not entirely void of interest for physics, is the following. Let $\mathcal{H} = \mathcal{L}^{(2)}(\mathbb{R})$ and $U(a)$ be unitary translation operators i.e. $(U(a)\psi)(x) = \psi(x-a)$; x, $a \in \mathbb{R}$. Let K be a *Hilbert-Schmidt operator*, i.e.

$$(K\psi)(x) = \int K(x,x')\psi(x')dx'$$

where the integral kernel is a square integrable function of x and x'. Obviously, for a sequence $\psi_a = U(a)\psi$ we have

$$\lim_{a\to\infty}\langle\psi_a|K|\psi_a\rangle = 0 \qquad \text{(III.2.51)}$$

The norm closure of the set of Hilbert-Schmidt operators is a C*-subalgebra of $\mathfrak{B}(\mathcal{H})$ without unit. It is the set of *compact operators*, a closed 2-sided *-ideal $\mathfrak{J}_C$ in $\mathfrak{B}(\mathcal{H})$. By (III.2.51) the sequence of vector states ψ_a, restricted to $\mathfrak{J}_C$, converges weakly to zero as $a \to \infty$. On the other hand, since $\langle\psi_a|\mathbb{1}|\psi_a\rangle = 1$ the sequence of states over $\mathfrak{B}(\mathcal{H})$ must have limit points in $\mathfrak{B}(\mathcal{H})^*$ which are states on $\mathfrak{B}(\mathcal{H})$. These states are, however, not normal states of the defining representation of $\mathfrak{B}(\mathcal{H})$.

W*-Algebras. A von Neumann algebra also has an abstract counterpart, a W*-algebra. This is a C*-algebra which, in addition, is the dual space of a Banach space V over the field of complex numbers. From the point of view of our application to physics W*-algebras arise naturally if we consider state-space and its structure rather than the algebra as the primary object. V is then interpreted as the complex linear span of the space of states (expectation functionals, not state vectors). It is a Banach space with a distinguished positive cone V^+ of which the elements with norm 1 are the states.

Its dual space V^* (the Banach space of bounded linear forms over V) is then equipped with a norm topology and the weak topology induced by V as described above. By theorem 2.2.11 it is also weakly closed. If a product can be defined within V^* satisfying (III.2.4), (III.2.5) so that V^* becomes a C*-algebra, with V^{*+} as its positive cone, then this algebra, equipped with the weak topology inherited from V is called a W*-algebra. When should one be able to introduce such a product in V^*? This is encoded in the facial structure of V^+ (see Chapter VII). The papers quoted there give a partial answer. If we write $V^* = \mathcal{A}$ then V is called the *predual* of $\mathcal{A}$ (which is another way of saying that $\mathcal{A}$ is the dual of V). The predual is denoted by a lower star: one writes $V = \mathcal{A}_*$. The set of all bounded linear forms over $\mathcal{A}$, the dual space $\mathcal{A}^*$, will be larger than $V = \mathcal{A}_*$. Thus a W*-algebra has a distinguished folium of states, the normalized elements of $\mathcal{A}_*^+$. They are called the *normal states*. If we use one of the normal states of a W*-algebra in the GNS-construction of a representation we obtain as the image a weakly closed operator algebra i.e. a von Neumann ring.

Thus one might also say that a W*-algebra arises from a C*-algebra together with a distinguished folium of states. It is isomorphic to the closure of the algebra in the weak topology determined by this folium. The distinction of a specific folium may be relevant in physics since it is not clear whether one can or should define the net of C*-algebras of local observables in such a way that all mathematical states are physically realizable. An example of physically unrealizable states would be states carrying infinite energy in a finite region.

References. The basic theory of von Neumann algebras was developed between 1936 and 1943 in a series of papers by F.J. Murray and J. von Neumann in the Annals of Mathematics. C*-algebras originated in the Moscow school of I.E. Gelfand in the early 40's (see the still highly recommendable book [*Naimark 1972*]). An independent approach was pioneered by I.E. Segal [Seg 47].

Standard mathematical texts covering this area are [*Dixmier 1981*], [*Dixmier 1982*], [*Sakai 1971*], [*Pedersen 1979*], [*Kadison and Ringrose 1983, 1986*]. A rather detailed exposition addressed to physicists is contained in [*Bratteli and Robinson, Vol I, 1979*].

III.3 The Net of Algebras of Observables

With the help of the mathematical concepts described in the last section we can elaborate the frame of local quantum theory in Minkowski space. In this section and in the next one we shall look at some specific structural properties which the net of algebras of local observables should (or might) have and describe some consequences. One objective is to place axiom **A** into a broader context. In the endeavour of formulating the assumptions concisely one has to balance between the need for keeping them general enough to encompass the physically interesting situations and the wish to make them as restrictive as possible. It should be borne in mind that the theory is not in a closed, final stage; hence there is some leeway.

III.3.1 Smoothness and Integration. Local Definiteness and Local Normality

In chapter II we frequently used integrals like $\int f(x)\,\alpha_x A\, d^4x$ with f a smooth function with fast decrease and α_x the automorphism corresponding to a translation by x. We have to see how such integrals can be defined. Let $A \in \mathfrak{A}$. Since $\| \alpha_x A \|$ is independent of x and the norm has the subadditivity property (III.2.3) the only question concerns the local integrability. If one has this then

$$\alpha_f A \equiv \int f(x)\,\alpha_x A\, d^4x \tag{III.3.1}$$

is defined for continuous $\mathcal{L}^1$-functions f as a Bochner integral (the generalization of the Riemann integral to integrands with values in a Banach space) and one has

$$\| \alpha_f A \| \leq \| A \| \, \| f \|_1, \tag{III.3.2}$$

where $\| f \|_1 = \int | f(x) |\, d^4x$ is the $\mathcal{L}^1$-norm of f. One has local integrability (in the sense of the Bochner integral) certainly for those elements $A \in \mathfrak{A}$ for which the function $x \to \alpha_x A$ is norm-continuous.

Definition 3.1.1
An element $A \in \mathfrak{A}$ is called differentiable if

$$\lambda^{-1} \| \alpha_{\lambda x} A - A \| \quad \text{converges as} \quad \lambda \to 0. \tag{III.3.3}$$

The differential quotients $\partial/\partial x^\mu \alpha_x A|_{x=0}$ are then defined and belong again to $\mathfrak{A}$. The element A is called *smooth* if it is infinitely often differentiable.

We argue now that it is possible and warranted to choose the algebras so that the action of the translation automorphisms on the elements is continuous in the norm topology. We shall denote such a choice of the algebras by $\mathfrak{A}_S$.

Assumption 3.1.2 (Smoothness)
We can choose for the algebra of local observables in $\mathcal{O}$ a C*-algebra $\mathfrak{A}_S(\mathcal{O})$ containing a norm-dense set of smooth elements. Then $\alpha_x A$ is continuous in the norm topology i.e.

$$\| \alpha_x A - A \| \to 0 \quad \text{as } x \to 0 \quad \text{for any } A \in \mathfrak{A}_S. \tag{III.3.4}$$

Let us present some arguments supporting this assumption.

Suppose we start from a net $\{\mathfrak{A}(\mathcal{O})\}$ not satisfying this assumption. Denote the representation obtained by the GNS-construction from the vacuum state by π_0 and take the von Neumann rings

$$\mathcal{R}(\mathcal{O}) = \pi_0(\mathfrak{A}(\mathcal{O}))''; \quad \mathcal{R} = \pi_0(\mathfrak{A})''. \tag{III.3.5}$$

Then α_x is implementable by a strongly continuous group of unitary operators $U(x)$ and the energy-momentum operators P^μ are defined in the representation. Let $\mathcal{H}_E = P_E\mathcal{H}$ be the spectral subspace of P^0 corresponding to spectral values below E and P_E the spectral projector. Then matrix elements

$$\langle \Psi' | U(x) X U^{-1}(x) | \Psi \rangle \tag{III.3.6}$$

are differentiable functions of x for any $X \in \mathcal{R}$ if Ψ and Ψ' belong to $\mathcal{H}_E$ because $P_E P^\mu = P^\mu P_E$ is bounded. For arbitrary vectors Ψ, Ψ' (III.3.6) is still continuous since the vectors can be strongly approximated by vectors from $\mathcal{H}_E$ as $E \to \infty$. Let $f(x)$ be a smooth function with support contained in a neighborhood $\mathcal{O}_0$ of the origin and let $A \in \mathfrak{A}(\mathcal{O}_1)$. Then the integral

$$\pi(A_f) \equiv \int f(x)\, \pi(\alpha_x A)\, d^4x \tag{III.3.7}$$

is defined in the sense of matrix elements yielding an operator in $\mathcal{R}(\mathcal{O}_1 + \mathcal{O}_0)$ with norm

$$\| \pi(A_f) \| \leq \| A \| \, \| f \|_1 \, . \tag{III.3.8}$$

Furthermore $\pi(A_f)$ is infinitely often differentiable in the norm topology because the differentiations can be shifted to the smooth function f. Since any open region $\mathcal{O}$ can be written as $\mathcal{O}_1 + \mathcal{O}_0$ with suitably chosen, open $\mathcal{O}_1$, $\mathcal{O}_0$ we see that there are smooth elements in every $\mathcal{R}(\mathcal{O})$ and we may define a (concrete) C*-algebra $\mathfrak{A}_S(\mathcal{O})$ as consisting of smooth elements of $\mathcal{R}(\mathcal{O})$ and their limits in the norm topology.

In this construction the folium of states in the vacuum representation was used. What happens if we consider other folia of states? If we adhere strictly to the dogma that all physical information is contained in a net of abstract C*-algebras then we must regard every mathematical state (normalized positive, linear functional on the net) as a physically realizable state. There are, however, good reasons for the following principle which suggests that we should soften the dogma and distinguish a subset of the mathematical states as the *physically realizable states.*

Principle 3.1.3 (Local Definiteness)
Let $\mathcal{O}_n$ be a directed set of regions shrinking to a point x, i.e. $\mathcal{O}_{n+1} \subset \mathcal{O}_n$; $\cap\, \mathcal{O}_n = x$. Any two physically realizable states ω_i ($i = 1, 2$) will become indistinguishable in restriction to $\mathfrak{A}(\mathcal{O}_n)$ in the limit $n \to \infty$. Specifically

$$\| (\omega_1 - \omega_2)|_{\mathfrak{A}(\mathcal{O}_n)} \| \to 0 \quad \text{as} \quad n \to \infty. \tag{III.3.9}$$

This principle is natural if we think of defining the theory in terms of relations in the small. There should remain no superselection rules if we restrict attention to sufficiently small regions of space-time. The distinction of two states by means of observations in a very small region requires very high energies and there remain no observables at a point.[1] If we believe, in additition, that among the realizable states we have states which are primary for the algebras of small regions then, appealing to theorem 2.2.16 we conclude from the principle 3.1.3 that for any two folia of physically allowed states and any point $x \in \mathcal{M}$ there will be a neighborhood of x such that the restrictions of the folia to the algebra of this neighborhood coincide and are primary.

Remark. We motivated the use of abstract algebras by the wish of understanding superselection rules in an intrinsic way, namely as arising from inequivalent representations of one and the same algebraic structure. This is indeed one good reason. However, the superselection rules encountered so far (various types of total charges, thermodynamic-hydrodynamic quantities of the infinite system) all relate to global aspects of the states and are not effective once we restrict attention to a finite region. This fits in with the above conclusion and suggests the following working hypothesis.

Tentative Assumption 3.1.4 (Local Normality)
In restriction to the algebra of a finite, contractible region all physically realizable states belong to a unique primary folium.

This makes the supporting argument for the smoothness assumption 3.1.2 independent of the use of the vacuum representation. We have then the following picture. There is the net $\mathfrak{A}_S(\mathcal{O})$ on which the automorphisms of translations act continously and to which we shall refer the notions of states and representations. On the other side we have the set $\mathfrak{S}$ of physically realizable states. Their restriction to an algebra $\mathfrak{A}_S(\mathcal{O})$ will give the set of *partial states on* $\mathcal{O}$, denoted by $\mathfrak{S}(\mathcal{O})$. For $\mathcal{O}_1 \subset \mathcal{O}$ there is the natural restriction map

$$\mathfrak{S}(\mathcal{O}) \to \mathfrak{S}(\mathcal{O}_1). \tag{III.3.10}$$

It defines an equivalence class in $\mathfrak{S}(\mathcal{O})$ with respect to $\mathcal{O}_1$, consisting of all partial states on $\mathcal{O}$ which have the same restriction to $\mathcal{O}_1$. So, in mathematical terms, the collection $\{\mathfrak{S}(\mathcal{O})\}$ is a presheaf. Let $\Sigma(\mathcal{O})$ denote the Banach space over $\mathbb{C}$ generated by $\mathfrak{S}(\mathcal{O})$. It is the complex linear span of $\mathfrak{S}(\mathcal{O})$ completed in the norm topology inherited from the norm in $\mathfrak{S}(\mathcal{O})$ (in precise terms Σ is a "base norm space"). The dual space of $\Sigma(\mathcal{O})$ is a W*-algebra which we denote by $\mathcal{R}(\mathcal{O})$

$$\mathcal{R}(\mathcal{O}) = \Sigma^*(\mathcal{O}); \;\; \Sigma(\mathcal{O}) = \mathcal{R}(\mathcal{O})_*. \tag{III.3.11}$$

For finite, contractible regions $\mathcal{R}(\mathcal{O})$ is isomorphic to the corresponding von Neumann algebra in the vacuum sector for which we shall use the same symbol.

[1] The relevance of this principle was recognized in the context of quantum field theory in curved space-time and used in the discussion of the Hawking temperature by Haag, Narnhofer and Stein [Haag 84]. If the metric structure of space-time is not (classically) given but determined by the prevailing physical state then the possibility of regarding the metric at each point as a superselection rule may be of interest; see [Fred 87] and [Bann 88].

The C*-algebra $\mathfrak{A}_S(\mathcal{O})$ is weakly dense in $\mathcal{R}(\mathcal{O})$. The net $\{\mathfrak{A}_S(\mathcal{O})\}$ is not needed in the discussion of superselection rules in the next chapters. What is important there is that the global algebra $\mathfrak{A}$ must be considered as a C*-algebra, not a W*-algebra. It is the completion of the union of local algebras in the norm topology in order to preserve the quasilocal character, irrespective of whether we take the local algebras as W*- or C*-algebras. Nevertheless a physically distinguished C*-net $\{\mathfrak{A}_S(\mathcal{O})\}$ carrying more information than the net $\{\mathcal{R}(\mathcal{O})\}$ may be expected to become important in a finer analysis, building up the theory from information in the small in analogy to the methods of differential geometry. An example may be the (not yet achieved) understanding of a local gauge principle directly in the quantum theoretic setting. For the definition of a physical distinguished net $\mathfrak{A}_S$ one needs, of course, more than just the demand that the translation automorphisms act continuously on it. It should satisfy for instance some minimality requirement.

III.3.2 Symmetries and Symmetry Breaking. Vacuum States

Definition 3.2.1
A group $\mathcal{G}$ is called a symmetry group if there is a realization of $\mathcal{G}$ by automorphisms of the net:

$$g \in \mathcal{G} \rightarrow \alpha_g \in \mathrm{Aut}\mathfrak{A} \tag{III.3.12}$$

with the additional requirement that the image of the algebra of a finite region under α_g shall be again the algebra of a finite region.

If ω is an invariant state, i.e. a normalized, positive linear functional over $\mathfrak{A}$ with

$$\omega(\alpha_g A) = \omega(A) \quad \text{for} \quad g \in \mathcal{G}, \quad A \in \mathfrak{A} \tag{III.3.13}$$

then in the GNS-representation built from ω the automorphisms α_g are certainly implementable. Denoting by $(\Omega,\ \pi,\ \mathcal{H})$ respectively the cyclic vector corresponding to ω, the representation and the Hilbert space on which it acts, the definition

$$U(g)\pi(A)\Omega = \pi(\alpha_g A)\Omega \tag{III.3.14}$$

provides a unitary representation $g \rightarrow U(g)$ of the group.[2]

Each group element has a "geometric part", a point transformation of Minkowski space conserving the causal structure. This is because we required that α_g transforms the algebra of a finite region into that of another finite region and α_g respects inclusion properties of regions and causal structure since

[2] One has to check that the definition is consistent i.e. if $\pi(A)\Omega = 0$ then also $\pi(\alpha_g A)\Omega = 0$. This follows from the isometry

$$\| U(g)\pi(A)\Omega \|^2 = (\Omega|\pi(\alpha_g A)^*\pi(\alpha_g A)|\Omega) = \omega(\alpha_g A^* A) = \omega(A^* A) = \| \pi(A)\Omega \|^2 .$$

Since Ω is a cyclic vector for $\pi(A)$ (III.3.14) defines $U(g)$ on a dense domain. The image is again a dense domain because α_g is invertible. So $U(g)$ is extendable to a unitary operator on $\mathcal{H}_\omega$. One easily checks the representation property $U(g_2)U(g_1) = U(g_2 g_1)$.

these manifest themselves in the algebraic structure of the net. $\mathcal{G}$ is therefore a semidirect product of a purely geometric symmetry group and an "internal symmetry group" which transforms the algebra of each region into itself:

$$\alpha_g \mathfrak{A}(\mathcal{O}) = \mathfrak{A}(\mathcal{O}) \quad \text{for} \quad g \in \mathcal{G}_{\text{int}}. \tag{III.3.15}$$

Since the geometric symmetry group must conserve the causal structure it could be at most the conformal group. We shall mostly take it to be the Poincaré group. In any case the geometric symmetry group is locally compact and has an invariant measure $d\mu(g)$. We assume that the same is true for the full symmetry group. Then we can construct $\mathcal{G}$-invariant states in the following way. Consider an increasing sequence of compact subsets: $S_k \subset \mathcal{G}$, $S_{k+1} \supset S_k$, $\cup S_k = \mathcal{G}$. Starting from an arbitrary state ω we define ω_k as[3]

$$\omega_k(A) = \mu(S_k)^{-1} \int_{S_k} \omega(\alpha_g A) d\mu(g). \tag{III.3.16}$$

Obviously ω_k is again a state since $\omega(\mathbb{1}) = 1$ and, using the invariance of the measure one gets

$$\omega_k(\alpha_g A) = \mu(S_k)^{-1} \int_{S_k} \omega_k(\alpha_{g'g} A) d\mu(g') = \mu(S_k)^{-1} \int_{S_k g} \omega(\alpha_{g''} A) d\mu(g'')$$

where $S_k g$ is the set resulting from S_k by right translation with g. One has

$$| \, \omega_k(\alpha_g A) - \omega_k(A) \, | \leq \| \, A \, \| \, \mu(S_k)^{-1} \left(\mu(\Delta_k') + \mu(\Delta_k'') \right),$$

where

$$\Delta_k' = S_k - (S_k g \cap S_k), \quad \Delta_k'' = S_k g - (S_k g \cap S_k).$$

By theorem 2.2.18 the set of all states over $\mathfrak{A}$ is weakly compact. Thus the sequence ω_k has weak limit points which are states. For the subgroups formed by space-time translations together with any compact group (but not for the Poincaré group!) the ratios $\mu(\Delta_k)/\mu(S_k)$ go to zero for fixed g and $k \to \infty$. So one can construct states which are invariant under translations and rotations by this procedure.

The set of invariant states is evidently a convex subset of state space (mixtures of invariant states are invariant); it is weakly closed and therefore weakly compact (the set of all states being weakly compact). Therefore, by the theorem of Krein and Milman (see [*Dunford and Schwartz 1958*]) all invariant states can be obtained as convex combinations (mixtures) of *extremal invariant states*, i.e. invariant states which cannot be decomposed any more into a convex combination of other invariant states. There is, however, no reason yet why an extremal invariant state should be pure or primary. It may be possible to decompose it into a mixture of other states which are not invariant. This is the scenario of spontaneous symmetry breaking.

[3]We assume here that α_g acts continuously on $\mathfrak{A}$ for the whole symmetry group, not only for the translation subgroup.

The spontaneous breakdown of symmetries is a well known phenomenon in solid state physics or, more generally, in the nonrelativistic quantum theory of the structure of matter, idealized as an infinitely extended medium with a finite mean density of particles. Translation invariance may be broken by crystalization, rotation invariance by spontaneous magnetization. Galilei invariance (replacing Lorentz invariance) is always broken if the mean particle density is finite because a state describing the matter at rest in the reference system and one describing a flow of matter with uniform velocity are "infinitely different" and thus lie certainly in disjoint folia.

The mathematical counterpart of these phenomena is the following. Let ω be an extremal invariant state with respect to the symmetry group and π_ω the GNS-representation arising from it. If ω is not primary then the center $\mathfrak{Z} = \pi_\omega(\mathfrak{A})'' \cap \pi_\omega(\mathfrak{A})'$ is not trivial. We may regard a self adjoint operator $Z \in \mathfrak{Z}$ as an idealized observable which commutes with all local observables (it is typically a question involving the behavior of the state at infinity). Thus Z may be considered as a classical quantity. If we do not know its value this may be considered as subjective ignorance. If we decompose the Hilbert space according to the simultaneous spectrum of the Abelian von Neumann algebra $\mathfrak{Z}$

$$\mathcal{H}_\omega = \int_{\mathcal{K}} \mathcal{H}_\kappa d\mu(\kappa) \tag{III.3.17}$$

we get a unique decomposition of ω as a convex combination of primary states ω_κ ($\kappa \in \mathcal{K}$, the spectrum of $\mathfrak{Z}$).[4] Each such primary component may then be regarded as an optimal objective description of the state of an individual system. In the case of crystalization the relevant elements of the center are of the form

$$Z = \lim_{\mathcal{V}\to\infty} \mathcal{V}^{-1} \int_{\mathcal{V}} \alpha_{\mathbf{x}} A\, f(\mathbf{x})\, d^3x \tag{III.3.18}$$

where f is a periodic function. In the case of magnetization it is of the same form with $f = 1$ and A a component of a vectorial observable (the magnetic moment per unit volume).

The physical significance of the central decomposition is emphasized by the following theorem.

Theorem 3.2.2
A primary state ω has the cluster property

$$|\, \omega(A\alpha_{\mathbf{x}}B) - \omega(A)\omega(\alpha_{\mathbf{x}}B)\,| \to 0 \quad \text{as} \quad |\,\mathbf{x}\,| \to \infty. \tag{III.3.19}$$

Proof. The von Neumann algebra $\pi_\omega(\mathfrak{A})''$ is the dual of a Banach space. Therefore the ball $\| X \| \leq c$, $X \in \pi_\omega(\mathfrak{A})''$ is weakly compact. A sequence $\pi_\omega(\alpha_{\mathbf{x}}B)$ with $|\,\mathbf{x}\,| \to \infty$ has weak limit points in $\pi_\omega(\mathfrak{A})''$. For large space translations the

[4]The measure theoretic intricacies of this "central decomposition" and of the decomposition of invariant states into extremal invariant states are discussed in [*Bratteli and Robinson 1979, chapter 4 of Vol. 1.*]

commutator of $\alpha_{\mathbf{x}}B$ with any fixed element of the quasilocal algebra tends to zero. Therefore the limit elements of $\pi_\omega(\alpha_{\mathbf{x}}B)$ lie also in the commutant $\pi_\omega(\mathfrak{A})'$. So they lie in the center. If ω is primary the center consists only of multiples of the unit operator. Thus

$$\pi_\omega(\alpha_{\mathbf{x}}B) - \omega(\alpha_{\mathbf{x}}B)\mathbb{1} \to 0 \quad \text{(weakly)}$$

which implies (III.3.19).

Let us illustrate this in the example of spontaneous magnetization. A primary state describes a sharp direction of magnetization. If the observable $\mathbf{M}(\mathbf{x})$ denotes the magnetic moment per unit volume around the point $\mathbf{x}$ and ω is primary then $\omega(\mathbf{M}(\mathbf{x})) = \mathbf{m}(\mathbf{x}) \neq 0$ but the correlation $\omega(\mathbf{M}(\mathbf{x})\mathbf{M}(\mathbf{y})) - \mathbf{m}(\mathbf{x})\mathbf{m}(\mathbf{y})$ tends to zero for large $| \mathbf{x} - \mathbf{y} |$. If we take instead the extremal invariant state $\overline{\omega}$ which results from ω, as in (III.3.16), by averaging over the rotation group then the 1-point function $\overline{\omega}(\mathbf{M}(\mathbf{x}))$ vanishes and the 2-point function has infinitely long range correlations.

Let us turn now to the notion of a vacuum state in relativistic, local theory. There is a strong analogy between quantum field theory and the treatment of an infinitely extended medium in nonrelativistic quantum theory. In the latter case we know, however, a priori what we mean by matter density and vacuum in terms of the mathematical objects from which the theory is constructed. In our case, where we define the theory by a net of algebras of local observables the meaning of "absence of matter" is not immediately evident. We can start from lemma 4.1.2 in chapter II which tells us that for any $A \in \mathfrak{A}$ and any function $f(x)$ whose Fourier transform has support in the momentum space region Δ the algebraic element

$$A(f) = \int f(x)\, \alpha_x A\, d^4x$$

affects an energy-momentum transfer within Δ. This suggests that the positive observables $A(f)^*A(f)$ with $\operatorname{supp} \tilde{f}(p)$ confined to negative values of the energy p^0 register the presence of matter, or, taking Lorentz invariance into account, that such observables are detectors of matter whenever $\operatorname{supp} \tilde{f}(p)$ lies outside the forward cone V^+. In the context of the GNS-construction we saw that a state ω defines a left ideal in the algebra, its *Gelfand ideal* or annihilator ideal, consisting of all elements $X \in \mathfrak{A}$ for which $\omega(X^*X) = 0$. Conversely, given any proper, closed left ideal $\mathfrak{J}$ there exists at least one state which has this ideal as its Gelfand ideal. This fact was used by Doplicher [Dopl 65] to give an algebraic formulation of axiom **A3**. We associate with each region Δ of momentum space the left ideal of $\mathfrak{A}$

$$\mathfrak{J}(\Delta) = \{X = BA(f)\} \quad \text{with} \quad B \in \mathfrak{A}, \quad A \in \mathfrak{A},\ \operatorname{supp} \tilde{f} \subset \Delta. \qquad \text{(III.3.20)}$$

Definition 3.2.3
A state whose Gelfand ideal (annihilator ideal) contains the "spectral ideal"

$$\mathfrak{J}(V^{+c}) = \mathfrak{J}(\cup\Delta), \quad \Delta \cap V^+ = \emptyset \qquad \text{(III.3.21)}$$

(the union is taken over all momentum space regions which do not intersect the forward cone) is called a *vacuum state.*

The existence of a vacuum corresponds to the algebraic property that $\mathfrak{J}(V^{+c})$ *is a proper left ideal i.e. that its closure is not the whole algebra* $\mathfrak{A}$.

Obviously so defined vacuum states form a Poincaré invariant set. So one can obtain translation invariant vacuum states by averaging. The GNS-construction from a translation invariant vacuum satisfies axiom **A3** of chapter II.

One can show now that translation invariance cannot be spontaneously broken in a vacuum state: the vacuum cannot be a crystal. Furthermore, an extremal invariant vacuum state is automatically pure. One has

Theorem 3.2.4 [Ara 64b]
Let $\mathcal{R}$ be a v. Neumann algebra and $U(x)$ a strongly continuous unitary representation of the space-time translation group such that the spectrum of the generators is contained in the forward cone and

$$\alpha_x A \equiv U(x) A U(x)^{-1}, \quad A \in \mathcal{R}$$

transforms $\mathcal{R}$ into itself. Then

(i) each element of the center $\mathfrak{Z} \equiv \mathcal{R} \cap \mathcal{R}'$ commutes with all $U(x)$.

(ii) If the Hilbert space contains a cyclic vector Ω which is invariant under $U(x)$ then $U(x) \in \mathcal{R}$.

Proof. For any $X \in \mathfrak{B}(\mathcal{H})$ and any class $\mathcal{S}$-function $f(x)$ we use the abbreviations $X(x) = U(x)XU(x)^{-1}$, $X(f) = \int f(x)X(x)d^4x$. If $Z \in \mathfrak{Z}$ then $Z^* \in \mathfrak{Z}$, $Z(f) \in \mathfrak{Z}$. We show that for any vector Ψ from a dense set in $\mathcal{H}$, any $Z \in \mathfrak{Z}$ and any f such that its Fourier transform $\tilde{f}$ has support in a region Δ of momentum space which excludes the origin, $Z(f)\Psi = 0$. This implies claim (i) of the theorem, because then, for any $\Phi \in \mathcal{H}$ and Ψ in this dense set $\langle \Phi | Z(x) | \Psi \rangle$ is a bounded function whose Fourier transform is a Laurent Schwartz distribution with point support at $p = 0$. Hence it is a constant and we have

$$Z(x) = Z, \tag{III.3.22}$$

which is part (i) of the theorem.

To show that $Z(f)\Psi = 0$ under the stated conditions we note that for central elements $Z^n\Psi = 0$ implies $Z\Psi = 0$ and $Z^*\Psi = 0$ because $Z^{*n}Z^n = (Z^*Z)^n = (ZZ^*)^n$ and Z^*Z is a positive operator. Next, if Δ_0 is some compact subset of V^+ then for any $p \in \mathcal{M}$ except the origin there is a neighborhood Δ such that for sufficiently large n either the region

$$\Delta_0 + n\Delta \equiv \{p = q + p_1 + p_2 + \cdots p_n : q \in \Delta_0,\ p_i \in \Delta \quad \text{for} \quad i = 1, \ldots n\}$$

or the region $\Delta_0 - n\Delta$ is disjoint from V^+. Thus for vectors Ψ with spectral support in Δ_0 either $Z(f)^n\Psi$ or $Z^*(f)^n\Psi$ vanish for large n and $\operatorname{supp} \tilde{f} \subset \Delta$

and therefore $Z(f)\Psi = 0$. Any $\tilde{f}$ with support disjoint from the origin is a sum of functions to which the above argument applies.

To prove part (ii) we observe that with $S \in \mathcal{R}'$ and $A \in \mathcal{R}$

$$\langle\Omega|S(x)A|\Omega\rangle = \langle\Omega|SU(-x)A|\Omega\rangle = \langle\Omega|AS(x)|\Omega\rangle = \langle\Omega|AU(x)S|\Omega\rangle.$$

Due to the spectral properties of $U(x)$ the Fourier transform of any matrix element of $U(x)$ has support in the forward cone. Thus the Fourier transform of the above function has support in the intersection of the forward and the backward cone, hence only at the point $p = 0$ i.e. the function is constant. We can write this

$$\langle\Psi|S(x)|\Omega\rangle = \langle\Psi|S|\Omega\rangle \quad \text{where} \quad \Psi = A^*\Omega.$$

Since Ω is cyclic for $\mathcal{R}$ this means $S(x)\Omega = S\Omega$ and ultimately $S(x) = S$. So S commutes with the translation operators i.e. $U(x) \in (\mathcal{R}')' = \mathcal{R}$. □

Part (i) of the theorem implies that the central decomposition of a translation invariant vacuum state respects translation invariance. It leads to translation invariant primary vacua. Thus there is no breaking of translation invariance possible in a vacuum state.

From theorem 3.2.2 one obtains immediately

Lemma 3.2.5

Let ω be a primary translation invariant state, Ω the corresponding cyclic vector in the GNS-representation π_ω. Then there is no discrete eigenvector of the spatial momentum operators $\mathbf{P}$ in $\mathcal{H}_\omega$ apart from the multiples of Ω.

Proof. From (III.3.19) together with the invariance of ω and the asymptotic commutativity of $\alpha_{\mathbf{x}}A$ with fixed elements $B, C \in \mathfrak{A}_S$ we get

$$\omega(B(\alpha_{\mathbf{x}}A)C) \to \omega(BC)\omega(A).$$

Since Ω is a cyclic vector in the GNS-representation this means that $\pi_\omega(\alpha_{\mathbf{x}}A)$ converges weakly to $\omega(A)\mathbb{1}$ for large space-like translations. Assume Ω' is a discrete eigenvector of $\mathbf{P}$ to eigenvalue $\mathbf{p}$ (possibly 0). Then

$$\langle\Omega'|\pi_\omega(\alpha_{\mathbf{x}}A)|\Omega\rangle = e^{i\mathbf{px}}\langle\Omega'|\pi_\omega(A)|\Omega\rangle.$$

If $\Omega' \neq c\,\Omega$ we may take Ω' orthogonal to Ω. Then the left hand side vanishes for all A in the limit $|\,\mathbf{x}\,| \to \infty$, the right hand side does not since Ω is cyclic.□

From part (ii) of theorem 3.2.4 and lemma 3.2.5 we get

Theorem 3.2.6

A primary vacuum state is pure.

Proof. Any operator from the commutant of $\pi_\omega(\mathfrak{A})$ will transform Ω into another eigenvector of P_μ to eigenvalue zero due to (ii) of theorem 3.2.4. Thus by lemma 3.2.5 the commutant must transform the ray of Ω into itself. The cyclicity of Ω implies then that the commutant can only contain multiples of the unit operator. □

This leaves the question as to whether Lorentz symmetry or internal symmetries may be spontaneously broken in vacuum states. For $\Lambda \in \mathfrak{L}$ the group relations give

$$\alpha_\Lambda \alpha_x = \alpha_{x'} \alpha_\Lambda; \quad x' = \Lambda x. \tag{III.3.23}$$

For $g \in \mathcal{G}_{\text{int}}$, if the translations commute with the internal symmetries,[5]

$$\alpha_g \alpha_x = \alpha_x \alpha_g. \tag{III.3.24}$$

We note that spontaneous symmetry breaking in a vacuum sector is synonymous with non-invariance of the vacuum state. If the vacuum is invariant then the automorphism is implementable (equ. (III.3.14)). If the vacuum is not invariant the automorphism cannot be implemented according to lemma 3.2.5 together with (III.3.23), (III.3.24).

Consider now the continuous part of $\mathcal{G}$. An element of the Lie algebra generates a 1-parameter subgroup. The corresponding automorphisms will be denoted by α_λ. $\lambda \in \mathbb{R}$ is the parameter, and $\lambda = 0$ corresponds to the identity. Assume that there is a dense set $\mathfrak{D}(O) \subset \mathfrak{A}(O)$ (norm dense in $\mathfrak{A}_S(\mathcal{O})$, weakly dense in $\mathcal{R}(\mathcal{O})$) for whose elements $\alpha_\lambda A$ are smooth. Then

$$\left.\frac{d}{d\lambda}\alpha_\lambda A\right|_{\lambda=0} = \delta A, \quad A \in \mathfrak{D} \equiv \cup \mathfrak{D}(\mathcal{O}) \tag{III.3.25}$$

defines a derivation δ on the dense set $\mathfrak{D}$. Translation invariance of the vacuum and relations (III.3.23) or (III.3.24) give

$$\omega_0(\delta \alpha_x A) = \omega_0(\delta A). \tag{III.3.26}$$

The condition for implementability is

$$\omega_0(\delta A) = 0, \quad A \in \mathfrak{D} \tag{III.3.27}$$

It has been recognized that spontaneous symmetry breaking is usually connected with the existence of zero mass particles ("Goldstone's theorem") [Gold 61]. Here we give the following variant of this.

[5] This is the "normal situation". An action of the translations on the internal symmetry group of the observables can be ruled out in a massive theory if there is sufficient interaction. See e.g. Coleman and Mandula [Col 67]. In a free field theory any Weyl operator $W(f) = e^{i\phi(f)}$ gives an internal symmetry for a free bose field ϕ since

$$\alpha_f A \equiv W(f) A W^*(f)$$

defines an automorphism transforming each $\mathfrak{A}(\mathcal{O})$ into itself because

$$\alpha_f \phi(g) = \phi(g) + c(f,g)\mathbb{1}.$$

Local gauge transformations and supersymmetries give other examples of an action by translations on internal symmetries.

Theorem 3.2.7
If the vacuum is separated by an energy gap $m \neq 0$ from the other states in its folium and if there is a uniform bound

$$| \omega_0(\delta A) | \leq \Phi(R) (\| A\Omega \| + \| A^*\Omega \|), \quad A \in \mathfrak{D}(\mathcal{O}_R), \tag{III.3.28}$$

within the diamond $\mathcal{O}_R$ with base radius R centered at the origin such that

$$\lim R^{-n} \Phi(R) \to 0 \quad \text{as } R \to \infty \quad \text{for some} \quad n > 0, \tag{III.3.29}$$

then one has (III.3.27) i.e. the symmetry is unbroken.

If the symmetry is related to a conserved current then the property (III.3.28), (III.3.29) follows. In that case one has, due to locality

$$\delta A = i[Q_R, A] \quad \text{for} \quad A \in \mathfrak{D}(\mathcal{O}_R) \tag{III.3.30}$$

with

$$Q_R = \int \varrho(y) f(y) d^4y \tag{III.3.31}$$

where $\varrho(x)$ denotes the zero component of the current density, $f(y) = 1$ in a small time slice covering the base of $\mathcal{O}_R$ and $f(y) = 0$ outside of some slightly larger region. This yields

$$| \omega_0(\delta A) | \leq \| \Omega_R \| (\| A\Omega \| + \| A^*\Omega \|); \quad A \in \mathfrak{D}(\mathcal{O}_R) \tag{III.3.32}$$

with $\Omega_R = Q_R\Omega$,

$$\| \Omega_R \|^2 = \int \langle \Omega | \varrho(x) \varrho(y) | \Omega \rangle \overline{f}(x) f(y) \, d^4x d^4y.$$

We may choose $\langle \Omega | \varrho(x) | \Omega \rangle = 0$. The 2-point function decreases fast for increasing space-like separation of the points if the theory has a lowest mass $m > 0$ so that one obtains

$$\| \Omega_R \| \leq c R^{3/2} \quad \text{for large} \quad R. \tag{III.3.33}$$

Thus one has (III.3.28), (III.3.29) for any $n > 3/2$. This observation was the starting point of the discussion of the Goldstone theorem by Kastler, Robinson and Swieca [Kast 66].

To prove the theorem we can now proceed as follows. Specializing in (III.3.26) α_x to time translations α_t, multiplying with a smooth function $g(t)$ and integrating we get

$$\omega_0(\delta A) \int g(t) dt = \omega_0(\delta A(g)) \tag{III.3.34}$$

with

$$A(g) = \int g(t) \, \alpha_t A \, dt. \tag{III.3.35}$$

One has

$$A(g)\Omega = \int g(t)e^{iHt}A\Omega\, dt = \tilde{g}(H)A\Omega, \tag{III.3.36}$$

where $\tilde{g}$ is the Fourier transform of g and H the Hamiltonian. Since $\delta\mathbb{1} = 0$ we can take A so that

$$\omega_0(A) = 0; \quad A \in \mathfrak{D}(\mathcal{O}_R) \tag{III.3.37}$$

Then, if the spectrum of the Hamiltonian in the subspace orthogonal to the vacuum is confined to values $\varepsilon \geq m$ we have

$$\| A(g)\Omega \| \leq \sup_{\varepsilon \geq m} | \tilde{g}(\varepsilon) | \, \| A\Omega \| \, . \tag{III.3.38}$$

We can use now the freedom in the choice of g. Setting $g(t) = T^{-1}\, h(t/T)$ and denoting the set of smooth functions $h(\tau)$ with support in the interval $[-1,+1]$ and $\tilde{h}(0) = 1$ by $\mathcal{F}$ we get

$$\| A(g)\Omega \| \leq F(mT) \, \| A\Omega \|, \tag{III.3.39}$$

where

$$F(\xi) = \inf_{h\in\mathcal{F}} \sup_{\omega\geq\xi} | \tilde{h}(\omega) | \, . \tag{III.3.40}$$

F is a universal function which decreases exponentially as shown in [Buch 92]. Analogously one gets a bound for $\|A(g)^*\Omega\|$. Applying the estimate (III.3.28) to the right hand side of (III.3.34) one obtains

$$| \omega_0(\delta A) | \leq c.\Phi(R+T)\, F(mT).$$

Letting $T \to \infty$ one obtains (III.3.27).

One concludes that if the generator of a 1-parameter group of symmetries satisfies condition (III.3.28), (III.3.29) and if there is a mass gap then the symmetry is implementable (unbroken) in the vacuum sector. The conclusion may be strengthened to the statement that spontaneous symmetry breaking is only possible if the mass spectrum contains zero as a discrete eigenvalue. A discussion using conserved currents is given in [Ez 67].[6]

Since we know that there exist photons and neutrinos the P^μ-spectrum has no gap; so, strictly speaking, we can conclude nothing from the above theorem in elementary particle physics. In particular, the masslessness of the photon could allow a spontaneous breaking of Lorentz invariance even in the vacuum sector so that there might be no Lorentz invariant (pure) vacuum state. If one considers an idealized theory of strong interactions only in which one assumes a mass gap then it follows that the vacuum is automatically Lorentz invariant and invariant under the continuous part of the internal symmetry group of the observables. The latter does not include the gauge groups since these do not act on the observables. But it does include the chiral symmetry expressed by an (algebraic) conservation law of an axial vector current. The idea that one has

[6]Precise conditions for the occurrence of spontaneous symmetry breaking with or without the existence of massless particles have been given by Buchholz, Doplicher, Longo and Roberts [Buch 92].

such a symmetry in the idealized theory and that it is spontaneously broken implies then that there must be a massless particle in this model, notably the pion. Since the pion mass is relatively small compared to other hadronic masses this picture appears reasonable and has led to the idea of a "partially conserved axial vector current" in the full theory.

In the many body problem the Goldstone theorem does play a relevant rôle (breaking of Galilei invariance, existence of phonons; see [Swieca 67]). In the presence of long range forces there is, however, another mechanism for spontaneous symmetry breaking. The Bardeen-Cooper-Schrieffer model of superconductivity and the Coulomb forces in a plasma provide the prime examples. Within the algebraic setting the former is discussed in [Haag 62b], the general situation by Morchio and Strocchi in [Morch 85, 87]. The idea that in relativistic field theory local gauge invariance provides an analogon to long range forces (irrespective of the presence or absence of a mass gap) and that the spontaneous breaking of gauge invariance may in fact, be a mechanism for generating an energy gap, parallel to the cases of the BCS-model and the plasma, was forwarded by Nambu and Higgs. It led to the Higgs-Kibble mechanism which became a central theme in the development of the standard model in elementary particle physics. We leave this important topic aside here because of the gaps remaining in our understanding of the principle of local gauge invariance in quantum physics (compare, however, the localization properties of charges in the BF-analysis described in Chapter IV, section 3).

III.3.3 Summary of the Structure

To conclude this section let us summarize the essential structure of the theory. We reserve here the symbol $\mathcal{O}$ for a finite, contractible, open region in Minkowski space. The discussion above suggests that the following structures are relevant:

(i) A net of C*-algebras with common unit

$$\mathcal{O} \rightarrow \mathfrak{A}_S(\mathcal{O}), \tag{III.3.41}$$

with the total C*-algebra (inductive limit)

$$\mathfrak{A}_S = \overline{\cup_{\mathcal{O}} \mathfrak{A}_S(\mathcal{O})}. \tag{III.3.42}$$

(The bar denotes the completion in the norm topology).

The action of α_x on $\mathfrak{A}_S$ is continuous in the norm topology.

(ii) A set $\mathfrak{S}$ of physical states over $\mathfrak{A}_S$ and the complex linear span of $\mathfrak{S}$, denoted by Σ.

The restrictions of $\mathfrak{S}$ and Σ to the local algebras $\mathfrak{A}_S(\mathcal{O})$ yield the presheafs $\{\mathfrak{S}(\mathcal{O})\}$, $\{\Sigma(\mathcal{O})\}$ of partial states or linear forms respectively.

(iii) The dual of the presheaf Σ is a net of W*-algebras with common unit

$$\mathcal{O} \to \mathcal{R}(\mathcal{O}) = \Sigma(\mathcal{O})^*. \tag{III.3.43}$$

$\mathcal{R}(\mathcal{O})$ is closed in the weak topology induced by Σ and $\mathfrak{A}_S(\mathcal{O})$ is weakly dense in $\mathcal{R}(\mathcal{O})$.

(iv) The symmetry group $\mathcal{G}$. Its elements have a geometric part, point transformations of space-time conserving the causal structure. They are realized by automorphisms of the net satisfying (III.1.7) where $\mathfrak{A}(\mathcal{O})$ may be taken as either $\mathcal{R}(\mathcal{O})$ or $\mathfrak{A}_S(\mathcal{O})$. The action of the continuous part of $\mathcal{G}$ on $\mathfrak{A}_S$ is assumed to be continuous in the norm topology.

Remark. The local normality assumption asserts that for a finite, contractible region $\mathcal{O}$ we have to deal with only one quasi-equivalence class of representations of $\mathfrak{A}_S(\mathcal{O})$. Further, that the restriction of any state $\omega \in \mathfrak{S}$ to $\mathfrak{A}_S(\mathcal{O})$ is primary and the GNS-construction leads to a von Neumann factor $\pi_\omega(\mathfrak{A}_S(\mathcal{O}))''$ which is isomorphic to $\mathcal{R}(\mathcal{O})$. On the other hand $\mathfrak{S}$ will still contain many disjoint folia of states over the total algebra $\mathfrak{A}_S$. The von Neumann algebras $\pi_\omega(\cup\mathfrak{A}_S(\mathcal{O}))''$ will not be isomorphic for all states. We call each primary folium of states over $\mathfrak{A}_S$ a *sector* of the theory. For the analysis of the sector structure it is not relevant whether we start from the net $\mathfrak{A}_S(\mathcal{O})$ or from $\mathcal{R}(\mathcal{O})$ but the total algebra must be taken as the C*-inductive limit of the local algebras conforming with the idea that we want to include only quasilocal elements. Examples of different folia of physical states are provided by hydrodynamic-thermodynamic states characterized by non vanishing matter and energy distribution at infinity (chapter V) and by the charge superselection sectors in elementary particle physics (chapter IV).

The stability requirement, formulated in II.1.2 as axiom **A2**, **A3**, may be expressed in algebraic form as the requirement that the left ideal $\mathfrak{J}(V^{+c})$ defined in (III.3.21) is a proper ideal. The causality requirement, contained in axiom **E** of II.1.2, is, of course, expressed by (III.1.4), where $\mathcal{A}_{\text{obs}}$ may be taken as either $\mathfrak{A}_S(\mathcal{O})$ or $\mathcal{R}(\mathcal{O})$. The remnant of axiom **B**, the additivity property of $\mathcal{A}$ suggested in (III.1.2), cannot be expected to hold in $\mathfrak{A}_S$ but will require $\mathcal{R}$ (see next section). The same will very probably be true for the dynamical law. We expect that it cannot be formulated within $\mathfrak{A}_S$ but requires also the states $\mathfrak{S}$ or, alternatively speaking, the net $\mathcal{R}$. So we should replace (III.1.10) by

$$\mathcal{R}(\hat{\mathcal{O}}) = \mathcal{R}(\mathcal{O}). \tag{III.3.44}$$

This feature appears also satisfactory in view of general relativity where even the metric, which is essential for the dynamical equations, depends on the state.

We have seen that the properties listed imply the existence of at least one pure, translation invariant ground state ω_0. In the GNS-representation induced by ω_0 the spectral condition **A3** is satisfied. If the theory has a lowest mass $m > 0$ then ω_0 is invariant under the full continuous part of $\mathcal{G}$. In particular

ω_0 is Poincaré invariant, axiom **A1** will apply and one can expect that the vacuum state is unique. If there is no mass gap then the Lorentz invariance of the vacuum is not guaranteed on general grounds.

III.4 The Vacuum Sector

We discuss here some further properties which the von Neumann algebras in the vacuum sector possess (or might possess), considering also algebras associated with infinitely extended or not contractible regions.

III.4.1 The Lattice of Causally Complete Regions

Definition 4.1.1
Let M be any set of points in Minkowski space. The *causal complement* of M is the set of all points which lie space-like to all points of M. It will be denoted by M'.

$(M')' \equiv M''$ is called the *causal completion* of M. M is called causally complete if $M'' = M$.

Proposition 4.1.2
(i) M' is always causally complete. One has

$$M' = (M')'' = (M'')' \equiv M'''. \tag{III.4.1}$$

(ii) The intersection of two causally complete regions is causally complete. If $M_k = M_k''$, $(k = 1,\ 2)$ then

$$M_1 \cap M_2 = (M_1 \cap M_2)'' \equiv M_1 \wedge M_2. \tag{III.4.2}$$

The symbol $\wedge$ is used to indicate that the set is the largest causally complete set contained in both M_1 and M_2.

(iii) There exists a lowest bound, denoted by $M_1 \vee M_2$, of the causally complete sets containing both M_1 and M_2. It is given by

$$M_1 \vee M_2 = (M_1 \cup M_2)'' = (M_1' \wedge M_2')'. \tag{III.4.3}$$

In short: The set of causally complete regions is an orthocomplemented lattice. The smallest element is $\emptyset$ (the empty set), the largest element is $\mathbb{R}^4$. We denote this lattice by $\mathcal{K}$.

Proof. (i). $(M')'' = ((M')')' = (M'')'$. Since causal completion cannot diminish a set one has $(M')'' \supset M'$ and $M'' \supset M$ which yields $(M'')' \subset M'$. Thus $M''' = M'$.

(ii) and (iii). The set of all points lying space-like to both M_1 and M_2 may be characterized either as $M_1' \cap M_2'$ or as $(M_1 \cup M_2)'$. Thus

$$M_1' \cap M_2' = (M_1 \cup M_2)'. \tag{III.4.4}$$

For causally complete sets we may put $M_k' = N_k$, $M_k = N_k'$ and obtain

$$N_1 \cap N_2 = (N_1' \cup N_2')'.$$

This shows, according to (i) that $N_1 \cap N_2$ is causally complete. Taking the causal complement one gets

$$(M_1 \cup M_2)'' = (M_1' \cap M_2')'.$$

Clearly this is the smallest causally complete set containing both M_1 and M_2.□

III.4.2 The von Neumann Rings in the Vacuum Sector

We consider now the von Neumann algebras resulting by the GNS-construction from the vacuum state ω_0. As long as we believe in the local normality assumption for all finite, contractible regions the von Neumann rings $\pi_0(\mathcal{R}(\mathcal{O}))$ and the W*-algebras $\mathcal{R}(\mathcal{O})$ will be algebraically and topologically isomorphic. So we can and shall omit the symbol π_0 and take the von Neumann rings in the vacuum sector as defining the net $\mathcal{R}$. Only for an infinitely extended or topologically non trivial region $\mathcal{U}$ the limitation to the vacuum sector becomes relevant. To such a region we associate the von Neumann algebra

$$\mathcal{R}(\mathcal{U}) = (\cup_{\mathcal{O}\subset\mathcal{U}} \mathcal{R}(\mathcal{O}))'', \tag{III.4.5}$$

where the union is taken over all finite, contractible regions contained in $\mathcal{U}$. Note that the double commutant refers to the vacuum representation. So this object is not universal for all sectors.

We may confine attention to causally complete regions (see (III.3.44)). One notes the parallelism between the lattice structure of $\mathcal{K}$ described in the previous subsection and that of systems of von Neumann rings where the commutant gives an orthocomplement in $\mathcal{K}$. The causality principle relates the orthocomplementation in $\mathcal{K}$ with that in the set of von Neumann rings. This is, in fact, the reason why we used a prime to denote the orthocomplement in $\mathcal{K}$. This parallelism makes it tempting to assume that in the vacuum sector, ignoring now the distinction between finite and infinite regions (as well as distinctions between other classes of regions), the structural properties of the net can be strengthened to the

Tentative Postulate 4.2.1
The vacuum sector of the theory is described by a homomorphism from the orthocomplemented lattice $\mathcal{K}$ of causally complete regions of Minkowski space into the lattice of von Neumann rings on a Hilbert space $\mathcal{H}$.

This summarizes and extends the previously mentioned properties of the net $\mathcal{R}$. Specifically then

$$K \in \mathcal{K} \to \mathcal{R}(K) \subset \mathfrak{B}(\mathcal{H}), \tag{III.4.6}$$

$$\mathcal{R}(K_1 \wedge K_2) = \mathcal{R}(K_1) \wedge \mathcal{R}(K_2), \tag{III.4.7}$$

$$\mathcal{R}(K_1 \vee K_2) = \mathcal{R}(K_1) \vee \mathcal{R}(K_2), \tag{III.4.8}$$

$$\mathcal{R}(K') = \mathcal{R}(K)', \tag{III.4.9}$$

$$\mathcal{R}(\emptyset) = \{\mathbb{1}\}. \tag{III.4.10}$$

It must be realized that postulate 4.2.1 is a considerable extrapolation from physically motivated assumptions. It is suggested by mathematical simplicity, motivated by the desire to impose the strongest restrictions which are not in conflict with known facts. It is not used in this sharp form in the previous literature nor in the subsequent chapters of this book. If one wants to be more cautious one can start from considerably weaker requirements. Thus, in the analysis of superselection sectors in chapter IV we shall assume the *duality relation* (III.4.9) only for particularly simple regions, the diamonds (double cones)

This duality relation extends the causality requirement which demands that

$$\mathcal{R}(K') \subset \mathcal{R}(K)'. \tag{III.4.11}$$

If this is satisfied we may ask whether we can add further elements to $\mathcal{R}(K)$ so that causal commutativity is still respected in the resulting augmented net. If this is possible the left hand side in (III.4.11) increases, the right hand side decreases. Duality means that one can make all local algebras maximal. (III.4.9) was suggested on such grounds by Haag and Schroer, [Haag 62a]. In the field theoretic setting it is supported by Borchers' argument on "transitivity of locality" (see II.5) and, more generally, by the work of Bisognano and Wichmann [Bis 75, 76]. Araki proved the duality relation for von Neumann rings associated with free scalar fields [Ara 63b, 64a]. However most of the supporting evidence for (III.4.9) relates to simple regions.

The intersection property (III.4.7) has been previously discussed for disjoint diamonds where $K_1 \wedge K_2$ is empty. In this restricted form it has been called *extended locality* and was shown to be a consequence of the standard requirements by L.J. Landau [Land 69].

The strength of the combination of (III.4.6) through (III.4.10) can be illustrated in the following examples. They show that postulate 4.2.1 is violated in some models constructed from free fields but that it may hold if an interaction satisfying the local gauge principle is present.

Take the free Dirac field ψ and consider the current density $j_\mu(x)$ as the observable field by which the algebras $\mathcal{R}(\mathcal{O})$ are generated.[1] Then take

[1] Explicitly, start from the polynomial algebras of the smeared out currents in the vacuum sector and go over to bounded functions of them e.g. by the polar decomposition of the unbounded operators.

$K = K_1 \vee K_2$ where K_k $(k = 1, 2)$ are space-like separated diamonds. K is disconnected. Each $\mathcal{R}(K_k)$ satisfies the duality relation (III.4.9). This was proved in [Dop 69a]. But if $\mathcal{R}(K)$ is defined by the additivity relation (III.4.8) then $\mathcal{R}(K)$ does not satisfy duality for the simple reason that objects like $\psi^*(f)\psi(g)$, where the support of f is in K_1 and the support of g in K_2, commute with all elements of $\mathcal{R}(K')$ and are not in $\mathcal{R}(K_1) \vee \mathcal{R}(K_2)$. One cannot construct $\psi(x)$ from the currents in a neighborhood. Thus we have duality only for connected regions in this model. But worse than that. Take K_1, K_2 as two concentric diamonds, the causal completions of the 3-dimensional balls $|\mathbf{x}| < r_k$ at $x_0 = 0$, $r_1 < r_2$. The quantity $Q = \int j_0(\mathbf{x}) f(\mathbf{x}) d^3x$ with smooth f and

$$f(\mathbf{x}) = \begin{cases} 1 & \text{for} \quad |\mathbf{x}| < r_1 \\ 0 & \text{for} \quad |\mathbf{x}| > r_2 \end{cases}$$

(or more precisely an expression like (III.3.31) is the approximate charge in the smaller region. It commutes with $\mathcal{R}(K_1)$ due to the global gauge invariance of the currents and with $\mathcal{R}(K_2')$ due to locality. By (III.4.9) it should be contained in $\mathcal{R}(K_1') \cap \mathcal{R}(K_2)$, and by (III.4.7) in $\mathcal{R}(K_1' \cap K_2)$, the causal completion of the spherical shell. This is not the case in the free Dirac theory. The situation is improved, however, in full quantum electrodynamics because Gauss' law allows the conversion of Q into an integral over the electric field strength in the region of the spherical shell. This observation was presented as an argument in favour of the validity and strength of the unrestricted duality relation [Haag 63]. Postulate 4.2.1 excludes the free Dirac theory and requires the currents to be accompanied by another field which allows the determination of the charge inside a region by measurements in a surface layer. Indeed, also the failure of (III.4.9) for disconnected regions in the free Dirac theory is remedied in full electrodynamics because there $\psi^*(f)\psi(g)$ is not in the algebra of observables; it is not invariant under local gauge transformations. Instead of $\psi^*(x)\psi(y)$ we must now take the gauge invariant objects (see I.5.34)

$$\varrho(x, y) = \psi^*(x) \exp ie \int_x^y A_\mu(x') dx'^\mu \psi(y) \tag{III.4.12}$$

as (improper) elements of the observable algebra. A last example is provided by the Maxwell field. The relation $\mathrm{div}\mathbf{B} = 0$ implies that the magnetic flux through two 2-dimensional surfaces having the same 1-dimensional boundary (for instance a circle) is the same. We may take small neighborhoods in $\mathbb{R}^4$ of the surfaces and call their causal completions K_1 resp. K_2. Then one can find an observable (essentially the magnetic flux through either one of the two surfaces) which belongs to $\mathcal{R}(K_1)$ and to $\mathcal{R}(K_2)$, hence by (III.4.7) to $\mathcal{R}(K_1 \cap K_2)$ which is a neighborhood of the common boundary of the two surfaces. In free Maxwell theory this is not possible if we require the additivity property (III.4.8). Ignoring the small 4-dimensional extension of the regions considered, which was introduced only to guarantee that the integrals give well defined observables, the flux is given by the line integral of the vector potential around the (common) boundary. But the vector potential is not an observable while its line integral around a closed curve is measurable namely as the magnetic flux. This latter cannot be constructed from the field strengths in the neighborhood of the boundary curve. In full electrodynamics we have, however, besides the field strengths the quantities (III.4.12) and the currents generating the observable algebra and one may hope to construct from them the line integral of A_μ around a closed curve using only a neighborhood of this curve. Unfortunately a verification of this hope with any degree of rigour is difficult due to the very singular character of the quantities involved. So this remark cannot be taken very seriously.

The discussion in the last paragraph is intended to show that postulate 4.2.1 may be reasonable in physically interesting theories and that it is very strong. It excludes free theories with non trivial charge structure while it *possibly* admits full electrodynamics. The postulate looks mathematically canonical so that it might be worth while to investigate its consequences for a mathematical classification of theories. Still, since such a study is lacking we shall subsequently adopt the cautious point of view and use the duality relation only for diamonds, the intersection property only for disjoint elements of $\mathcal{K}$ ("extended locality"). An immediate consequence of postulate 4.2.1 is that $\mathcal{R}(K)$ is a factor for any $K \in \mathcal{K}$. If we adopt the cautious version then we shall only assume that $\mathcal{R}(K)$ is a factor if K is a diamond.

IV. Charges, Global Gauge Groups and Exchange Symmetry

IV.1 Charge Superselection Sectors

By focusing attention on observables and algebras we have kept only those parts of the axioms of chapter II which have a direct, immediately clear physical significance. Other parts, reflecting traditional field theoretic formalism, such as the assumption of a Bose-Fermi alternative, or the founding of the theory on fields transforming according to finite dimensional representations of the Lorentz group do not appear. The understanding of internal symmetries is also modified. In the development of elementary particle theory during the past two decades gauge symmetries have acquired a more and more prominent position and it may well be that all internal symmetries should properly be regarded as gauge symmetries, formulated in the field theoretic frame by transformations of unobservable fields leaving all observables unchanged.[1] Conversely, if in the algebraic approach we take the extreme position and claim that the net of observable algebras defines the theory completely without need for any additional specification, then we cannot allow any internal symmetry of the observable net in the sense of (III.3.15) with $\alpha_g \neq \text{id}$. So this would also lead to the conclusion that internal symmetries should always be gauge symmetries. The most salient observable consequence of internal symmetries is the set of charge quantum numbers which serves to distinguish different species of particles and to characterize their properties. In the case of gauge symmetries these quantum numbers manifest themselves through the existence of superselection rules for the states over the observable algebra. The charge structure with its composition laws as well as the exchange symmetry ("statistics") are encoded in the structure of the net of algebras of observables. By studying how they are encoded we get a natural approach to these questions, unbiased by traditional formalism.

Strange Statistics. The question as to whether every particle must be either a boson or a fermion and the wish for a deeper understanding of the Pauli principle have been a recurrent theme of discussion on various levels of the

[1] Of course this need not apply in models in which approximate symmetries are idealized as exact ones.

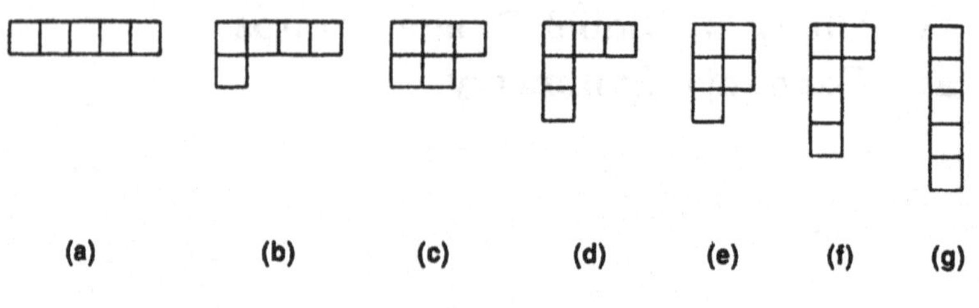

Fig. IV.1.1.

theory. One generalization of the Bose-Fermi alternative was suggested by H.S. Green [Green 53]. He pointed out that certain trilinear commutation relations of creation-annihilation operators are in harmony with the equation of motion and transformation properties of free fields. He showed then that this algebra can be decomposed into disjoint components, "the parastatistics of order p" (p a natural number) and for each p one has a bosonic and a fermionic version. As $p \to \infty$ the two versions converge towards "Boltzmann statistics". Physical consequences of the scheme have been studied by several authors, e.g. Volkov [Volk 60], Greenberg and Messiah [Greenb 64, 65]. The contents can be most simply described in the wave mechanical setting where states of n identical particles are characterized by wave functions $\psi(\xi_1, \ldots, \xi_n)$; here ξ stands for position and spin orientation. In the Hilbert space spanned by such wave functions one has a unitary representation of the permutation group S_n resulting from permutations of the arguments in the wave function. This representation may be decomposed into irreducible parts. An irreducible representation of S_n is labeled by a Young tableau, an array of boxes in rows and columns of non-increasing length as pictured for the case $n = 5$ in the figure (IV.1.1).[2] The length of a row indicates the degree of symmetrizability, the length of a column the degree of antisymmetrizability. Thus a 5-particle wave function of symmetry type (c) can be symmetrized in 3 arguments and then again in the remaining 2. Alternatively it can be antisymmetrized in 2 arguments and then again in two others but not more. If ψ is an n-particle wave function of para-bosons of order p then, for arbitrary n, precisely all such symmetry types are admitted which have not more than p rows (i.e. for which ψ cannot be antisymmetrized in more than p variables). Correspondingly, for para-fermions of order p the number of columns is limited to at most p.

A permutation of the order of arguments in a wave function of indistinguishable particles leads to the same physical state. Since for $p > 1$ we encounter representations of S_n which are more than 1-dimensional this implies that pure states are now no longer described by rays in Hilbert space but by subspaces with the dimensionality of the representation of S_n belonging to the Young tableau. The specific limitations for the allowed Young tableaux in parastatis-

[2] See any text book on group representations.

tics can be traced to the requirement that if some particles are very far away then (asymptotically) the others can be regarded as an independent system [Stolt 70], [Lands 67].

One may notice that the situation is analogous to the one encountered in the treatment of the fine structure of atomic spectra by F. Hund. There, building up wave functions from products of position space wave functions and spin functions, the two factors need not be totally antisymmetric. In order to obtain an antisymmetric total wave function the Young tableau of the spatial wave function must be the mirror image of the tableau for the spin wave function. The Young tableau of a spin wave function can have at most two rows because the electron spin has only two orthogonal states. The tableau of the spatial wave function can therefore have, at most, two columns. Thus, if the spin were unobservable, one could say that the electron is a para-fermion of order 2. This suggests that parastatistics of order p can be replaced by ordinary Bose or Fermi statistics, respectively, if one introduces a hidden degree of freedom ("generalized isospin") with p possible orientations or, alternatively speaking, if one has a non-Abelian unbroken global gauge group (isospin group). This has been elaborated by Ohnuki and Kamefuchi [Ohnu 68, 69] and by Drühl, Haag and Roberts [Drühl 70].

Other types of exotic statistics are encountered in models on lower dimensional position space. In these the exchange symmetry cannot be described in terms of representations of the permutation group. Thus the full intrinsic significance of exchange symmetry is not seen if one starts with multiparticle wave functions, claiming that a permutation of the order of arguments leads to the same state, nor if one bases the discussion on commutation relations between unobservable quantities.

We shall show in the next sections how exchange symmetry arises from the causality principle for observables and describe the possible types of exchange symmetry which ensue.

Charges. Superselection rules arise because the abstract net has inequivalent Hilbert space representations. But only a small subset of these representations will concern us in this chapter, namely those whose states have vanishing matter density at space-like infinity. These are the states of direct interest to elementary particle physics, corresponding to the idealization that we have empty space far away.[3] The natural formulation of this idealization would be to say that a state ω is relevant to elementary particle physics if

$$\omega(\alpha_x B) \to 0 \tag{IV.1.1}$$

as x goes to space-like infinity when B is any element of the Doplicher ideal $\mathfrak{J}(V^{+c})$ (see section III. Def.3.2.3). The analysis of superselection structure

[3] Other representations, corresponding to thermal equilibrium (without boundaries) will be discussed in chapter V.

within the set of states restricted only by (IV.1.1) has not been carried through. Borchers proposed instead to study the following set of representations.

Borchers' Selection Criterion (Positive Energy). We consider all representations of the observable algebra in which the translation symmetry can be implemented and the P^μ-spectrum lies in $\overline{V}^+$.

The relation of this criterion to the idealization of absence of matter at infinity is qualitatively seen in the following way. If the operators P_μ can be defined at all in a representation of $\mathfrak{A}$ then, starting from a state with finite energy we may lower the energy by taking matter away - i.e. by acting on the state vectors with quasilocal elements from $\mathfrak{J}(V^-)$. If there is an infinite reservoir of matter extending to spatial infinity then this procedure can be repeated indefinitely often and the energy can have no lower bound. The first systematic attempt to analyse the superselection structure was made by Borchers [Borch 65a, 65b] and based on this criterion. Unfortunately it contained some errors which were hard to trace but led to conclusions to which counterexamples could be constructed.[4] An analysis based on Borchers' criterion without any further restrictive assumptions is burdened by complications arising from long range forces and infrared clouds when the mass spectrum has no gap. Therefore the first objective was to understand the situation in massive theories. With this regime in view Doplicher, Haag and Roberts started from the following criterion.[5]

DHR Selection Criterion. The representations considered are those which, in restriction to the causal complement of any diamond of sufficiently large diameter, become unitarily equivalent to the vacuum representation. In symbols

$$\pi\mid_{\mathfrak{A}(\mathcal{O}')} \cong \pi_0\mid_{\mathfrak{A}(\mathcal{O}')}, \tag{IV.1.2}$$

if $\mathcal{O}$ is a sufficiently large diamond.

This requirement is closely related to the common (vacuum-like) appearance with respect to measurements in very distant regions of all the states considered.

Lemma 1.1
Take a directed sequence of diamonds exhausting space-time i.e. $\mathcal{O}_{n+1} \supset \mathcal{O}_n$, $\bigcup \mathcal{O}_n = \mathbb{R}^4$. If ω is a state in the folium of a representation satisfying (IV.1.2) then

$$\lim_{n\to\infty} \left\| (\omega - \omega_0)\mid_{\mathcal{O}'_n} \right\| = 0 . \tag{IV.1.3}$$

Here ω_0 denotes the vacuum, $\varphi\mid_{\mathcal{O}'}$ the restriction of the linear form φ to the subalgebra $\mathfrak{A}(\mathcal{O}')$.

---

[4] This provided, in fact, the motivation for [Dopl 69a].

[5] In the present chapter we refer to this work as DHR 1, 2, 3, 4. This corresponds to [Dopl 69a, 69b, 71, 74] in the author index.

Proof. By (IV.1.2) the folia of π and of π_0 coincide when the states are restricted to an algebra $\mathfrak{A}(\mathcal{O}'_n)$ and n is large enough. Therefore $\varphi = \omega - \omega_0$ is a normal linear form on the von Neumann algebra $\mathcal{R}(\mathcal{O}'_n) = \pi_0(\mathfrak{A}(\mathcal{O}'_n))''$. Consider a sequence $B_n \in \mathcal{R}(\mathcal{O}'_n)$ with $\| B_n \|= 1$. Since the unit ball of a von Neumann algebra is compact in the weak *-topology this sequence has weak limit points in $\mathfrak{B}(\mathcal{H})$ and each such limit B commutes with all observables. Since the vacuum representation is irreducible B is a multiple of the identity and $\varphi(B) = 0$. □

The converse of lemma 1.1 is also true under rather mild assumptions. If the restrictions of π and π_0 to $\mathfrak{A}(\mathcal{O}'_n)$ are primary then (IV.1.3) implies that these restrictions are quasiequivalent for large n. For type III factors quasiequivalence is the same as unitary equivalence. It will be shown in chapter V that the relevant von Neumann algebras are indeed of type III. We can therefore consider (IV.1.3) as an alternative formulation of the selection criterion (IV.1.2) .

Condition (IV.1.3) is a much more stringent restriction than (IV.1.1) because it demands for the states considered a uniform approach of the expectation values for all observables far away to the vacuum expectation values (irrespective of the extension of their support region). Thus it excludes from consideration states with electric charge because, by Gauss' law, the electric charge in a finite region can be measured by the flux of the field strength through a sphere of arbitrarily large radius. This is due to the long range character of electric forces which, in turn, is tied to the vanishing of the photon mass. In a massive theory forces are expected to decrease exponentially with distance so that the charges become shielded. However, a careful analysis by Buchholz and Fredenhagen [Buch 82a] showed that even in a purely massive theory there can exist charges accompanied by correlation effects which are discernable at arbitrarily large distances. They found that if one starts from a sector containing single particle states but not the vacuum in a massive theory then Borchers' criterion is equivalent to a requirement of the form (IV.1.2) but with $\mathcal{O}$ replaced by $\mathcal{C}$, an infinitely extended cone around some arbitrarily chosen space-like direction. It is this modification which allows topological charges. The adjective "topological" is appropriate because a particular state does not determine the direction of the cone which has to be excluded. It is only important that a large sphere surrounding the charge must be punctured somewhere. We shall describe this in section 3 (BF-analysis).

The DHR-criterion excludes such "topological charges" from consideration. It is therefore too narrow. It is instructive to see that this criticism goes deeper and touches some of the pillars of general quantum field theory. The main argument in favour of (IV.1.2) as a reasonable selection criterion in a massive theory was the fact that it is implicitly assumed in the standard treatment of collision processes described in II.4. This, in turn, was based on the axioms in II.1 but continues to work in a more general setting where the Bose-Fermi alternative is eliminated from the assumptions. Specifically, let us consider the frame sketched in III.1. We have a Hilbert space $\mathcal{H}$, a net of operator algebras

$\mathcal{A}(\mathcal{O})$ whose elements are interpreted as representing physical "operations" in the respective region. We may imagine that it is generated by fields, some of which are charge carrying and unobservable. Therefore we call $\mathcal{A}$ the *field net*. We have furthermore a unitary representation of $\overline{\mathfrak{P}}$ transforming the field net according to equ. (III.1.3). There is a ground state (vacuum). Inside $\mathcal{A}(\mathcal{O})$ we have the observable part $\mathfrak{A}(\mathcal{O})$, concretely realized as an operator algebra on $\mathcal{H}$. The set of all unitary operators which commute with all observables and represent an internal symmetry of the field net is called the *global gauge group* $\mathcal{G}$. The observables are assumed to obey causal commutation relations but the commutation relations between unobservable quantities are left open. The standard construction of incoming particle states works if the vacuum expectation values of fields have the cluster property

$$\lim \omega_0 \left(F_1(\alpha_x F_2)F_3\right) = \omega_0(F_1F_3)\ \omega_0(F_2); \quad F_i \in \mathcal{A}_{\text{loc}} \tag{IV.1.4}$$

for x moving to space-like infinity and if the fields commute with the observables at space-like distances. Under these circumstances, as shown in [Dopl 69a], the vector states in $\mathcal{H}$ satisfy the criterion (IV.1.2). The decomposition of $\mathcal{H}$ with respect to the center of $\mathfrak{A}''$ (which is also the center of $\mathcal{G}''$ because $\mathcal{G}'' = \mathfrak{A}'$) gives the decomposition into charge superselection sectors. If the DHR-criterion is too narrow then also the standard collision theory is too narrow. The topological charges encountered in the BF-analysis cannot be created by quasilocal operators of a field net.

The Program and the Results. Having settled for one of the criteria singling out the states of interest the remaining task is clear. Starting with the abstract algebra of observables we have to classify the equivalence classes of its irreducible representations conforming with the criterion. We shall call each of these a *charge superselection sector* and the labels distinguishing sectors *charge quantum numbers*. This terminology is appropriate if the set of sectors is discrete and this will turn out to be the case in massive theories under physically transparent conditions.[6] Still the term charge is used in a wide sense. In specific models it may be baryon number, lepton number, magnitude of generalized isospin ...

I give a brief preview of the results. Two properties of charge quantum numbers follow from the general setting: a composition law and a conjugation. Tied to this is the exchange symmetry of identical charges ("statistics"). If the theory is based on Minkowski space (or higher dimensional space times) the total structure is remarkably simple and not affected by the presence or absence of topological charges. The charge quantum numbers correspond in one-to-one fashion to the labels of (equivalence classes of) irreducible representations of a compact group, *the global gauge group*. The composition law of charges corresponds to the tensor product of representations of this group, charge conjugation to the

[6] In DHR1 it corresponds to the compactness of the global gauge group which was shown there to follow from the absence of infinite degeneracy of particle types with equal mass and the completeness of scattering states (*asymptotic completeness*).

complex conjugate representation. In addition, intrinsically attached to each type of charge, there is a sign, determining whether the charge is of bosonic or fermionic nature. If the global gauge group is Abelian then the charge quantum numbers themselves form a group, the Pontrjagin dual of the gauge group. The exchange symmetry gives ordinary Bose or Fermi statistics. If it is not Abelian the composition of two charges leads to a direct sum of several charge sectors corresponding to the Clebsch-Gordan decomposition of the tensor product of representations of the gauge group. The statistics in the sectors arising from n-fold composition of one type of charge may be described in two ways. One may either say that it is para-Bose (resp. para-Fermi) of order d, where d is the dimension of the irreducible representation of the gauge group in the charge 1-sector. Or one may take an extension of the observable algebra, adding "hidden elements", and consider the non-Abelian part of the gauge group as an internal symmetry group of this enlarged algebra. The statistics reduces then again to the ordinary Bose-Fermi alternative. A simple example illustrating these different descriptions is the Yukawa type theory of pions and nucleons neglecting electromagnetic and weak interactions, the "charge independent theory". There SU(2)×U(1) is the global gauge group, SU(2) corresponding to charge independence, U(1) to baryon number conservation. One is unable to distinguish a neutron from a proton or the different electric charges of a π-meson. Instead one has a nucleon, belonging to isospin 1/2 as a para-fermion of order 2 and a meson, belonging to isospin 1, as a para-boson of order 3. Enlarging the observable algebra we come to the usual description with two distinguishable types of nucleons, three types of mesons, which are ordinary fermions and bosons respectively. If we do this without giving up charge independence then the SU(2)-group appears as an internal symmetry group of the (extended) observable algebra and only the U(1) part remains as a gauge group. Practically, of course, the very fact that states with different electric charge can be experimentally distinguished means that the SU(2) symmetry is lost.

In recent years there has been growing interest in models based on 2- or 3-dimensional space times. The motivation has come from various sources: the existence of solvable or partly solvable models in lower dimensions, the information about non-relativistic statistical mechanics at critical points which may be drawn from conformal invariant field theories in 2-dimensional space-time and the fact that such a model serves as a point of departure for theories with strings and superstrings. It is therefore noteworthy that the discussion of the superselection structure and the exchange symmetry is less simple in low dimensions. The permutation group is replaced by the braid group, the gauge group by a more complicated structure ("quantum group" or worse). In the presence of topological charges the richer structure appears already in $(2+1)$-dimensional space-time, for localizable charges in $(1+1)$-dimensional space-time. We shall discuss this in section 5 of the present chapter.

IV.2 The DHR-Analysis

Although, as mentioned, the DHR-criterion is too restrictive it is worthwhile to present the DHR-analysis. It illustrates the essential features. Modifications will be discussed in sections 3 and 5 of this chapter and, concerning infrared problems, in Chapter VI.

In this section the symbol $\mathcal{O}$ shall always denote a diamond (as the prototype of a finite, causally complete region) and $\mathcal{O}'$ its causal complement. We write Rep_L for the set of representations of the observable algebra conforming with (IV.1.2). The index L stands for "localizable charges". $\mathcal{R}(\mathcal{O}) = \pi_0\left(\mathfrak{A}(\mathcal{O})''\right), \mathcal{R}(\mathcal{O}') = \pi_0\left(\mathfrak{A}(\mathcal{O}')\right)''$ denote the von Neumann algebras in the vacuum representation. The duality relation

$$\mathcal{R}(\mathcal{O}')' = \mathcal{R}(\mathcal{O}) \tag{IV.2.1}$$

will play an important rôle. All representations from Rep_L are *locally equivalent* i.e.

$$\pi\left(\mathfrak{A}(\mathcal{O})\right) \cong \pi_0\left(\mathfrak{A}(\mathcal{O})\right); \quad \pi \in Rep_L\ . \tag{IV.2.2}$$

We meet, in restriction to finite regions, only one equivalence class of representations, the normal representations of the von Neumann algebras $\mathcal{R}(\mathcal{O})$. The same holds, by (IV.1.2), for the infinite regions $\mathcal{O}'$ at least if $\mathcal{O}$ is large enough. So the charge sectors differ essentially only in the way how the representations of $\mathcal{R}(\mathcal{O})$ and $\mathcal{R}(\mathcal{O}')$ are coupled. We may, for our purposes here, take the net $\{\mathcal{R}(\mathcal{O})\}$ instead of $\{\mathfrak{A}(\mathcal{O})\}$ as defining the theory, write A instead of $\pi_0(A)$ and consider the elements $A \in \cup\mathcal{R}(\mathcal{O})$ as operators acting in $\mathcal{H}_0$. The algebra of all approximately local observables $\mathfrak{A}$ is the C*-inductive limit of the net $\mathcal{R}$

$$\mathfrak{A} = \overline{\bigcup_{\mathcal{O}} \mathcal{R}(\mathcal{O})}\ . \tag{IV.2.3}$$

IV.2.1 Localized Morphisms

The language is simplified if we describe all representations in the same Hilbert space, the space $\mathcal{H}_0$ on which the vacuum representation is defined by the operator algebras $\mathcal{R}(\mathcal{O})$. If $\pi \in Rep_L$ acts on $\mathcal{H}_\pi$ then we may replace it by an equivalent representation acting on $\mathcal{H}_0$

$$\varrho(A) = V\,\pi(A)\,V^{-1}; \quad A \in \mathfrak{A} \tag{IV.2.4}$$

where V is any unitary mapping $V : \mathcal{H}_\pi \to \mathcal{H}_0$. Such mappings exist because $\mathcal{H}_0$ and $\mathcal{H}_\pi$ are separable Hilbert spaces. By (IV.1.2) we can choose V so that it establishes the equivalence between π and π_0 for some $\mathcal{O}'$. This means

$$\varrho(A) = A \quad \text{for} \quad A \in \mathfrak{A}(\mathcal{O}'). \tag{IV.2.5}$$

Note that on the one hand $A \to \varrho(A)$ gives a representation of $\mathfrak{A}$ and of the net $\mathcal{R}$ (in the equivalence class of π) by operator algebras in $\mathfrak{B}(\mathcal{H}_0)$. On the other hand we may regard ϱ as a map from $\mathfrak{A}$ into $\mathfrak{A}$ with the property

$$\varrho\, \mathcal{R}(\mathcal{O}_1) \subset \mathcal{R}(\mathcal{O}_1) \quad \text{for} \quad \mathcal{O}_1 \supset \mathcal{O}\,. \tag{IV.2.6}$$

Property (IV.2.6) is a consequence of the duality relation (IV.2.1). Take $A \in \mathcal{R}(\mathcal{O}_1)$. Then by (IV.2.4), (IV.2.5) $\varrho(A)$ commutes with $\mathcal{R}(\mathcal{O}_1')$ since $\mathcal{O}_1' \subset \mathcal{O}'$. Thus (IV.2.1) implies (IV.2.6). Further, the map ϱ is isometric

$$\|\varrho(A)\| = \|A\|\,. \tag{IV.2.7}$$

This follows because we can obtain an equivalent representation ϱ_1, starting from a different choice of the region $\mathcal{O}$ in the above construction as long as the radius of the region is large enough. In particular we can choose it space-like to the support of any element $A \in \mathfrak{A}_{\text{loc}}$ which we consider. Then (IV.2.5) for ϱ_1 will imply (IV.2.7) for ϱ. Since $\mathfrak{A}_{\text{loc}}$ is norm dense in $\mathfrak{A}$ one sees that ϱ indeed maps $\mathfrak{A}$ into $\mathfrak{A}$. Finally, since $A \to \varrho(A)$ is a representation, the map ϱ respects the algebraic relations. Thus we may regard ϱ as an endomorphism of the C*-algebra $\mathfrak{A}$ with the additional features (IV.2.5) and (IV.2.6). We call such an endomorphism of $\mathfrak{A}$ a *localized morphism* with localization region (or support) $\mathcal{O}$ because ϱ acts trivially in the causal complement of $\mathcal{O}$ and transforms $\mathcal{R}(\mathcal{O})$ into itself. We have

Proposition 2.1.1
For any $\pi \in Rep_L$ and any diamond $\mathcal{O}$ of sufficiently large diameter we can obtain a representation in the equivalence class of π by operators acting on $\mathcal{H}_0$ applying a localized morphism with support $\mathcal{O}$ to the net $\mathcal{R}$.

Definitions 2.1.2
(i) Two localized morphisms ϱ_1, ϱ_2 are called equivalent if the representations $\varrho_1(\mathfrak{A}), \varrho_2(\mathfrak{A})$ are unitarily equivalent. We denote the equivalence class of ϱ by $[\varrho]$. The morphism ϱ is called irreducible if the representation $\varrho(\mathfrak{A})$ is irreducible.

(ii) A localized morphism with support in $\mathcal{O}$ is called *transportable* if there exist equivalent morphisms for any support region arising by a Poincaré transformation from $\mathcal{O}$. The set of transportable, localized morphisms will be denoted by Δ, the subset of irreducible ones by Δ_{irr}, the set of those with support in $\mathcal{O}$ by $\Delta(\mathcal{O})$.
(iii) A unitary $U \in \mathfrak{A}_{\text{loc}} = \bigcup \mathcal{R}(\mathcal{O})$ defines a transportable, localized automorphism σ_U by $\sigma_U A \equiv UAU^{-1}$, $A \in \mathfrak{A}$. The set of these (the inner automorphisms of $\mathfrak{A}_{\text{loc}}$) is denoted by $\mathcal{I}$.

Lemma 2.1.3
Two localized morphisms ϱ_1, ϱ_2 are equivalent if and only if

$$\varrho_2 = \sigma \varrho_1; \quad \sigma \in \mathcal{I} \tag{IV.2.8}$$

Proof. The equivalence means that there exists a unitary $U \in \mathfrak{B}(\mathcal{H}_0)$ such that $\varrho_2(A) = U\varrho_1(A)U^{-1}$. There exists a diamond $\mathcal{O}$ containing the supports of both ϱ_1 and ϱ_2 so that ϱ_1 and ϱ_2 act trivially on $\mathcal{R}(\mathcal{O}')$. Hence U must commute with $R(\mathcal{O}')$. By duality $U \in \mathcal{R}(\mathcal{O})$ and $\sigma_U \in \mathcal{I}$. □

The important point about describing representations by morphisms is that morphisms can be composed. We show that the product of morphisms from Δ respects the class division so that we also obtain a product of equivalence classes.

Notation. Considering ϱ as a representation we have written $\varrho(A)$ for the representor of the algebraic element A. Considering ϱ as a mapping it is more convenient to write ϱA instead. Since we now have two multiplications, one between morphisms, one between algebraic elements, some consideration has to be given to the setting of brackets. We shall omit them in a straight product of morphisms and algebraic elements. Thus $\varrho_2\varrho_1 AB$ means that the two morphisms are successively applied to the algebraic product of A and B, a bracket indicates the end of the effect of a morphism. Examples: $\varrho AB = (\varrho A)(\varrho B)$,

$$\varrho\, \sigma_U\, A = \varrho\, UAU^{-1} = (\varrho U)(\varrho A)(\varrho U^{-1}) = \sigma_W\, \varrho\, A, \qquad \text{(IV.2.9)}$$
$$W = \varrho U.$$

Lemma 2.1.4

If $\varrho_i \in \Delta$, $i = 1, 2$ then the equivalence class of the product depends only on the equivalence classes of the factors. Therefore a product of classes is defined

$$[\varrho_2\varrho_1] = [\varrho_2][\varrho_1]\,. \qquad \text{(IV.2.10)}$$

Proof. If $\varrho_i' \cong \varrho_i$ i.e. $\varrho_i' = \sigma_{U_i}\varrho_i$ then, using (IV.2.9), we have $\varrho_2'\varrho_1' = \sigma_W\varrho_2\varrho_1$ with $W = U_2\varrho_2 U_1$. □

Next we note that the causal structure carries over from the observables to the morphisms.

Lemma 2.1.5

Morphisms from Δ commute if their supports lie space-like to each other.

Proof. Let $\varrho_i \in \Delta(\mathcal{O}_i)$, $i = 1, 2$ and $A \in \mathcal{R}(\mathcal{O})$ where $\mathcal{O}_1$ lies spacelike to $\mathcal{O}_2$ and $\mathcal{O}$ is arbitrary. As indicated in the figure (IV.2.1) we can achieve, by translating the regions $\mathcal{O}_1, \mathcal{O}_2$ respectively to $\mathcal{O}_3, \mathcal{O}_4$, that $\mathcal{O}$ is space-like to both $\mathcal{O}_3$ and $\mathcal{O}_4$, that $\mathcal{O}_2$ is space-like to a diamond $\mathcal{O}_5$ containing both $\mathcal{O}_1$ and $\mathcal{O}_3$ and $\mathcal{O}_1$ is space-like to a diamond $\mathcal{O}_6$ containing both $\mathcal{O}_2$ and $\mathcal{O}_4$. There are morphisms $\varrho_3 = \sigma_{U_{31}}\varrho_1, \varrho_4 = \sigma_{U_{42}}\varrho_2$, localized in $\mathcal{O}_3, \mathcal{O}_4$ respectively and equivalent to ϱ_1, resp. ϱ_2. Then

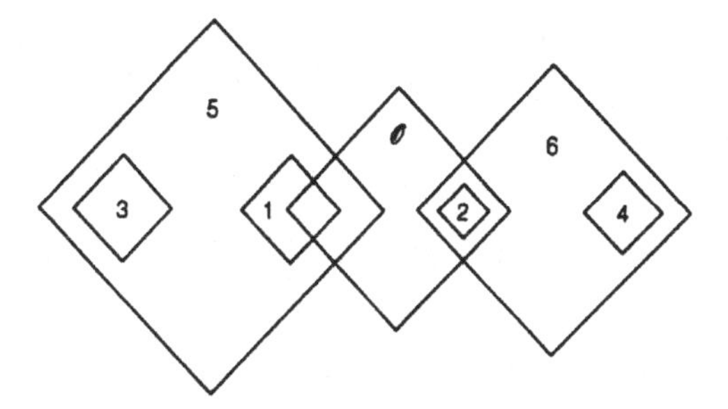

Fig. IV.2.1.

$$\varrho_4\varrho_3 A = \varrho_3\varrho_4 A = A \tag{IV.2.11}$$

By (IV.2.1) (duality) the localization of the unitaries which transport the morphisms is given by $U_{42} \in \mathcal{R}(\mathcal{O}_6), U_{31} \in \mathcal{R}(\mathcal{O}_5)$. Therefore $\varrho_1 U_{42} = U_{42}, \varrho_2 U_{31} = U_{31}$; $U_{31}U_{42} = U_{42}U_{31}$, and (IV.2.11) implies $\varrho_2\varrho_1 A = \varrho_1\varrho_2 A$. □

As a consequence we have

Proposition 2.1.6
(i) The charge quantum numbers are in one to one correspondence with the elements of $\Delta_{\text{irr}}/\mathcal{I}$ (the set of classes).
(ii) Δ is a semigroup, $\Delta/\mathcal{I}$ is an Abelian semigroup.

IV.2.2 Intertwiners and Exchange Symmetry (Statistics)

The concept of intertwiners between non-disjoint representations was described in III.2. In our present context, where the representations are obtained by localized morphisms of a local net, there results some additional structure. We denote the set of intertwining operators from a morphism ϱ to a morphism ϱ' by $i(\varrho', \varrho)$. An element of this set is an operator $R \in \mathfrak{B}(\mathcal{H}_0)$ such that

$$(\varrho' A)R = R(\varrho A); \quad A \in \mathfrak{A} \tag{IV.2.12}$$

It follows then from duality that $R \in \mathfrak{A}_{\text{loc}}$. We shall use the term *intertwiner* for the triple (ϱ', R, ϱ) standing in the relation (IV.2.12) and denote it by $\mathbf{R}$. The morphism ϱ will be called its source, ϱ' its target. The adjoint

$$\mathbf{R}^* = (\varrho, R^*, \varrho') \tag{IV.2.13}$$

is an intertwiner in the opposite direction. If $R_1 \in i(\varrho_1, \varrho), R_2 \in i(\varrho_2, \varrho_1)$ then $R_2 R_1 \in i(\varrho_2, \varrho)$. Thus, if the target of $\mathbf{R}_1$ coincides with the source of $\mathbf{R}_2$, there is an obvious composition product of these intertwiners

$$\mathbf{R}_2 \circ \mathbf{R}_1 = (\varrho_2, R_2 R_1, \varrho)\,. \tag{IV.2.14}$$

The product of morphisms introduces another kind of product between intertwiners. Given $R_i \in i(\varrho_i', \varrho_i), (i = 1, 2)$ then one checks that $R = R_2\varrho_2 R_1 = (\varrho_2' R_1)R_2$ is an intertwining operator from $\varrho_2\varrho_1$ to $\varrho_2'\varrho_1'$. We write

$$\mathbf{R}_2 \times \mathbf{R}_1 = (\varrho_2'\varrho_1', R_2\varrho_2 R_1, \varrho_2\varrho_1) . \tag{IV.2.15}$$

One checks that this cross product is associative, that

$$(\mathbf{R} \times \mathbf{S})^* = \mathbf{R}^* \times \mathbf{S}^* , \tag{IV.2.16}$$

and, if $\mathbf{R}_i' = (\varrho_i'', R_i', \varrho_i'), \mathbf{R}_i = (\varrho_i', R_i, \varrho_i), i = 1, 2$ then

$$(\mathbf{R}_2' \circ \mathbf{R}_2) \times (\mathbf{R}_1' \circ \mathbf{R}_1) = (\mathbf{R}_2' \times \mathbf{R}_1') \circ (\mathbf{R}_2 \times \mathbf{R}_1) . \tag{IV.2.17}$$

The cross product inherits causal commutativity if both the sources and the targets of the factors have causally disjoint supports.

Lemma 2.2.1
Let $\mathbf{R}_i = (\varrho_i', R_i, \varrho_i), i = 1, 2$. If the supports of ϱ_1, ϱ_2 as well as those of ϱ_1', ϱ_2' are causally disjoint then

$$\mathbf{R}_2 \times \mathbf{R}_1 = \mathbf{R}_1 \times \mathbf{R}_2 . \tag{IV.2.18}$$

Warning: this lemma does not hold in $(1+1)$-dimensional space-time.

Proof. We have to show that under the stated conditions

$$R_1(\varrho_1 R_2) = R_2(\varrho_2 R_1) \tag{IV.2.19}$$

This is trivially true if for instance

$$\text{supp } \varrho_1 = \text{supp } \varrho_1' = \mathcal{O}_1; \quad \text{supp } \varrho_2 = \text{supp } \varrho_2' = \mathcal{O}_2 \tag{IV.2.20}$$

and $\mathcal{O}_1, \mathcal{O}_2$ are causally disjoint. Consider the change of both sides of (IV.2.19) when we make a small shift of ϱ_2 to ϱ_3 so that the smallest diamond containing the supports of both ϱ_2 and ϱ_3 remains space-like to the support of ϱ_1 (see fig. (IV.2.2)). This replaces $\mathbf{R}_2$ by $\mathbf{R}_3 = \mathbf{R}_2 \circ \mathbf{U}$ where

$$\mathbf{U} = (\varrho_2, U, \varrho_3); \quad \varrho_3 = \sigma_{U^*}\varrho_2$$

and supp U is space-like to supp ϱ_1, thus $\varrho_1 U = U$. Then

$$\begin{aligned} \mathbf{R}_1 \times \mathbf{R}_3 &= (\varrho_1'\varrho_2', R_{13}, \varrho_1\varrho_3) \quad \text{with} \\ R_{13} &= R_1\varrho_1 R_3 = R_1\varrho_1 R_2 U \\ &= R_1(\varrho_1 R_2)(\varrho_1 U) = R_1(\varrho_1 R_2)U . \end{aligned}$$

In the same way one gets

$$\begin{aligned} \mathbf{R}_3 \times \mathbf{R}_1 &= (\varrho_2'\varrho_1', R_{31}, \varrho_3\varrho_1) \quad \text{with} \\ R_{31} &= R_2(\varrho_2 R_1)U . \end{aligned}$$

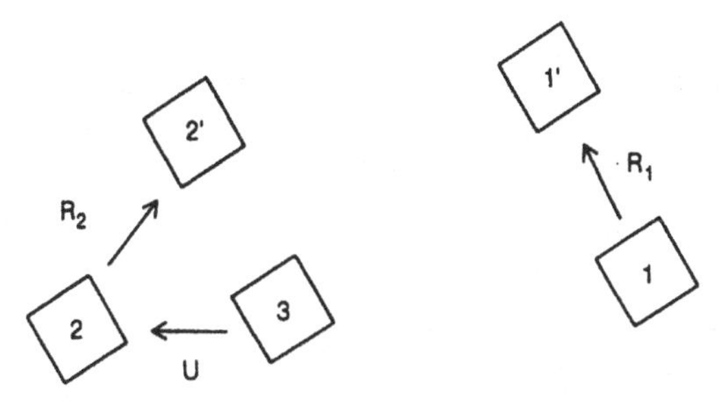

Fig. IV.2.2.

Thus, by the change from ϱ_2 to ϱ_3 the two sides of (IV.2.19) change by the same factor U. If they were equal before they remain equal. By a succession of such small shifts we can change ϱ_2 to any morphism in its class with support space-like to $\mathcal{O}_1$. Then, analogously, we can change ϱ_1' by continuous deformation to any morphism in its class with support space-like to $\mathcal{O}_2$ and thus reach the general placement of all the supports specified in the lemma. Note that in these deformations we always have to keep the support of the moving morphism space-like to that of the fixed one. In order to reach an arbitrary final configuration from the initial one we need a space with at least two dimensions (at fixed time). □

The next topic is the exchange symmetry. We start from a charge $\xi \in \Delta_{\text{irr}}/\mathcal{I}$ and consider the folium ξ^2 of "doubly charged states". To characterize state vectors we pick a reference morphism ϱ in the equivalence class of ξ and a reference state $\omega = \omega_0 \circ \varrho^2$ in the folium (i.e. $\omega(A) = \omega_0(\varrho^2 A)$). Other doubly charged states are obtained by taking other morphisms

$$\varrho_i = \sigma_{U_i}\varrho; \quad i = 1, 2 \tag{IV.2.21}$$

in the class and forming

$$\omega_{21}(A) = \omega_0(\varrho_2\varrho_1 A) \ . \tag{IV.2.22}$$

Such a state may be pictured as one where a charge ξ has been placed into the support regions of ϱ_1 and of ϱ_2, leaving the vacuum everywhere else. Having picked the unitary intertwiners from ϱ to ϱ_i in the singly charged sector

$$\mathbf{U}_i = (\varrho_i, U_i, \varrho) \tag{IV.2.23}$$

we get the intertwiner from ϱ^2 to $\varrho_2\varrho_1$ as

$$\mathbf{U}_2 \times \mathbf{U}_1 = (\varrho_2\varrho_1, U_{21}, \varrho^2); \quad U_{21} = U_2\varrho U_1 \tag{IV.2.24}$$

and we get as the natural assignment of a state vector to the state ω_{21} in the representation ϱ^2 the vector

$$\Omega_{21} = U_{21}^* \Omega \qquad \text{(IV.2.25)}$$

where Ω denotes the reference state vector corresponding to $\omega_0 \circ \varrho^2$ which is the vacuum state vector in $\mathcal{H}_0$. Now, if the supports of ϱ_1, ϱ_2 are causally disjoint then, by lemma 2.1.5, $\omega_{21} = \omega_{12}$ and we might equally well have assigned to this state the state vector

$$\Omega_{12} = U_{12}^* \Omega; \quad U_{12} = U_1 \varrho U_2 \, . \qquad \text{(IV.2.26)}$$

The two assignments differ by a unitary $\varepsilon_\varrho = U_{12}^* U_{21}$ whose properties are described in the next lemma.

Key Lemma 2.2.2
Let $\varrho_1, \varrho_2, \varrho$ be equivalent morphisms from Δ. So

$$\varrho_i = \sigma_{U_i} \varrho, \quad i = 1, 2, \quad \sigma_{U_i} \in \mathcal{I} \, , \qquad \text{(IV.2.27)}$$

and let the support of ϱ_1 be space-like to that of ϱ_2. Then the unitary operator (the *statistics operator*)

$$\varepsilon_\varrho = (\varrho U_2^{-1}) U_1^{-1} U_2 (\varrho U_1) \qquad \text{(IV.2.28)}$$

has the following properties

a) ε_ϱ commutes with all observables in the representation ϱ^2, i.e.

$$\varepsilon_\varrho \in (\varrho^2 \mathfrak{A})' \, . \qquad \text{(IV.2.29)}$$

b) ε_ϱ depends only on ϱ, not on ϱ_1, ϱ_2 as long as they vary in the class $[\varrho]$ and have causally disjoint supports. Nor does it depend on the choice of the unitary operators U_i which implement the inner automorphisms σ_{U_i} in the singly charged sector.

c)

$$\varepsilon_\varrho^2 = \mathbb{1} \, . \qquad \text{(IV.2.30)}$$

d) If $\varrho' = \sigma_W \varrho$ then $\varepsilon_{\varrho'} = \sigma_V \varepsilon_\varrho$ with $V = W(\varrho W)$.

Warning: parts b) and c) do not hold in 2-dimensional space-time.

Proof. Part a) expresses the fact that ε_ϱ is an intertwining operator from ϱ^2 to ϱ^2:

$$\varepsilon_\varrho = (\mathbf{U}_1 \times \mathbf{U}_2)^* \circ (\mathbf{U}_2 \times \mathbf{U}_1) = (\varrho^2, \varepsilon_\varrho, \varrho^2) \, . \qquad \text{(IV.2.31)}$$

Part b). The independence of ε_ϱ from the choice of the ϱ_i follows from lemma 2.2.1. Let ϱ_1', ϱ_2' be two other morphisms in the class $[\varrho]$ with causally disjoint supports, related to ϱ_1, ϱ_2 respectively by the intertwiners

$$\mathbf{W}_i = (\varrho_i', W_i, \varrho_i), \quad i = 1, 2$$

(see fig. (IV.2.3)). Then the intertwiners from ϱ^2 to $\varrho_2' \varrho_1'$, respectively to ϱ_1', ϱ_2' are

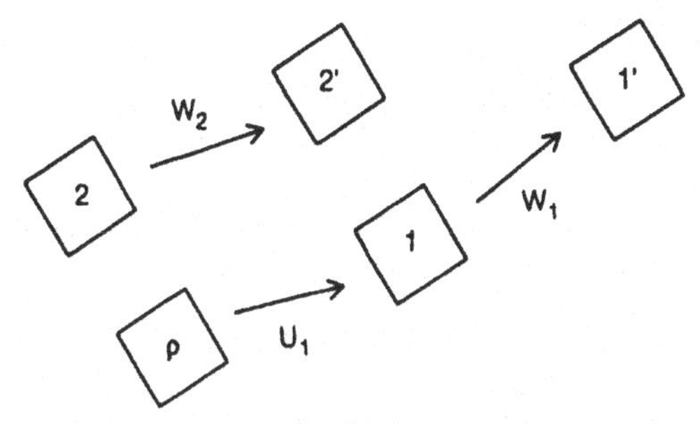

Fig. IV.2.3.

$$\begin{aligned}\mathbf{U}'_{21} &= (\mathbf{W}_2 \circ \mathbf{U}_2) \times (\mathbf{W}_1 \circ \mathbf{U}_1) &= (\mathbf{W}_2 \times \mathbf{W}_1) \circ (\mathbf{U}_2 \times \mathbf{U}_1) \\ \mathbf{U}'_{12} &= (\mathbf{W}_1 \circ \mathbf{U}_1) \times (\mathbf{W}_2 \circ \mathbf{U}_2) &= (\mathbf{W}_1 \times \mathbf{W}_2) \circ (\mathbf{U}_1 \times \mathbf{U}_2)\end{aligned} \tag{IV.2.32}$$

where we have used (IV.2.17). Since the sources as well as the targets of $\mathbf{W}_1$ and $\mathbf{W}_2$ are causally disjoint one has $\mathbf{W}_2 \times \mathbf{W}_1 = \mathbf{W}_1 \times \mathbf{W}_2$ by lemma 2.2.1. Inserting (IV.2.32) in the definition (IV.2.31) for ε'_ϱ the W-factors cancel and one obtains $\varepsilon'_\varrho = \varepsilon_\varrho$. The independence of ε_ϱ from the choice of the intertwining operators U_i in the representation ϱ follows as the special case $\varrho'_i = \varrho_i$. A direct proof of part b), avoiding the use of lemma 2.2.1, is given in DHR3. It should be kept in mind that the argument depends on the possibility of shifting space-like separated diamonds continuously to arbitrary other space-like separated positions without having to cross their causal influence zones. This homotopic property needs a space-time dimension of at least 3. Part c) follows from b) because an interchange of ϱ_1 and ϱ_2 in (IV.2.31) changes ε_ϱ into ε_ϱ^{-1}. Part d) follows from elementary computation. □

The generalization of this lemma to the sector $[\varrho]^n$, the states with n identical charges, is given in the following theorem.

Theorem 2.2.3

Let $\varrho_1, \ldots \varrho_n$ be morphisms from Δ, with causally disjoint supports and all equivalent to a reference morphism ϱ with

$$\mathbf{U}_i = (\varrho_i, U_i, \varrho), \quad i = 1, \ldots, n$$

intertwiners from ϱ to ϱ_i. For each permutation

$$P = \begin{pmatrix} 1 & \ldots & n \\ P(1) & \ldots & P(n) \end{pmatrix}$$

define

$$\begin{aligned}\mathbf{U}_\varrho(P) &= \mathbf{U}_{P^{-1}(1)} \times \mathbf{U}_{P^{-1}(2)} \times \ldots \mathbf{U}_{P^{-1}(n)} \\ &= \left(\varrho_{P^{-1}(1)} \ldots \varrho_{P^{-1}(n)}, U_\varrho(P), \varrho^n\right)\end{aligned} \tag{IV.2.33}$$

and

$$\begin{aligned} \mathcal{E}_\varrho^{(n)}(P) &= \mathbf{U}_\varrho^*(P) \circ \mathbf{U}_\varrho(e) \\ &= \left(\varrho^n, \varepsilon_\varrho^{(n)}(P), \varrho^n\right) \end{aligned} \tag{IV.2.34}$$

where e denotes the unit element of the permutation group. Then

a) $\varepsilon_\varrho^{(n)}(P)$ commutes with all observables in the representation ϱ^n, i.e.

$$\varepsilon_\varrho^{(n)}(P) \in (\varrho^n \mathfrak{A})' . \tag{IV.2.35}$$

b) $\varepsilon_\varrho^{(n)}(P)$ is independent of the choice of $\varrho_1, \ldots, \varrho_n$ and of the choice of the intertwining operators U_i as long as the supports of the ϱ_i are causally disjoint.

c) The operators $\varepsilon_\varrho^{(n)}(P)$ form a unitary representation of the permutation group:

$$\varepsilon_\varrho^{(n)}(P_2)\varepsilon_\varrho^{(n)}(P_1) = \varepsilon_\varrho^{(n)}(P_2 P_1) . \tag{IV.2.36}$$

d) If τ_m denotes the transposition of m and $m+1, (m < n)$,

$$\mathcal{E}_\varrho^{(n)}(\tau_m) = \mathbf{I}_\varrho^{m-1} \times \varepsilon_\varrho \times \mathbf{I}_\varrho^{n-m-1} \tag{IV.2.37}$$

where $\mathbf{I}_\varrho^r$ stands for $(\varrho^r, \mathbb{1}, \varrho^r)$ and $\varepsilon_\varrho = (\varrho^2, \varepsilon_\varrho, \varrho^2)$ is defined in lemma 2.2.2 (the special case for $n = 2$); in terms of the operators

$$\varepsilon_\varrho^{(n)}(\tau_m) = \varrho^{m-1}\varepsilon_\varrho . \tag{IV.2.38}$$

Proof. Parts a) and b) are the generalizations of the corresponding statements in the key lemma from $n = 2$ to arbitrary n and the proof proceeds in complete analogy. For part c) pick two permutations P' and P'' and set $k' = P'(k), k'' = P''(k), k = 1, \ldots, n$. An intertwiner from $\varrho_{k'}$ to $\varrho_{k''}$ is

$$\mathbf{U}_{k''k'} = (\varrho_{k''}, U_{k''}, \varrho) \circ (\varrho, U_{k'}^*, \varrho_{k'}) .$$

By lemma 2.2.1 the cross product over all k

$$\mathbf{V} = \times_{k=1}^{k=n} \mathbf{U}_{k''k'}$$

is independent of the order of factors since all sources are causally disjoint and all targets also. Hence $\mathbf{V}$ does not change if P' and P'' are replaced by $P'P'''$ and $P''P'''$ respectively. On the other hand we have by (IV.2.17) and (IV.2.33)

$$\mathbf{V} = \mathbf{U}_\varrho(P''^{-1}) \circ \mathbf{U}_\varrho(P'^{-1})^* .$$

Thus

$$\mathbf{U}_\varrho(P''^{-1}) \circ \mathbf{U}_\varrho(P'^{-1})^* = \mathbf{U}_\varrho(e) \circ \mathbf{U}(P''P'^{-1})^* .$$

With the definition of $\varepsilon_\varrho^{(n)}$, putting $P''P'^{-1} = P_1, P''^{-1} = P_2$, this gives (IV.2.36). Part d) is evident from the definitions. □

We note that, quite generally, the equivalence class of a product of morphisms does not depend on the order of factors. One can construct natural intertwiners between products of morphisms in different order. This is described in the following generalization of the preceding theorem.

Proposition 2.2.4
Given $\varrho_k \in \Delta$ which may be in different equivalence classes; $k = 1, \ldots, n$. Choose n morphisms $\varrho_k^{(0)}$ with mutually space-like supports, equivalent respectively to ϱ_k so that there exist unitary intertwiners $\mathbf{U}_k = (\varrho_k^{(0)}, U_k, \varrho_k)$. The cross product of these intertwiners in the order determined by the permutation P as in (IV.2.33) is denoted by $\mathbf{U}(P)$ and

$$\varepsilon(\varrho_1, \ldots, \varrho_n; P) = \mathbf{U}^*(P) \circ \mathbf{U}(e) . \tag{IV.2.39}$$

Then

a) $\varepsilon(\varrho_1, \ldots, \varrho_n; P)$ is independent of the choice of the $\varrho_k^{(0)}$ and U_k within the specified limitations.

b) $\varepsilon(\varrho_1, \ldots, \varrho_n; P) = \mathbb{1}$ if the ϱ_k have mutually space-like supports.

c) Given intertwiners $\mathbf{R}_k = (\varrho'_k, R_k, \varrho_k)$ then

$$\mathbf{R}(P) \circ \varepsilon(\varrho_1, \ldots, \varrho_n; P) = \varepsilon(\varrho'_1, \ldots, \varrho'_n; P) \circ \mathbf{R}(e) \tag{IV.2.40}$$

The computations establishing these claims can be found in DHR3.

Discussion. The significance of the operators $\varepsilon_\varrho^{(n)}$ has been previously described for the case $n = 2$. Similarly, in the folium of states with n identical charges we pick a reference state $\omega_0 \circ \varrho^n$ and assign to it the state vector Ω in the representation ϱ^n. The morphisms ϱ_k may be interpreted as charge creation in the support regions $\mathcal{O}_k$. If these regions are causally disjoint then the state $\omega_o \circ \varrho_{P^{-1}(1)} \cdots \varrho_{P^{-1}(n)}$ does not depend on the order of the factors ϱ_k i.e. it is independent of the permutation P. The natural assignment of a state vector to this state depends, however, on the order "in which the charges are created". For the above order it is

$$\Psi_P = U_\varrho(P)^* \Omega . \tag{IV.2.41}$$

This vector in the representation ϱ^n may be regarded as the product[1] of the single charge state vectors $U_k^* \Omega$ from the representation ϱ, applied in the order in which the factors U_k appear in $U_\varrho(P)$. The $\varepsilon_\varrho^{(n)}$ change this order

$$\varepsilon_\varrho^{(n)}(P') \Psi_P = \Psi_{P'P} . \tag{IV.2.42}$$

The state is independent of the order because the $\varepsilon_\varrho^{(n)}$ commute with all observables in the representation ϱ^n. These properties of the permutation operators ε

[1]Product in the sense of the symbol $\otimes^t$ used in the collision theory of chapter II. The time t corresponds now to a space-like surface through the regions $\mathcal{O}_k$.

make them completely analogous to place permutations of product wave functions of n identical particles. By part d) of the theorem the $\varepsilon_\varrho^{(n)}$ for arbitrary n can be constructed from ε_ϱ of lemma 2.2.2.

Remark. If ϱ is an automorphism (if the image of $\mathfrak{A}$ under ϱ is all of $\mathfrak{A}$) then ϱ^n is irreducible and thus all $\varepsilon_\varrho^{(n)}$ are multiples of the unit operator. This leaves only the alternative between the completely symmetric representation of the permutation group (Bose case) and the completely antisymmetric one (Fermi case); ε_ϱ of lemma 2.2.2 is either $+1$ or -1 and this sign is an intrinsic property of the charge $[\varrho]$. In this context the following lemma is of interest.

Lemma 2.2.5
Let $\varrho \in \Delta_{\text{irr}}$. The following conditions are equivalent

a) ϱ is an automorphism,

b) ϱ^2 is irreducible,

c) $\varepsilon_\varrho = \pm\mathbb{1}$.

d) The representation of the permutation group in the sector ϱ^n is either the completely symmetric or the completely antisymmetric one.

e) The representation ϱ satisfies the duality condition

$$(\varrho\mathcal{R}(\mathcal{O}'))' = \varrho\mathcal{R}(\mathcal{O}) . \qquad \text{(IV.2.43)}$$

The proof is rather simple (see DHR3, lemmas 2.2, 2.7 and equation (IV.2.37) above). We shall record here only the argument leading from e) to a). It suffices to show that ϱ maps $\mathcal{R}(\mathcal{O})$ onto itself when $\mathcal{O}$ contains the support of ϱ. In that case, however, ϱ acts trivially on $\mathcal{R}(\mathcal{O}')$ so one can omit ϱ on the left hand side of (IV.2.43). Then, using the duality in the vacuum sector, we get $\mathcal{R}(\mathcal{O}) = \varrho\mathcal{R}(\mathcal{O})$.

The fact that e) entails a) is of interest because it suggests an alternative approach to the construction of a global gauge group, simpler than that of section 4 below and in line with the qualitative arguments at the end of section 1. It should be possible to show that within the set of representations considered one has "essential duality" namely that there exists an extension of the algebra of observables which is maximal i.e. satisfies the duality relation. This requires a generalization of the theorem of Bisognano and Wichmann (see chapter V). For the extended algebra one has then the Bose-Fermi alternative in all charge sectors. The automorphism group of the extended algebra which leaves the observables element-wise invariant is the non-Abelian part of the gauge group. It may be regarded as an internal symmetry group for the extended algebra. The parastatistics for the observable algebra arises from the ordinary statistics of the extended algebra by ignoring the "hidden", unobservable degrees of freedom as described at the end of section 1.

IV.2.3 Charge Conjugation, Statistics Parameter

We show next that all morphisms $\varrho \in \Delta$ can be obtained as limits of sequences of inner automorphisms, the charge transfer chains. Let $\varrho \in \Delta(\mathcal{O})$ and take a sequence $\mathcal{O}_k$, space-like translates of $\mathcal{O}$ moving to infinity as $k \to \infty$. For each k take a morphism ϱ_k with support in $\mathcal{O}_k$ in the equivalence class of ϱ, so

$$\varrho_k A = \sigma_{U_k} \varrho A = U_k(\varrho A)U_k^{-1} \tag{IV.2.44}$$

with U_k a unitary from $\mathcal{R}(\mathcal{O}) \vee \mathcal{R}(\mathcal{O}_k)$. If $A \in \mathfrak{A}_{\text{loc}}$ then $\varrho_k A = A$ for sufficiently large k. Therefore

$$\varrho A = \lim_{k\to\infty} \sigma_{U_k^{-1}} A = \lim U_k^* A U_k \tag{IV.2.45}$$

where the convergence is understood in the uniform topology of $\mathfrak{B}(\mathcal{H}_0)$. If ϱ is an automorphism then also the sequence $U_k A U_k^*$ will converge uniformly to $\varrho^{-1} A$. If ϱ is only an endomorphism the convergence of this sequence is not evident. Under some additional conditions it has been shown that this sequence converges weakly in $\mathfrak{B}(\mathcal{H}_0)$. These conditions, which cover the cases of principal physical interest are [Buch 82a]

BF-Conditions 2.3.1
The automorphisms of space-time translations are implementable in the sector $[\varrho]$. The energy spectrum is positive, the theory is purely massive and the sector $[\varrho]$ contains single particle states. In other words, there is an isolated mass hyperboloid in the P_μ-spectrum of the sector.

Lemma 2.3.2[2]
Let U_k be a sequence of charge transfer operators as described above with $\mathcal{O}_k$ moving to space-like infinity as $k \to \infty$. If ϱ satisfies the BF-conditions then $\sigma_{U_k} A$ converges weakly. The limit is independent of the chain ϱ_k (though not of the starting point ϱ) and defines a positive map from $\mathfrak{A}$ to $\mathfrak{A}$:

$$\Phi_\varrho A = \mathrm{w} - \lim \sigma_{U_k} A \tag{IV.2.46}$$

with the properties

(i)

$$\begin{aligned} \Phi_\varrho A &= A \quad \text{for} \quad A \in \mathfrak{A}(\mathcal{O}') \,; \\ \Phi_\varrho \mathfrak{A}(\mathcal{O}_1) &\subset \mathfrak{A}(\mathcal{O}_1) \quad \text{if} \quad \mathcal{O}_1 \supset \mathcal{O} \,, \end{aligned} \tag{IV.2.47}$$

(ii) Φ_ϱ is a *left inverse* of ϱ, i.e.

$$\Phi_\varrho(\varrho A) B(\varrho C) = A(\Phi_\varrho B) C \,, \tag{IV.2.48}$$

$$\Phi_\varrho(\mathbb{1}) = \mathbb{1} \,. \tag{IV.2.49}$$

[2] In DHR3 the weak convergence of $\sigma_{U_k} A$ was not used. The analysis is then more complicated but the conclusions remain essentially the same except that the case of "infinite statistics" i.e. $\lambda_\xi = 0$ in (IV.2.55) can then not be excluded and that in this pathological case the left inverse is not unique and the conjugate sector cannot be defined.

The hard part in the proof of this lemma concerns the weak convergence and uniqueness. We defer it to the next section. Claim (i) then follows from the fact that σ_{U_k} acts trivially on those elements of $\mathfrak{A}$ whose support is causally disjoint from both $\mathcal{O}$ and $\mathcal{O}_k$, together with the duality in the vacuum sector. For part (ii) we note that due to the uniform convergence in (IV.2.45) $\sigma_{U_k}\varrho A$ converges uniformly to A and

$$\sigma_{U_k}(\varrho A)B(\varrho C) = (\sigma_{U_k}\varrho A)(\sigma_{U_k}B)(\sigma_{U_k}\varrho C)\ .$$

Remark. Φ_ϱ is not a morphism of $\mathfrak{A}$. It does not respect the multiplicative structure i.e. $\Phi_\varrho AB \neq (\Phi_\varrho A)(\Phi_\varrho B)$ unless one of the factors belongs to $\varrho\mathfrak{A}$. Φ_ϱ is, however, a positive linear map of $\mathfrak{A}$ into $\mathfrak{A}$. This, together with (IV.2.49) implies that

$$\omega_0(\Phi_\varrho A) \equiv \overline{\omega}_\varrho(A) \tag{IV.2.50}$$

is a state. It determines a representation $\overline{\pi}$ via the GNS-construction. One finds (see below) that, if $\varrho \in \Delta_{\text{irr}}$ satisfies the BF-condition, the automorphisms of space-time translations are implementable in $\overline{\pi}$ with an energy-momentum spectrum coinciding with that of the representation ϱ and that $\overline{\pi}$ is irreducible. In particular this representation satisfies the selection criterion (IV.1.2) and is thus unitarily equivalent to a representation $A \to \overline{\varrho}A$. Obviously this representation must be interpreted as the charge conjugate to ϱ.

Charge conjugation entails further properties of the statistics operators $\varepsilon_\varrho, \varepsilon_\varrho^{(n)}(P)$. Since ε_ϱ commutes with $\varrho^2\mathfrak{A}$, $\Phi\varepsilon_\varrho$ commutes with $\varrho\mathfrak{A}$ and is therefore a multiple of the identity operator if $\varrho \in \Delta_{\text{irr}}$.

Lemma 2.3.3
Associated with each charge $\xi \in \Delta_{\text{irr}}/\mathcal{I}$ there is a number λ_ξ, called the *statistics parameter* of the charge, given by

$$\Phi_\varrho\varepsilon_\varrho = \lambda_\xi \mathbb{1}\ , \tag{IV.2.51}$$

where ϱ is any morphism in the class ξ.

One easily checks the invariance of λ_ξ i.e.

$$\Phi_\varrho\varepsilon_\varrho = \Phi_{\varrho'}\varepsilon_{\varrho'} \quad \text{if} \quad \varrho \cong \varrho'\ .$$

Similarly one finds

Lemma 2.3.4
With $\varrho \in \Delta_{\text{irr}}$, $[\varrho] = \xi$ one has

$$\Phi_\varrho^{(n-1)}\varepsilon_\varrho^{(n)}(P) = \lambda_\xi^{(n)}(P)\mathbb{1}. \tag{IV.2.52}$$

The numbers $\lambda_\xi^{(n)}(P)$ define, for fixed ξ and n, a normalized, positive linear form over the group algebra of the permutation group S_n.

Comment. The $\varepsilon_{\varrho}^{(n)}(P)$ form a representation of S_n by unitary operators from $\mathfrak{A} \subset \mathfrak{B}(\mathcal{H}_0)$. Their complex linear combinations form a concrete C*-algebra homomorphic to the abstract group algebra of S_n. The fact that Φ is a positive linear map with $\Phi(\mathbb{1}) = \mathbb{1}$ implies that $\lambda_{\xi}^{(n)}(P)$ defines a state on this C*-algebra. From (IV.2.37) this state can be computed explicitly in terms of λ_{ξ} which shows that it depends only on ξ. We give only the results and refer to DHR3 for the computation.

Lemma 2.3.5
Let $E_s^{(n)}, E_a^{(n)}$ denote the totally symmetric, respectively the totally antisymmetric projectors in the group algebra of S_n. Then

$$\lambda_{\xi}^{(n)}(E_s^{(n)}) = n!^{-1}(1+\lambda_{\xi})(1+2\lambda_{\xi})\dots(1+(n-1)\lambda_{\xi}), \quad \text{(IV.2.53)}$$

$$\lambda_{\xi}^{(n)}(E_a^{(n)}) = n!^{-1}(1-\lambda_{\xi})(1-2\lambda_{\xi})\dots(1-(n-1)\lambda_{\xi}) \, . \quad \text{(IV.2.54)}$$

This limits the allowed values of λ_{ξ} to the set

$$\lambda_{\xi} = 0, \pm d_{\xi}^{-1} \quad \text{where } d_{\xi} \text{ is a natural number.} \quad \text{(IV.2.55)}$$

Comment. E_s and E_a are positive elements of the group algebra. For any value of λ_{ξ} outside of the set (IV.2.55) either (IV.2.53) or (IV.2.54) will become negative for large n.

We shall call the integer d_{ξ} the *statistics dimension* of the charge. It corresponds precisely to the *order of the parastatistics* discussed in section 1. To show this one notes first that the representation of S_n in the sector ξ^n is quasi-equivalent to that obtained by the GNS-construction from the state $\lambda_{\xi}^{(n)}$. For $\lambda_{\xi} \neq 0$ the explicit expression for this state is described in the following lemma.

Lemma 2.3.6
Let $\mathfrak{h}$ be a d-dimensional Hilbert space (d finite) and $\mathfrak{h}^{\otimes n}$ the n-fold tensor product of $\mathfrak{h}$ with itself. The natural representation of S_n on $\mathfrak{h}^{\otimes n}$ which acts by permuting the order of factors in product vectors is denoted by π; the representation π' is defined as $\pi'(P) = \operatorname{sign} P \cdot \pi(P)$. Then

$$\lambda_{\xi}^{(n)}(P) = \begin{cases} d^{-n} \ \operatorname{tr} \pi(P) & \text{for} \quad \lambda_{\xi} > 0, \\ d^{-n} \ \operatorname{tr} \pi'(P) & \text{for} \quad \lambda_{\xi} < 0, \end{cases} \quad d = \mid \lambda_{\xi}^{-1} \mid \, . \quad \text{(IV.2.56)}$$

The product vectors in $\mathfrak{h}^{\otimes n}$ can be antisymmetrized in at most d factors while there is no limitation for the symmetrization. This yields

Proposition 2.3.7
If $\lambda_{\xi} > 0$ the representation of S_n in the sector ξ^n contains all irreducible representations whose Young tableaux have not more than d_{ξ} rows, if $\lambda_{\xi} < 0$ it contains all those whose Young tableaux have not more than d_{ξ} columns.

Thus, if the statistics parameter of a charge is positive it means that the multiply charged states obey para-Bose statistics of order d, if it is negative they obey para-Fermi statistics of order d. There remains the case $\lambda_\xi = 0$ (or $d = \infty$). This case, called *infinite statistics*, can, however, not occur if we consider sectors in which the translation group is implementable with positive energy spectrum and where the charge ξ has a conjugate charge $\overline{\xi}$ characterized by the property that $\xi\overline{\xi}$ contains the vacuum sector (see the appendix in DHR4). Thus, in particular, the case of infinite statistics is excluded if the charge satisfies the BF-conditions. The proof uses an analyticity argument. We shall return to this in the subsections 2.4, 2.5 and 3.4.

The close relation between the exclusion of infinite statistics and the existence of charge conjugation is also seen by the following direct construction of the conjugate sector (see DHR4). Let $E_s^{(n)}, E_a^{(n)}$ denote the totally symmetric, respectively totally antisymmetric projectors in the group algebra of S_n and, for $|\lambda_\xi| = d^{-1}, d > 1$

$$E = \varepsilon_\varrho^{(d)}(E_a^{(d)}); \quad E' = \varepsilon_\varrho^{(d-1)}(E_a^{(d-1)}) \qquad \text{if} \quad \lambda_\xi > 0$$
$$E = \varepsilon_\varrho^{(d)}(E_s^{(d)}); \quad E' = \varepsilon_\varrho^{(d-1)}(E_s^{(d-1)}) \qquad \text{if} \quad \lambda_\xi < 0 .$$

In words: in the para-Bose case E and E' are the representors of the antisymmetrizers in the representation spaces ϱ^d respectively ϱ^{d-1}. In the para-Fermi case they are the representors of the symmetrizers. Then one finds that the representation of the observable algebra $E\varrho^d$ has "statistical dimension" 1. By lemma 2.2.5 it is thus equivalent to a localized *auto*morphism γ. The representation $E'\varrho^{d-1}$ is equivalent to a morphism ϱ' with the same statistics parameter as ϱ. Since γ has an inverse one can define

$$\overline{\varrho} = \varrho'\gamma^{-1} \tag{IV.2.57}$$

and verify that the sector $\overline{\varrho}$ is indeed the conjugate to ϱ.

Theorem 2.3.8
If $\varrho \in \Delta_{\text{irr}}$ has finite statistics, $[\varrho] = \xi$, there exists a unique conjugate charge $\overline{\xi} = [\overline{\varrho}]$ characterized by the property that $\varrho\overline{\varrho}$ contains the vacuum sector. It is obtained by (IV.2.57). Moreover $\varrho\overline{\varrho}$ contains the vacuum sector precisely once and

$$\lambda_{\overline{\xi}} = \lambda_\xi . \tag{IV.2.58}$$

If R is an intertwining operator from the vacuum representation to $\overline{\varrho}\varrho$ then

$$\overline{R} = \text{sign}(\lambda_\varrho)\, \varepsilon(\overline{\varrho}, \varrho)R \tag{IV.2.59}$$

is an intertwining operator from the vacuum sector to $\varrho\overline{\varrho}$. R may be chosen so that

$$\overline{R}^*\varrho R = \mathbb{1}; \quad R^*\overline{\varrho}\overline{R} = \mathbb{1}; \quad R^*R = \overline{R}^*\,\overline{R} = d(\varrho)\mathbb{1} . \tag{IV.2.60}$$

For the proof see DHR4. The relations (IV.2.59), (IV.2.60) hold also when ϱ is reducible. In that case d is the sum of the dimensions of the irreducible parts contained in ϱ. If ϱ is irreducible then R is uniquely determined by (IV.2.59), (IV.2.60) up to a phase.

IV.2.4 Covariant Sectors and Energy-Momentum Spectrum

Let us assume now that the sector ϱ is covariant i.e. that the covering group of the Poincaré group is implementable by unitary operators $U_\varrho(g), g = (x,\alpha) \in \tilde{\mathfrak{P}}$. Then the operators

$$T_\varrho(g) = U_0(g)U_\varrho^*(g) \qquad \text{(IV.2.61)}$$

are charge shifting operators as considered in (IV.2.44). Specializing to translations $g = x$ we have

$$T_\varrho(x)(\varrho A)T_\varrho(x)^* = \varrho_x A \qquad \text{(IV.2.62)}$$

where ϱ_x denotes the "shifted" morphism

$$\varrho_x = \alpha_x \varrho \alpha_{-x} \qquad \text{(IV.2.63)}$$

The key lemma 2.2.2 gives then, with $U_1 = \mathbb{1}, U_2 = T_\varrho(x)$, $\operatorname{supp} \varrho \subset \mathcal{O}$,

$$\varepsilon_\varrho = (\varrho T_\varrho(x)^*)T_\varrho(x) \quad \text{if} \quad \mathcal{O} + x \quad \text{is space-like to} \quad \mathcal{O}\,. \qquad \text{(IV.2.64)}$$

Applying the left inverse Φ we get for such x

$$\Phi T_\varrho(x) = \lambda_\varrho T_\varrho(x). \qquad \text{(IV.2.65)}$$

In the case of infinite statistics ($\lambda_\varrho = 0$) this would give

$$\Phi T_\varrho(x) = 0 \quad \text{for large space-like} \quad x\,. \qquad \text{(IV.2.66)}$$

This relation indicates the essential reason why λ_ϱ cannot vanish. Combined with the positivity of the energy in the representations under consideration one finds by analyticity arguments that it would entail the identical vanishing of some two-point functions.

The properties of covariance and finiteness of statistics are conserved if one takes products of morphisms, subrepresentations or conjugates (see DHR4). In a massive theory all sectors of interest should be of this type. We list the most relevant consequences.

Lemma 2.4.1
If ϱ is irreducible then $U_\varrho(g)$ is uniquely determined and

(i) $U_\varrho(g) \in (\varrho\mathfrak{A})''$,

(ii) if $\mathbf{R}$ is an intertwiner from ϱ_1 to ϱ_2 then it also intertwines from U_{ϱ_1} to U_{ϱ_2}.

We denote the P_μ-spectrum in the representation ϱ by Spect U_ϱ. The additivity of the spectrum, discussed at the end of chapter II, now becomes

Theorem 2.4.2

(i) Spect $U_{\varrho_2\varrho_1} \supset$ Spect U_{ϱ_2} + Spect U_{ϱ_1} .

(ii) If ϱ_1 and ϱ_2 are irreducible and ϱ is a subrepresentation of $\varrho_2\varrho_1$ then also

$$\text{Spect } U_\varrho \supset \text{Spect } U_{\varrho_1} + \text{Spect } U_{\varrho_2}\,.$$

The proof of part (i) is a simple adaptation of the proof given for the field theoretic case in section 5 of chapter II. Part (ii) follows if one uses in addition part (ii) of lemma 2.4.1. For details see DHR4.

One would expect that in analogy to the field theoretic case one can construct a CPT-operator Θ, giving an antiunitary mapping between the representations ϱ and $\overline{\varrho}$ and that, as a consequence, Spect U_ϱ and Spect $U_{\overline{\varrho}}$ are the same. It must be remembered, however, that an essential ingredient in Jost's construction of Θ was the assumption of fields with a finite number of components obeying a transformation law specified in axiom **D** (section 1 of chapter II). This assumption is now absent. Still, H. Epstein has shown that the essential features of Jost's conclusions remain valid in a theory based on a net of local algebras, i.e. in the frame described in section 1 of chapter III [Ep 67]. One can construct a CPT-operator for the single particle states and for the collision states constructed from them and one retains the spin-statistics connection. The method of Epstein was adapted to the present analysis in DHR4. One has

Theorem 2.4.3

a) If the sector ξ contains single particle states with mass m, spin s then the conjugate sector contains single particle states with the same mass, spin and multiplicity.

b) Spin-statistics connection:

The sign of the statistics parameter is $(-1)^{2s}$.

IV.2.5 Fields and Collision Theory

The construction of an S-matrix in the general collision theory, described in II.4 used charged local fields. It can be applied without change in the algebraic setting of III.1 but its adaptation to the present analysis, where instead of a field algebra we have localized morphisms, is less evident. The point is the distinction between states and state vectors. It is obvious how to construct states (expectation functionals) corresponding to specific configurations of incoming or outgoing particles. They determine transition probabilities and collision cross sections. The S-matrix contains in addition to this also the phases of scalar products of state vectors. For this we must use the complex linear structure of the Hilbert spaces of state vectors in the various sectors and the "charged fields" are a convenient tool to express this. The construction of a field algebra from the observables and localized morphisms will be discussed in section 4. For the purposes of collision theory a more primitive concept, called *field bundle* suffices. We sketch this here because the structure has some interest in its own right and is needed in cases where a field algebra cannot be constructed (e.g. section 5) .

We have described all representations by operators acting in the same Hilbert space $\mathcal{H}_0$. Thus, picking a vector in $\mathcal{H}_0$ determines a state on $\mathfrak{A}$ only if we also specify the representation in which it is to be understood. Let us therefore consider as a generalized state vector a pair $\{\varrho; \Psi\}$ where the first member is a covariant morphism and the second a vector in $\mathcal{H}_0$. Correspondingly we consider a set $\mathcal{B}$ of operators acting on these generalized vectors. An element of $\mathcal{B}$ is a pair $\mathbf{B} = \{\varrho; B\}$ with $\varrho \in \Delta$ and $B \in \mathfrak{A}$. It acts on $\Psi = \{\varrho'; \Psi\}$ by

$$\mathbf{B}\Psi = \{\varrho'\varrho; (\varrho' B)\Psi\} \ . \tag{IV.2.67}$$

This leads to the associative multiplication law within $\mathcal{B}$

$$\mathbf{B}_2\mathbf{B}_1 = \{\varrho_1\varrho_2; (\varrho_1 B_2)B_1\} \quad \text{when} \quad \mathbf{B}_k = \{\varrho_k; B_k\} \ . \tag{IV.2.68}$$

The action of $\overline{\mathfrak{P}}$ on these objects is then defined as

$$\mathbf{U}(g)\{\varrho; \Psi\} = \{\varrho; U_\varrho(g)\Psi\}, \quad g \in \overline{\mathfrak{P}}, \tag{IV.2.69}$$

$$\alpha_g\{\varrho; B\} = \left\{\varrho; U_\varrho(g) B U_0(g)^{-1}\right\} \ . \tag{IV.2.70}$$

The description is, of course, redundant because a representation may be replaced by an equivalent one without changing the physical significance. This means that

$$\{\varrho; \Psi\} \sim \{\sigma_U\varrho; U\Psi\}, \tag{IV.2.71}$$

$$\{\varrho; B\} \sim \{\sigma_U\varrho; UB\} \tag{IV.2.72}$$

when U is a unitary from $\mathfrak{A}_{\text{loc}}$. The reader is invited to check that the appearance of UB rather than $\sigma_U B$ on the right hand side of (IV.2.72) is no printing mistake! One may then consider the set of generalized state vectors and $\mathcal{B}$ as bundles whose base are the equivalence classes $[\varrho]$ and where the unitary group in $\mathfrak{A}_{\text{loc}}$ is the structure group. One can use the equivalence relation to endow $\mathcal{B}$ with a local structure:

Definition 2.5.1
$\mathbf{B} = \{\varrho; B\} \in \mathcal{B}(\mathcal{O})$ if there exists a unitary $U \in \mathfrak{A}_{\text{loc}}$ such that $\sigma_U\varrho$ has support in $\mathcal{O}$ and $UB \in \mathfrak{A}(\mathcal{O})$

A conjugation is defined by

$$\mathbf{B} = \{\varrho; B\} \longrightarrow \mathbf{B}^\dagger = \{\bar{\varrho}; (\bar{\varrho}B^*)R\} \tag{IV.2.73}$$

where R is an intertwiner from the vacuum sector to $\bar{\varrho}\varrho$. Note that in the case of infinite statistics the conjugation could not be defined and relation (IV.2.66) would yield

$$\langle \mathbf{B}\Omega \mid \mathbf{U}(x) \mid \mathbf{B}\Omega \rangle = 0 \tag{IV.2.74}$$

if $\mathbf{B} \in \mathcal{B}(\mathcal{O})$ and $\mathcal{O} + x$ space-like to $\mathcal{O}$.
This relation could be extended by analyticity to all x and one would arrive at a contradiction.

With the help of these bundles the construction of state vectors for configurations of incoming or outgoing particles can be carried through in analogy to II.4. For the explicit discussion and details of the algorithm concerning $\mathcal{B}$ the reader is referred to DHR4.

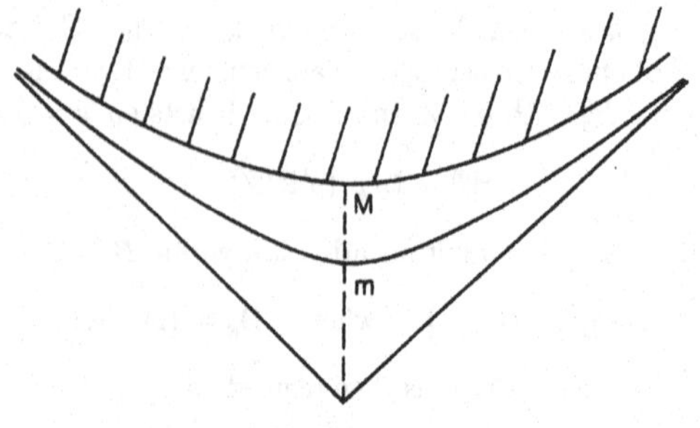

Fig. IV.3.1.

IV.3 The Buchholz-Fredenhagen (BF)-Analysis

IV.3.1 Localized Single Particle States

The DHR-criterion (IV.1.2) aimed at singling out the subset of states with vanishing matter density at infinity within a purely massive theory. With the same aim in mind Buchholz and Fredenhagen start from the consideration of a charge sector in which the space-time translations are implementable by unitary operators $U(x)$ and the energy-momentum spectrum is as pictured in fig. (IV.3.1). It shall contain an isolated mass shell of mass m (single particle states) separated by a gap from the remainder of the spectrum which begins at mass values above $M > m$. In this situation they construct states with spectral support on a bounded part of the single particle mass shell which may be regarded as strongly localized in the following sense. Let π denote the representation of the observable algebra for this sector, $\mathcal{H}$ the Hilbert space on which it acts and Ψ the state vector of such a localized state. Then the effect of a finite translation on Ψ can be reproduced by the action of an almost local operator on Ψ, or, in terms of the infinitesimal generators (the energy-momentum operators in this representation) by

$$P_\mu \Psi = B_\mu \Psi; \quad B_\mu = B_\mu^* \in \pi(\mathfrak{A}_{\text{a.l.}}). \tag{IV.3.1}$$

Here $\mathfrak{A}_{\text{a.l.}}$ denotes the almost local part of $\mathfrak{A}$ i.e. the set of elements which can be approximated by local observables in a diamond of radius r with an error decreasing in norm faster than any inverse power of r.

To see what is involved in the claim (IV.3.1) let us first suppose that Lorentz transformations are implementable in the sector considered. Then one can adapt the argument of Wigner (chapter I section 3.2) and introduce for the state vectors of single particles of mass m an improper basis $|\,\mathbf{p}, \xi\rangle$, with

$$\begin{aligned} P^\mu \,|\,\mathbf{p}, \xi\rangle &= p^\mu \,|\,\mathbf{p}, \xi\rangle \quad ; \; p^0 = (\mathbf{p}^2 + m^2)^{1/2}, \\ |\,\mathbf{p}, \xi\rangle &= U(\beta(p))\,|\,\mathbf{0}, \xi\rangle \end{aligned}$$

(see equ. (I.3.18), (I.3.20)). ξ labels a basis in the degeneracy space $\mathfrak{h}$. A general state vector Ψ in $\mathcal{H}^{(1)}$ can be described by a wave function $\tilde{\Psi}(\mathbf{p})$ or, in position space, by

$$\Psi(\mathbf{x}) = \int \tilde{\Psi}(\mathbf{p}) e^{i\mathbf{p}\cdot\mathbf{x}} d^3p$$

which is determined by its values on the space-like surface $x^0 = 0$. It may be considered as a function of $\mathbf{x}$ taking values in the little Hilbert space $\mathfrak{h}$.

An "almost localized" state in $\mathcal{H}^{(1)}$ is naturally characterized as one for which $\| \Psi(\mathbf{x}) \|_{\mathfrak{h}}$ decreases faster than any inverse power as $| \mathbf{x} | \to \infty$. This corresponds to the Newton-Wigner concept of localization for single particle states. If Ψ and Φ describe localized states in this sense then $\langle \Phi \mid U(x) \mid \Psi \rangle$ decreases fast to zero when x moves to space-like infinity. The construction of such states is illustrated by the following argument. Let Δ_1, Δ_2 be two disjoint regions of momentum space intersecting the spectrum only on the mass shell m. If C is an almost local operator from $\pi(\mathfrak{A})$ with momentum transfer Δ so that $\Delta_1 + \Delta$ lies in Δ_2, Φ a state vector with momentum support in Δ_1 and $\Psi = C\Phi$ then we expect that

$$\Psi(\mathbf{x}) = \int K(\mathbf{x}, \mathbf{x}') \phi(\mathbf{x}') d^3x' \qquad \text{(IV.3.2)}$$

with an integral kernel K with values in $\mathfrak{B}(\mathfrak{h})$ and such that $\| K(\mathbf{x}, \mathbf{x}') \|$ tends to zero fast when either $\mathbf{x}$ or $\mathbf{x}'$ move to infinity. This must be so if we believe that the notion of localization of Newton and Wigner agrees roughly with the one determined by the interpretation of local algebras. In fact C, being almost local, can produce a change essentially only in a finite neighborhood of the origin, but it must produce a momentum change from Δ_1 to Δ_2 and so cannot depend on the parts of $\phi(\mathbf{x})$ with large $\mathbf{x}$. In this way we can produce an almost localized state Ψ from an arbitrary one by momentum space restrictions using an almost local operator. Clearly, if Ψ is almost localized and has bounded spectral support then the same will be true for $P^\mu\Psi$. The last question in establishing (IV.3.1) is therefore whether the set of almost local operators in $\pi(\mathfrak{A})$ is rich enough to transform any almost local state in $\mathcal{H}^{(1)}$ into any other such state. If $\mathfrak{h}$ is finite dimensional this poses no problem.

In their analysis [Buch 82a], quoted as BF in this section, the authors dispense with the assumption that the Lorentz symmetry is implementable and they do not use the Newton-Wigner notion of localization. Their construction of localized states follows the heuristic idea underlying (IV.3.2), starting from a single particle state Φ and an almost local operator C with the mentioned restrictions of their momentum space supports. Their proof that such states satisfy (IV.3.1) uses only the shape of the spectrum as pictured in fig. (IV.3.1) and the relation $P^0 = (\mathbf{P}^2 + m^2)^{1/2}$ on $\mathcal{H}^{(1)}$. In fact they derive a stronger result. Instead of P^μ one may take in (IV.3.1) any smooth function of the P^μ. Their proof is, however, rather technical and we do not reproduce it here.

In the following analysis we shall assume that π is irreducible[1] and that there is a state vector satisfying (IV.3.1).

Theorem 3.1.1
The representation π determines a pure, translationally invariant state ω_0 (vacuum) by

$$\mathrm{w}-\lim \pi(\alpha_x A) = \omega_0(A)\mathbb{1}; \quad A \in \mathfrak{A} \tag{IV.3.3}$$

as x goes to space-like infinity in any direction. The GNS-representation π_0 constructed from ω_0 has a P^μ-spectrum consisting of the isolated point $p = 0$ which is a nondegenerate eigenvalue corresponding to the vacuum state and the other part contained in $p^0 > 0, p^2 \geq (M - m)^2$.

Proof. Let $A \in \mathfrak{A}(\mathcal{O})$ and $\mathcal{O}$ a diamond centered at the origin. Since the unit ball of $\mathfrak{B}(\mathcal{H})$ is weakly compact there exist weak limit points of $\pi(\alpha_x A)$ as x moves to space-like infinity. Each such limit is in $\pi(\mathfrak{A})'$ and hence, as π was assumed irreducible, it is a multiple of the identity. The coefficient, denoted by $\omega_0(A)$ above, can be calculated if we take the expectation value of (IV.3.3) for an arbitrary state vector $\Psi \in \mathcal{H}$. For Ψ we choose a vector satisfying (IV.3.1). Then

$$\begin{aligned}\partial_\mu \langle \Psi \mid \pi(\alpha_x A) \mid \Psi\rangle &= i\langle \Psi \mid [P_\mu, \pi(\alpha_x A)] \mid \Psi\rangle \\ &= i\langle \Psi \mid [B_\mu, \pi(\alpha_x A)] \mid \Psi\rangle .\end{aligned}$$

Since B_μ is almost local and A local, the commutator decreases fast as x moves to space-like infinity. Integrating from x to y along some path connecting these points we get

$$|\langle \Psi \mid \pi(\alpha_y A) \mid \Psi\rangle - \langle \Psi \mid \pi(\alpha_x A) \mid \Psi\rangle| < h(r) \parallel A \parallel , \tag{IV.3.4}$$

with $r = R - d$ where R is the minimal (space-like) Lorentz distance from the origin to an arc connecting x and y and d is the diameter of the diamond $\mathcal{O}$. $h(r)$ decreases faster than any inverse power of r. Thus $\pi(\alpha_x A)$ converges weakly for $A \in \mathfrak{A}_{\mathrm{loc}}$ and hence also for $A \in \mathfrak{A}$ and the limit is independent of the way how x moves to infinity. (The last statement is, of course, no longer true if we have only one space dimension. There we might have to distinguish between a "right hand vacuum " and a "left hand vacuum", a phenomenon characteristic for solitons).

ω_0 is a state on $\mathfrak{A}$ since it is a positive, linear functional on $\mathfrak{A}_{\mathrm{loc}}$ with $\omega_0(\mathbb{1}) = 1$, it is translation invariant and continuous with respect to the norm topology of $\mathfrak{A}$. Therefore we have in the representation π_0, arising from ω_0 by the GNS-construction, certainly an implementation of α_x by unitary operators $U_0(x)$ (see equ. (III.3.14)).

The additivity of the energy-momentum spectrum discussed in II.5 generalizes to a relation between the spectra of U_0 in π_0 and U in π. Suppose $\Phi \in \mathcal{H}$ has

[1] This is done only to simplify some arguments. In BF π is only assumed to be primary.

spectral support (with respect to U) in the momentum space region Δ and f is a test function with momentum space support Δ_1, fast decreasing in x-space, $A \in \mathfrak{A}_{\text{loc}}, A(f) = \int f(x)\alpha_x A d^4x$. Then $\pi(A(f)) \mid \Phi\rangle$ has spectral support in $\Delta + \Delta_1$ and the same is true for $\pi(\alpha_y A(f)) \mid \Phi\rangle$ with arbitrary y. The norm square of this vector converges to $\omega_0(A(f)^* A(f))$ for $\mid \mathbf{y} \mid \rightarrow \infty$ which does not vanish for all A if $\Delta_1 \in \text{Spect } U_0$. Therefore

$$\text{Spect } U \supset \text{Spect } U + \text{Spect } U_0 \,. \tag{IV.3.5}$$

Given Spect U as in fig. (IV.3.1) the spectrum of U_0 can therefore consist only of the point $p = 0$ and momentum vectors with positive energy and mass above $(M - m)$. So π_0 is a vacuum representation with mass gap of (at least) $M - m$. Next one shows that ω_0 has the cluster property

$$\omega_0(A_2 \alpha_x A_1) \rightarrow \omega_0(A_2)\omega_0(A_1) \quad \text{as} \quad \mid \mathbf{x} \mid - \mid x^0 \mid \rightarrow \infty \,. \tag{IV.3.6}$$

Consider the function

$$F(x, y) = \langle \Psi \mid \pi\left(\alpha_y(A_2 \alpha_x A_1)\right) \mid \Psi\rangle \tag{IV.3.7}$$

for $A_1, A_2 \in \mathfrak{A}(\mathcal{O})$, y and x space-like to the origin. Take the radial straight line from y to infinity and its parallel beginning at $x + y$. The minimal space-like distance from these lines to the origin shall be R. Then, by the same argument leading to (IV.3.4) we get

$$|F(x, y) - \omega_0(A_2 \alpha_x A_1)| \; < h(r) \parallel A_2 \parallel \parallel A_1 \parallel \tag{IV.3.8}$$

where h is a fast decreasing function of $r = R - d$, independent of A_1, A_2 in $\mathfrak{A}(\mathcal{O})$. On the other hand

$$\lim_x F(x, y) = \langle \Psi \mid \pi(\alpha_y A_2) \mid \Psi\rangle \, \omega_0(A_1) \underset{y \rightarrow \infty}{\longrightarrow} \omega_0(A_2)\omega_0(A_1) \,. \tag{IV.3.9}$$

The cluster property of ω_0 implies that ω_0 is the only translation invariant state in the folium of π_0, so the GNS-vector Ω_0 is a non-degenerate eigenstate of P^μ. Since, by theorem III.3.2.4, $U_0(x)$ is in the weak closure of $\pi_0(\mathfrak{A})$ and Ω_0 is cyclic for π_0 the commutant of $\pi_0(\mathfrak{A})$ must be trivial i.e. π_0 is irreducible, ω_0 is a pure state. □

IV.3.2 BF Topological Charges

Consider in a space-like plane the region $\mathcal{C}$ (the shaded region in fig. (IV.3.2)). Starting with a ball around the origin with radius r we draw a straight line from the origin to infinity and, around the point at distance r' from the origin on this line we take a ball with radius $r + \gamma r'$; $\gamma > 0$. $\mathcal{C}$ is the union of all these balls for $0 \leq r' < \infty$. We denote by $\mathfrak{A}^c(\mathcal{C})$ the *relative commutant* of $\mathfrak{A}(\mathcal{C})$ i.e. the set of all elements of $\mathfrak{A}$ which commute with every element of $\mathfrak{A}(\mathcal{C})$.

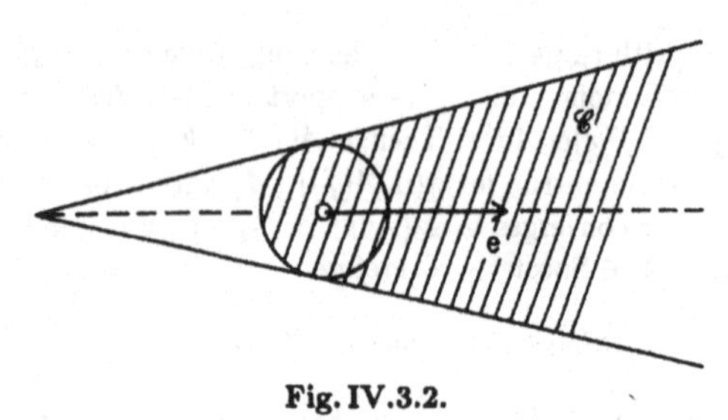

Fig. IV.3.2.

Lemma 3.2.1
Let $\omega(A) = \langle \Psi \mid \pi(A) \mid \Psi \rangle$ with Ψ satisfying (IV.3.1), ω_0 as defined in theorem 3.1.1 and C as described above. Then the restrictions of the states ω, ω_0 to the relative commutant $\mathfrak{A}^c(C)$ satisfy

$$\left\| (\omega - \omega_0) \restriction_{\mathfrak{A}^c(C)} \right\| < \varepsilon \,, \qquad \text{(IV.3.10)}$$

where ε does not depend on the direction of the center line of C and vanishes fast as $r \to \infty$.

Proof. Writing $\Psi_x = U(x)\Psi, B_\mu(x) = U(x)B_\mu U^*(x)$ one obtains from (IV.3.1)

$$P_\mu \Psi_x = B_\mu(x)\Psi_x$$

and from theorem 3.1.1, with $\mathbf{e}$ denoting the unit vector in direction of the center line of C

$$\begin{aligned} |(\omega - \omega_0)(A)| &= \left| \int_0^\infty \frac{\partial}{\partial x'_\mu} \langle \Psi_{x'} \mid \pi(A) \mid \Psi_{x'} \rangle \, e^\mu dr' \right| \\ &= \left| \int \langle \Psi_{x'} \mid [e^\mu B_\mu(x'), \pi(A)] \mid \Psi_{x'} \rangle \, dr' \right| \,, \end{aligned}$$

with $x' = (0, r'\mathbf{e}); e = (0, \mathbf{e})$.
$e^\mu B_\mu(x')$ can be approximated by an element of the algebra of a diamond with center x' and diameter d' up to an error with norm $f(d')$ where f is a fast decreasing function. For $A \in \mathfrak{A}^c(C)$, choosing $d' = r + \gamma r'$, this implies that the commutator under the integral above has norm less than $2 \parallel A \parallel f(r + \gamma r')$. Thus

$$\begin{aligned} |(\omega - \omega_0)(A)| &\leq \parallel A \parallel \int f(r + \gamma r') dr' \\ &\leq \parallel A \parallel g(r, \gamma) \,, \end{aligned} \qquad \text{(IV.3.11)}$$

where g is fast decreasing with r. □

Definition 3.2.2
Let $\mathcal{O}$ be a diamond space-like separated from the origin and $a \in \mathbb{R}^4$ an arbitrary point. The region

$$S = a + \cup_\lambda \lambda\mathcal{O}; \quad 0 \leq \lambda \leq \infty \tag{IV.3.12}$$

is called a space-like cone with apex a.

Theorem 3.2.3
Let π be a representation of $\mathfrak{A}$ as described above[2], acting in the Hilbert space $\mathcal{H}$, and $(\pi_0, \mathcal{H}_0)$ the corresponding vacuum representation (according to theorem 3.1.1). Then, for any space-like cone S one has a unitary mapping V from $\mathcal{H}$ onto $\mathcal{H}_0$ such that

$$V\pi(A) = \pi_0(A)V \quad \text{for} \quad A \in \mathfrak{A}^c(S) \,. \tag{IV.3.13}$$

In other words: the restrictions of π and π_0 to the relative commutant of the algebra of any space-like cone are unitarily equivalent.

Proof. If the theorem is true for one position of the apex it will be true for any apex position since the translations are implementable in the representation π. The region C in lemma 3.2.1 is contained in a space-like cone S with apex $a = -\gamma^{-1}re$ and central axis along e. Therefore lemma 3.2.1 says that for arbitrarily small $\varepsilon > 0$ we can choose r sufficiently large so that there is a state ω in the folium of π which in restriction to $\mathfrak{A}^c(S)$ approximates ω_0 with an error of norm less than ε. Since the folium of states in a representation is complete in the norm topology the folium of π contains a state $\widehat{\omega}$ which in restriction to $\mathfrak{A}^c(S)$ coincides with ω_0. One shows that under the prevailing circumstances there is a vector $\widehat{\Omega} \in \mathcal{H}$ inducing the state $\widehat{\omega}$ on $\mathfrak{A}^c(S)$ and cyclic for this subalgebra, i.e.

$$\left\langle \widehat{\Omega} \mid \pi(A) \mid \widehat{\Omega} \right\rangle = \widehat{\omega}(A) = \omega_0(A) \quad \text{for} \quad A \in \mathfrak{A}^c(S) \tag{IV.3.14}$$

$$\overline{\pi(\mathfrak{A}^c(S))\widehat{\Omega}} = \mathcal{H} \,. \tag{IV.3.15}$$

Accepting this, the intertwining operator in (IV.3.13) is defined by

$$V\pi(A)\widehat{\Omega} = \pi_0(A)\Omega \quad \text{for} \quad A \in \mathfrak{A}^c(S) \,, \tag{IV.3.16}$$

where Ω is the vacuum state vector in $\mathcal{H}_0$. To show that a vector $\widehat{\Omega}$ with the properties (IV.3.14), (IV.3.15) exists and that (IV.3.16) gives a consistent definition of an operator V mapping $\mathcal{H}$ onto $\mathcal{H}_0$ one may use the following specialization of [*Sakai 1971, theorem 2.7.97*]:

[2]The P^μ-spectrum shall be as in fig. (IV.3.1) and, for simplicity, we also assume irreducibility.

Theorem 3.2.4
Let $\mathcal{N}$ be a von Neumann algebra acting on a Hilbert space $\mathcal{H}$, $\hat{\omega}$ a normal state on $\mathcal{N}$. If there exists a vector $\Psi \in \mathcal{H}$ which is cyclic for $\mathcal{N}'$ then there is a vector $\hat{\Omega} \in \mathcal{H}$ inducing $\hat{\omega}$. If $\hat{\omega}$ is faithful on $\mathcal{N}$ and Ψ cyclic for $\mathcal{N}$ then $\hat{\Omega}$ is cyclic.

For our application $\mathcal{N} = \pi(\mathfrak{A}^c(S))''$ and Ψ may be taken as the vector from which the construction started. It was chosen to have bounded spectral support and is therefore cyclic for both $\mathcal{N}$ and $\mathcal{N}'$ due to the (generalized) Reeh-Schlieder theorem. By the same theorem Ω is cyclic for $\pi_0(\mathfrak{A}^c(S))$ and by a slight variant of this theorem one shows that $\hat{\omega}$ is faithful on $\mathcal{N}$.

Replacing A by AB in (IV.3.16) one sees that V satisfies (IV.3.13). The isometry of V follows from (IV.3.14).

This completes the proof for the case of cones with an axis e having vanishing time component. But lemma 3.2.1 holds in any Lorentz frame. □

Comment. Lemma 3.2.1 comes close to the DHR-selection criterion (IV.1.3) since we can cover the outside of a ball of radius r by the complements of two regions C_i of the shape depicted in fig.(IV.3.2). The step from lemma 3.2.1 to theorem 3.2.3 is analogous to the one from (IV.1.3) to (IV.1.2). It is therefore at first sight surprising that one cannot infer (IV.1.3) from lemma 3.2.1 using additivity (III.4.8) for the local algebras and the cluster property for the states which, in a massive theory, demands a fast decrease of correlations with increasing distance. The following rudimentary model may show how the difference between lemma 3.2.1 and equ. (IV.1.3) can arise due to homotopy properties of regions. Consider an Abelian subnet on 3-dimensional space at a sharp time. The net shall be generated by unitaries $\mathcal{V}(\mathbf{f})$ where $\mathbf{f}$ is a real, 3-vector-valued function of $\mathbf{x} \in \mathbb{R}^3$. $\mathcal{V}(\mathbf{f})$ is assigned to the algebra of the support region of $\mathbf{f}$ and

$$\mathcal{V}(\mathbf{f}_1 + \mathbf{f}_2) = \mathcal{V}(\mathbf{f}_1)\mathcal{V}(\mathbf{f}_2)\,.^3 \qquad \text{(IV.3.17)}$$

This guarantees additivity because, given an open covering of the support region of $\mathbf{f}$, we can write $\mathbf{f}$ as a sum of functions $\mathbf{f}$ with supports in the subregions and then apply (IV.3.17). Now let

$$\omega_0\left(\mathcal{V}(\mathbf{f})\right) = e^{-(\mathbf{f},K\mathbf{f})}\,, \qquad \text{(IV.3.18)}$$
$$\omega\left(\mathcal{V}(\mathbf{f})\right) = e^{i(\mathbf{h},\mathbf{f})}\omega_0\left(\mathcal{V}(\mathbf{f})\right)\,, \qquad \text{(IV.3.19)}$$

where $\mathbf{h}$ is a fixed function with curl $\mathbf{h} = 0$ and the support of div $\mathbf{h}$ is contained in the ball $|\,\mathbf{x}\,| < r_0$. We may interpret $\mathbf{h}$ as the electric field of an external charge with density $\varrho = \text{div}\,\mathbf{h}$.

The linear combinations of elements $\mathcal{V}(\mathbf{f})$ with complex coefficients form a *-algebra and ω_0 and ω define states on it if $\mathbf{K}$ is a positive operator in the test function space. The states have the cluster property if in

$$(\mathbf{f}\,,\mathbf{K}\,\mathbf{f}) \equiv \int \mathbf{f}(\mathbf{x}')\mathbf{K}(\mathbf{x}'-\mathbf{x})\mathbf{f}(\mathbf{x})d^3x'd^3x \qquad \text{(IV.3.20)}$$

[3] We may put $\mathcal{V}(\mathbf{f}) = e^{i\mathbf{E}(\mathbf{f})}$ and consider $\mathbf{E}$ as an analogue to the electric field strength.

the kernel $\mathbf{K}$ is fast decreasing with $|\mathbf{x}' - \mathbf{x}|$. Now $(\mathbf{h}, \mathbf{f}) = 0$ if supp $\mathbf{f}$ is in the complement of any cone containing the support of div $\mathbf{h}$. One sees this most easily if one decomposes $\mathbf{f}$ into a longitudinal and a transversal part:

$$\mathbf{f} = \text{grad}\, \varphi + \text{curl}\, \mathbf{g}\,. \tag{IV.3.21}$$

Then, since $\mathbf{h}$ is curl free

$$(\mathbf{h}, \mathbf{f}) = -\int \varphi \,\text{div}\, \mathbf{h}\, d^3x \tag{IV.3.22}$$

where φ is given as the solution of the potential equation

$$\Delta\varphi = \text{div}\, \mathbf{f}$$

with the boundary condition that φ vanishes at infinity. If the support of div $\mathbf{h}$ (the ball $|\mathbf{x}| < r_0$) is connected to infinity by paths avoiding the support of div $\mathbf{f}$ then $\varphi = 0$ in the support region of div $\mathbf{h}$ and hence $(\mathbf{h}, \mathbf{f}) = 0$. Therefore ω and ω_0 coincide on $\mathcal{V}(\mathbf{f})$ whenever supp $\mathbf{f}$ is in the complement of an arbitrary cone containing the support of div $\mathbf{h}$. On the other hand let

$$\mathbf{f}(\mathbf{x}) = \alpha\,(|\,\mathbf{x}\,|) \cdot \mathbf{x} \tag{IV.3.23}$$

with α having support in the spherical shell $r_1 < |\,\mathbf{x}\,| < r_2$ with $r_1 > r_0$. Then, for $r < r_1$, φ is a nonvanishing constant $\varphi_0 = -\int \alpha(r) r dr$ and

$$\omega\,(\mathcal{V}(\mathbf{f})) = e^{-iq\varphi_0}\omega_0\,(\mathcal{V}(\mathbf{f}))\,; \quad q = \int \text{div}\, \mathbf{h}\, d^3x\,. \tag{IV.3.24}$$

We can choose r_1 arbitrarily large keeping the value of φ_0 fixed. So there are elements in the observable algebra of a spherical shell of arbitrarily large inner radius for which ω_0 and ω differ markedly. If we decompose $\mathbf{f}$ in (IV.3.23) smoothly

$$\mathbf{f} = \mathbf{f}_1 + \mathbf{f}_2$$

where $\mathbf{f}_1$ has its support limited to polar angles $\vartheta < \vartheta_1$ and $\mathbf{f}_2$ to polar angles $\vartheta > \vartheta_2$ $(\vartheta_2 < \vartheta_1)$ then

$$\omega\,(\mathcal{V}(\mathbf{f}_i)) = \omega_0\,(\mathcal{V}(\mathbf{f}_i))\,, \quad i = 1, 2\,,$$

but $|\omega_0\,(\mathcal{V}(\mathbf{f})) - \omega_0\,(\mathcal{V}(\mathbf{f}_1))\,\omega_0\,(\mathcal{V}(\mathbf{f}_2))|$ need not be small in spite of the fact that $\mathbf{f}_1$ and $\mathbf{f}_2$ have only a small overlap and $\mathbf{K}(\mathbf{x}' - \mathbf{x})$ is fast decreasing with increasing distance. Suppose as a typical example

$$(\mathbf{f}\,, \mathbf{K}\,\mathbf{f}) = \int (\text{curl}\, \mathbf{f})^2 d^3x\,.$$

Then, for $\mathbf{f}$ given by (IV.3.23), $\omega_0(\mathcal{V}(\mathbf{f})) = 1$ because $\mathbf{f}$ is curl free. But $\mathbf{f}_1$ arising from $\mathbf{f}$ by multiplication with a function of ϑ which decreases from 1 to 0 in the interval $\vartheta_2 < \vartheta < \vartheta_1$, is not curl free. In fact, the closer one takes ϑ_1 to ϑ_2 the larger becomes $(\mathbf{f}\,, \mathbf{K}\,\mathbf{f})$. Thus, no matter how one makes the decomposition, $\omega_0(\mathcal{V}(\mathbf{f}_i)) < 1$.

IV.3.3 Composition of Sectors and Exchange Symmetry

Again, as in section 2, it is convenient to replace the representation π by an equivalent representation acting in $\mathcal{H}_0$. Choosing an arbitrary space like cone S we map $\mathcal{H}$ onto $\mathcal{H}_0$ by the unitary operator V of (IV.3.16) and define the representation $A \to \varrho A \in \mathfrak{B}(\mathcal{H}_0)$ by

$$\varrho A = V\pi(A)V^*, \quad A \in \mathfrak{A} . \tag{IV.3.25}$$

Again we may consider the vacuum representation as the defining representation and omit the symbol π_0. The map ϱ from $\mathfrak{A}$ to $\mathfrak{B}(\mathcal{H}_0)$ acts trivially on $\mathfrak{A}^c(S)$ and, in fact, the triviality extends to the weak closure $\mathfrak{A}(S)''$. Thus ϱ maps $\mathcal{R}(S) = \mathfrak{A}(S)''$ into itself and one may say that ϱ is "localized in S". For general elements $A \in \mathfrak{A}$ the image ϱA will, however, not necessarily belong to $\mathfrak{A}$. Thus ϱ need not be a morphism of $\mathfrak{A}$ and therefore the composition $\varrho_2\varrho_1$ is not directly defined because $\varrho_1 A$ may lie outside the domain of definition of ϱ_2. Now, for $A \in \mathfrak{A}(\mathcal{O})$, ϱA will be in the commutant of $\mathfrak{A}(S' \cap \mathcal{O}')$ because, if $\mathcal{O}_1$ is space-like to both S and $\mathcal{O}$ and $B \in \mathfrak{A}(\mathcal{O}_1)$ then

$$B\varrho A = (\varrho B)(\varrho A) = \varrho BA = \varrho AB = (\varrho A)(\varrho B) = (\varrho A)B .$$

If we take an element $A \in \mathfrak{A}(\mathcal{O})$ and two mappings ϱ_k localized respectively in S_k so that S_2 is space-like to both S_1 and $\mathcal{O}$ then $\varrho_2\varrho_1 A = \varrho_1 A$ because $\varrho_1 A$ is in $\mathfrak{A}(S_1' \cap \mathcal{O}')'$ which is contained in $\mathfrak{A}(S_2)'$ where ϱ_2 acts trivially. To define a composition of two maps ϱ_k which are localized in cones with arbitrary mutual placement one introduces a suitably placed auxiliary cone. In BF the auxiliary cone is used to define an extended algebra which is stable under the action of the maps so that the ϱ_k become morphisms of the extended algebra. A somewhat simpler way to define the composition product was described by Doplicher and Roberts [Dopl 90]. Let S_1 be the support of ϱ_1 and $A \in \mathfrak{A}(\mathcal{O})$. Choose an auxiliary cone with support in S_3, space-like to both S_1 and $\mathcal{O}$ Then there is a representation ϱ_3 with support in S_3 and equivalent to ϱ_2 with unitary intertwiner V_3:

$$\varrho_2 B = V_3(\varrho_3 B)V_3^*; \quad B \in \mathfrak{A} .$$

Since we have

$$\varrho_3\varrho_1 A = \varrho_1 A \quad \text{for} \quad A \in \mathfrak{A}(\mathcal{O})$$

this suggests the definition

$$\varrho_2\varrho_1 A = V_3(\varrho_1 A)V_3^*; \quad A \in \mathfrak{A}(\mathcal{O}) . \tag{IV.3.26}$$

The location of the auxiliary cone has to vary as the support of A varies. One shows, however, that (IV.3.26) defines a representation depending only on ϱ_1, ϱ_2, independent of the choice of ϱ_3 and V_3 within the specified limitations.

Once the composition of sectors is defined the analysis of exchange symmetry, the construction of a field bundle and of collision states proceeds in analogy to DHR (see subsection 2.5). The discussion becomes somewhat more tedious because one has to verify the independence from the choice of the auxiliary cone at all steps. We shall not repeat it here but refer to BF and [Dopl 90]. In a theory based on Minkowski space the conclusions remain the same as those described in section 2. It is, however, already intuitively clear that now all 4 dimensions are needed since the morphisms are now homotopically no longer points but half lines extending to infinity and one needs three such lines for the construction of ε_ϱ. So one needs at least three space-like dimensions to effect the interchange of two mutually space-like morphisms by continuous shifts without entering into time-like configurations. Thus the relevance of the braid group instead of the permutation group (section 5) begins already in 3-dimensional space-time for BF topological charges.

IV.3.4 Charge Conjugation and Absence of Infinite Statistics

Let S be the support cone of ϱ and e a space-like direction so that $S + \lambda e$ becomes space-like to every diamond as $\lambda \to +\infty$. We have

$$U_\varrho(x)(\varrho A)U_\varrho^*(x) = \varrho\alpha_x A, \quad A \in \mathfrak{A}\ .$$

Consider now the sequence $U_\varrho(x)AU_\varrho^*(x), A \in \mathfrak{A}$ with $x = \lambda e, \lambda \to +\infty$. (Note that $A = \pi_0(A) \in \mathfrak{B}(\mathcal{H}_0)$). Due to the weak compactness of the unit ball of $\mathfrak{B}(\mathcal{H}_0)$ it has weak limit points in $\mathfrak{B}(\mathcal{H}_0)$. Each such limit is in the commutant of $\varrho\mathfrak{A}$. To see this take A and $B \in \mathfrak{A}_{\mathrm{loc}}$. Then

$$\left[U_\varrho(x)AU_\varrho^*(x), \varrho B\right] = U_\varrho(x)\left[A, \varrho\alpha_{-x}B\right]U_\varrho^*(x)\ .$$

But $\alpha_{-x}B$ moves out of the support of ϱ as $\lambda \to \infty$. Then ϱ acts trivially and the commutator vanishes for large enough λ. Since ϱ was assumed irreducible the limit points are multiples of the identity and we can evaluate them by taking the expectation value in any state. For this we choose the state vector Ψ which satisfies (IV.3.1). Note that with the present notation P^μ is the generator of $U_\varrho^*(x)$ and B_μ in (IV.3.1) should be written as ϱB_μ with $B_\mu \in \mathfrak{A}_{\mathrm{a.l.}}$ The same argument as used in the proof of theorem 3.1.1 shows then that the sequence converges weakly and one has

Lemma 3.4.1
With e as above

$$\mathrm{w} - \lim_{\lambda\to\infty} U_\varrho(\lambda e)AU_\varrho^*(\lambda e) = \overline{\omega}(A)\mathbb{1}; \quad A \in \mathfrak{A} \tag{IV.3.27}$$

The state $\overline{\omega}$ satisfies

$$\overline{\omega}(\varrho A) = \omega_0(A). \tag{IV.3.28}$$

The relation (IV.3.28) follows directly from theorem 3.1.1. It shows that $\overline{\omega}$ is a state in the conjugate sector.

We can also construct a left inverse of ϱ. Defining the charge shifting operator

$$T_\varrho(x) = U_0(x)U_\varrho^*(x) \tag{IV.3.29}$$

(compare (IV.2.61), (IV.2.62)) we get for large λ

$$\begin{aligned}\left\langle C\Omega_0 \mid \sigma_{T_\varrho(\lambda e)}A \mid B\Omega_0\right\rangle &= \left\langle \Omega_0 \mid (\alpha_{-\lambda e}C^*)U_\varrho^*(\lambda e)AU_\varrho(\lambda e)(\alpha_{-\lambda e}B) \mid \Omega_0\right\rangle \\ &= \left\langle \Omega_0 \mid U_\varrho^*(\lambda e)(\varrho C^*)A(\varrho B)U_\varrho(\lambda e) \mid \Omega_0\right\rangle \\ &\overset{\lambda\to\infty}{\longrightarrow} \quad \overline{\omega}\left((\varrho C^*)A\varrho B\right).\end{aligned}$$

We have used that the vacuum vector Ω_0 is invariant under $U_0(x)$, that $\alpha_{-\lambda e}\mathfrak{A}_{\text{loc}}$ moves outside the support of ϱ so that we can insert the action of ϱ on the respective elements, and finally (IV.3.28). This gives

Lemma 3.4.2

$$\begin{aligned} \mathrm{w} - \lim_{\lambda\to\infty} \sigma_{T_\varrho(\lambda e)}A &= \Phi(A); \quad A \in \mathfrak{A}, &\text{(IV.3.30)}\\ \langle C\Omega_0 \mid \Phi(A) \mid B\Omega_0\rangle &= \overline{\omega}((\varrho C^*)A\varrho B)\,. &\text{(IV.3.31)}\end{aligned}$$

Φ is a left inverse of ϱ.

One can use the left inverse Φ to define the conjugate sector, the statistics parameter $\lambda_\varrho = \Phi(\varepsilon_\varrho)$ and show, as in section 2 that the possible values of λ_ϱ are limited to $0, \pm d^{-1}$. The final aim is to exclude $\lambda_\varrho = 0$ (infinite statistics). This was achieved by Fredenhagen [Fred 81a]. He shows that $\lambda_\varrho = 0$ implies (IV.2.74) for a non void open set of points $x \in \mathcal{M}$ and that one has sufficient analyticity to conclude that (IV.2.74) must hold then for all x. This is in conflict with the positivity of the Hilbert space metric.

A proof of the spin-statistics connection in the case of BF-charges has been given by Buchholz and Epstein [Buch 85].

IV.4 Global Gauge Group and Charge Carrying Fields

In the traditional field theoretic approach (which in algebraic guise was summarized in III.1) we deal with a Hilbert space $\mathcal{H}^{(u)}$ which contains all superselection sectors as subspaces (the index u was added to indicate "universal") and we deal with a larger algebra $\mathfrak{F}$ (denoted by $\mathcal{A}$ in III.1) acting irreducibly on $\mathcal{H}^{(u)}$. It contains besides the observables also operators mapping from one coherent subspace to a different one ("charge carrying fields"). Furthermore there is a faithful, strongly continuous representation of a compact group, the global gauge group $\mathcal{G}$, by unitary operators in $\mathfrak{B}(\mathcal{H}^{(u)})$, singling out the observables in $\mathfrak{F}$ as the invariant elements under the action induced by $\pi^{(u)}(\mathcal{G})$. In other words

$$\pi^{(u)}(\mathfrak{A}) = \mathfrak{F} \cap \pi^{(u)}(\mathcal{G})'\,. \tag{IV.4.1}$$

This poses the question as to whether $\mathcal{G}$ and the algebraic structure of $\mathfrak{F}$ are determined in a natural fashion by the physical content of the theory i.e. by $\mathfrak{A}$ and its superselection structure as described in the last sections. There are several facets to this question.

Implementation of Endomorphisms. If we start from a representation $\pi(\mathfrak{A})$ in a Hilbert space $\mathcal{H}$ in which each sector occurs at least once (we omit now the index $^{(u)}$ on π and $\mathcal{H}$) then, for $\varrho \in \Delta(\mathcal{O})$ there will exist intertwining operators ψ between $\pi(\mathfrak{A})$ and $\pi(\varrho\mathfrak{A})$:

$$\psi\pi(A) = \pi(\varrho A)\psi; \quad \psi^*\pi(\varrho A) = \pi(A)\psi^*. \tag{IV.4.2}$$

We may choose ψ as a partial isometry, an isometric mapping from a subspace $E_1\mathcal{H}$ onto a subspace $E_2\mathcal{H}$ where the projectors E_i are given by

$$E_1 = \psi^*\psi; \quad E_2 = \psi\psi^*. \tag{IV.4.3}$$

The subspaces must be chosen so that each equivalence class of subrepresentations of $\pi(\mathfrak{A})E_1$ appears in $\pi(\varrho\mathfrak{A})E_2$ with the same multiplicity. If the operators ψ and ψ' both satisfy (IV.4.2) then we have

$$\psi'^*\psi\pi(A) = \pi(A)\psi'^*\psi; \quad \text{i.e.} \quad \psi'^*\psi \in \pi(\mathfrak{A})'. \tag{IV.4.4}$$

Now the commutant $\pi(\mathfrak{A})'$ contains only global quantities, not referring to any particular region. On the other hand, if $\varrho \in \Delta(\mathcal{O})$ then $\varrho A = A$ for $A \in \mathfrak{A}(\mathcal{O}')$. Hence ψ and ψ' commute with $\pi(\mathfrak{A}(\mathcal{O}'))$. This suggests that we associate ψ and ψ' with a field algebra of the region $\mathcal{O}$;

$$\psi \in \mathfrak{F}(\mathcal{O}); \quad \mathfrak{F}(\mathcal{O}) \subset \pi(\mathfrak{A}(\mathcal{O}'))'. \tag{IV.4.5}$$

We do not want $\mathfrak{F}(\mathcal{O})$ to contain any non trivial global quantities. Since $\psi'^*\psi \in \mathfrak{F}(\mathcal{O}) \cap \pi(\mathfrak{A})'$ we should demand

$$\psi'^*\psi = c\mathbb{1}. \tag{IV.4.6}$$

The set of operators satisfying (IV.4.2) and (IV.4.6) forms a linear space in $\mathfrak{B}(\mathcal{H})$. It is in fact a Hilbert space (of operators) since we can define a numerical sesquilinear positive definite scalar product $\langle\psi' \mid \psi\rangle$ by

$$\psi'^*\psi = \langle\psi' \mid \psi\rangle\mathbb{1}. \tag{IV.4.7}$$

Let us denote this Hilbert space by $\mathfrak{h}$ and assume that it has finite dimension d. The elements of unit length are isometries i.e. their source projector $E_1 = \mathbb{1}$. We may choose an orthonormal basis ψ_k in $\mathfrak{h}$:

$$\psi_i^*\psi_k = \delta_{ik}\mathbb{1}. \quad i, k = 1, \ldots d \tag{IV.4.8}$$

If (and only if) the range projectors $\psi_k\psi_k^*$ add up to $\mathbb{1}$, i.e.

$$\sum \psi_k \psi_k^* = \mathbb{1} \tag{IV.4.9}$$

we can implement the endomorphism ϱ in the representation π and obtain from (IV.4.2)

$$\pi(\varrho\mathfrak{A}) = \sum \psi_k \pi(\mathfrak{A}) \psi_k^*. \tag{IV.4.10}$$

The relations (IV.4.8), (IV.4.9) may be considered as the defining relations of an abstract *-algebra generated by the ψ_k, ψ_k^*. Cuntz has shown that there is a unique C^*-norm on this algebra and that the completion of the algebra in this norm yields a simple C^*-algebra, the Cuntz algebra O_d.

One checks that with $\psi \in \mathfrak{h}$ and $C \in O_d$

$$\psi C = (\sigma_{\mathfrak{h}} C)\psi\ , \tag{IV.4.11}$$

where $\sigma_{\mathfrak{h}}$ is the "inner" endomorphism of O_d

$$C \in O_d \longrightarrow \sigma_{\mathfrak{h}} C = \sum \psi_k C \psi_k^*\ . \tag{IV.4.12}$$

Note that $\sigma_{\mathfrak{h}}$ depends only on $\mathfrak{h}$, not on the choice of the basis ψ_k. We call $\sigma_{\mathfrak{h}}$ the canonical endomorphism of O_d. Thus $\sigma_{\mathfrak{h}}$ extends the endomorphism ϱ to the part of $\mathfrak{F}(\mathcal{O})$ consisting of the representation of the Cuntz algebra connected with ϱ. One anticipates that the dimension d of $\mathfrak{h}$ must be the statistics dimension of ϱ. To show this one needs the full machinery involving the conjugate charge and permutation operators (see theorem 4.2 below).

Charges with d = 1. Here the charges form an Abelian group, the conjugate charge being the inverse. For each charge ξ we choose some morphism ϱ_ξ in its class and a representation π_ξ of $\mathfrak{A}$ acting in a Hilbert space $\mathcal{H}_\xi$, equivalent to $\varrho_\xi \mathfrak{A}$ in $\mathcal{H}_0$. So we have a unitary map V_ξ from $\mathcal{H}_0$ to $\mathcal{H}_\xi$ with

$$\pi_\xi(A) = V_\xi(\varrho_\xi A)V_\xi^*.$$

It appears natural to form the direct sums

$$\mathcal{H}^{(u)} = \sum_{\widehat{\mathcal{G}}}{}^{\oplus} \mathcal{H}_\xi; \quad \pi^{(u)} = \sum{}^{\oplus} \pi_\xi\ . \tag{IV.4.13}$$

Then each sector appears in $\mathcal{H}^{(u)}$ exactly once. Since each π_ξ is irreducible the commutant of $\pi^{(u)}(\mathfrak{A})$ reduces to multiples of the identity on each $\mathcal{H}_\xi$ and thus consists of multiplication operators by functions $F(\xi)$. A basis system of functions, in terms of which general functions may be expanded, is provided by the characters of the group (harmonic analysis). We recall that a character may be defined as a 1-dimensional, unitary representation i.e. it is a function $\chi(\xi)$ with values in T (the unit circle in the complex plane) with the property

$$\chi(\xi_2)\chi(\xi_1) = \chi(\xi_2\xi_1)\ . \tag{IV.4.14}$$

The set of characters of an Abelian group forms itself a group, the so called "dual group", by

$$(\chi_2\chi_1)(\xi) = \chi_2(\xi)\chi_1(\xi); \quad \chi^{-1}(\xi) = (\chi(\xi))^{-1} . \tag{IV.4.15}$$

We denote it by $\mathcal{G}$. It is the (global) gauge group. By the Pontrjagin duality theorem the charge group is then the dual group of $\mathcal{G}$. The representation of $\mathcal{G}$ in $\mathcal{H}^{(u)}$ is

$$\pi^{(u)}(\chi)\Psi = \chi(\xi)\Psi \quad \text{for} \quad \Psi \in \mathcal{H}_\xi . \tag{IV.4.16}$$

Next we construct unitary operators in $\mathfrak{B}(\mathcal{H}^{(u)})$ which implement the automorphisms $\varrho \in \Delta$. Let $[\varrho] = \xi$ then

$$\pi^{(u)}(\varrho A) \mid_{\mathcal{H}_{\xi'}} = V_{\xi'} V^*_{\xi'\xi} \pi_{\xi'\xi}(\sigma_W A) V_{\xi'\xi} V^*_{\xi'} \tag{IV.4.17}$$

where W is an intertwining operator between $\varrho_{\xi'}\varrho$ and $\varrho_{\xi'\xi}$

$$\varrho_{\xi'}\varrho = \varrho_{\xi'\xi}\sigma_W . \tag{IV.4.18}$$

Introducing an operator ψ_ϱ which lowers the charge by ξ and acts on $\mathcal{H}_{\xi'}$ by

$$\psi_\varrho \mid_{\mathcal{H}_{\xi'}} = V_{\xi'} V^*_{\xi'\xi} \pi_{\xi'\xi}(W) \tag{IV.4.19}$$

we have

$$\pi^{(u)}(\varrho A) = \psi_\varrho \pi^{(u)}(A) \psi^*_\varrho . \tag{IV.4.20}$$

The adjoint (and inverse) ψ^*_ϱ raises the charge by ξ.

The computation of W for arbitrary ϱ and ξ' and hence the commutation properties of $\psi_\varrho, \psi^*_\varrho$ depend on some conventions. We refer to DHR2 for a general discussion of the case of charges with $d = 1$. The result is that there remains some freedom but that it is always possible to construct a field algebra with normal commutation relations i.e. such that at space-like distances two fields carrying fermionic charges ($\lambda = -1$) anticommute, a field carrying bosonic charge commutes with all other fields.

Endomorphisms and non Abelian Gauge Group. New aspects occur when the charge structure corresponds to localized endomorphisms, the general situation described in section 2.

It was shown there that $\varrho \in \Delta_{\text{irr}}$ leads to parastatistics of some finite order d in the multiply charged sectors ϱ^n. The qualitative discussion of section 1 suggests that there is an alternative description in which the net of observable algebras is embedded in a larger net $\breve{\mathfrak{A}}$ which still has causal commutativity but has a non Abelian internal symmetry group $\mathcal{G}_0$ realized by automorphism $g \in \mathcal{G}_0 \to \alpha_g \in \text{Aut}\, \breve{\mathfrak{A}}$ [1] so that the observables are the invariant elements of $\breve{\mathfrak{A}}$ under α_g. It is then a matter of taste whether one regards $\breve{\mathfrak{A}}$ as as the "true" algebra of observables and $\mathcal{G}_0$ as an internal symmetry or $\mathfrak{A}$ as the observable algebra and $\mathcal{G}_0$ as (part of) the global gauge group. The latter point of view is more appropriate if the symmetry is not broken. The question is now whether the structure encountered in section 2 can always be interpreted in this way.

[1] i.e. α_g transforms each $\breve{\mathfrak{A}}(\mathcal{O})$ into itself.

This would mean that the superselection structure determines a group such that the pure charges $\xi \in \Delta_{\rm irr}/\mathcal{I}$ are in one-to-one correspondence with the equivalence classes of irreducible representations of the group, the composition $\xi_2\xi_1$ corresponds to the tensor product of the group representations, charge conjugation to the complex conjugate representation and the decomposition of $\xi_2\xi_1$ into irreducibles mirrors the Clebsch-Gordan decomposition of the tensor product. In short: the structure of the semigroup $\Delta/\mathcal{I}$ should be recognizable as the dual object of a group.

To say simply that the answer is affirmative would not do justice to the work invested by Doplicher and Roberts in the study of this question and the insights gained through this investigation. It has added a new chapter to the mathematical theory of group duality. In particular it has shown that a semigroup of endomorphisms of a C*-algebra together with its intertwiners, possessing precisely all the properties elaborated in section 2 (apart from the sign of the statistics parameter), may be regarded in a natural fashion as the (abstract) dual of a compact group. Further that a concrete group dual is obtained by embedding the structure in a larger algebra (the algebra $\breve{\mathfrak{A}}$ mentioned above or, ultimately, the field algebra).

It is beyond the scope of this book to present a full account of this work. I shall outline the central ideas and state (without proof) results. This may facilitate the study of the original papers for the interested reader.

The natural language is provided by category theory. This need not be a deterrent for a theoretical physicist of our days. The basic concepts are simple and natural. A category consists of "objects" carrying some mathematical structure (e.g. algebras, representations of a group ...) and "arrows" between objects corresponding to maps conserving the structure. In the examples relevant here the objects are Hilbert spaces (possibly equipped with a representation of a group or algebra). This implies that the set of arrows from object α to object β, which will be denoted by $i(\alpha,\beta)$[2], is a Banach space: linear combinations of arrows in this set are naturally defined and so is a norm on these arrows. In any category one has a composition of arrows provided that the target of the first coincides with the source of the second. If $\mathbf{S} \in i(\alpha,\beta), \mathbf{S}' \in i(\beta,\gamma)$ then $\mathbf{S}' \circ \mathbf{S} \in i(\alpha,\gamma)$. Here $\| \mathbf{S}' \circ \mathbf{S} \| \leq \| \mathbf{S}' \| \| \mathbf{S} \|$; one has an adjoint $\mathbf{S}^* \in i(\beta,\alpha)$ and $\| \mathbf{S}^* \| = \| \mathbf{S} \|, \| \mathbf{S}^* \circ \mathbf{S} \| = \| \mathbf{S} \|^2$. This justifies the name C*-category for such a structure. It stands in the same relation to a C*-algebra as a groupoid to a group (the algebraic operations being only defined if the elements fit together). For each object α the arrows $i(\alpha,\alpha)$ form a C*-algebra. Its unit will be denoted by $\mathbb{1}_\alpha$. This allows the definition of subobjects and direct sums of objects. The category is said to have subobjects if, for every projector $E \in i(\alpha,\alpha)$ there is an object γ (called a subobject of α) and an arrow $\mathbf{S} \in i(\alpha,\gamma)$ such that $\mathbf{S} \circ \mathbf{S}^* = \mathbb{1}_\gamma, \mathbf{S}^* \circ \mathbf{S} = E$. The direct sum of two objects α, β is defined if there is an object γ (called the direct sum of α and β) such that one has arrows $\mathbf{V} \in i(\gamma,\alpha), \mathbf{W} \in i(\gamma,\beta)$ with $\mathbf{V} \circ \mathbf{V}^* = \mathbb{1}_\alpha, \mathbf{W} \circ \mathbf{W}^* = \mathbb{1}_\beta$

[2] In order to conform with standard usage we have now written the source to the left, the target to the right in the bracket. The opposite convention was used in section 2.

and $\mathbf{V}^* \circ \mathbf{V} + \mathbf{W}^* \circ \mathbf{W} = \mathbb{1}_\gamma$. The C*-categories we deal with here are closed under direct sums and subobjects.

Two such categories concern us in our context. One is the representation theory of a compact Lie group $\mathcal{G}$, the "dual of $\mathcal{G}$". We may take as objects finite dimensional unitary representations of $\mathcal{G}$ and as arrows the homomorphisms between them. It suffices if the set of objects contains at least one sample of each equivalence class of representations. We denote this category by $\mathcal{T}_\mathcal{G}$. The other category is the superselection structure discussed in section 2. The objects are the localized morphisms from Δ, the arrows are the intertwiners. We denote it by $\mathcal{T}_\mathfrak{A}$.

In both cases the categories have the following additional structure. There is a product of objects which entails an (associative) product of arrows. In $\mathcal{T}_\mathcal{G}$ this is the tensor product of representations accompanied by the tensor product of homomorphisms. In $\mathcal{T}_\mathfrak{A}$ it is the product of morphisms accompanied by the cross product of intertwiners. The relations (IV.2.16), (IV.2.17) hold in both cases. There is a unit object, denoted by ι. It is the trivial representation of $\mathcal{G}$ (respectively the vacuum representation or the identity morphism of $\mathfrak{A}$). A category with such a product structure is called a *strict monoidal category*. The categories are also *symmetric*. To each pair of objects α, β there is a unitary $\varepsilon(\alpha,\beta) \in i(\alpha\beta, \beta\alpha)$ interchanging the order of factors of arrows such that for $\mathbf{S} \in i(\alpha,\beta), \mathbf{S}' \in i(\alpha',\beta')$ one has

$$\varepsilon(\beta',\beta) \circ (\mathbf{S}' \times \mathbf{S}) = (\mathbf{S} \times \mathbf{S}') \circ \varepsilon(\alpha',\alpha) \ . \qquad \text{(IV.4.21)}$$

This leads, as discussed in section 2 to a unitary representation of the permutation group S_n in the algebra $i(\alpha^n, \alpha^n)$ for each α.

The final element of structure is the existence of conjugates. To each object α one can pick an object $\overline{\alpha}$ and arrows

$$\mathbf{R} \in i(\iota, \overline{\alpha}\alpha), \quad \overline{\mathbf{R}} = \varepsilon(\overline{\alpha},\alpha) \circ \mathbf{R} \in i(\iota, \alpha\overline{\alpha}) \ , \qquad \text{(IV.4.22)}$$

with[3]

$$(\overline{\mathbf{R}}^* \times \mathbb{1}_\alpha) \circ (\mathbb{1}_\alpha \times \mathbf{R}) = \mathbb{1}_\alpha; \quad (\mathbf{R}^* \times \mathbb{1}_{\overline{\alpha}}) \circ (\mathbb{1}_{\overline{\alpha}} \times \overline{\mathbf{R}}) = \mathbb{1}_{\overline{\alpha}}. \qquad \text{(IV.4.23)}$$

In the categories $\mathcal{T}_\mathcal{G}$ and $\mathcal{T}_\mathfrak{A}$ the algebra $i(\iota,\iota)$ reduces to $\mathbb{C}$ (the complex numbers). For $\mathcal{T}_\mathfrak{A}$ this follows from the irreducibility of the vacuum representation. Let us now call, for short, a category with all the structure mentioned above a *DR-category*. It is a strict monoidal C*-category, closed under subobjects and direct sums, symmetric and with conjugates and we shall also require

[3]Note that the relations (IV.4.23) differ by the sign of the statistics parameter from (IV.2.59). This finds its natural expression by saying that $\mathcal{T}_\mathfrak{A}$ should be regarded as a "$\mathbb{Z}_2$ graded category" [Dopl 88]. In keeping with the objective mentioned at the beginning of the section we may also concern ourselves first with the subcategory formed by the bosonic charges and construct $\mathcal{G}_0$ and $\breve{\mathfrak{A}}$. This algebra will still have localized automorphisms containing a charge group $\mathbb{Z}_2$; in a second step one can then apply the methods for the case $d = 1$ to construct the full gauge group $\mathcal{G}$ and the field algebra $\mathfrak{F}$. We shall not take up the Bose-Fermi problem here again.

$i(\iota, \iota) = \mathbb{C}$. We list some important properties of DR-categories. The proofs may be found in [Dopl 89b].

Lemma 4.1
The conjugate is defined up to unitary equivalence i.e. if $\overline{\alpha}_1$ and $\overline{\alpha}_2$ are possible choices for a conjugate to α and $\mathbf{R}_1, \mathbf{R}_2$ are the respective intertwiners of (IV.4.22), (IV.4.23) then there is a unitary intertwiner $\mathbf{U} \in i(\overline{\alpha}_1, \overline{\alpha}_2)$ so that

$$\mathbf{R}_2 = (\mathbf{U} \times \mathbb{1}_\alpha) \circ \mathbf{R}_1 .$$

Theorem 4.2
There is a dimension function on the objects $\alpha \to d(\alpha)$ defined by

$$d(\alpha) = \mathbf{R}^* \circ \mathbf{R} \in i(\iota, \iota)$$

where $\mathbf{R}$ is given by (IV.4.22), (IV.4.23) and $d(\alpha)$ is independent of the choice of $\overline{\alpha}$ (due to lemma 4.1). It satisfies

(i) $d(\alpha_1\alpha_2) = d(\alpha_1)d(\alpha_2)$,

(ii) $d(\alpha_2 \oplus \alpha_1) = d(\alpha_2) + d(\alpha_1)$,

(iii) the possible values of $d(\alpha)$ are the natural numbers.

Definition 4.3
The subobject of $\alpha^{d(\alpha)}$ corresponding to the completely antisymmetric subspace will be called the *determinant* of α and denoted by det α. The objects with determinant ι will be called *special objects.*

Theorem 4.4

(i) det α has dimension 1.

(ii) det $(\alpha \oplus \overline{\alpha}) = \iota$.

(iii) Every finite set $\{\alpha_1, \ldots, \alpha_n\}$ of objects is "dominated" by a special object, i.e. all α_k $(k = 1, \ldots n)$ are subobjects of some α with det $\alpha = \iota$.

Consider now the symmetric, monoidal C*-category whose objects are the tensor powers of a d-dimensional Hilbert space $\mathfrak{h}_d$ (not equipped with any additional structure) and their subspaces. The arrows are the isometries between these spaces (or, more conveniently, all linear maps between them). It is obvious how this can be embedded in the Cuntz algebra. $\mathfrak{h}_d$ corresponds to the basic Hilbert space $\mathfrak{h}$ in O_d; $\mathfrak{h}_d^{\otimes n}$ to the linear span of products of n elements $\psi \in \mathfrak{h} \subset O_d$; $i(\mathfrak{h}^{\otimes n+k}, \mathfrak{h}^{\otimes n})$ to the linear span of monomials with $l + k$ factors ψ on the left and l factors ψ^* on the right where $l \leq n$, $l + k \geq 0$ and $k \in \mathbb{Z}$. In the last correspondence we have identified arrows $\mathbf{T}$ and $\mathbf{T} \times \mathbb{1}_{\mathfrak{h}_d}$ with the same element of the Cuntz algebra; then tensoring on the left with $\mathbb{1}_{\mathfrak{h}_d}$ corresponds to the canonical endomorphism of O_d.

If $\mathcal{G}$ is a compact Lie group it has a faithful unitary representation $g \in \mathcal{G} \to u(g)$ with determinant 1 in some d-dimensional space and such that every (finite dimensional) representation of $\mathcal{G}$ is equivalent to one contained in some tensor power of u. The category generated by u (the objects being the tensor powers of u, their subrepresentations and direct sums, with the trivial representation as the element ι; the arrows being homomorphisms of these representations) may be regarded as a dual of $\mathcal{G}$. We denote it by $\mathcal{T}_{\mathcal{G}}$. It is a DR-category in which u is a special object and it is naturally embedded in the Cuntz algebra, equipped now with an automorphism group $g \in \mathcal{G} \to \alpha_g \in \mathrm{Aut}\, O_d$ where α_g is defined by its action on $\mathfrak{h}$

$$\alpha_g \psi = u(g)\psi \,. \tag{IV.4.24}$$

The arrows correspond to invariant elements of O_d under the action of α_g.

Theorem 4.5
The DR-category $\mathcal{T}_\varrho$ generated by a single object ϱ of dimension d and determinant ι is isomorphic to $\mathcal{T}_{\mathcal{G}}$ for some compact Lie group $\mathcal{G}$. Both can be naturally embedded in O_d , equipped with an automorphism group α_g ($g \in \mathcal{G}$) as described in (IV.4.24) such that the arrows in $\mathcal{T}_\varrho$ are mapped onto a subalgebra $O_{\mathcal{G}} \subset O_d$, the fixed points under α_g in O_d. Let $T \in O_{\mathcal{G}}$ be the image of the arrow $\mathbf{T}$ under this map then T is also the image of $\mathbf{T} \times \mathbb{1}_\varrho$ whereas $\mathbb{1}_\varrho \times \mathbf{T}$ is mapped on $\sigma_{\mathfrak{h}} T$ (see (IV.4.12)).

Returning to the original problem: we have the observable algebra $\mathfrak{A}$, the set Δ of its localized morphisms and the intertwiners, implemented by elements of $\mathfrak{A}$. Considering first the bosonic morphisms Δ_{bos} (positive statistics parameter) one has

Theorem 4.6
The structure $(\mathfrak{A}, \Delta_{\mathrm{bos}})$ can be embedded in a C*-algebra $\breve{\mathfrak{A}}$ with causal net structure and equipped with an automorphism group $\mathcal{G}_0$ such that to each $\varrho \in \Delta_{\mathrm{bos}}$ there is a subspace $\mathfrak{h}_\varrho$ of isometries in $\breve{\mathfrak{A}}$ (basis $\psi_1, \ldots, \psi_d$) implementing ϱ by

$$\varrho A = \sum \psi_k A \psi_k^*, \quad A \in \mathfrak{A}, \tag{IV.4.25}$$

and $\mathfrak{A}$ consists of the invariant elements of $\breve{\mathfrak{A}}$ under the automorphisms $\alpha_g \in \mathcal{G}_0$. If the semigroup of charges is generated by a finite set then $\mathcal{G}_0$ is isomorphic to a compact Lie group.[4] For the full set $(\mathfrak{A}, \Delta)$ one can find an embedding in a C*-algebra $\mathfrak{F}$ with normal Bose-Fermi causal net structure and a compact automorphism group $\mathcal{G}$ such that $\mathfrak{A}$ is the fixed point subalgebra of $\mathfrak{F}$ under α_g and (IV.4.25) holds with $\psi_k \in \mathfrak{F}$.

References The survey given in this section is based on the original papers [Dopl 88], [Dopl 89a], [Dopl 89b], [Dopl 90].

[4] $\mathcal{G}$ need not be connected and it may possibly be zero dimensional i.e. discrete.

IV.5 Low Dimensional Space-Time and Braid Group Statistics

Let us follow the steps of the DHR-analysis for a quantum field theory in 2-dimensional space-time. The results of subsection 2.1 remain unchanged but lemma 2.2.1 and the key lemma 2.2.2 are modified. Consider the set of ordered pairs $(\mathcal{O}_1, \mathcal{O}_2)$ with space-like separation. If the dimension of space-time is larger than 2 then this set is connected; we can continuously shift a space-like configuration of two points to any other such configuration without crossing their causal influence zone. In 2-dimensional space-time we have two disconnected components. If $\mathcal{O}_1$ lies to the left of $\mathcal{O}_2$ then $\mathcal{O}_1(s)$ will have to remain on the left of $\mathcal{O}_2(s)$ for any continuous family of pairs $\mathcal{O}_k(s)$ with space-like separation. In the proof of lemma 2.2.1 we used the possibility of continuously moving the supports of the pair (ϱ_1, ϱ_2) to the supports of the pair (ϱ_1', ϱ_2'). In 2-dimensional space-time this is only possible if these (ordered) pairs of supports lie in the same connectivity component. The consequence for the key lemma is that ε_ϱ is not completely independent of the choice of the morphisms ϱ_1, ϱ_2 but we may obtain two different operators ε_ϱ , depending on whether in the construction (IV.2.27), (IV.2.28) we choose the support of ϱ_1 to the left of the support of ϱ_2 or to the right. Let us adopt the convention of defining ε_ϱ by the first mentioned choice of the supports. Then one finds that the opposite choice leads to ε_ϱ^{-1} in (IV.2.28). The relation (IV.2.30) is lost. Therefore ε_ϱ does not correspond to a permutation of two elements but generates a braiding of two strands. An illustration is afforded by two strands of hair of a young lady. If they are originally parallel then the position of the loose end points may be interchanged in two inequivalent fashions, rotating by 180° around the center line in a clockwise or in an anticlockwise sense. ε_ϱ corresponds to the one, ε_ϱ^{-1} to the other operation. Repetition of one of the procedures does not lead back to the original situation but to the beginning of a braid.

The braid group B_n for n strands is generated by such operations on neighboring strands. Let σ_k $(k = 1, \ldots, n-1)$ denote the interchange of the ends of the k-th and $(k+1)$-th strands by clockwise rotation. Then, as Artin has shown, the only independent relations which the σ_k will satisfy are

$$\sigma_k \sigma_{k+1} \sigma_k = \sigma_{k+1} \sigma_k \sigma_{k+1}\,, \quad k = 1, \ldots, n-2 \tag{IV.5.1}$$

$$\sigma_i \sigma_k = \sigma_k \sigma_i \quad \text{if} \quad |\, i - k \,| \geq 2\,. \tag{IV.5.2}$$

The elements of B_n are then the products of the σ_k and their inverses. Since for $n > m$ the group B_m is naturally embedded in B_n one may define B_∞ as the braid group for an unspecified finite number of strands.

Next we look at the adaptation of theorem 2.2.3, the generalization of the key lemma to the sector ϱ^n . In (IV.2.37) we have only to change τ_m to σ_m (corresponding to the choice that supp ϱ_m lies to the left of supp ϱ_{m+1}) and obtain as the operator representing σ_m in B_n (see equ. (IV.2.38))

$$\varepsilon_\varrho^{(n)}(\sigma_m) = \varrho^{m-1}\varepsilon_\varrho. \tag{IV.5.3}$$

One checks that this respects the group relations (IV.5.1), (IV.5.2):

$$\varepsilon_\varrho^{(n)}(\sigma_m)\varepsilon_\varrho^{(n)}(\sigma_k) = \varepsilon_\varrho^{(n)}(\sigma_k)\varepsilon_\varrho^{(n)}(\sigma_m) \quad \text{for} \quad |\, m-k \,| \geq 2 \tag{IV.5.4}$$

and

$$\varepsilon_\varrho^{(n)}(\sigma_m)\varepsilon_\varrho^{(n)}(\sigma_{m+1})\varepsilon_\varrho^{(n)}(\sigma_m) = \varepsilon_\varrho^{(n)}(\sigma_{m+1})\varepsilon_\varrho^{(n)}(\sigma_m)\varepsilon_\varrho^{(n)}(\sigma_{m+1}) \,. \tag{IV.5.5}$$

Thus (IV.5.3) generates a unitary representation of B_n . The relation (IV.5.4) follows directly from the fact that ε_ϱ commutes with $\varrho^2\mathfrak{A}$. This is part a) of lemma 2.2.2 which remains unaffected by the dimensionality of space-time. To show (IV.5.5) we start from the definition (IV.2.28) for ε_ϱ which we may simplify by choosing $\varrho_2 = \varrho, U_2 = \mathbb{1}$ to

$$\varepsilon_\varrho = U_1^{-1}\varrho U_1 \tag{IV.5.6}$$

if supp ϱ_1 is chosen in the left space-like complement of supp ϱ. Then

$$\begin{aligned} \varepsilon_\varrho\varrho\varepsilon_\varrho &= U_1^{-1}\varrho^2 U_1 \,, \\ (\varrho\varepsilon_\varrho)\varepsilon_\varrho(\varrho\varepsilon_\varrho) &= (\varrho\varepsilon_\varrho)U_1^{-1}\varrho^2 U_1 \\ &= U_1^{-1}(\varrho_1\varepsilon_\varrho)(\varrho^2 U_1) \\ &= U_1^{-1}\varepsilon_\varrho(\varrho^2 U_1) \\ &= \varepsilon_\varrho(\varrho\varepsilon_\varrho)\varepsilon_\varrho \,, \end{aligned} \tag{IV.5.7}$$

where we have used that U_1 is an intertwining operator from ϱ to ϱ_1, then that ε_ϱ is localized in supp ϱ which is space-like to supp ϱ_1, implying $\varrho_1\varepsilon_\varrho = \varepsilon_\varrho$ and finally (IV.2.29). Applying ϱ^{m-1} to (IV.5.7) yields (IV.5.5).

We note that in (IV.5.3) the right hand side is independent of n, the number of strands used, as long as $n > m$. Therefore we get a unitary representation of B_∞, which we denote by ε^ϱ

$$b \in B_\infty \to \varepsilon^\varrho(b) \tag{IV.5.8}$$

generated from

$$\varepsilon^\varrho(\sigma_m) = \varrho^{m-1}\varepsilon_\varrho \,. \tag{IV.5.9}$$

The next question concerns the characterization of the representation (IV.5.8) of B_∞ which is associated with an irreducible sector $[\varrho]$. For this, as in subsection 2.3, the conjugate sector $[\bar{\varrho}]$ and the left inverse Φ_ϱ of ϱ are important. Φ_ϱ is defined by (IV.2.46) and has the properties (IV.2.47), (IV.2.48), (IV.2.49). Due to (IV.2.48) and (IV.2.29) $\Phi_\varrho\varepsilon_\varrho$ commutes with $\varrho\mathfrak{A}$ and thus, for irreducible ϱ, it is a multiple of the identity

$$\Phi_\varrho\varepsilon_\varrho = \lambda_\varrho\mathbb{1}, \quad (\lambda_\varrho \in \mathbb{C}) \,. \tag{IV.5.10}$$

As in (IV.2.51) the "statistics parameter" λ_ϱ depends only on the class $[\varrho]$. But it may now be a complex number. We decompose it into a phase and an absolute value

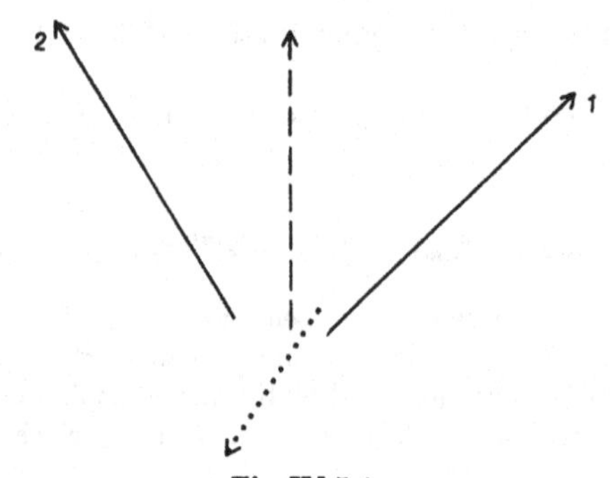

Fig. IV.5.1.

$$\lambda_\varrho = \omega_\varrho \cdot d_\varrho^{-1}; \quad |\, \omega_\varrho \,| = 1, \quad d_\varrho \geq 1, \tag{IV.5.11}$$

and call ω_ϱ the "statistics phase", d_ϱ the "statistics dimension". In contrast to (IV.2.55) ω_ϱ is not limited to the values ± 1 and d_ϱ need not be an integer.

In the simplest case, when ϱ is an automorphism, ε_ϱ is itself a multiple of the identity, the representation ε^ϱ of B_∞ is 1-dimensional, $d_\varrho = 1$ but ω_ϱ may be any phase factor. We do not have the Bose-Fermi alternative. Models with particles carrying such a charge have probably first been described by Streater and Wilde [Streat 70]. The term "anyon" was introduced for such a particle in [Wilc 82] to emphasize the fact that such particles may carry any statistics phase. For particles whose statistics dimension differs from 1 (i.e. where ϱ is an endomorphism) the term "plekton" (for Greek "braided") was suggested by Fredenhagen, Rehren, Schroer in [Fred 89a].

At this stage it may be remarked that in 3-dimensional space-time we meet the same situation if instead of the compactly localized charges of DHR we consider the string-like localized charges of the BF-analysis. This results from the fact that in order to define the composition of two charges localized in cones S_1, S_2 we need an auxiliary cone S_r which is space-like to both S_1 and S_2 (see section 3). There are two disconnected possibilities for the choice of S_r (see fig. (IV.5.1) where we have represented the cones by their center lines in the space-like plane $x^0 = 0$ and marked two inequivalent choices of S_r by a dotted and a dashed line respectively.)

As far as the statistics is concerned one arrives at the same conclusions as in the case of compactly localized charges in 2-dimensional space-time. There is, however, one further feature. In 3-dimensional space-time we may have spin. Adapting the analysis by Wigner (subsection I.3.2) we have to look at the stability group of a momentum vector. This is, in the case of a massive particle, the Abelian rotation group in a plane which has $\mathbb{R}$ as its covering group. An irreducible representation of this is 1-dimensional and given by $x \in \mathbb{R} \to e^{isx}$ where s is an arbitrary real number labelling the representation and corresponding to the spin. There is a relation between the spin s and the statistics phase

$$e^{4\pi i s} = \omega_\varrho^2 , \tag{IV.5.12}$$

and an addition theorem for the spin of plektons with statistics phases ω_1, ω_2, spins s_1, s_2 combining to a pure state with statistics phase ω, spin s

$$\frac{e^{2\pi i s}}{e^{2\pi i(s_1+s_2)}} = \frac{\omega}{\omega_1 \cdot \omega_2}. \tag{IV.5.13}$$

For the derivation see [Fred 89b], [Fröh 89].

Continuing with the discussion of statistics we can define a complex valued function φ on B_∞ using the definitions (IV.5.8), (IV.5.9), setting

$$\lim_{n\to\infty} \Phi_\varrho^n \varepsilon^\varrho(b) = \varphi(b)\ \mathbb{1} . \tag{IV.5.14}$$

This is analogous to (IV.2.52). We have used the fact that any $b \in B_\infty$ belongs to some B_m for finite m. If the power n of Φ_ϱ in (IV.5.14) exceeds m then by (IV.2.49) $\Phi_\varrho^n \varepsilon^\varrho$ is a multiple of the identity and does not change any more if n is increased. So we become in the limit independent of the number of strands in b. The function φ defines a state on the group algebra of B_∞ and it has properties analogous to lemma 2.3.6.

$$\varphi(b_1 b_2) = \varphi(b_2 b_1) , \tag{IV.5.15}$$

and, if $b_1 \in B_n$ and b_2 is generated by the σ_k with $k \geq n$,

$$\varphi(b_1 b_2) = \varphi(b_1)\varphi(b_2) . \tag{IV.5.16}$$

For the derivation see [Fred 89a]. A function on B_∞ with these properties is called a "strong Markov trace".

The representation ε^ϱ is obtained from φ by the GNS-construction. Similar to proposition 2.3.7 the statistics dimension d_ϱ limits the irreducible components which can occur in the restriction of ε^ϱ to B_n.

The finite dimensional irreducible representations of B_n have not yet been classified. However, an important advance in this direction has been initiated by the work of Jones [Jones 83]. He considered inclusions of von Neumann factors of type II_1, showed that such an inclusion defines a number with remarkable invariance properties, called the "index of the inclusion" and he gave a formula for the possible values this index can take. Kosaki as well as Pimsner and Popa generalized the notion of the Jones index to inclusions of arbitrary factors [Kos 86], [Pim 86]. Longo showed that in the case of the inclusion $\varrho\mathcal{R}(\mathcal{O}) \subset \mathcal{R}(\mathcal{O})$ which we discussed above the Jones index is just the square of the statistics dimension [Longo 89]

$$\mathrm{Ind}\ \varrho = d_\varrho^2 . \tag{IV.5.17}$$

The recognition by Jones, Ocneanu and others in the mid 80's that the theory of W*-inclusions leads to Markov traces on B_∞ and to invariants for knots and links spurred a considerable mathematical activity in this area. One interesting result is that the allowed values for the statistics dimension consist of a discrete spectrum in the interval between 1 and 2 with an accumulation point at 2 and a subset of $[2, \infty)$.[1] The discrete spectrum below 2 is given by

$$d = 2\cos\frac{\pi}{m}; \quad m \quad \text{a positive integer and} \quad m \geq 3 . \tag{IV.5.18}$$

For these values of d the Markov trace is uniquely determined by the statistics parameter λ.

There are cross connections to several other much discussed topics both in mathematics and physics. Among them are "quantum groups" and the Yang-Baxter equations in models of statistical mechanics and exactly solvable quantum field theory models in 2-dimensional space-time. For reviews and guides to the literature see [de Vega 89], [Fröh 88], [Fadd 90].

Choosing one representative morphism ϱ_α in each class of irreducible morphisms the composition law of the charges can be written as

$$\varrho_\alpha \varrho_\gamma = \sum N^\gamma_{\alpha\beta} \varrho_\beta, \tag{IV.5.19}$$

where $N^\gamma_{\alpha\beta}$ is the multiplicity with which the sector $[\varrho_\beta]$ occurs in the decomposition of the product $\varrho_\alpha\varrho_\gamma$. $N^\gamma_{\alpha\beta}$ is a non negative integer and N^γ is called the fusion matrix. Keeping $\varrho_\gamma = \varrho$ fixed one considers diagrams with the ϱ_α as vertices and paths between them allowed by the fusion rules, (see [Ver 88]). For each pair α, β we have $N^\gamma_{\alpha\beta}$ edges and a general path from an initial vertex to a final vertex is composed of such edges. The set of intertwiners from ϱ_β to $\varrho_\alpha\varrho$ is an $N^\gamma_{\alpha\beta}$-dimensional linear space in which an hermitean scalar product is defined because, if T_1 and T_2 are two such intertwiners then the product $T_1^* T_2$ is a multiple of the identity and we may set

$$T_1^* T_2 = \langle T_1 \mid T_2 \rangle \, \mathbb{1} . \tag{IV.5.20}$$

Picking an orthonormal basis in this space of intertwiners and picturing the basis elements as the edges in the diagrammatic description we get for each path ξ of length n, starting at ϱ_α and ending at ϱ_δ an intertwiner T_ξ from ϱ_δ to $\varrho_\alpha\varrho^n$, namely the product of the intertwiners associated with the edges. One has an action of the braid group B_n on the set of intertwiners $\{T_\xi\}$ belonging to paths of length n from α to δ. $\varepsilon^\varrho(b)$ intertwines from ϱ^n to ϱ^n; so $\varrho_\alpha\varepsilon^\varrho(b)$ intertwines from $\varrho_\alpha\varrho^n$ to $\varrho_\alpha\varrho^n$ and $(\varrho_\alpha\varepsilon^\varrho(b))T_\xi$ from ϱ_δ to $\varrho_\alpha\varrho^n$. Thus

$$(\varrho_\alpha\varepsilon^\varrho(b))\, T_\xi = \sum_\eta R_{\xi\eta}(b) T_\eta. \tag{IV.5.21}$$

Such R-matrices, labelled by paths, have previously been encountered in models of 2-dimensional conformal invariant quantum field theories. They are representation matrices of B_∞ and as such satisfy the group relations (IV.5.1), (IV.5.2).

[1] Recently Longo showed that there must be a gap in the spectrum above $d = 2$ [Longo 92]. The spectrum of d below $\sqrt{6}$ has been completely determined by Rehren [Rehr 95].

One approach to the construction of charge carrying fields starts from a Hilbert space

$$\mathcal{H} = \sum{}^{\oplus} \mathcal{H}_\alpha \,, \tag{IV.5.22}$$

where each $\mathcal{H}_\alpha$ can be identified with $\mathcal{H}_0$ but carries the representation $\varrho_\alpha \mathfrak{A}$ of the observable algebra. Then one introduces operators, called "exchange fields" or "generalized vertex operators" leading from $\mathcal{H}_\alpha$ to $\mathcal{H}_\beta$ and induced by a morphism ϱ_γ as in the composition (IV.5.19). One defines localization for such exchange fields in the same way as it was done for the elements of the "field bundle" in subsection 2.5. Products of space-like separated exchange fields satisfy commutation relations in which the R-matrices enter. This structure is the "exchange algebra" of Rehren and Schroer [Rehr 89]. It corresponds to the field bundle rather than to the field algebra described in the last section. One may ask whether a more economical net of algebras $\mathfrak{F}(\mathcal{O})$ can be constructed so that there are no non trivial common elements of $\mathfrak{F}(\mathcal{O}_1)$ and $\mathfrak{F}(\mathcal{O}_2)$ for disjoint regions and one has (IV.4.6). In the last section this succeeded by taking instead of (IV.5.22) each sector with a suitable multiplicity. Here it seems improbable that this can be achieved (see [Longo 95], [Fred 92]).

There remains the question of finding the analogue of the gauge group. This problem has not yet been satisfactorily resolved. Although the representation theory of a "quantum group" as defined by Drinfeld [Drin 87] or Woronowicz [Wor 87] shares some features with the charge structure encountered in braid group statistics there seem to be some differences. So the dual object of the superselection structure in 2-dimensional field theories is not yet known. In the situation discussed in [Longo 95] it is a "paragroup" in the sense of Ocneanu [Ocn 88]. Other examples are discussed by Buchholz, Mack, Todorov [Buch 88], Mack and Schomerus [Mack 90, 91a, 91b], Fröhlich and Kerler [Frö 91], Hadjiivanov, Paunov and Todorov [Hadj 90], Majid [Maj 89], Rehren [Rehr 91].

The classification of braid group statistics remains largely open. An interesting treatment of the relation between spin and statistics which emphasizes the algebraic aspect is due to Guido and Longo [Guido 95a,b].

V. Thermal States and Modular Automorphisms

V.1 Gibbs Ensembles, Thermodynamic Limit, KMS-Condition

V.1.1 Introduction

In chapter IV we focused attention on states which, with respect to local observations, differ from the vacuum essentially only in some finite space region at a given time. This is an appropriate idealization for elementary particle physics but not for the study of properties of bulk matter, the "many body problem" (many $\sim 10^{24}$). There the simplification leading to the deduction of laws of thermodynamics and hydrodynamics from statistical physics is achieved by the idealization that matter fills all space with finite density and, instead of the vacuum, we have as the simplest states of interest the thermodynamic equilibrium states. To avoid misinterpretation: of course the realistic material systems in which we are interested have finite extension and the standard approach in statistical mechanics starts with a system of N particles enclosed in a box of volume $\mathcal{V}$ with total energy E. But in order to give concepts like "temperature", "phase transition" an unambiguous meaning the *thermodynamic limit* $N \to \infty$, $\mathcal{V} \to \infty$, $E \to \infty$ with $N/\mathcal{V}$ and $E/\mathcal{V}$ finite must be performed (or is implicitly understood). By basing the theory on the algebra $\mathfrak{A}$ of local observables with its net structure we have the advantage that the box is dispensable. Equilibrium states in the thermodynamic limit are good states over $\mathfrak{A}$ and can be directly characterized. In the terminology of thermodynamics the elements of $\mathfrak{A}$ are "intensive quantities".

To arrive at such a description of equilibrium states let us recall the rules for computing them in non-relativistic quantum mechanics for a 1-component system (a system containing only one type of particles) confined to a box of volume $\mathcal{V}$. If the total number of particles is given one considers the Hilbert space $\mathcal{H}_N$ of totally antisymmetric (resp. totally symmetric) N-particle wave functions. Choosing suitable boundary conditions on the walls of the box one has a Hamilton operator H. A general state is described by a *density matrix* ϱ (a positive operator with trace 1); the expectation value of an observable $A \in \mathfrak{B}(\mathcal{H}_N)$ is given by

$$\omega(A) = \operatorname{tr} \varrho A. \tag{V.1.1}$$

To an equilibrium state with inverse temperature $\beta = (kT)^{-1}$ (k is the Boltzmann constant, T the absolute temperature) one assigns $\varrho = \varrho_\beta$

$$\varrho_\beta = Z^{-1} e^{-\beta H}; \quad Z = \operatorname{tr} e^{-\beta H}. \tag{V.1.2}$$

This is the adaptation of Gibbs' canonical ensemble to quantum mechanics. Note that $e^{-\beta H}$ is a trace class operator in the case of finite $\mathcal{V}$ since H then has discrete spectrum with a level density increasing less than exponentially for large energies and H is positive. The internal energy in the sense of thermodynamics is then given by

$$E = \operatorname{tr} \varrho_\beta H. \tag{V.1.3}$$

One may remark that if, instead of β, one prescribes E one may choose for ϱ other positive functions of H. Putting $\varrho = F(H)$ with $\operatorname{tr} F(H) = 1$ and $\operatorname{tr} F(H)H = E$ one observes that the shape of the function F plays no important rôle. The "microcanonical ensemble" where F is chosen to have support in a neighborhood of E is a familiar example. The difference between the consequences of different choices disappear in the thermodynamic limit if $E/\mathcal{V}$ is related to β by (V.1.3). The canonical ensemble is, however, the most convenient choice for F in computations.

Even more convenient, with the thermodynamic limit in view, is the *grand canonical ensemble.* Here we do not fix the number N but consider the Fock space $\mathcal{H}_F = \oplus_{N=0}^{\infty} \mathcal{H}_N$. The "observable" N (the particle number) is then an operator in $\mathcal{H}_F$. One considers then the algebra of operations, generated from bosonic and fermionic creation and annihilation operators. We shall denote it again by $\mathcal{A}$ to distinguish it from the algebra of observables $\mathfrak{A}$. With H denoting the Hamiltonian in $\mathcal{H}_F$ one takes ϱ as

$$\varrho_{\beta,\mu} = G^{-1} e^{-\beta(H-\mu N)}; \quad G = \operatorname{tr} e^{-\beta(H-\mu N)}. \tag{V.1.4}$$

β is again the inverse temperature, μ is the "chemical potential". By

$$E = \operatorname{tr} \varrho_{\beta,\mu} H; \quad \langle N \rangle = \operatorname{tr} \varrho_{\beta,\mu} N \tag{V.1.5}$$

the expectation values of energy and particle number become functions of β, μ and $\mathcal{V}$. In the thermodynamic limit $\mathcal{V}^{-1}E$, $\mathcal{V}^{-1}\langle N \rangle$ and $\Theta = \mathcal{V}^{-1} \ln G$ become functions of β and μ alone. One finds $\Theta = \beta p$, where p is the pressure. All thermodynamic quantities may be obtained from Θ by differentiation with respect to β and μ. As μ increases the matter density increases. For fermions μ can take all values beween $-\infty$ and $+\infty$, whereas for bosons $\mu \leq 0$.[1]

The Gibbs states (V.1.1), (V.1.2) have a property first pointed out by Kubo [Kubo 57] and used to define "thermodynamic Green's functions" by Martin and Schwinger [Mart 59]. Many applications are described in [*Kadanoff and Baym 1962*]. If A is any observable then

[1]For the preceeding claims see any standard text book on statistical mechanics.

$$\alpha_t(A) = e^{iHt} A e^{-iHt} \tag{V.1.6}$$

is its time translate. Due to the invariance of the trace under cyclic permutations one has for $A, B \in \mathfrak{B}(\mathcal{H})$ and ω_β defined by (V.1.1), (V.1.2)

$$\omega_\beta((\alpha_t A)B) = Z^{-1} \mathrm{tr}\, e^{-\beta H} e^{iHt} A e^{-iHt} B = Z^{-1} \mathrm{tr}\, B e^{iH(t+i\beta)} A e^{-iHt}$$

$$= \omega_\beta(B e^{iH(t+i\beta)} A e^{-iH(t+i\beta)}).$$

So

$$\omega_\beta((\alpha_t A)B) = \omega_\beta(B \alpha_{t+i\beta} A), \tag{V.1.7}$$

where we have written, replacing t by a complex variable z

$$\alpha_z A = e^{iHz} A e^{-iHz}. \tag{V.1.8}$$

Introducing for each pair $A, B \in \mathfrak{B}(\mathcal{H})$ the two functions of z

$$\begin{aligned} F_{A,B}^{(\beta)}(z) &= \omega_\beta(B(\alpha_z A)) \\ G_{A,B}^{(\beta)}(z) &= \omega_\beta((\alpha_z A)B) \end{aligned} \tag{V.1.9}$$

we see that with $z = t + i\gamma$

$$F_{A,B}^{(\beta)}(z) = Z^{-1} \mathrm{tr}\, B e^{iHt} e^{-\gamma H} A e^{-iHt} e^{-(\beta-\gamma)H}$$

is an analytic function of z in the strip

$$0 < \gamma < \beta, \quad \gamma = \mathrm{Im}\, z, \tag{V.1.10}$$

because $He^{-H\alpha}$ is a trace class operator for $\alpha > 0$. For real values of z, F and G are bounded, continuous functions of t and one obtains $G(t)$ as the boundary value of $F(z)$ for $z \to t + i\beta$.

$$G_{A,B}^{(\beta)}(t) = F_{A,B}^{(\beta)}(t + i\beta). \tag{V.1.11}$$

The same relation results in the case of the grand canonical ensemble, replacing β by β, μ and H by $H(\mu)$, α_t in (V.1.6) by α_t^μ, $\mathfrak{A}$ by $\mathcal{A}$, where

$$H(\mu) = H - \mu N, \quad \alpha_t^\mu A = e^{iH(\mu)t} A e^{-iH(\mu)t}. \tag{V.1.12}$$

The importance of relation (V.1.11) stems from the fact that it survives the thermodynamic limit. Specifically we may regard A and B as local quantities, $\omega_{\beta,\mu}$ as the state (normalized, positive, linear form over $\mathcal{A}$) corresponding to equilibrium with inverse temperature β, chemical potential μ in unlimited space. We consider H and N in (V.1.4) as generators of symmetries which are realized by automorphism groups on $\mathcal{A}$, namely the time translations α_t and the U(1)-gauge transformations

$$\gamma_\varphi A = e^{iN\varphi} A e^{-iN\varphi}. \tag{V.1.13}$$

$H(\mu)$ is an element in the Lie algebra of the symmetry group and

$$\alpha_t^{\mu} = \alpha_t \gamma_{-\mu t}. \tag{V.1.14}$$

We define, for each pair $A,\ B \in \mathcal{A}$ the functions

$$F_{A,B}^{\beta,\mu}(t) = \omega_{\beta,\mu}(B\alpha_t^{\mu}A) - \omega_{\beta,\mu}(A)\omega_{\beta,\mu}(B), \tag{V.1.15}$$

$$G_{A,B}^{\beta,\mu}(t) = \omega_{\beta,\mu}((\alpha_t^{\mu}A)B) - \omega_{\beta,\mu}(A)\omega_{\beta,\mu}(B). \tag{V.1.16}$$

Then the above discussion suggests that $F(t)$ is the boundary value on the real axis of a function $F(z)$, analytic in the strip (V.1.10) and that $G(t)$ is obtained as the boundary value for $\operatorname{Im} z = \beta$ as in (V.1.11).

With this adaptation to the algebraic setting of the work of Kubo and of Martin and Schwinger quoted above Haag, Hugenholtz and Winnink [Haag 67] postulated the relation

$$G_{A,B}^{\beta,\mu}(t) = F_{A,B}^{\beta,\mu}(t + i\beta), \tag{V.1.17}$$

with the definitions (V.1.15), (V.1.16) and the analyticity requirement for F as the defining property of an equilibrium state with parameters $\beta,\ \mu$ and called it the KMS-condition (for Kubo, Martin, Schwinger). This paper will be referred to in the following as HHW. The KMS-condition implies that $\omega_{\beta,\mu}$ is an invariant state with respect to α_t^{μ} [2]

$$\omega_{\beta,\mu}(\alpha_t^{\mu}A) = \omega_{\beta,\mu}(A). \tag{V.1.18}$$

The subtraction of $\omega(A)\omega(B)$ in (V.1.15), (V.1.16) (which could be omitted) takes away the uncorrelated part in the expectation value of the product. This has the advantage that then we may also assume in the infinite volume limit

$$\lim_{t\to\pm\infty} F(t) = \lim G(t) = 0, \tag{V.1.19}$$

because in a system without boundaries the correlations between local quantities at different times are expected to tend to zero as the time difference increases to infinity.[3]

An alternative characterization of the KMS-condition is the following; (for details of the argument see HHW).

---

[2]Putting $B = \mathbb{1},\ t = 0$ one has $\omega_{\beta,\mu}(\alpha_{i\beta}^{\mu}A) = \omega_{\beta,\mu}(A)$. Due to the analyticity and boundedness in the strip $0 < \operatorname{Im} z < \beta$ $\omega_{\beta,\mu}(\alpha_z^{\mu}A)$ is then a periodic function in the direction of the imaginary axis, hence analytic in the whole complex plane and bounded. Thus, by Liouville's theorem, it is constant and one has (V.1.18).

[3]For a system enclosed in a box this will not be so because the Hamiltonian then has a discrete spectrum and so $F_{A,B}(t)$ will generically be an almost periodic function. In classical mechanics this corresponds to the Poincaré recurrence. In contrast, in an unbounded medium the effect of a local operation will spread and dissipate so that it is not felt in a finite region after a very long time.

Lemma 1.1.1
The KMS-condition is equivalent to the requirement that the Fourier transforms of the functions F and G, defined in (V.1.15), (V.1.16) are related by a Boltzmann factor:

$$\tilde{G}(\varepsilon) = e^{-\beta\varepsilon}\tilde{F}(\varepsilon). \tag{V.1.20}$$

Here the Fourier transform is taken as

$$\tilde{F}(\varepsilon) = \int F(t)e^{-i\varepsilon t}dt$$

and considered as a distribution over smooth test functions $f(\varepsilon)$ with compact support.

V.1.2 Equivalence of KMS-Condition and Gibbs Canonical Ensembles for Finite $\mathcal{V}$

Does the KMS-condition suffice to characterize an equilibrium state? Let us first look at the case of the system enclosed in a box. The standard way to describe such a system with an arbitrary number of (identical) particles in non relativistic quantum theory uses creation and annihilation operators acting in a Fock space $\mathcal{H}_F$.

To fix the ideas[4] let us think of spinless bosons interacting by 2-body forces given by a potential $V(\mathbf{x}-\mathbf{x}')$. Denoting by $a(\mathbf{x})$ the (improper) annihilation operator of a particle at the point $\mathbf{x} \in \mathcal{V}$ (the region of $\mathbb{R}^3$ enclosed by the box), by $a^*(\mathbf{x})$ the creation operator, we have the commutation relations

$$[a(\mathbf{x}), a^*(\mathbf{x}')] = \delta^3(\mathbf{x}-\mathbf{x}'); \quad [a(\mathbf{x}), a(\mathbf{x}')] = 0. \tag{V.1.21}$$

We have a vacuum state vector $\Omega \in \mathcal{H}_F$ characterized by

$$a(\mathbf{x})\Omega = 0 \quad \text{for all} \quad \mathbf{x} \in \mathcal{V}. \tag{V.1.22}$$

The Hamiltonian H and the particle number operator N are formally given by

$$H = \frac{1}{2m}\int(\nabla a^*)(\nabla a)d^3x + \frac{1}{2}\int a^*(\mathbf{x}')a^*(\mathbf{x})V(\mathbf{x}-\mathbf{x}')a(\mathbf{x})a(\mathbf{x}')d^3x d^3x', \tag{V.1.23}$$

$$N = \int a^* a \, d^3x. \tag{V.1.24}$$

In order to make them well defined self adjoint operators the formal expressions have to be supplemented by boundary conditions on the walls of the box defining the domains of H and N. These conditions reflect the way in which the walls influence the system. They have no counterpart in the case of the infinite system. The question of the sensitivity of the thermodynamic limit to the choice of the boundary conditions will concern us later. The operators H, N and $H(\mu) =$

[4] The reader insisting on mathematical rigour is invited to skip this passage. It could be made precise but the effort would only serve to deviate attention from the essentials.

$H - \mu N$ yield the automorphism groups α_t, γ_φ and α_t^μ acting on $\mathfrak{B}(\mathcal{H}_F)$ as described before.

For the finite system we can phrase the question of the equivalence between the KMS-condition and the Gibbs prescription as follows. Given the 1-parametric automorphism group α_t^μ. Does the requirement that ω is a normal state on $\mathfrak{B}(\mathcal{H}_F)$ satisfying the KMS-condition imply that ω is given by (V.1.1), (V.1.4)? The answer is yes. First, any normal state on $\mathfrak{B}(\mathcal{H}_F)$ is described by a density matrix as in (V.1.1). Secondly, all automorphisms of $\mathfrak{B}(\mathcal{H}_F)$ are inner; so α_t^μ defines an implementing unitary $U^\mu(t) \in \mathfrak{B}(\mathcal{H}_F)$ up to a phase factor and, given adequate continuity with respect to t, it defines a generator $H(\mu)$ up to an additive constant which plays no rôle in (V.1.4). If we insert in (V.1.15), (V.1.16) for A an invariant element i.e. one which commutes with all $U^\mu(t)$ then $F_{A,B}(z)$ and $G_{A,B}(z)$ are independent of z and the KMS-condition says that F and G are equal in that case, which implies for the density matrix

$$\mathrm{tr}\,[\varrho, A]B = 0 \quad \text{for all} \quad B \in \mathfrak{B}(\mathcal{H}_F)$$

i.e. $[\varrho, A] = 0$ if A commutes with $H(\mu)$. Thus $\varrho \in \{\cup_t U^\mu(t)\}''$ and ϱ must be a bounded function of $H(\mu)$. That this function is of the form $ce^{-\beta H(\mu)}$ follows then by choosing for A and B operators which have nonvanishing matrix elements only between two vectors Ψ_1, Ψ_2 which are simultaneous eigenvectors of $H(\mu)$ and ϱ. Note that since ϱ is a trace class operator it must have discrete spectrum. This implies, incidentally, that a KMS-state over $\mathfrak{B}(\mathcal{H}_F)$ for positive β exists only if $H(\mu)$ has discrete spectrum, bounded below, with a density of eigenvalues increasing less than $e^{\beta\varepsilon}$ for energy $\varepsilon \to \infty$.

V.1.3 The Arguments for Gibbs Ensembles

The line of argument justifying Gibbs ensembles as the proper characterization of equilibrium for finite systems is the following. An equilibrium state is stationary. Therefore the density matrix (or, in the classical case the distribution function in phase space) should be a function of the constants of motion. There are some constants of motion we know a priori; they relate to the symmetries (see section 4 of Chapter I). In the non relativistic case for a system with a single species of particles in a fixed box there remain among those only two: energy and particle number.[5] The essential question is whether there are no "hidden" constants of motion, stationary quantities not tied to generally known symmetries, which could enter in the distinction of equilibrium states. It was the original objective of ergodic theory to show that this worry can be excluded for "almost all" dynamical systems. In classical systems the situation is, however, much less simple. I refer to the famous KAM-theorem.[6] For infinitely extended quantum systems we shall come back to this problem in Section 3.

[5]The box destroys spatial translation invariance, Galilei invariance and also rotation invariance unless it is of special symmetric shape. In some problems (rotating stars, spin systems..) a non vanishing angular momentum must, of course, be considered.

[6]Kolmogorov, Arnold, Moser. See e.g. the book by [*Arnold 1978*] and references therein.

Assuming that there are no "hidden" constants of motion one concludes that the density matrix for an equilibrium state of a 1-component system must be a function of H and N. Which function? As mentioned earlier the thermodynamic consequences are not sensitive to the choice of this function (within reasonable limits) as long as it yields the right expectation values for the energy density u and the particle number density n:

$$u = \mathcal{V}^{-1}\mathrm{tr}\,\varrho H; \quad n = \mathcal{V}^{-1}\mathrm{tr}\,\varrho N. \tag{V.1.25}$$

To describe equilibrium states for different values of u and n one needs a 2-parametric family of functions of H and N (for fixed $\mathcal{V}$). The grand canonical ensemble (V.1.4) parametrized by β and μ, the canonical ensemble (V.1.2), parametrized by β and n, the microcanonical ensemble, parametrized by u and n are the standard choices. They, as well as other reasonable choices of the function of H and N, all lead to the same family of states over the local algebras in the thermodynamic limit, namely to KMS-states with respect to a 1-parameter subgroup of the automorphism group of symmetries, which in the case of the 1-component system is generated by α_t of (V.1.6) and γ_φ of (V.1.13). In section 3 we shall give direct justifications (independent of the above arguments) for the claim that in an infinite medium the KMS-condition is the appropriate characterization of equilibrium.

Another road to the Gibbs ensembles starts from the principle of maximal entropy. For a state over $\mathfrak{B}(\mathcal{H})$, described by a density matrix ϱ, the entropy is defined as

$$S = -\mathrm{tr}\,\varrho \ln \varrho. \tag{V.1.26}$$

It is some measure for the lack of information, the impurity of the state.[7] An equilibrium state may be defined as one having maximal entropy among the states with the same (prescribed) expectation values for energy and particle number. An infinitesimal variation of ϱ by $\delta\varrho$ changes S by

$$\delta S = -\mathrm{tr}\,\delta\varrho(\ln\varrho + 1).$$

Note that in spite of the fact that $\delta\varrho$ need not commute with ϱ one has the formula $\delta\mathrm{tr}\,F(\varrho) = \mathrm{tr}\,\delta\varrho F'(\varrho)$ which may be justified for instance by using first order Schrödinger perturbation theory for the eigenvalues of $F(\varrho+\delta\varrho)$. Taking the auxiliary conditions $\delta\mathrm{tr}\,\varrho = 0$, $\delta\mathrm{tr}\,\varrho H = 0$, $\delta\mathrm{tr}\,\varrho N = 0$ into account by Lagrange's multiplier method the extremality requirement for S yields

$$\mathrm{tr}\,\delta\varrho(\ln\varrho + 1 + \alpha + \beta H + \nu N) = 0$$

for essentially arbitrary $\delta\varrho$.[8] The multipliers α, β and ν are numerical constants. So S is extremal under the stated auxiliary conditions if and only if the factor

[7] For a pure state $S = 0$. For a mixture of n pure states corresponding to linearly independent state vectors we have $0 < S \leq \ln n$ and the maximum value of S is attained for $\varrho = n^{-1}P_n$ where P_n is the projection operator on the subspace spanned by the state vectors. The state of maximal entropy is the one which prefers no direction in the accessible subspace.

[8] The only remaining restriction for $\delta\varrho$ is that $\varrho + \varepsilon\,\delta\varrho$ should remain positive for $\varepsilon \to 0$. If ϱ is not on the boundary of the convex set of states this gives no restriction for $\delta\varrho$.

of $\delta\varrho$ under the trace vanishes. This means

$$\varrho = c\, e^{-\beta H - \nu N}$$

which is the grand canonical ensemble.

Comment. Before proceeding further let us comment on the terminology used in connection with (V.1.13). We have tied the conservation of particle number to the invariance under a U(1) gauge group. If the term "gauge group" is understood in the sense of Chapter IV this elevates the conservation of particle number to a superselection rule. The algebra $\mathcal{A}$ then corresponds to the "field algebra" whereas the "algebra of observables" is the subalgebra of "gauge invariant elements", i.e. operations which do not change the number of particles. In the case of a system with several components[9] the "gauge group" is a direct product of U(1)-groups, one factor for each component. Correspondingly we need for the parametrization of equilibrium states a chemical potential for each component. This illustrates that we are dealing here with an approximation, an idealization appropriate to a specific regime, since there may be reactions which are not forbidden in principle but strongly inhibited at the energies considered. If we would have to take the possibility of nuclear reactions into account we would be left with three components, the total electric charge, the total baryon number and the lepton number. Strictly speaking the conservation laws do not refer to particle number but to charges and this is indeed tied to the invariance under a global gauge group as discussed in Chapter IV. One simple and well known illustration of the importance of this distinction is the case of a "photon gas". There is no charge associated with a photon, no superselection rule between the vacuum and states with photons. Therefore one has no chemical potential and there is only the single parameter β needed to distinguish equilibrium states of a photon gas, a fact of paramount historical importance for the discovery of Planck's constant in black body radiation.

Still, also in a regime where some reactions which are possible in principle are so strongly inhibited that one can treat them as forbidden, it is appropriate to consider the relevant approximate conservation laws as tied to a gauge group and restrict attention to the "observable algebra" composed of the gauge invariant elements. The rôle of the chemical potentials as a distinguishing feature for equilibrium states on the observable algebra must then be understood in analogy to the appearance of charge superselection sectors in Chapter IV. In the direct characterization of equilibrium states of an infinite medium which does not start from the Gibbs Ansatz for finite systems this is indeed necessary and we shall deal with this in subsection 3.4.

As mentioned before the advantage of the KMS-condition over the Gibbs prescription (V.1.1), (V.1.4) is that it carries over to the infinitely extended medium. There $\omega_{\beta,\mu}$ can no longer be described by a density operator in $\mathcal{H}_F$ but

[9] a system composed of different species of particles where we count as components only a basic set of chemically independent species, omitting to count those which can be reached by chemical reactions from the basic set.

as a state over the abstract C*-algebra $\mathcal{A}$, equipped with its local net structure and the automorphism groups describing the symmetries. This eliminates the box, the boundary conditions, the consideration of the thermodynamic limit and applies with equal ease to the relativistic case. An equilibrium state is characterized as one satisfying the KMS-condition with respect to a 1-parameter subgroup of automorphisms whose geometric correspondence in Minkowski space is a time-like motion for all points.[10]

V.1.4 The Representation Induced by a KMS-State

Let ω be a KMS-state over $\mathcal{A}$. It is natural to look at the representation π of $\mathcal{A}$ resulting from ω by the GNS-construction. The states in this folium are those which deviate from the equilibrium state ω only by essentially local excitations. As shown in HHW the representation π has some remarkable properties. It is instructive to exhibit them first for the case of the system in a box where $\mathcal{A} = \mathfrak{B}(\mathcal{H}_F)$ and ω is given by (V.1.1), (V.1.4). We drop the indices β, μ during this discussion. The density operator ϱ is positive. Therefore

$$\kappa_0 = \varrho^{1/2} \tag{V.1.27}$$

is well defined and of Hilbert-Schmidt class, that is in the set $\{\kappa : \operatorname{tr} \kappa^* \kappa < \infty,\ \kappa \in \mathfrak{B}(\mathcal{H}_F)\}$. We denote this set here by $\mathcal{H}$ because it is a Hilbert space with respect to the scalar product

$$\langle \kappa | \kappa' \rangle = \operatorname{tr} \kappa^* \kappa', \tag{V.1.28}$$

complete with respect to the norm $\| \kappa \|_H = (\operatorname{tr} \kappa^* \kappa)^{1/2}$. It is also a *-algebra (though not a C*-algebra because $\| \kappa \|_H$ is not the operator norm of κ) and it is a 2-sided ideal in $\mathfrak{B}(\mathcal{H}_F)$:

$$\kappa \in \mathcal{H} \quad \text{and} \quad A \in \mathfrak{B}(\mathcal{H}_F) \quad \text{implies} \quad A\kappa \in \mathcal{H} \quad \text{and} \quad \kappa A \in \mathcal{H}. \tag{V.1.29}$$

Since ϱ has finite trace, $\kappa_0 \in \mathcal{H}$ and since all spectral values of ϱ and hence of κ_0 are non vanishing, we have

$$A\kappa_0 \neq 0; \quad \kappa_0 A \neq 0; \quad \omega(A^* A) \equiv \operatorname{tr} \varrho A^* A \neq 0 \quad \text{for} \quad A \in \mathfrak{B}(\mathcal{H}_F), \quad A \neq 0. \tag{V.1.30}$$

Thus ω is a faithful state; it has no Gelfand ideal. Consider now the following representation of $\mathcal{A} = \mathfrak{B}(\mathcal{H}_F)$ by operators acting on $\mathcal{H}$.

$$\pi_l(A)|\kappa\rangle = |A\kappa\rangle; \quad \kappa \in \mathcal{H}, \quad A \in \mathfrak{B}(\mathcal{H}_F). \tag{V.1.31}$$

[10] This limits the relevant symmetries to a combination of positive time-like translations and internal symmetries. If one takes a generator in the Lie algebra of symmetries which brings in rotations and Lorentz transformations then the orbits of points are not positive time-like everywhere and global KMS-states with respect to such subgroups will either not exist at all or have no physical significance. There is, however, the interesting case of causally complete subregions, invariant under such groups which we shall meet in connection with the Bisognano-Wichmann theorem and the Hawking temperature of black holes.

Here we used for distinction the notation $|\kappa\rangle$ if the Hilbert-Schmidt operator κ is to be considered as a vector of the Hilbert space $\mathcal{H}$. One has

$$\langle\kappa_0|\pi_l(A)|\kappa_0\rangle = \omega(A). \tag{V.1.32}$$

Due to (V.1.30) $|\kappa_0\rangle$ is a cyclic and separating vector for the representation π_l. From (V.1.31) and (V.1.32) one recognizes that π_l is isomorphic to the GNS-representation induced by ω and that $|\kappa_0\rangle$ is the state vector corresponding to ω. We shall therefore drop the index l and write π instead of π_l.

There is another natural mapping $A \in \mathfrak{B}(\mathcal{H}_F) \to \pi_r(A) \in \mathfrak{B}(\mathcal{H})$:

$$\pi_r(A)|\kappa\rangle = |\kappa A^*\rangle. \tag{V.1.33}$$

It gives a conjugate linear representation of $\mathcal{A}$, namely

$$\pi_r(AB) = \pi_r(A)\pi_r(B);\ \ \pi_r(A^*) = (\pi_r(A))^*;\ \ \pi_r(cA) = \bar{c}\pi_r(A). \tag{V.1.34}$$

From the definitions one sees directly that $\pi_r(A)$ commutes with $\pi(B)$ for any $A,\ B \in \mathcal{A}$. One finds

Theorem 1.4.1

(i) The operator norms of $\pi(A)$ and $\pi_r(A)$ are equal.

$$\| \pi(A) \| = \| \pi_r(A) \| . \tag{V.1.35}$$

(ii) The commutant of $\pi(\mathcal{A})$ is precisely the weak closure of $\pi_r(\mathcal{A})$

$$(\pi(\mathcal{A}))' = (\pi_r(\mathcal{A}))'' . \tag{V.1.36}$$

(iii) κ_0 is a cyclic vector for both $\pi(\mathcal{A})$ and $\pi_r(\mathcal{A})$ and

$$\omega(A) = \langle\kappa_0|\pi(A)|\kappa_0\rangle = \overline{\langle\kappa_0|\pi_r(A)|\kappa_0\rangle}. \tag{V.1.37}$$

(iv) π and π_r are transformed into each other by an antiunitary operator J defined by

$$J|\kappa\rangle = |\kappa^*\rangle. \tag{V.1.38}$$

One has

$$J\pi(A)J = \pi_r(A);\ \ J^2 = \mathbb{1};\ \ J|\kappa_0\rangle = |\kappa_0\rangle. \tag{V.1.39}$$

We omit the proof which is simple and may be found in HHW.

Since ω is invariant under α_t^μ we can implement this automorphism group in the GNS-representation by unitary operators $U^\mu(t)$ in the standard way putting

$$U^\mu(t)\pi(A)|\kappa_0\rangle = \pi(\alpha_t^\mu A)|\kappa_0\rangle. \tag{V.1.40}$$

Similarly, if ω is invariant under α_t and γ_φ separately, we have unitaries $U(t),\ V(\varphi)$ implementing these automorphisms. The generators of U, V, U^μ

may be regarded as the Hamiltonian H, the particle number operator N and $H(\mu) = H - \mu N$, which all are operators acting in $\mathcal{H}$ and satisfy

$$H|\kappa_0\rangle = 0; \quad N|\kappa_0\rangle = 0. \tag{V.1.41}$$

These operators are, however, clearly not the representors of H, N considered in (V.1.23), (V.1.24). Rather, if we denote the latter by H_F, N_F and regard $U_F(t) = \exp iH_F t$ as elements of $\mathcal{A}$ then

$$U(t) = \pi(U_F(t))\pi_r(U_F(t)), \tag{V.1.42}$$

or, symbolically

$$H = \pi(H_F) - \pi_r(H_F), \tag{V.1.43}$$

with analogous expressions for U^μ, $H(\mu)$ and for $V(\varphi)$, N. Since $\pi_r(\mathcal{A})$ commutes with $\pi(\mathcal{A})$ the second factor in (V.1.42) has no effect on the automorphism and

$$\frac{\partial}{\partial t}\pi(\alpha_t A)\Big|_{t=0} = i[H, \pi(A)] = i[\pi(H_F), \pi(A)]. \tag{V.1.44}$$

The subtraction of $\pi_r(H_F)$ in (V.1.43) becomes, however, essential in the thermodynamic limit. The total energy operator $\pi(H_F)$ becomes meaningless in the limit. Its expectation value and its fluctuations become infinite as $\mathcal{V} \to \infty$. The second term in (V.1.43) cancels the infinities so that the equilibrium state vector becomes an eigenvector of H to eigenvalue zero as seen in (V.1.41). The mechanism for this can be seen by noting that

$$\kappa_0 = Z^{-1/2} e^{-\beta/2\, H_F}.$$

We have

$$\pi(H_F)|\kappa_0\rangle = Z^{-1/2}|H_F e^{-\beta/2\, H_F}\rangle \equiv \Psi,$$

Whereas, according to (V.1.43)

$$H|\kappa_0\rangle = Z^{-1/2}|\,[H_F, e^{-\beta/2\, H_F}]\,\rangle = 0.$$

The norm of Ψ increases to infinity as the volume $\mathcal{V} \to \infty$. We also see[11] that

$$e^{-\beta/2\, H}|A\kappa_0\rangle = |\kappa_0 A\rangle = J|A^*\kappa_0\rangle.$$

This gives the relation

$$\pi(A^*)|\kappa_0\rangle = J e^{-\beta/2\, H}\pi(A)|\kappa_0\rangle, \tag{V.1.45}$$

which also survives the thermodynamic limit and whose key rôle will become apparent in the mathematical theory of Tomita and Takesaki described below.

[11] Note that due to the conjugate linearity of π_r we get from (V.1.42) for imaginary values of t

$$e^{-1/2\beta H} = \pi(e^{-1/2\beta H_F})\pi_r(e^{1/2\beta H_F}) \tag{V.1.42a}$$

Summing up: for the system in a box we have two equivalent descriptions of equilibrium states. The traditional one uses an irreducible representation of $\mathcal{A}$ in $\mathcal{H}_F$ (in fact we have taken $\mathcal{A} = \mathfrak{B}(\mathcal{H}_F)$). The impure state $\omega_{\beta,\mu}$ is then described by the density operator (V.1.4). The other uses a reducible representation π in $\mathcal{H}$. The state $\omega_{\beta,\mu}$ is then described by a vector in $\mathcal{H}$, an eigenvector of the modified Hamiltonian $H(\mu)$. The commutant of π is obtained from π by conjugation with an antiunitary operator J (see (V.1.36), (V.1.39)). The first description is no longer possible in the thermodynamic limit but the second one survives and the essential features remain.

Let $\mathcal{V}_n$ be a set of increasing volumes exhausting space

$$\mathcal{V}_{n+1} \supset \mathcal{V}_n; \quad \cup \mathcal{V}_n = \mathbb{R}^3.$$

For each box $\mathcal{V}_n$ we have a Fock space $\mathcal{H}_F^{(n)}$, operators H_n, N_n defining automorphisms $\alpha_t^{(n)}$, $\gamma_\varphi^{(n)}$, $\alpha_t^{\mu,(n)}$ of $\mathcal{A}_n = \mathfrak{B}(\mathcal{H}_F^{(n)})$ and statistical operators $\varrho_{\beta,\mu}^{(n)}$ as described before. The algebra $\mathcal{A}_m$ is naturally identified as a subalgebra of $\mathcal{A}_n$ when $n > m$ and the inductive limit of this directed set of algebras (as a C*-algebra) is denoted by $\mathcal{A}$. Then one has

Theorem 1.4.2
Assume that

a) for each positive β and a certain range of values of μ (independent of n) $\varrho_{\beta,\mu}^{(n)}$ is a trace class operator on $\mathcal{H}_F^{(n)}$,

b)
$$\| \alpha_t^{\mu,(n)}(A) - \alpha_t^{\mu,(m)}(A) \| \to 0 \quad \text{as} \quad n, m \to \infty \tag{V.1.46}$$

for any $A \in \mathcal{A}_k$ and any $t \in \mathbb{R}$ (A and t being kept fixed in the limiting process),

c) The numerical sequence $\omega_{\beta,\mu}^{(n)}(A) = \operatorname{tr} \varrho_{\beta,\mu}^{(n)} A$, $A \in \mathcal{A}_k$ converges as $n \to \infty$.[12]

Then

(i)
$$\alpha_t^\mu A = \lim \alpha_t^{\mu,(n)} A \tag{V.1.47}$$

defines an automorphism group on $\mathcal{A}$,

$$\omega_{\beta,\mu}(A) = \lim \omega_{\beta,\mu}^{(n)}(A) \tag{V.1.48}$$

defines a state on $\mathcal{A}$ which satisfies the KMS-condition for α^μ (with parameter β).

[12]This assumption concerns only the uniqueness of the limit state. It may be dropped and, in fact it does not hold in many cases of interest (see the comments below).

(ii) The GNS-representation π of $\mathcal{A}$ induced by $\omega_{\beta,\mu}$ has properties analogous to those listed in theorem 1.4.1. Specifically there is a self-adjoint operator $H(\mu)$ and an antiunitary operator J on the representation space $\mathcal{H}$ defined by

$$\pi(\alpha_t^\mu A) = e^{iH(\mu)t}\pi(A)e^{-iH(\mu)t}; \quad H(\mu)\Omega = 0, \tag{V.1.49}$$

$$\pi(A^*)\Omega = Je^{-\beta/2\, H(\mu)}\pi(A)\Omega; \quad J\Omega = \Omega, \tag{V.1.50}$$

where $\Omega \in \mathcal{H}$ is the vector representing $\omega_{\beta,\mu}$.

$$J = J^{-1}. \tag{V.1.51}$$

Defining a conjugate representation π_r by

$$\pi_r(A) = J\pi(A)J, \tag{V.1.52}$$

one has (V.1.35) through (V.1.37) with κ_0 replaced by Ω.

The proof can be skipped here. It is obtained from the stated assumptions in a straightforward way using lemma 1.1.1. For details see HHW. Of particular interest will be part (ii) which establishes the connection between thermodynamic equilibrium states and the Tomita-Takesaki theory of modular automorphisms of von Neumann algebras, a mathematical structure which will be discussed in the next section.

Comments. Conditions for the interparticle forces which guarantee the assumptions a) and b) have been extensively studied. The first concerns "saturation". For fixed volume the ground state energy of an N-particle system should not drop towards $-\infty$ faster than linearly with increasing N. Assumption b) means that the effect of the boundary of the box on the time translation of a well localized quantity during a fixed time interval should become negligible when the box is taken sufficiently large. The worst case for both a) and b) is that of long range, attractive forces. For certain classes of models the validity of the assumptions has been rigorously established. They include lattice systems with short range interaction (see [Streat 68], [*Ruelle 1969*], [*Bratteli and Robinson, Vol. II 1981*]) and Galilei invariant theories in which the interaction decreases not only with the distance but also with the relative momentum of the particles (see Narnhofer and Thirring [Narn 91]). For Coulomb forces between an equal number of positive and negative charges saturation has been proved by Dyson and Lenard [Dy 67, 68] and, with significantly improved estimates, by Lieb and Thirring [Lieb 75 a,b]. See also [*Thirring, Vol.4 1983*]. Thermodynamics in cases of non saturating forces has been discussed by Hertel, Narnhofer and Thirring [Hert 72], Narnhofer and Sewell [Narn 81].

Granted the assumptions a) and b) the problem of the existence of the thermodynamic limit (assumption c)) is essentially a question of choosing a consistent set of boundary conditions as the size of the box increases. This relates to the question of uniqueness of KMS-states with given parameters β, μ. We know that at a phase transition point or in connection with spontaneous breaking of a symmetry (e.g. spontaneous magnetization) the KMS-state cannot be unique and this has its counterpart in the sensitivity of the states $\omega_{\beta,\mu}^{(n)}$ on the boundary conditions, on the precise definition of the box Hamiltonian. This leads to the next topic, the decomposition theory of KMS-states in the infinite medium.

V.1.5 Phases, Symmetry Breaking and Decomposition of KMS-States

Let K_β be the set of KMS-states over the C*-algebra $\mathfrak{A}$ with respect to an automorphism group α_t and parameter β. Mixtures of KMS-states with fixed α_t and β again satisfy the KMS-condition. So K_β is a convex set. Furthermore it is closed in the weak topology induced by $\mathfrak{A}$ in state space and hence weakly compact. Therefore K_β has extremal elements (states which are not decomposable as mixtures of other states in K_β). General states in K_β may be regarded as mixtures of the extremal states in K_β. The basic fact is now

Theorem 1.5.1

(i) An extremal state of K_β is primary; a primary KMS-state is extremal.

(ii) K_β is a simplex i.e. the decomposition of a general state in K_β into extremals is unique.

Proof. Part (ii) follows from (i) because the central decomposition of a state is unique. It is obtained by considering the restriction of the state ω on the center of the von Neumann algebra $\pi_\omega(\mathfrak{A})''$. The center is Abelian and the state space of an Abelian algebra is a simplex (the situation of classical physics).

For part (i) we may start from a state ω in the interior of the convex set K_β and look at the GNS-representation induced by it, obtaining thus a Hilbert space $\mathcal{H}$, a von Neumann algebra $\mathcal{R}$ with a cyclic, separating vector Ω and a group of unitaries $U(t)$ implementing the automorphisms α_t. Any other state $\omega_1 \in K_\beta$ is then dominated by ω and can be obtained as

$$\omega_1(A) = \omega(AT') \tag{V.1.53}$$

where T' is a positive operator in the commutant i.e. $T' \in \mathcal{R}'$. The requirement that ω_1 is again in K_β implies that T' must commute with the unitaries $U(t)$ because, with $A, B \in \mathcal{R}$

$$\langle A\Omega | U(t)T'U^*(t) | B\Omega\rangle = \langle \Omega | U(t)T'U^*(t)A^*B|\Omega\rangle$$

$$= \langle \Omega | T'U(-t)A^*BU^*(-t)|\Omega\rangle = \omega_1(\alpha_{-t}A^*B) = \omega_1(A^*B) = \langle A\Omega|T'|B\Omega\rangle.$$

Here we have used the fact that $U(t)T'U^*(t)$ as well as T' belong to the commutant, that $U(t)\Omega = \Omega$ and that ω_1, being a KMS-state with respect to α_t, is invariant under α_t. So we get

$$U(t)T'U^*(t) = T' \tag{V.1.54}$$

because $\mathcal{R}\Omega$ is dense in $\mathcal{H}$.

The proof is completed now by the following lemma

Lemma 1.5.2
Let $(\mathcal{H}, \mathcal{R}, \Omega, U(t))$ be the representation induced by a KMS-state ω with the notation as defined above and J the conjugation defined in (V.1.50). If $T' \in \mathcal{R}'$ commutes with the unitaries $U(t)$ then

(i) $T = JT'J \in \mathcal{R}$ also commutes with the $U(t)$ and

$$T'\Omega = T\Omega; \quad \omega_1(A) \equiv \omega(AT') = \omega(AT). \tag{V.1.55}$$

(ii) If furthermore the state ω_1, defined by (V.1.53) is in K_β then

$$T = T' \in \mathcal{R} \cap \mathcal{R}' = \mathfrak{Z}. \tag{V.1.56}$$

Proof. Part (i) follows directly from (V.1.50), (V.1.52). For part (ii) we use the KMS-condition for both ω and ω_1 in the form

$$\omega_k(AB) = \omega_k((\alpha_{-i\beta}B)A); \quad \omega_k \in K_\beta, \quad A, \ B \in \mathcal{R}. \tag{V.1.57}$$

One has, with $A, \ B, \ C, \ T \in \mathcal{R}$

$$\omega(ATBC) = \omega((\alpha_{-i\beta}BC)AT) = \omega_1((\alpha_{-i\beta}B)(\alpha_{-i\beta}C)A)$$

$$= \omega_1((\alpha_{-i\beta}C)AB) = \omega((\alpha_{-i\beta}C)ABT) = \omega(ABTC).$$

Thus the matrix elements of $[T, B]$ between a dense set of state vectors vanish for all $B \in \mathcal{R}$ i.e. $T \in \mathcal{R}'$. So T is in the center $\mathfrak{Z} = \mathcal{R} \cap \mathcal{R}'$. □

The theorem suggests the physical interpretation that an extremal KMS-state should correspond to a pure thermodynamic phase. K_β consists of more than one element if and only if there are several coexisting phases possible at the parameter value β. The term "phase" used in this context needs some qualification. If we think of the example of a liquid and gaseous phase which may coexist at special values of β and μ (the latter entering in the choice of the automorphism group α_t) then, at these parameter values there is also an equilibrium state in which we have a phase boundary, say the plane $x^3 = 0$, such that the half space $x^3 > 0$ is occupied by the gas, the half space $x^3 < 0$ by the liquid. This is not a statistical mixture of two homogeneous states (gas, resp. liquid in all space) but again an extremal state of K_β. Thus the pure phases in the thermodynamic sense should be understood as the macroscopically homogeneous, extremal KMS-states. In the example we have only two homogeneous, extremal KMS-states (at the special values of β, μ) but an infinity of inhomogeneous extremal KMS-states at these parameter values. The interior points of the convex set K_β result if we average the latter over the position of the phase boundary. The existence of two distinct homogeneous extremal KMS-states is not related in this example to any symmetry but analogous to an accidental degeneracy, yielding the same value for the pressure in two entirely different configurations. Therefore it is not expected to happen for generic values of the equilibrium parameters but at most on a submanifold of these of codimension 1. This leads to Gibbs' phase rule for the maximal number of coexisting phases of this type (compare the introduction by Wightman to [*Israel 1979*]).

Another reason for the appearance of several phases is spontaneous symmetry breaking. Clearly, if γ is an automorphism which commutes with α_t then the transform under γ of an extremal KMS-state i.e. the state ω', defined as $\omega'(A) = \omega(\gamma A)$, is again an extremal KMS-state to the same parameter values. If $\omega' \neq \omega$ one says that the symmetry is spontaneously broken and then K_β consists of more than one point. This phenomenon, being not "accidental" may occur in a subset of the manifold of equilibrium parameters with the full dimensionality.

We cannot enter into a discussion of conditions for phase transitions and spontaneous symmetry breaking in states with finite matter density. This fascinating subject is so rich that any serious attempt to include it would unbalance this book. It has been a central topic in statistical mechanics for decades. Some results using the algebraic approach and the KMS-condition are given in the books [*Ruelle 1969*], [*Bratteli and Robinson, Vol. II, 1981*], [*Sewell 1989*].

A relativistic version of the KMS-condition which involves all space-time variables and is stronger than the standard form has been proposed by Bros and Buchholz [Bros 94]. For finite temperatures Lorentz invariance is spontaneously broken [Ojima 86].

V.1.6 Extremality and Autocorrelation Inequalities

The maximal entropy principle for equilibrium states of a system in a finite volume can be cast in another form.

Proposition 1.6.1
The grand canonical ensemble to parameters β, μ is distinguished among all density matrices ϱ as the one yielding the maximal value for the functional

$$\Phi_{\beta,\mu}(\varrho) = \beta^{-1}S(\varrho) - \operatorname{tr} \varrho H(\mu); \quad H(\mu) = H - \mu N. \tag{V.1.58}$$

The maximal value of Φ, achieved for $\varrho = \varrho_{\beta,\mu}$, equals $\beta^{-1}\ln G = p\mathcal{V}$ where G is the grand partition function, p the equilibrium pressure.

Proof. By (V.1.4)

$$\beta H(\mu) = -\ln \varrho_{\beta,\mu} - \ln G_{\beta,\mu}\mathbb{1}.$$

So

$$\ln G_{\beta,\mu} - \beta\Phi_{\beta,\mu}(\varrho) = \operatorname{tr}(\varrho \ln \varrho - \varrho \ln \varrho_{\beta,\mu}), \tag{V.1.59}$$

$$\Phi_{\beta,\mu}(\varrho_{\beta,\mu}) = \beta^{-1} \ln G_{\beta,\mu}. \tag{V.1.60}$$

Now we note that for a positive trace class operator B and any unit vector Ψ one has

$$\langle\Psi|\ln B|\Psi\rangle \leq \ln\langle\Psi|B|\Psi\rangle, \tag{V.1.61}$$

where we do not exclude the value $-\infty$ for the left hand side. This is seen by writing for B its spectral resolution $B = \sum b_i\,|\Psi_i\rangle\langle\Psi_i|$. Then

$$\langle\Psi|\ln B|\Psi\rangle = \sum \lambda_i \ln b_i, \quad \langle\Psi|B|\Psi\rangle = \sum \lambda_i b_i. \tag{V.1.62}$$

$$\lambda_i = |\langle\Psi_i|\Psi\rangle|^2 \geq 0, \quad \sum \lambda_i = 1. \tag{V.1.63}$$

Since the function $\ln x$ is convex one has

$$\sum \lambda_i \ln b_i \leq \ln \sum \lambda_i b_i,$$

which gives the inequality (V.1.61).

Next, let A be a positive trace class operator with the spectral resolution $A = \sum a_i\,|\Phi_i\rangle\langle\Phi_i|$; thus $a_i \geq 0$ and $\{\Phi_i\}$ a complete, orthonormal system. Then, using (V.1.61)

$$\operatorname{tr} A \ln B = \sum a_i \langle\Phi_i|\ln B|\Phi_i\rangle \leq \sum a_i \ln\langle\Phi_i|B|\Phi_i\rangle,$$

$$\operatorname{tr} A \ln A - \operatorname{tr} A \ln B \geq \sum a_i(\ln a_i - \ln\langle\Phi_i|B|\Phi_i\rangle).$$

Now, for positive numbers x and y

$$x(\ln x - \ln y) \geq x - y.$$

Thus we get, for positive trace class operators A, B the inequality

$$\operatorname{tr} A \ln A - \operatorname{tr} A \ln B \geq \operatorname{tr} A - \operatorname{tr} B. \tag{V.1.64}$$

In the derivation of (V.1.64) we have interchanged the order of summation of infinite series and this is legitimate only if the series are absolutely convergent which means that also $A \ln A$ and $A \ln B$ should be trace class operators. Putting $A = \varrho$ and $B = \varrho_{\beta,\mu}$ (V.1.64) with (V.1.59), (V.1.60) gives

$$\Phi_{\beta,\mu}(\varrho_{\beta,\mu}) - \Phi_{\beta,\mu}(\varrho) \geq 0. \tag{V.1.65}$$

This is the claim of the proposition. □

The practical advantage of proposition 1.6.1 as compared to the principle of maximal entropy is that the maximality of $\Phi_{\beta,\mu}$ is no longer subject to auxiliary conditions. It gives a powerful tool for the derivation of inequalities for the functions $\omega_{\beta,\mu}(A^*\alpha^{\mu}_{i\gamma}A)$, $\gamma \in [0,\beta]$ i.e. the functions $F_{A,A^*}(z)$ of equation (V.1.15) with imaginary argument. Such functions are called autocorrelation functions (of A) and the inequalities play an important rôle in the study of phase transitions. The derivation and applications of such inequalities are extensively discussed in [*Bratteli and Robinson, Vol. II, 1981*]. We shall collect two of them in the following theorem and note that - in contradistinction to proposition 1.6.1 - they remain valid in the case of an infinite medium; they may be regarded as a replacement of the extremality principles in this case and, moreover, they give an alternative characterization of KMS-states. The following theorem evolved from the work of Roepstorff, Araki and Sewell, Fannes and Verbeure; [Roep 76, 77], [Ara 77a], [Fann 77a and b]. For a proof and further references see Bratteli and Robinson loc. cit..

Theorem 1.6.2
Let $\mathfrak{A}$ be a C*-algebra, α_t a 1-parameter subgroup of automorphisms, δ the generator of α_t i.e. $\delta A = \partial \alpha_t A / \partial t|_{t=0}$. Note that δ is a (possibly unbounded) derivation, defined on a dense domain $D(\delta) \subset \mathfrak{A}$ so that $\delta A \in \mathfrak{A}$ for $A \in D(\delta)$. Let ω be a state on $\mathfrak{A}$ and put

$$u = \omega(A^*A), \quad v = \omega(AA^*). \tag{V.1.66}$$

Then the following conditions are equivalent

a) ω is a KMS-state with respect to the group α_t with parameter β;

b) $$-i\beta\omega(A^*\delta A) \geq u \ln \tfrac{u}{v} \quad \text{for all} \quad A \in D(\delta); \tag{V.1.67}$$

c) $$\beta^{-1} \int_0^\beta \omega(A^* \alpha_{i\gamma} A) d\gamma \leq (u - v)(\ln u - \ln v)^{-1} \quad \text{for all} \quad A \in D(\delta). \tag{V.1.68}$$

In short: the inequalities (V.1.67), (V.1.68) are consequences of the KMS-condition and either one of these inequalities implies the KMS-condition.

V.2 Modular Automorphisms and Modular Conjugation

This section is a mathematical digression. It is devoted to the Tomita-Takesaki theorem and consequences thereof. This theorem is a beautiful example of "prestabilized harmony" between physics and mathematics. On the one hand it is intimately related to the KMS-condition. On the other hand it initiated a significant advance in the classification theory of von Neumann algebras and led to powerful computational techniques. The results will be used in the discussion of physical questions in the subsequent sections.

V.2.1 The Tomita-Takesaki Theorem

It was known before that any W*-algebra which has a faithful representation on a separable Hilbert space is (algebraically and topologically) isomorphic to a von Neumann algebra $\mathcal{R}$ in "standard form". "Standard" means that $\mathcal{R}$ acts on a Hilbert space $\mathcal{H}$ possessing a cyclic and separating vector Ω. In other words the domains $D = \mathcal{R}\Omega \subset \mathcal{H}$ and $D' = \mathcal{R}'\Omega \subset \mathcal{H}$ are both dense in $\mathcal{H}$. If

$(\mathcal{R}, \mathcal{H}, \Omega)$ is standard then one may consider the conjugate linear operator S from D to D

$$S A\Omega = A^* \Omega; \quad A \in \mathcal{R}. \tag{V.2.1}$$

Similarly one has a conjugate linear operator F from D' to D'

$$F A'\Omega = A'^* \Omega; \quad A' \in \mathcal{R}'. \tag{V.2.2}$$

One finds that S and F are closeable. This means that if Ψ_n is a sequence in D converging strongly to $\Psi \in \mathcal{H}$ and if at the same time $S\Psi_n$ converges strongly to some vector $\Phi \in \mathcal{H}$ then Φ is uniquely determined by Ψ and we may extend the domain of S by defining $S\Psi = \Phi$. Otherwise put, if $\Psi_n \in D$, $\| \Psi_n \| \to 0$ and $S\Psi_n$ converges strongly then $\lim S\Psi_n = 0$. We use the letters S and F for the closures of the operators defined by (V.2.1), (V.2.2). A closed operator has a unique polar decomposition which we write

$$S = J\Delta^{1/2}. \tag{V.2.3}$$

Δ is a self adjoint, positive operator (unbounded in general) and J is a conjugate linear isometry, in our case J is an antiunitary operator, since D is dense in $\mathcal{H}$. From the definitions one finds the relations

$$FS = \Delta; \quad SF = \Delta^{-1}; \tag{V.2.4}$$

$$J = J^* = J^{-1}; \tag{V.2.5}$$

$$J\Delta^{1/2} A\Omega = \Delta^{-1/2} J A\Omega = A^*\Omega; \tag{V.2.6}$$

$$\Delta\Omega = \Omega; \quad J\Omega = \Omega. \tag{V.2.7}$$

Defining the 1-parameter unitary group

$$U(t) = \Delta^{it} \tag{V.2.8}$$

one has

$$J\Delta J = \Delta^{-1}; \quad JU(t)J = U(t). \tag{V.2.9}$$

The key theorem is now

Theorem 2.1.1 (Tomita-Takesaki).
Let $\mathcal{R}$ be a von Neumann algebra in standard form, Ω a cyclic and separating vector and Δ, J, $U(t)$ as defined above. Then

$$J\mathcal{R}J = \mathcal{R}', \tag{V.2.10}$$

$$\begin{aligned} U(t)\mathcal{R}U^*(t) &= \mathcal{R}, \\ U(t)\mathcal{R}'U^*(t) &= \mathcal{R}' \end{aligned} \tag{V.2.11}$$

for all real t.

Thus the map σ_t defined by

$$\sigma_t A \equiv U(t) A U^*(t), \quad A \in \mathcal{R} \tag{V.2.12}$$

is an automorphism group of $\mathcal{R}$. It is called the group of modular automorphisms of the state ω on the algebra $\mathcal{R}$. Correspondingly J is called the modular conjugation and Δ the modular operator of $(\mathcal{R}, \Omega)$.

We shall skip the proof of these claims. It can be found in many books e.g. [*Takesaki, 1970*], [*Bratteli and Robinson, Vol. I, 1979*], [*Kadison and Ringrose, Vol. II, 1987*], [*Stratila and Zsido 1978*], [*Stratila 1981*]. Let us try to elucidate the significance of the theorem.

Writing

$$\Delta = e^{-K}, \quad \text{thus} \quad U(t) = e^{-itK}, \tag{V.2.13}$$

we call K the "modular Hamiltonian". K is a self adjoint operator whose spectrum will extend in general from $-\infty$ to $+\infty$ and which has Ω as an eigenvector to eigenvalue zero. The analogy of equations (V.2.1) through (V.2.12) with the relations discussed in theorems 1.4.1 and 1.4.2 in connection with the representation induced by a KMS-state is striking. Indeed, a simple calculation shows that ω satisfies the KMS-condition with respect to the automorphism group σ_t for the parameter value $\beta = -1$:

$$\begin{aligned}
\omega\left((\sigma_t A)B\right) &= \langle \Omega | A U^*(t) B | \Omega \rangle = \langle J\Delta^{1/2} A\Omega | U^*(t) J \Delta^{1/2} B^* \Omega \rangle \\
&= \langle J\Delta^{1/2} A\Omega | J U^*(t) \Delta^{1/2} B^* \Omega \rangle = \langle U^*(t) \Delta^{1/2} B^* \Omega | \Delta^{1/2} A \Omega \rangle \\
&= \langle \Omega | B \Delta^{1/2} U(t) \Delta^{1/2} A \Omega \rangle \\
&= \langle \Omega | B e^{-K-iKt} A | \Omega \rangle = \langle \Omega | B \sigma_{t-i} A \Omega \rangle .
\end{aligned} \tag{V.2.14}$$

We have used $U(z)\Omega = \Omega$, relations (V.2.6) and (V.2.9) and the antiunitarity of J, i.e. $\langle J\Psi | J\Phi \rangle = \langle \Phi | \Psi \rangle$. Comparison with (V.1.7) shows that (V.2.14) is the KMS-condition if we put

$$\sigma_t = \alpha_{-\beta t}. \tag{V.2.15}$$

The (unfortunate) change of sign of the parameter results from the definition (V.2.8), (V.2.12) due to which the parameter of the modular group becomes proportional to negative time in statistical mechanics.

The striking conclusion for physics is:
An equilibrium state with inverse temperature β may be characterized as a faithful state over the observable algebra whose modular automorphism group σ_τ (as a group) is the time translation group, the parameter τ being related to the time t by $t = -\beta\tau$. The extension of the state to the algebra of operations, the field algebra in the sense of Chapter IV, is a faithful state whose modular automorphism group is a 1-parameter subgroup of the symmetries composed of time translations and gauge transformations.

One salient feature of the KMS-property, and therefore also of modular automorphisms, is the fact that it generalizes the positive energy requirement.

While the latter corresponds to the analyticity of $\omega(B\alpha_t A)$ in the whole upper half plane $\mathrm{Im}\, t > 0$ (Chapter II.2) we have, for an equilibrium state with inverse temperature β, analyticity in the strip $0 < \mathrm{Im}\, t < \beta$ and for a faithful state with modular automorphism group σ_t analyticity of $\omega(B\sigma_\tau A)$ in the strip $-1 < \mathrm{Im}\, \tau < 0$. The modular Hamiltonian (V.2.13) is not positive but, with respect to $\mathcal{R}$, its negative part is "suppressed" while, with respect to $\mathcal{R}'$, its positive part is suppressed in the following sense: Let $E_\kappa^{(-)}$ be the spectral projection of K for the interval $[-\infty, -\kappa]$, $E_\kappa^{(+)}$ the spectral projector for the interval $[\kappa, \infty]$, taking κ positive in both cases. Then

$$\| E_\kappa^{(+)} \Delta \| \leq e^{-\kappa}; \quad \| E_\kappa^{(-)} \Delta^{-1} \| \leq e^{-\kappa}; \quad J E_\kappa^{(+)} J = E_\kappa^{(-)}. \tag{V.2.16}$$

Therefore we get from (V.2.6)

$$\| E_\kappa^{(-)} A\Omega \| \leq e^{-\kappa/2} \| A^*\Omega \| \leq e^{-\kappa/2} \| A \|, \quad \text{for} \quad A \in \mathcal{R}. \tag{V.2.17}$$

Similarly

$$\| E_\kappa^{(+)} A'\Omega \| \leq e^{-\kappa/2} \| A' \|, \quad \text{for} \quad A' \in \mathcal{R}'. \tag{V.2.18}$$

Thus any vector in $\mathcal{R}\Omega$ has exponentially decreasing components in the negative part of the spectrum of K. The vectors in $\mathcal{R}\Omega$ are in the domain of Δ^α for $0 \leq \alpha \leq 1/2$, the vectors in $\mathcal{R}'\Omega$ are in the domain of $\Delta^{-\alpha}$ for $0 \leq \alpha \leq 1/2$.

V.2.2 Vector Representatives of States. Convex Cones in $\mathcal{H}$

Another important fact is that any normal state φ of a W*-algebra has a representative state vector $\Phi \in \mathcal{H}$ in the standard form representation

$$\varphi(A) = \langle \Phi | A | \Phi \rangle; \quad \Phi \in \mathcal{H}, \quad A \in \mathcal{R}. \tag{V.2.19}$$

Thus the folium of normal states is already provided by the vector states in this representation and we do not have to resort to density matrices. Of course the correspondence $\varphi \to \Phi$ is one too many. One has

Lemma 2.2.1
Two vectors Φ_1, Φ_2 are representatives of the same state iff there is a partial isometry $V \in \mathcal{R}'$ from a subspace $E_1\mathcal{H}$ to a subspace $E_2\mathcal{H}$ such that

$$\Phi_2 = V\Phi_1.$$

This means, more explicitly, there are projectors $E_i \in \mathcal{R}'$ $(i = 1, 2)$ satisfying

$$E_1 = V^*V; \quad E_2 = VV^*; \quad V \in \mathcal{R}'. \tag{V.2.20}$$

The subspaces $E_i\mathcal{H}$ are the "cyclic components" of Φ_i i.e.

$$\overline{\mathcal{R}\Phi_i} = E_i\mathcal{H}.$$

One can make the correspondence $\varphi \to \Phi$ unique if one subjects Φ to further conditions. There are the two important relations

$$\langle\Omega|A\Delta^{\nu}B|\Omega\rangle = \langle\Omega|B\Delta^{1-\nu}A|\Omega\rangle; \quad 0 \leq \nu \leq 1, \quad A,\, B \in \mathcal{R}. \tag{V.2.21}$$

$$\langle\Omega|A^{*}A\Delta^{1/2}B^{*}B|\Omega\rangle \geq 0. \tag{V.2.22}$$

The first follows from

$$\langle\Omega|A\Delta^{\nu}B|\Omega\rangle = \langle\Delta^{-1/2}JA\Omega|\Delta^{\nu}|\Delta^{-1/2}JB^{*}\Omega\rangle = \langle JA\Omega|J\Delta^{1-\nu}B^{*}|\Omega\rangle$$
$$= \langle\Delta^{1-\nu}B^{*}\Omega|A\Omega\rangle = \langle\Omega|B\Delta^{1-\nu}A|\Omega\rangle.$$

The second by

$$\langle\Omega|A^{*}A\Delta^{1/2}B^{*}B|\Omega\rangle = \langle\Omega|A^{*}AJB^{*}B\Omega\rangle = \langle\Omega|X^{*}X|\Omega\rangle \geq 0$$

with

$$X = A\jmath(B)$$

where we have used the abbreviation

$$\jmath(B) \equiv JBJ, \tag{V.2.23}$$

and the fact that $\jmath$ maps $\mathcal{R}$ on the commutant so that we can shift $\jmath(B^{*})$ to the left.

One introduces now for each α in the interval $[0, 1/2]$ a convex cone of vectors in $\mathcal{H}$

$$\mathcal{P}^{\alpha}_{\Omega} \equiv \overline{\Delta^{\alpha}\mathcal{R}^{+}|\Omega\rangle}, \quad 0 \leq \alpha \leq \frac{1}{2}, \tag{V.2.24}$$

where $\mathcal{R}^{+}$ is the set of positive elements of $\mathcal{R}$ and the bar denotes the closure in the norm topology of Hilbert space. Equation (V.2.22) shows that for $\Psi \in \mathcal{P}^{\alpha}_{\Omega}$, $\Phi \in \mathcal{P}^{1/2-\alpha}_{\Omega}$ we have $\langle\Psi|\Phi\rangle \geq 0$. A closer study shows that $\mathcal{P}^{1/2-\alpha}_{\Omega}$ is precisely the polar (or dual) cone to $\mathcal{P}^{\alpha}_{\Omega}$ i.e. the set

$$\mathcal{P}^{1/2-\alpha}_{\Omega} = \{\Phi : \langle\Phi|\Psi\rangle \geq 0 \quad \text{for all} \quad \Psi \in \mathcal{P}^{\alpha}_{\Omega}\}. \tag{V.2.25}$$

Since

$$\Delta^{\alpha}A^{*}A|\Omega\rangle = \Delta^{\alpha-1/2}\jmath(A^{*}A)|\Omega\rangle$$

the cones may also be obtained as

$$\mathcal{P}^{\alpha}_{\Omega} = \overline{\Delta^{\alpha-1/2}\mathcal{R}'^{+}|\Omega\rangle}. \tag{V.2.26}$$

In particular

$$\mathcal{P}^{1/2}_{\Omega} = \overline{\mathcal{R}'^{+}|\Omega\rangle}. \tag{V.2.27}$$

With these definitions one obtains now the following theorems which evolved mainly through the work of Araki and Connes. We state them without giving the proofs. They are given in the book by Bratteli and Robinson Vol. II where also the references to the original papers may be found.

Theorem 2.2.2
Every normal state on $\mathcal{R}$ has precisely one vector representative in each of the cones $\mathcal{P}^{\alpha}_{\Omega}$ for $0 \leq \alpha \leq 1/4$. A state φ which is dominated by ω in the sense $\lambda\omega - \varphi > 0$ for some $\lambda > 0$ has a (unique) vector representative also in the cones with $1/4 < \alpha \leq 1/2$.

Of special significance is the cone $\mathcal{P}^{1/4}_{\Omega}$, called the *natural cone*, which we shall denote simply by $\mathcal{P}_{\Omega}$. It is self dual. We list the main properties of $\mathcal{P}_{\Omega}$ in the next theorem. We shall omit indicating the reference state ω and its vector representative Ω to which the objects J, Δ, $U(t)$, $\mathcal{P}$ are related.

Theorem 2.2.3
Let Ψ, $\Phi \in \mathcal{P}$. Then

$$\text{(i)} \quad \langle\Psi|\Phi\rangle = \langle\Phi|\Psi\rangle \geq 0, \tag{V.2.28}$$

$$\text{(ii)} \quad U(t)\Psi \in \mathcal{P}, \tag{V.2.29}$$

$$\text{(iii)} \quad J\Psi \in \mathcal{P}. \tag{V.2.30}$$

Conversely, if $\Psi \in \mathcal{H}$ and $J\Psi = \Psi$ then

$$\Psi = \Psi_+ - \Psi_- \quad \text{with} \quad \Psi_+ \in \mathcal{P}, \Psi_- \in \mathcal{P} \quad \text{and} \quad \langle\Psi_+|\Psi_-\rangle = 0. \tag{V.2.31}$$

(iv) $\mathcal{P}$ may also be characterized as the closure of the set

$$Aj(A)|\Omega\rangle, \quad A \in \mathcal{R}. \tag{V.2.32}$$

(v) If $\Phi \in \mathcal{P}$ is separating for $\mathcal{R}$ then it is also cyclic for $\mathcal{R}$ and vice versa.
(vi) Let Φ_1, Φ_2 be the (unique) vector representatives of the states φ_1, φ_2 then

$$\| \Phi_1 - \Phi_2 \|^2 \leq \| \varphi_1 - \varphi_2 \|^2 \leq \| \Phi_1 - \Phi_2 \| \, \| \Phi_1 + \Phi_2 \| . \tag{V.2.33}$$

Most remarkable is the property (vi). It means that the map from the set of normal states over $\mathcal{R}$ onto the set of representative vectors in $\mathcal{P}$ is a homeomorphism with respect to the norm topology in state space on the one side and the Hilbert space norm topology on $\mathcal{P}$. Of course, if two state vectors are close together then also the corresponding states will be close together. This is the second of the inequalities (V.2.33); it does not need $\mathcal{P}$. The converse, however, the fact that closeness of states implies the closeness of their vector representatives is special to the natural cone $\mathcal{P}$. It does not hold if the vector representatives are chosen on any one of the cones $\mathcal{P}^{\alpha}$ with $\alpha \neq 1/4$.

Another remarkable consequence of theorem 2.2.3 is

Theorem 2.2.4
Every automorphism α of a W*-algebra is implementable in a standard form representation by a unitary $U(\alpha)$ i.e.

$$\alpha A = U(\alpha) A U^*(\alpha); \quad A \in \mathcal{R}, \quad U(\alpha) \quad \text{unitary from} \quad \mathfrak{B}(\mathcal{H}). \tag{V.2.34}$$

$U(\alpha)$ may be chosen so that

$$U(\alpha)\mathcal{P} = \mathcal{P}; \quad JU(\alpha) = U(\alpha)J. \tag{V.2.35}$$

Comment. Theorem 2.2.4 shows that a W*-algebra is a very rigid object. If an automorphism of a C*-algebra $\mathfrak{A}$ cannot be implemented in a representation π then it cannot be extended from $\mathfrak{A}$ to the von Neumann algebra $\pi(\mathfrak{A})''$.

Let us describe the construction of $U(\alpha)$ implementing the automorphism α as stated in theorem 2.2.4. The state ω' defined as

$$\omega'(A) = \omega(\alpha^{-1}A)$$

is faithful. Hence it has a vector representative $\Psi \in \mathcal{P}_\Omega$ which is separating for $\mathcal{R}$ and thus, by theorem 2.2.3 (v), also cyclic. Set

$$U(\alpha)A|\Omega\rangle = (\alpha A)|\Psi\rangle. \tag{V.2.36}$$

Due to the cyclic and separating property of both Ω and Ψ (V.2.36) defines $U(\alpha)$ on a dense domain and the range is dense too. Then

$$\| U(\alpha)A|\Omega\rangle \|^2 = \| \alpha A|\Psi\rangle \|^2 = \omega'(\alpha A^* A) = \omega(A^*A) = \| A|\Omega\rangle \|^2 .$$

Therefore $U(\alpha)$ can be extended to a unitary operator on $\mathcal{H}$. From the definition (V.2.36) one checks the relations (V.2.34), (V.2.35).

V.2.3 Relative Modular Operators and Radon-Nikodym Derivatives

We study next the dependence of modular automorphisms on the reference state. Let ω, ω' be faithful normal states and Ω, Ω' vector representatives which are cyclic and separating for $\mathcal{R}$. Then one can define the conjugate linear operator $S_{\Omega',\Omega}$ by

$$S_{\Omega',\Omega}A|\Omega\rangle = A^*|\Omega'\rangle; \quad A \in \mathcal{R}. \tag{V.2.37}$$

It is closeable and hence has a polar decomposition

$$S_{\Omega',\Omega} = J_{\Omega',\Omega}(\Delta_{\Omega',\Omega})^{1/2}. \tag{V.2.38}$$

The positive operator $\Delta_{\Omega',\Omega}$ is called the relative modular operator of the pair Ω', Ω. If Ω' is on the natural cone of Ω then

$$J_{\Omega',\Omega} = J_{\Omega,\Omega} = J. \tag{V.2.39}$$

Defining the unitaries

$$U_{\Omega',\Omega}(t) = (\Delta_{\Omega',\Omega})^{it} \tag{V.2.40}$$

one finds for any pair of cyclic and separating vectors

$$U_{\Omega',\Omega}(t)AU^*_{\Omega',\Omega}(t) = U_{\Omega'}(t)AU^*_{\Omega'}(t); \quad A \in \mathcal{R}, \tag{V.2.41}$$

$$U_{\Omega',\Omega}(t)A'U^*_{\Omega',\Omega}(t) = U_{\Omega}(t)A'U^*_{\Omega}(t); \quad A' \in \mathcal{R}' \tag{V.2.42}$$

where $U_{\Omega}(t) = U_{\Omega,\Omega}(t)$ denotes the unitary defined in (V.2.8). Defining for a triple of such vectors

$$u_{\Omega',\Omega}(t) \equiv U_{\Omega',\Omega''}(t)U^*_{\Omega,\Omega''}(t), \tag{V.2.43}$$

$$u'_{\Omega',\Omega}(t) \equiv U^*_{\Omega'',\Omega'}(t)U^*_{\Omega'',\Omega}(t), \tag{V.2.44}$$

one sees from (V.2.41), (V.2.42) that $u_{\Omega',\Omega}$ commutes with $\mathcal{R}'$ and does not depend on Ω'' (as indicated in the notation). Thus

$$u_{\Omega',\Omega}(t) \in \mathcal{R}, \tag{V.2.45}$$

$$u'_{\Omega',\Omega}(t) \in \mathcal{R}'. \tag{V.2.46}$$

Furthermore the unitaries (V.2.43), (V.2.44) are independent of the choice of the vector representatives of the states ω, ω'. In their dependence on t they are strongly continuous.

To illustrate the significance, let us look at the example afforded by Gibbs states of a finitely extended system as described in section 1. For such a state with density matrix ϱ the modular operator factors into a part from $\mathcal{R}$ and one from $\mathcal{R}'$

$$\Delta_{\Omega} = \varrho_{\Omega}\, j(\varrho_{\Omega}^{-1}). \tag{V.2.47}$$

Writing

$$\varrho_{\Omega} = e^{-H_{\Omega}} \tag{V.2.48}$$

we see from (V.2.43) that all the operators $U_{\Omega',\Omega}$, $\Delta_{\Omega',\Omega}$ also factor. We get

$$U_{\Omega',\Omega}(t) = e^{-itH_{\Omega'}} e^{it\, j(H_{\Omega})}, \tag{V.2.49}$$

$$u_{\Omega',\Omega}(t) = e^{-itH_{\Omega'}} e^{itH_{\Omega}}, \tag{V.2.50}$$

a form familiar from collision theory in quantum mechanics. Introducing the *relative Hamiltonian*

$$h_{\Omega',\Omega} = H_{\Omega'} - H_{\Omega} \tag{V.2.51}$$

one may compute $u_{\Omega',\Omega}$ and $\varrho_{\Omega'}$ by standard perturbation expansion with respect to $h_{\Omega',\Omega}$ as described in subsection 2.4 of Chapter II.

The special feature here, the factorization of Δ, corresponds to the fact that in this case the modular automorphisms σ_t^{ω} are *inner*, i.e. they may be implemented by the unitaries $\exp(-iH_{\Omega}t) \in \mathcal{R}$. This feature is lost in general, for instance already in the thermodynamic limit of Gibbs states. Still there always remain the unitaries $u_{\Omega',\Omega}(t) \in \mathcal{R}$. Relation (V.2.50) is replaced by the *cocycle identity*

$$u_{\Omega',\Omega}(t_1 + t_2) = u_{\Omega',\Omega}(t_1)\left(\sigma^{\omega}_{t_1} u_{\Omega',\Omega}(t_2)\right), \tag{V.2.52}$$

from which, if σ_t^ω is given, $u_{\Omega',\Omega}(t)$ may be computed by perturbation expansion with respect to the relative Hamiltonian

$$h_{\Omega',\Omega} = i\frac{\partial}{\partial t}u_{\Omega',\Omega}(t)\Big|_{t=0} . \tag{V.2.53}$$

Thus any pair of faithful, normal states on $\mathcal{R}$ determines a cocycle of unitaries in $\mathcal{R}$. If σ_t^ω is inner then this cocycle is of the form (V.2.50); it is a coboundary in the terminology of group cohomology.

The cocycle $u_{\Omega',\Omega}(t)$ which characterizes the relation between the two states is called the *Radon-Nikodym cocycle* and is written as

$$(D\omega' : D\omega)(t) \equiv u_{\Omega',\Omega}(t) \in \mathcal{R}. \tag{V.2.54}$$

Besides the cocycle identity (V.2.52) it satisfies the *chain rule*

$$(D\omega_1 : D\omega_2)(t)\,(D\omega_2 : D\omega_3)(t) = (D\omega_1 : D\omega_3)(t), \tag{V.2.55}$$

has the intertwining property

$$(D\omega_1 : D\omega_2)(t)\;(\sigma_t^{\omega_2}A) = (\sigma_t^{\omega_1}A)\;(D\omega_1 : D\omega_2)(t), \tag{V.2.56}$$

and the initial condition

$$(D\omega_1 : D\omega_2)(0) = \mathbb{1}. \tag{V.2.57}$$

V.2.4 Classification of Factors

If, for a particular value of t and a particular faithful, normal state ω the modular automorphism σ_t^ω is inner (i.e. can be implemented by a unitary from $\mathcal{R}$, which implies that $U_\Omega(t)$ factors into a part from $\mathcal{R}$ and one from $\mathcal{R}'$) then (V.2.56) shows that for any other normal state ω' also $\sigma_t^{\omega'}$ is inner. Therefore the set of t-values

$$\mathcal{T} = \{t : \sigma_t^\omega \text{ is inner}\} \tag{V.2.58}$$

is a property of $\mathcal{R}$, independent of the choice of ω. If $\mathcal{R}$ is not a factor then $\mathcal{T}$ is the intersection of the sets $\mathcal{T}_\kappa$ attached to the factors $\mathcal{R}_\kappa$ occurring in the central decomposition of $\mathcal{R}$. Therefore $\mathcal{T}$ does not have much information unless $\mathcal{R}$ is a factor. We specialize to this now. Obviously always $0 \in \mathcal{T}$ and, with $t_1,\ t_2 \in \mathcal{T}$ also $t_1 \pm t_2 \in \mathcal{T}$. So $\mathcal{T}$ is a subgroup of $\mathbb{R}$, the group of real numbers (in additive notation).

There are three "simple" possibilities for $\mathcal{T}$:

a) $\mathcal{T} = \mathbb{R}$,
b) $\mathcal{T} = n\,t_0, \quad n \in \mathbb{Z}$,
c) $\mathcal{T} = \{0\}$.

But $\mathcal{T}$ could also be some dense subset of $\mathbb{R}$ since, so far, we have not equipped $\mathcal{T}$ with a topology. But, as Connes has shown, $\mathcal{T}$ is related to the spectrum of the modular operators Δ_ω [Conn 73, 74]. Defining the *spectral invariant*

$$S(\mathcal{R}) = \cap_\omega \mathrm{Spect}\Delta_\omega, \tag{V.2.59}$$

(where ω ranges over all normal states of $\mathcal{R}$) and the set

$$\Gamma(\mathcal{R}) = \{\lambda \in \mathbb{R} : e^{i\lambda t} = 1 \quad \text{for all} \quad t \in \mathcal{T}\}. \tag{V.2.60}$$

Connes showed that

$$\Gamma(\mathcal{R}) \supset \ln(S(\mathcal{R}) \setminus 0) \tag{V.2.61}$$

and that $S(\mathcal{R}) \setminus 0$ is a closed subgroup of the multiplicative group of positive real numbers. The three cases listed above correspond to

a) $S(\mathcal{R}) = \{1\}$,
b) $S(\mathcal{R}) = \{0 \cup \lambda^n\}, \quad n \in \mathbb{Z}; \quad 0 < \lambda < 1$,
c) $S(\mathcal{R})$ is the set of all non – negative reals.

The remaining possibility is

d) $S(\mathcal{R}) = \{0,\ 1\}$.

In case a) one can define a semifinite trace over $\mathcal{R}$ [*Takesaki 1970*]. If σ_t^ω is inner for all t then Δ_ω factors as in (V.2.47) where ϱ is a selfadjoint operator affiliated with $\mathcal{R}$ and commuting with Δ. This yields the following formal relations (provided that the expressions occurring exist)

$$\omega\left(\varrho^\alpha A\right) = \omega\left(\varrho^{\alpha-\beta} A \varrho^\beta\right), \tag{V.2.62}$$

$$\omega\left(\varrho^{-1}AB\right) = \langle\Omega|Aj\left(\varrho^{-1}\right)B|\Omega\rangle = \langle\Omega|A\Delta\varrho^{-1}B|\Omega\rangle = \langle\Omega|\varrho^{-1}BA|\Omega\rangle, \tag{V.2.63}$$

where we have used (V.2.20) and finally (V.2.21). Setting

$$\mathrm{tr}\, A = \omega\left(\varrho^{-1}A\right) = \omega\left(\varrho^{-1/2}A\varrho^{-1/2}\right), \tag{V.2.64}$$

one sees that this defines a positive linear functional over a dense set in $\mathcal{R}$ with the property

$$\mathrm{tr}\, AB = \mathrm{tr}\, BA.$$

So (V.2.64) defines a semifinite trace. Therefore case a) contains the factors of type I and II in von Neumann's classification. Case b) corresponds to the *Powers factors* [Pow 67]. They are denoted as type III_λ $(0 < \lambda < 1)$. Case c) are the factors denoted as type III_1. Case d) is denoted as type III_0 (one cannot quarrel with notations).

We shall see that case c) (type III_1) is of paramount interest for us. The representation of the quasilocal algebra induced by an equilibrium state as well as the local algebras of relativistic theory in the vacuum sector are of type

III_1. In fact we shall see in Section 6 that these algebras are isomorphic as W*-algebras to the (unique) hyperfinite factor of type III_1.

The special feature of type III_1 is that in this case the spectrum of each Δ_ω is maximal (all of $\mathbb{R}^+$). $S(\mathcal{R})$ is already given by the spectrum of the modular operator of any single state. If one can exhibit a faithful primary state ω such that SpectΔ_ω is minimal in the folium i.e. that $S(\mathcal{R})$ is already given by SpectΔ_ω then the possibilities are very limited. One can exclude case d) because it would mean that zero is an isolated, discrete eigenvalue. Furthermore case a) would mean then that $\Delta_\omega = \mathbb{1}$ i.e. that ω is a trace state. This special situation prevails only for equilibrium states at infinite temperature ($\beta = 0$). In this limiting case $\mathcal{R}$ is of type II_1. Apart from this and the very special case of Powers factors this leaves as the generic situation only type III_1.

In this context the following observation of Araki [Ara 72] and Størmer [Størm 72] is important.

Theorem 2.4.1

Let $\mathfrak{A}$ be a C*-algebra which is *asymptotically Abelian* with respect to an automorphism group, ω an invariant, primary, faithful state over $\mathfrak{A}$, π_ω the GNS-representation of $\mathfrak{A}$ induced by ω and $\mathcal{R} = \pi_\omega(\mathfrak{A})''$. If the extension of ω to $\mathcal{R}$ is faithful then

$$\text{Spect}\Delta_\omega = S(\mathcal{R}). \tag{V.2.65}$$

Comment. Asymptotically Abelian shall mean that there is an automorphism α of $\mathfrak{A}$ such that

$$\| [B, \alpha^n A] \| \rightarrow 0 \quad \text{as} \quad n \rightarrow \infty; \quad A,\ B \in \mathcal{R}. \tag{V.2.66}$$

Invariance of ω:

$$\omega(\alpha A) = \omega(A). \tag{V.2.67}$$

Typically, if $\mathfrak{A}$ is the quasilocal algebra and α corresponds to space translations by some 3-vector then (V.2.66) is satisfied. If ω is an extremal KMS-state for some finite temperature which is homogeneous (or at least invariant under some subgroup of translations as in the case of a crystal) then we have (V.2.67) and ω is also faithful on $\mathcal{R}$.

To prove the theorem one has to show that for any state ω' in the folium of ω the spectrum of $\Delta_{\omega'}$ contains the spectrum of Δ_ω. This results from the fact that under large translations α^n ($n \rightarrow \infty$) every state in the folium tends to a multiple of ω. The arguments are reminiscent of the discussion of the structure of the energy-momentum spectrum (theorem 5.4.1 in Chapter II). In fact one can see rather directly that SpectΔ_ω is a group and thus, since the spectrum is closed and Δ_ω is positive, a closed subgroup of the non-negative reals. Namely, we know that $1 \in \text{Spect}\Delta_\omega$, that with $\lambda \in \text{Spect}\Delta_\omega$, also $\lambda^{-1} \in \text{Spect}\Delta_\omega$, (this follows from (V.2.9)). It remains to show that SpectΔ_ω, is closed under multiplication. This corresponds to the additivity of the spectrum of the modular Hamiltonian K. The faithfulness of ω and its invariance under α imply that α commutes with the modular automorphism group. Therefore the additivity of the K-spectrum can be established with the technique used in the proof of theorem 5.4.1 in Chapter II.

V.3 Direct Characterization of Equilibrium States

V.3.1 Introduction

In the last section we argued that an equilibrium state of an infinitely extended medium can be characterized as a state over the algebra of local operations satisfying the KMS-condition at some value of β for some 1-parameter subgroup of automorphisms generated by time translations and gauge transformations. Obviously this characterization is far from "first principles". It resulted from the Gibbs Ansatz by passing to the thermodynamic limit. The Gibbs Ansatz in turn was justified by combining the requirement of stationarity of the state with some assumptions about the ergodicity of the dynamics or, alternatively, by the principle of maximal entropy for states having specified expectation values for the relevant constants of motion.

The objective of the present section is to give a direct and natural definition of equilibrium for the infinitely extended medium and derive the KMS-condition from it. Again we shall have two roads. The first uses a concept of *dynamical stability*, the other, starting from the second law of thermodynamics, the notion of *passivity*. Both lead to the result that an equilibrium state is a KMS-state over the algebra of observables with respect to the time translation automorphism group.

There remains the second task, to elucidate the rôle of the chemical potentials. We shall see that the existence of charge quantum numbers in the sense of Chapter IV implies, generically, the existence of a continuum of different KMS-states over the observable algebra with respect to α_t at the same value of the temperature; each such state has an extension to the algebra of operations (the field algebra in the sense of Chapter IV) and these extensions are KMS-states with respect to a modified automorphism group α_t^μ over the algebra of operations.

V.3.2 Stability

Let $\mathfrak{A}$ be the C*-algebra of quasilocal observables and α_t the automorphism group representing time translations. A (pure phase) equilibrium state is certainly a primary state over $\mathfrak{A}$ which is stationary (invariant under time translations)

$$\omega(\alpha_t A) = \omega(A). \tag{V.3.1}$$

But this alone may not suffice to distinguish equilibrium from other states. An equilibrium state should not be too sensitive to small changes in the dynamical law (adding a few grains of dust). This old idea was formalized by Haag, Kastler and Trych-Pohlmeyer [Haag 74] as a requirement of "dynamical stability". Suppose the dynamical law is slightly changed, corresponding to a change of the Hamiltonian by λh, where the "coupling constant" λ shall ultimately tend to zero and h is a fixed element of $\mathfrak{A}$. This is expressed by replacing α_t by $\alpha_t^{\lambda h}$ defined by

$$\frac{d}{dt}\alpha_t^{-1}\alpha_t^{\lambda h}A\bigg|_{t=0} = i\lambda[h, A], \tag{V.3.2}$$

$$\alpha_{t+t'}^{\lambda h} = \alpha_t^{\lambda h}\alpha_{t'}^{\lambda h}; \quad \alpha_t^{\lambda h}\big|_{t=0} = \text{id}. \tag{V.3.3}$$

From these relations one can compute $\alpha_t^{\lambda h}$ in terms of α_t and h, for instance by a norm convergent perturbation expansion as described in [*Bratteli and Robinson 1981*]. Introducing the cocycle of automorphisms (compare (V.2.52))

$$\beta_t^{\lambda h} = \alpha_t^{-1}\alpha_t^{\lambda h}, \tag{V.3.4}$$

one has

$$\frac{d}{dt}\beta_t^{\lambda h}A = i\lambda\alpha_t^{-1}[h, \alpha_t^{\lambda h}A], \tag{V.3.5}$$

$$\frac{d}{dt}\left(\beta_t^{\lambda h}\right)^{-1}A = -i\lambda\left(\alpha_t^{\lambda h}\right)^{-1}[h, \alpha_t A]. \tag{V.3.6}$$

If ω is a primary state, stationary with respect to the dynamical law α_t, then we say that ω is *dynamically stable with respect to the perturbation h* if, for sufficiently small λ, there exists a state $\omega^{\lambda h}$ in the primary folium of ω which is stationary with respect to the perturbed dynamics $\alpha_t^{\lambda h}$ and such that the family $\omega^{\lambda h}$ depends continuously on λ and tends to ω for $\lambda \to 0$. To simplify the arguments we shall even require differentiability. Thus $\varphi(A) = d/d\lambda\, \omega^{\lambda h}(A)\,|_{\lambda=0}$ shall be a normal linear form in the folium of ω. With this terminology we shall define a (pure phase) equilibrium state as a primary state over $\mathfrak{A}$ which is

i) *stationary with respect to* α_t,

ii) *stable with respect to any perturbation of the dynamics by* $h \in D$, *where* D *is a norm dense set in* $\mathfrak{A}$.

This definition is effective in the case of an infinite medium, not for a system bounded by walls. In the latter case things are more complicated due to recurrences caused by reflection from the walls. This is analogous to the situation in collision theory. The concept of an S-matrix can be introduced only if infinitely extended space is available. Otherwise, though the effects of walls far away may be insignificant, they prevent this idealization. In fact, the analogy is very close. The essential rôle in the subsequent arguments will be played by asymptotic Abelianness of $\mathfrak{A}$ with respect to time translations. The physical picture for this is the following. The effect of an operation in a finite space region at time t_0 on the outcome of later observations in another fixed, finite space region should decrease in strength as the time difference increases. Thus the commutator $[A, \alpha_t B]$ should tend to zero as $t \to \infty$. Specifically we shall base the discussion on

Assumption 3.2.1 ($\mathcal{L}^{(1)}$ Asymptotic Abelianness in Time).
There is a norm dense subset $D \subset \mathfrak{A}$ such that $\| [A, \alpha_t B] \|$ is an absolutely integrable function of t when $A,\ B \in D$.

Comments. According to the intuitive picture one should think of D as the set of sharply localized observables. For a system on non-interacting particles $\| [A, \alpha_t B] \|$ decreases generically like $|t|^{-3/2}$ and the same can be expected in the case of purely repulsive interactions. The assumption imposes limitations on the type of attractive interactions (e.g. saturation properties as mentioned at the end of section 1) and it may be that the assumption does not cover all cases of physical interest. We shall not discuss the possibility of replacing assumption 3.2.1 by weaker conditions. Some elaborations on the status of the assumption are given in [*Bratteli and Robinson 1981*]. Note that the assumption concerns an algebraic property, a property of the system not referring to any particular state. It implies that for every primary state ω the correlation functions tend to zero for large time differences. Thus, if ω is primary and stationary and assumption 3.2.1 holds, one conludes (by the same reasoning as in the proof of theorem 3.2.2 of Chapter III) that

$$|\omega(A\alpha_t B)| - \omega(A)\omega(B)| \to 0 \quad \text{as} \quad |t| \to \infty. \tag{V.3.7}$$

This does not specify the rate of decrease. The system may possess different types of equilibrium states (e.g. phase transition points of different order) in which the correlation functions have different rates of decrease for large separation in space and time and only the differences of the correlation functions $F - G$ are absolutely integrable.

From (V.3.6) we get for a state which is invariant under $\alpha^{\lambda h}$

$$\omega^{\lambda h}\left(\left(\beta_t^{\lambda h}\right)^{-1} A\right) = \omega^{\lambda h}(A) - i\lambda \int_0^t \omega^{\lambda h}\left([h, \alpha_{t'} A]\right) dt'. \tag{V.3.8}$$

By assumption 3.2.1 the limits for $t \to \pm\infty$ exist for $A \in D$. The normalized positive linear forms on D

$$\omega_\pm(A) = \lim_{t \to \pm\infty} \omega^{\lambda h}\left(\left(\beta_t^{\lambda h}\right)^{-1} A\right) \tag{V.3.9}$$

define, by extension, states on $\mathfrak{A}$. Since

$$\left(\beta_t^{\lambda h}\right)^{-1} \alpha_{t'} = \alpha_{t'}^{\lambda h} \left(\beta_{t+t'}^{\lambda h}\right)^{-1}, \tag{V.3.10}$$

the states $\omega_\pm$ are invariant under α_t i.e.

$$\omega_\pm(\alpha_t A) = \omega_\pm(A). \tag{V.3.11}$$

Now we argue that due to the dynamical stability requirement the states ω_+, ω_-, $\omega^{\lambda h}$, ω all must lie in the same primary folium for sufficiently small λ since two primary states with norm distance less than 2 lie in the same folium by theorem 2.2.16 of Chapter III. On the other hand a primary folium can contain at most one state which is invariant under an automorphism group with respect to which $\mathfrak{A}$ is aysymptotically Abelian. This follows from the argument establishing lemma 3.2.5 in Chapter III. Therefore we conclude

$$\omega_+ = \omega_- = \omega. \tag{V.3.12}$$

Combining (V.3.8), (V.3.9), (V.3.12) and $\| \omega^{\lambda h} - \omega \| \rightarrow 0$ we get the following expression of the stability requirement

$$\int_{-\infty}^{\infty} \omega\left([h, \alpha_t A]\right) dt = 0, \quad h, A \in D. \tag{V.3.13}$$

Remarks. (i) Narnhofer and Thirring [Narn 82] showed that the condition (V.3.13) may also be interpreted as the condition for adiabatic invariance of ω (instead of dynamical stability). One considers the effect on ω when the dynamical law is changed by a local perturbation which is slowly switched on and, as $t \rightarrow \infty$, slowly switched off again. Adiabatic invariance means that the state returns to its original form at the end of this procedure.

(ii) It is also instructive to compare (V.3.13) with the treatment of a metastable state in quantum mechanics. One standard approach starts from an approximate Hamiltonian H_0 which has a discrete eigenvalue inside a continuous part of the spectrum. Computing the spectral projectors of the true Hamiltonian H by perturbation expansion with respect to the difference $h = H - H_0$ one finds generically that the discrete eigenvalue disappears; it is dissolved in the continuum. A discrete eigenvalue survives only if h is such that (V.3.13) holds for sufficiently many $A \in \mathfrak{B}(\mathcal{H})$ (where ω denotes the original discrete eigenstate of H_0 and $\alpha_t A = e^{iH_0 t} A e^{-iH_0 t}$). In our case the equilibrium state of an infinitely extended medium is a discrete eigenstate of the effective Hamiltonian in the representation π_ω and it should remain generically a discrete eigenstate if the dynamics is changed slightly in some finite region. Hence the requirement of (V.3.13) for all $h \in D$.

In the notation of (V.1.9), (V.1.20) condition (V.3.13) becomes

$$\widetilde{F}_{A,h}(0) = \widetilde{G}_{A,h}(0), \tag{V.3.14}$$

i.e. it is the (Fourier transform of the) KMS-condition at the value $\varepsilon = 0$. It is now very remarkable that (V.3.13) together with the requirements that ω shall be primary and stationary implies already that ω is a KMS-state with respect to α_t for some value of the inverse temperature β. The proof presented here uses the assumption 3.2.1 and combines arguments from [Haag 74] with the method described in [Haag 77, appendix]. For more details and variants of the assumptions see [*Bratteli and Robinson 1981*]. Let us set for h and A in (V.3.13) respectively

$$h = h_1 \alpha_\tau h_2, \quad A = A_1 \alpha_\tau A_2.$$

This is legitimate for finite τ since $h,\ A \in D$ when $h_k,\ A_k \in D$ $(k = 1,\ 2)$. We obtain

$$0 = \int \omega(I + II + III + IV)dt, \tag{V.3.15}$$

$$\begin{aligned} I &= h_1[\alpha_\tau h_2, \alpha_t A_1]\alpha_{t+\tau} A_2, \\ II &= h_1(\alpha_t A_1)[\alpha_\tau h_2, \alpha_{t+\tau} A_2], \\ III &= [h_1, \alpha_t A_1](\alpha_{t+\tau} A_2)\alpha_\tau h_2, \\ IV &= (\alpha_t A_1)[h_1, \alpha_{t+\tau} A_2]\alpha_\tau h_2. \end{aligned} \tag{V.3.16}$$

All four integrals are finite, continuous functions of τ and we may consider their limits as $\tau \to \infty$. For the contribution of I we get, setting $t = \tau + t'$ and using the invariance of ω

$$\left|\int \omega(I)dt\right| \leq$$

$$\left|\int_{-T}^{T} \omega\left((\alpha_{-\tau} h_1)[h_2, \alpha_{t'} A_1]\alpha_{t'+\tau} A_2\right) dt'\right| + \| h_1 \| \| A_2 \| \int_{|t'|>T} \| [h_2, \alpha_{t'} A_1] \| \, dt'$$

In the first term we may take the limit $\tau \to \infty$ under the integral sign. For primary ω it approaches (due to asymptotic Abelianness and the cluster property)

$$\omega(h_1)\omega(A_2) \int_{-T}^{T} [h_2, \alpha_{t'} A_1] dt'.$$

Thus it vanishes in the limit $T \to \infty$ due to (V.3.13). The second term also approaches zero as $T \to \infty$ since the integrand is an $\mathcal{L}^{(1)}$-function. Thus $\lim \int \omega(I)dt = 0$ as $\tau \to \infty$ and the same is true for $\lim \int \omega(IV)dt$. The norms of II and III are bounded by $\mathcal{L}^{(1)}$-functions of t which are independent of τ. Hence the limit may be taken under the integral sign. This yields

$$\lim_{\tau \to \infty} \int_{-\infty}^{\infty} \omega(II + III)dt$$

$$= \int \left(\omega(h_1 \alpha_t A_1)\omega([h_2, \alpha_t A_2]) + \omega([h_1, \alpha_t A_1])\omega(\alpha_t A_2)h_2)\right) dt$$

$$= \int \left(F_1(t)F_2(t) - G_1(t)G_2(t)\right) dt,$$

with the abbreviations

$$F_k \equiv F_{A_k, h_k}; \quad G_k \equiv G_{A_k, h_k}. \tag{V.3.17}$$

Due to (V.3.13) we may subtract in F and G the uncorrelated parts (see the remarks in connection with (V.1.19)). (V.3.15) becomes

$$\int \left(F_1(t)F_2(t) - G_1(t)G_2(t)\right) dt = 0 \tag{V.3.18}$$

and, repeating the process with three factors we get

$$\int (F_1(t)F_2(t)F_3(t) - G_1(t)G_2(t)G_3(t))\, dt = 0. \tag{V.3.19}$$

We can write (V.3.18) as

$$\int \widetilde{F}_1(-\varepsilon)\widetilde{F}_2(\varepsilon)d\varepsilon = \int \widetilde{G}_1(-\varepsilon)\widetilde{G}_2(\varepsilon)d\varepsilon, \tag{V.3.20}$$

and

$$F_{A,h}(t) = \langle \Omega|hU(t)A|\Omega\rangle - \langle \Omega|h|\Omega\rangle\langle \Omega|A|\Omega\rangle, \quad U(t) = e^{iHt}, \tag{V.3.21}$$

where $U(t)$ implements α_t. Applying the standard technique (see Chapter II, lemma 4.1.2) we can choose A_2, h_2 so that $\widetilde{F}_2$ is non-vanishing only in an arbitrary small interval around a chosen point ε_0. For this purpose set

$$A_2 = \int f(t')\alpha_{t'} B dt', \quad h_2 = A_2^*, \tag{V.3.22}$$

$$f(t') = \int \widetilde{f}(\varepsilon')e^{-i\varepsilon' t'} d\varepsilon'. \tag{V.3.23}$$

Then $\widetilde{F}_2(\varepsilon)$ and $\widetilde{G}_2(\varepsilon)$ vanish if ε is outside the support of $\widetilde{f}$. If ε_0 is in the spectrum of H, we can choose B so that $\widetilde{F}_2$ is non-vanishing inside the support of $\widetilde{f}$. The choice (V.3.22) makes $\widetilde{F}_2(\varepsilon)$ positive; we can arrange that $\int \widetilde{F}_2(\varepsilon)d\varepsilon = 1$ and shrink the support. In other words we can, by suitable choice of B and $\widetilde{f}$ let $\widetilde{F}_2$ approximate a δ-function arbitrarily well. Simultaneously $\widetilde{G}_2(\varepsilon)$ will approach a multiple of a δ-function (possibly 0). Assuming that $\widetilde{F}_1$ is continuous[1] we conclude from (V.3.20), for $\varepsilon_0 \in \operatorname{Spect} H$

$$\widetilde{F}_{A_1,h_1}(-\varepsilon_0) = c\widetilde{G}_{A_1,h_1}(-\varepsilon_0),$$

where c results from the ratio of the integrals of $\widetilde{G}_2$ and $\widetilde{F}_2$ in the limit. The essential point is that c is independent of A_1 and h_1 but will depend on ε_0 since it results from the choice of A_2, h_2 as described above. If both ε_0 and $-\varepsilon_0$ are in the spectrum of H then, for suitable A_1, h_1, both $\widetilde{F}_1$ and $\widetilde{G}_1$ are nonvanishing. We have

$$\widetilde{F}_{A,h}(\varepsilon) = \Phi(\varepsilon)\widetilde{G}_{A,h}(\varepsilon) \quad \text{if } \varepsilon \text{ and } -\varepsilon \text{ are in } \operatorname{Spect} H. \tag{V.3.24}$$

From the definition of F and G follows

$$\widetilde{F}_{A,B}(\varepsilon) = \widetilde{G}_{B,A}(-\varepsilon), \tag{V.3.25}$$

[1] This will be the case if the correlation function $F_1(t)$ and not only the difference $F - G$ is square integrable. The conclusion of the subsequent argument remains valid under weaker assumptions. For instance, if for certain states (such as at a critical point) the correlation functions should decrease more slowly at infinity then one can expect that derivatives become $\mathcal{L}^{(1)}$-functions and so $\widetilde{F}(\varepsilon)$ is continuous with the exception of the point $\varepsilon = 0$. This does not affect the main conclusion.

which leads to

$$\Phi(-\varepsilon) = \Phi(\varepsilon)^{-1}. \tag{V.3.26}$$

Further information about the function $\Phi(\varepsilon)$ is obtained if one applies the same technique once more to the relation (V.3.19). This yields

$$\Phi(\varepsilon)\Phi(\varepsilon' - \varepsilon) = \Phi(\varepsilon'), \tag{V.3.27}$$

provided that $\pm\varepsilon$, $\pm\varepsilon'$ and $\pm(\varepsilon' - \varepsilon)$ are in the spectrum. Therefore, if Spect H is the whole real line

$$\Phi(\varepsilon) = e^{\beta\varepsilon} \qquad \text{for some value of } \beta. \tag{V.3.28}$$

This is the KMS-condition.

Next we show that assumption 3.2.1 implies that Spect H is either all of $\mathbb{R}$ or it is one-sided i.e. a subset of $\{0 \cup \mathbb{R}^+\}$ or of $\{0 \cup \mathbb{R}^-\}$. One has

Lemma 3.2.2
Assume 3.2.1. Then in a primary folium

(i) $\varepsilon_1,\ \varepsilon_2 \in \text{Spect}\, H$ implies $\varepsilon_1 + \varepsilon_2 \in \text{Spect}\, H$.

(ii) There can be no isolated point in spect H besides (possibly) $\varepsilon = 0$.

Proof. Part (i) is the frequently encountered additivity of the spectrum which follows from the assumption as in theorem 5.4.1 of Chapter II. Part (ii) follows from the argument proving lemma 3.2.5 in Chapter III, using asymptotic Abelianess for time translations instead of space translations.

Theorem 3.2.3
If ω is a primary, stationary state and assumption 3.2.1 holds then there remain the following three alternatives for Spect H

a) Spect $H = \mathbb{R}$;

b) Spect $H \subset \{0 \cup \mathbb{R}^+\}$, ω is a ground state;

b) Spect $H \subset \{0 \cup \mathbb{R}^-\}$, ω is a ceiling state.

Proof. We must only show that if Spect H contains two values of opposite sign, say a and $-b$ ($a > 0,\ b > 0$) then every point in $\mathbb{R}$ is in Spect H. By part (i) of lemma 3.2.2 $na - mb \in \text{Spect}\, H$ for all positive integers $n,\ m$. Since Spect H is closed, part (ii) of the lemma tells us that we may assume that a/b is not a rational number. Then the set $na - mb$ is dense in $\mathbb{R}$ and Spect $H = \mathbb{R}$.

We have finally as a consequence of assumption 3.2.1

Theorem 3.2.4
A stationary, primary state ω which is dynamically stable satisfies the KMS-condition with respect to α_t for some value of the parameter β unless it is a

ground state (which may be regarded as the limiting case for $\beta = \infty$) or a ceiling state (the limiting case $\beta = -\infty$).

Comment. One should take note that no information about the sign of β is provided by theorem 3.2.4. On the other hand, given a dynamical system $(\mathfrak{A}, \alpha_t)$, the set of values $\{\beta\}$ for which KMS-states exist is determined. It appears that one fundamental property of the net of local algebras and their geometric automorphisms must be that no KMS-states to negative values of β exist. This may be regarded as an extension or generalization of axiom **A3** in Chapter II which plays such an essential rôle in quantum field theory. Compare also the concept of ***passivity*** discussed in the next subsection. It is probable that this feature can be reduced to a principle concerning the structure of the net in the small (local definiteness, together with a characterization of the germ of the theory such as attempted via a concept of "local stability" by Haag, Narnhofer, Stein [Haag 84]).

What is the significance of the parameter β (the inverse temperature in physics) in this context? From

$$\omega^{\lambda h}(A) = \lim_{t\to\infty} \omega\left(\beta_t^{\lambda h} A\right) \tag{V.3.29}$$

and (V.3.5) we get

$$\frac{d}{d\lambda}\omega^{\lambda h}(A)\bigg|_{\lambda=0} = i\int_0^\infty \omega\left([h, \alpha_t A]\right) dt = i\int_0^\infty \left(F_{A,h}(t) - G_{A,h}(t)\right) dt.$$

Due to the analytic properties of $F(z)$ this may be replaced by an integral along the imaginary axis from 0 to $i\beta$

$$\frac{d}{d\lambda}\omega^{\lambda h}(A)\bigg|_{\lambda=0} = -\int_0^\beta \left(F_{A,h}(i\gamma)\right) d\gamma. \tag{V.3.30}$$

Therefore the change of the state due to a small local perturbation of the dynamics is - at least for small β (high temperature) - proportional to β. One may regard β as a stability parameter: The sensitivity of an equilibrium state to a change in the dynamical law decreases with decreasing β. At infinite temperature ($\beta = 0$) the state is completely insensitive.

Let us add one last remark to the topic of stability and equilibrium. It is intuitively suggested that an equilibrium state which is not at a phase transition point possesses higher stability. One expects that it will satisfy (V.3.13) even for homogeneous perturbations i.e. even if h is no longer in $\mathfrak{A}$. Therefore it is of interest to study the degree of stability of different states, considering changes of the dynamics of the form

$$h = \int f(\mathbf{x})\alpha_{\mathbf{x}} h' d^3x,$$

where $h' \in \mathfrak{A}$ and $\lim |\mathbf{x}|^n |f(\mathbf{x})| = 0$ as $|\mathbf{x}| \to \infty$. For $n > 3$ we have essentially local perturbations, $n = 0$ allows almost homogeneous perturbations. The value of n for which the stability breaks down relates to the rate of decrease of correlation functions in space-time and is one characteristic of different types of phase transitions.

V.3.3 Passivity

Another direct characterization of equilibrium states was given by Pusz and Woronowicz [Pusz 78]. It implies that an equilibrium state is a KMS-state over $\mathfrak{A}$ (with respect to time translation) for *some non-negative value of the parameter* β. One may start from the observation

Lemma 3.3.1
Let ω be a KMS-state over the C*-algebra $\mathfrak{A}$ with respect to the automorphism group α_t and parameter β. Denote by

$$\delta A = \frac{d}{dt}\alpha_t A\Big|_{t=0} \tag{V.3.31}$$

the infinitesimal generator of α_t, defined on a dense domain $D(\delta) \in \mathfrak{A}$.[2] Then, for any unitary $U \in D(\delta)$ and for any self-adjoint $A \in D(\delta)$ one has

$$-i\omega(U^*\delta U) \begin{cases} \geq 0 & \text{for} \quad \beta > 0 \\ \leq 0 & \text{for} \quad \beta < 0 \end{cases} \tag{V.3.32}$$

$$-i\omega(A\delta A) \begin{cases} \geq 0 & \text{for} \quad \beta > 0 \\ \leq 0 & \text{for} \quad \beta < 0 \end{cases} \tag{V.3.33}$$

Proof. Take the GNS-representation $(\pi_\omega,\ \Omega,\ \mathcal{H})$ in which α_t is implemented by the unitaries $U(t) = e^{iHt}$ and

$$\pi_\omega(\delta A) = i[H, \pi_\omega(A)], \quad A \in \mathfrak{A}, \tag{V.3.34}$$

$$H\Omega = 0. \tag{V.3.35}$$

The modular operator of Ω is $\Delta = e^{-\beta H}$. The basic relation following from the KMS-property (see (V.1.50)) is

$$Je^{-\beta/2\,H}\pi_\omega(A)|\Omega\rangle = \pi_\omega(A^*)|\Omega\rangle. \tag{V.3.36}$$

For unitary $U \in \mathfrak{A}$ this yields

$$1 = \langle\pi_\omega(U^*)\Omega|\pi_\omega(U^*)\Omega\rangle = \langle\pi_\omega(U)\Omega|e^{-\beta H}\pi_\omega(U)\Omega\rangle$$

or

$$\langle\Omega|\pi_\omega(U^*)\left(\mathbb{1} - e^{-\beta H}\right)\pi_\omega(U) \mid \Omega\rangle = 0. \tag{V.3.37}$$

Since, for a real variable x

[2] δ is an unbounded derivation on $\mathfrak{A}$.

$$x \geq 1 - e^{-x}$$

we get for $U \in D(\delta)$, using (V.3.35)

$$\langle \Omega \mid \pi_\omega(U^*)\beta H \pi_\omega(U) \mid \Omega\rangle \equiv -i\beta\langle\Omega|\pi_\omega(U^*\delta U)|\Omega\rangle \geq 0,$$

$$-i\beta\omega(U^*\delta U) \geq 0. \qquad \text{(V.3.38)}$$

For self-adjoint $A \in D(\delta)$ one finds similarly

$$-i\beta\omega(A\delta A) \geq 0. \qquad \text{(V.3.39)}$$

□

There is a converse to this lemma.

Theorem 3.3.2
If $\mathfrak{A}$ is asymptotically Abelian with respect to α_t and ω is a primary state satisfying either

$$-i\omega(U^*\delta U) \geq 0 \quad \text{for every unitary} \quad U \in D(\delta) \qquad \text{(V.3.40)}$$

or

$$-i\omega(A\delta A) \geq 0 \quad \text{for every self-adjoint} \quad A \in D(\delta) \qquad \text{(V.3.41)}$$

then ω is either a ground state or it is a KMS-state with respect to α_t for some *positive* value of β.

Note that (V.1.67) for $A^* = A$ reduces to

$$\operatorname{sign}\beta\,(-i\omega(A\delta A)) \geq 0. \qquad \text{(V.3.42)}$$

So the theorem tells that in the case where $\mathfrak{A}$ is asymptotically Abelian under α_t and ω is primary this special case of (V.1.67) suffices to prove that ω is a KMS-state (including the limiting situation $\beta = \infty$).

Definition 3.3.3
A primary state ω satisfying either (V.3.40) or (V.3.41) is called a passive state.

Let us make a few remarks concerning the proof of theorem 3.3.2. For a complete proof the reader is referred to the original paper by Pusz and Woronowicz or to volume II of Bratteli and Robinson.

First, one sees easily that (V.3.40) implies (V.3.41) by using for the unitaries a family $e^{i\varepsilon A}$ and expanding in powers of ε. Next one sees that (V.3.41) implies that ω is stationary: If $A^* = A$ then $(A\,\delta A)^* = \delta A\,A$ and, if $\omega(A\,\delta A)$ is purely imaginary then $\omega(A\,\delta A)^* = -\omega(A\delta A)$. So one has in this case

$$\omega(A\,\delta A) + \omega(\delta A\,A) \equiv \omega(\delta A^2) = 0,$$

which implies $\omega(\alpha_t A^2) = \omega(A^2)$. Since linear combinations of positive elements in $D(\delta)$ are dense in $\mathfrak{A}$ one has

$$\omega(\alpha_t B) = \omega(B) \quad \text{for all} \quad B \in \mathfrak{A}. \tag{V.3.43}$$

Next one forms the GNS-representation induced by ω yielding $(\pi,\ \Omega, \mathcal{H},\ U(t),\ H)$ with the relations (V.3.34), (V.3.35). By theorem 3.2.3 ω may either be a ground state or $\operatorname{Spect} H = \mathbb{R}$. The third possibility, that ω may be a ceiling state, is excluded by (V.3.41). If $\operatorname{Spect} H = \mathbb{R}$ then one shows that ω must be faithful. So one can apply the machinery of the Tomita-Takesaki theory. From (V.3.43) it follows that $U(t)$ commutes with the modular operator Δ. The last step is to show that then, under the assumptions of the theorem, the modular Hamiltonian K must be a multiple of H (possibly zero). This problem is similar to the one discussed in the last subsection (with dynamical stability replaced by passivity). We shall not reproduce here the somewhat lengthy argument.

The most interesting aspect of theorem 3.3.2 is its relation to the second law of thermodynamics. Suppose we can change the dynamical law, e.g. by applying external fields in a controlled way. Studying the behaviour of the system under different dynamical laws we should generalize the concept of "observable" to that of "observation procedure".[3] Recalling the discussion at the beginning of section 3 in Chapter I, a procedure specifies on the one hand the intrinsic construction of the measuring apparatus (the workshop drawing) and on the other hand the placement of the apparatus within the space-time reference frame of the laboratory. For the present purpose we shall combine the intrinsic construction and all placement parameters *with the exception of the time of measurement* into one symbol A but specify the time separately. Thus $(A,\ t)$ denotes a procedure and A will be mathematically represented by a self-adjoint element of an algebra $\mathfrak{A}$ which we might call the kinematical algebra and which we may also identify with $(\mathfrak{A},\ 0)$, the measurement procedures at $t = 0$. Given a dynamical law one has an equivalence between procedures at different times: To every procedure $(A,\ t_1)$ there is a procedure $(B,\ t_2)$ so that in every state the measuring results obtained by either one of the two procedures coincide. This may be expressed by saying that a dynamical law defines a 1-parameter family τ_t of automorphisms of $\mathfrak{A}$ and we can represent the procedure $(A,\ t)$ by the algebraic element $\tau_t A \in \mathfrak{A}$. If the dynamical law is not homogeneous in time then τ_t is not a group. We shall start, however, from a basic homogeneous dynamical law τ_t^0, denoted as before by α_t (absence of external fields) and consider changes of this law of the form

$$\frac{d}{dt}\tau_t A = \tau_t\left(\delta A + i[h(t), A]\right); \quad h(t) \in \mathfrak{A},\ A \in D(\delta), \tag{V.3.44}$$

where δ is the generator of α_t (see lemma 3.3.1).

Remark. In the case of a finite system we could write

$$\delta A = i[H_0, A], \tag{V.3.45}$$

[3]That this distinction is important was emphasized by H. Ekstein in connection with his concept of "presymmetry" [Ek 69].

where H_0 is affiliated with $\mathfrak{A}''$ and interpreted as the energy operator of the unperturbed system. Further, writing

$$\tau_t A = U(t)AU^*(t), \tag{V.3.46}$$

$$\frac{d}{dt}U(t) = i(\tau_t H(t))U(t) = iU(t)H(t); \quad \frac{d}{dt}\tau_t A = i\tau_t[H(t), A], \tag{V.3.47}$$

$$H(t) = H_0 + h(t). \tag{V.3.48}$$

The operator $\tau_t H(t)$, representing the procedure $(H(t),\ t)$, is the energy operator for the system at time t including the then prevailing external influence $(h(t),\ t)$. If

$$h(t) = 0 \quad \text{for} \quad t \leq 0 \quad \text{and} \quad t \geq T \tag{V.3.49}$$

then the energy transferred to the system in the time interval during which the external influence was acting is

$$\Delta E = \tau_T H_0 - H_0 = \int_0^T \frac{d}{dt}\tau_t H_0 dt. \tag{V.3.50}$$

This can be rewritten in another form. By (V.3.46), (V.3.47)

$$\frac{\partial}{\partial t}\tau_t H(t')\bigg|_{t'=t} = 0 = \frac{d}{dt}\tau_t H_0 + \frac{\partial}{\partial t}\tau_t h(t')\bigg|_{t'=t} \tag{V.3.51}$$

which is the adaptation of energy conservation to the case of time dependent influences. The energy transfer (V.3.50) can then be written, taking the condition (V.3.49) into account

$$\Delta E = -\int_0^T \left(\frac{d}{dt}\tau_t\right) h(t)\, dt = \int_0^T \tau_t \frac{d}{dt}h(t)\, dt. \tag{V.3.52}$$

On the other hand, defining

$$\Gamma(t) = U(t)e^{-iH_0 t} \tag{V.3.53}$$

(V.3.50) gives

$$\Delta E = \Gamma(T)H_0\Gamma(T)^* - H_0 = -i\Gamma(T)\delta\Gamma(T)^*. \tag{V.3.54}$$

In the infinite system H_0 and $U(t)$ cannot be defined but $\Gamma(t)$ remains a well defined unitary in $\mathfrak{A}$, given by the differential equation

$$\frac{d}{dt}\Gamma(t) = i\Gamma(t)\alpha_t h(t), \tag{V.3.55}$$

with initial condition $\Gamma(0) = \mathbb{1}$. The above argument remains valid if T is finite and $h(t)$ is affiliated with the algebra of a bounded region when we replace H_0 by the integral of the energy density over a sufficiently large region. So we can write for the energy transfer

$$\Delta E = -i\Gamma(T)\delta\Gamma(T)^* = \int_0^T \tau_t \frac{dh(t)}{dt}\, dt. \tag{V.3.56}$$

This justifies the term "passive" in the definition 3.3.3. We have

Proposition 3.3.4
A state ω is *passive* if, for any smooth family of self-adjoint elements $h(t) \in \mathfrak{A}$ satisfying (V.3.49)

$$\Delta E = \omega\left(\int_0^T \tau_t \frac{dh(t)}{dt}\, dt\right) \geq 0. \tag{V.3.57}$$

If we interpret the change of the dynamical law from α_t to τ_t during the interval $0 \leq t \leq T$ as a cyclic process caused by external forces we see that from a passive state we cannot extract energy by such a cyclic process.

One may then identify the equilibrium states of the dynamical system ($\mathfrak{A}$, α_t) *as the passive states. By theorem 3.3.2. the passive states are the KMS-states with respect to* α_t *and non-negative parameter* β *(including the limiting case* $\beta = \infty$*, the ground state).*

V.3.4 Chemical Potential

The characterization of equilibrium states by either one of the two properties: dynamical stability or passivity led to the conclusion that an equilibrium state is a KMS-state of the observable algebra $\mathfrak{A}$ with respect to time translations α_t. On the other hand, starting from the grand canonical ensemble for a finite system, the thermodynamic limit leads to a KMS-state over the "algebra of operations", the "field algebra" $\mathcal{A}$ with respect to a 1-parameter subgroup of time translations and gauge transformations. If we consider for simplicity here a 1-component system then this subgroup is characterized by the two parameters β, μ (see (V.1.14)). Of course, since the observables are invariant under γ_φ, the state restricted to $\mathfrak{A}$ is a KMS-state with respect to α_t and parameter β. But it is clear that the states $\omega_{\beta,\mu}$ over $\mathcal{A}$ with different values of μ yield different states over $\mathfrak{A}$ by this restriction. The problem of extending (α_t, β)-KMS-states over $\mathfrak{A}$ to states over $\mathcal{A}$ has been considered by Kastler [Kast 76], Araki and Kishimoto [Ara 77b], and most completely by Araki, Haag, Kastler, Takesaki [Ara 77c] which is the main source of the subsequent discussion.

Focusing attention on the observable algebra and ignoring for the moment the embedding algebra $\mathcal{A}$, the appearance of a distinguishing parameter μ within the set of (α_t, β)-KMS-states over $\mathfrak{A}$ has the same root as the appearance of charge quantum numbers in the set of states satisfying one of the selection criteria of Chapter IV. It is the existence of localized morphisms of $\mathfrak{A}$ which are not inner. In our simplest example (Abelian gauge group) we are dealing with an equivalence class of *automorphisms* modulo the inner automorphisms.

Let ϱ be an automorphism in the class which adds one unit of charge. If ω is a state over $\mathfrak{A}$ then

$$\omega_\varrho(A) \equiv \omega(\varrho A) \tag{V.3.58}$$

defines a state differing from ω by one additional charge localized in some finite volume at time zero.

$$\varrho_t = \alpha_t \varrho \alpha_t^{-1} \tag{V.3.59}$$

describes the automorphism shifted by a time translation i.e. the adding of one unit of charge the same way at a later time t. ϱ and ϱ_t are in the same class; so they are related by an inner automorphism[4] and we can write

$$\varrho_t = \varrho \operatorname{Ad} v_t, \quad \text{or explicitly} \quad \varrho_t A = \varrho(v_t A v_t^*), \tag{V.3.60}$$

where v_t is a unitary element of $\mathfrak{A}$. $\operatorname{Ad} v_t$ is determined by ϱ and α_t. Correspondingly, v_t is determined up to a phase factor which can be fixed by a convention. First, by (V.3.59), we see that we can choose the family v_t so that it satisfies the cocycle identity

$$v_{t+t'} = v_t \alpha_t v_{t'}, \tag{V.3.61}$$

or, equivalently, the differential equation

$$i\frac{d}{dt} v_t = v_t \alpha_t h; \quad h = -i\, \frac{d}{dt} v_t \bigg|_{t=0} . \tag{V.3.62}$$

Here h is a self-adjoint, uniquely determined by ϱ and α_t up to $c\mathbb{1}$ where c is an arbitrary real constant. This fixes v_t up to a factor e^{ict}. The important point is, however, that once we have adopted a convention for choosing c this cannot be changed any more in the subsequent consideration of equilibrium states with different chemical potential.

Next, let ω be a primary $(\alpha_t,\ \beta)$-KMS-state over $\mathfrak{A}$ and ω_ϱ as in (V.3.58). In contradistinction to Chapter IV, where we discussed states with vanishing matter density at infinity and the addition of a charge led to a different superselection sector, now ω describes a state which has already infinitely many particles and the addition of one particle[5] is not a drastic change; it does not lead out of the folium. Therefore ω_ϱ is again a vector state in the GNS-representation π_ω of $\mathfrak{A}$. We may compare the modular automorphism groups σ_t and σ_t^ϱ of the states ω and ω_ϱ. One sees easily from the defining relations (V.2.1) through (V.2.12) that

$$\sigma_t^\varrho = \varrho^{-1} \sigma_t \varrho. \tag{V.3.63}$$

The Radon-Nikodym cocycle

$$W_t \equiv D\omega_\varrho : D\omega \tag{V.3.64}$$

is a family of unitaries in $\pi_\omega(\mathfrak{A})''$ and the intertwining property (V.2.56) gives

[4] Concerning the notation: in Chapter IV the inner automorphism induced by a unitary U was denoted by σ_U. Here we shall write Ad U instead of σ_U in order to avoid confusion with modular automorphisms.

[5] In the non-relativistic context the "charge" we talk about is synonymous with particle number.

$$\sigma_t^{\varrho} = (\mathrm{Ad}\, W_t)\sigma_t, \tag{V.3.65}$$

or, by (V.3.63)

$$\sigma_t \varrho \sigma_t^{-1} = \varrho \,\mathrm{Ad}\, W_t. \tag{V.3.66}$$

But the modular group of ω is

$$\sigma_t = \alpha_{-\beta t}. \tag{V.3.67}$$

So, by (V.3.59)

$$\varrho_{-\beta t} = \varrho \,\mathrm{Ad}\, W_t. \tag{V.3.68}$$

This implies that Ad W_t is actually an inner automorphism of $\mathfrak{A}$ and the comparison with (V.3.60) gives

$$\mathrm{Ad}\, W_t = \mathrm{Ad}\, v_{-\beta t}. \tag{V.3.69}$$

Since the center of $\pi_\omega(\mathfrak{A})''$ consists only of multiples of the identity W_t and $\pi_\omega(v_{-\beta t})$ agree up to a phase factor and, since W_t satisfies the cocycle relation (V.2.52), this factor is a function $e^{i\lambda t}$. So we can write

$$W_t = e^{i\beta\mu t}\pi_\omega(v_{-\beta t}), \tag{V.3.70}$$

where the parameter μ depends on the state ω.

This parameter is the chemical potential (up to a constant c which depends on the convention used in fixing v_t). This is seen by looking at the extension of ω to a state $\widehat{\omega}$ over the field algebra $\mathcal{A}$. Let us suppose that this extension is an (α_t^μ, β)-KMS-state (see (V.1.14)) so that the modular group of $\widehat{\omega}$ is

$$\widehat{\sigma}_t = \alpha_{-\beta t}\gamma_{\beta\mu t}. \tag{V.3.71}$$

The extension $\widehat{\varrho}$ of the charge creation automorphism to $\mathcal{A}$ is implemented by a unitary $u \in \mathcal{A}$. For distinction we shall denote the elements of $\mathcal{A}$ by lower case letters $a, b, \ldots$, those of $\mathfrak{A}$ by capital letters $A, B, \ldots$. Then

$$\widehat{\varrho} a = u a u^*, \quad a \in \mathcal{A}, u \in \mathcal{A}, \tag{V.3.72}$$

with

$$\gamma_\varphi u = e^{i\varphi} u. \tag{V.3.73}$$

As above we have

$$\widehat{\varrho}_t \equiv \alpha_t \widehat{\varrho} \alpha_t^{-1} = \widehat{\varrho}\,\mathrm{Ad}\, \widehat{v}_t, \tag{V.3.74}$$

and find from (V.3.72) that $\widehat{v}_t$ can now be chosen as

$$\widehat{v}_t = u^* \alpha_t u. \tag{V.3.75}$$

The modular automorphism groups of $\widehat{\omega}_\varrho$ and $\widehat{\omega}$ are related as in (V.3.63) and the Radon-Nikodym cocycle is now

$$\widehat{W}_t \equiv D\widehat{\omega}_\varrho : D\widehat{\omega} = u^* \widehat{\sigma}_t u, \tag{V.3.76}$$

which becomes, inserting (V.3.71) and (V.3.73), according to (V.3.75)

$$\widehat{W}_t = e^{i\beta\mu t}\widehat{v}_{-\beta t}. \tag{V.3.77}$$

We have omitted writing the symbol $\pi_{\widehat{\omega}}$ since all the objects are understood to be taken in this GNS-representation. The restriction of (V.3.77) to $\mathfrak{A}$ just means that we drop the marking $\widehat{}$ and then (V.3.77) becomes (V.3.70).

One should still understand more directly why and how the extension of an (α_t, β)-KMS-state ω over $\mathfrak{A}$ leads to a state $\widehat{\omega}$ over $\mathcal{A}$ which is a KMS-state with respect to a modified automorphism group. Let us start from the following setting. $\mathcal{A}$ is the C*-algebra of quasilocal operations which may be constructed from the observable algebra $\mathfrak{A}$ and its localized morphisms as described in Chapter IV. The full symmetry group of $\mathcal{A}$ consists of the geometric symmetries and a (global) gauge group $\mathcal{G}$. Of the former we shall use here only the time translations α_t and the powers of some automorphism τ with respect to which $\mathcal{A}$ is asymptotically Abelian[6]

$$\lim_{n\to\infty}[a, \tau^n b] = 0. \tag{V.3.78}$$

τ may be interpreted as a space translation or, under the assumption 3.2.1, also as a time translation. We shall further assume that $\mathcal{G}$ is a compact Lie group (possibly non Abelian). It is represented by automorphisms of $\mathcal{A}$ denoted by γ_g, $g \in \mathcal{G}$ which commute with α_t and τ. We shall consider states over $\mathfrak{A}$ which are extremal τ-invariant i.e.

$$\omega(\tau A) = \omega(A),$$

and such that ω is not a mixture (convex combination) of other τ-invariant states. The asymptotic Abelianness of $\mathfrak{A}$ with respect to τ implies that extremal τ-invariance is equivalent to "weak clustering" (see Doplicher, Kadison, Kastler, Robinson [Dopl 67]):

$$\lim_{N\to\infty}(2N)^{-1}\sum_{-N}^{N}\omega(a\tau^n b) = \omega(a)\omega(b). \tag{V.3.79}$$

This is the main structural property used in the following.

Theorem 3.4.1
An extremal τ-invariant state over $\mathfrak{A}$ has extensions to extremal τ-invariant states over $\mathcal{A}$. Two such extensions φ_1, φ_2 can differ only by a gauge transformation, i.e. there is an element $g \in \mathcal{G}$ such that

$$\varphi_2(a) = \varphi_1(\gamma_g a). \tag{V.3.80}$$

[6] Since $\mathcal{A}$ contains fermionic and bosonic elements the commutator in (V.3.78) should be replaced by the "graded commutator" $[a, b]_\pm$. This can be taken into account by a straightforward adaptation of the subsequent argument. We shall omit it here; it is given in [Ara 77c].

The first part of the theorem, the existence of extremal τ-invariant extensions follows from general abstract arguments (Hahn-Banach theorem, averaging over the group τ^n and Krein-Milman theorem). The uniqueness of such an extension up to a gauge transformation follows from the consideration of the set of continuous functions over $\mathcal{G}$ obtained from a state φ by

$$f_a(g) = \varphi(\gamma_g a); \quad a \in \mathcal{A}. \tag{V.3.81}$$

Obviously this set (for a ranging through $\mathcal{A}$ and fixed φ) is a linear space, containing with every function also its complex conjugate. We define the norm of f as the supremum of $|f(g)|$ and denote the completion of this set in the norm topology by $\mathcal{C}_\varphi$. If φ is weakly clustering then (V.3.79) implies that with $f,\ f' \in \mathcal{C}_\varphi$ also the product $ff' \in \mathcal{C}_\varphi$. So $\mathcal{C}_\varphi$ is an Abelian C*-algebra. It may be that φ is invariant under some subgroup of $\mathcal{G}$ which will be called the *stability group* or the *stabilizer* of φ and denoted by $\mathcal{G}_\varphi$. By the definition (V.3.81) then

$$f_a(g) = f_a(hg) \quad \text{for all} \quad h \in \mathcal{G}_\varphi. \tag{V.3.82}$$

So the functions in $\mathcal{C}_\varphi$ are constant on each right coset of $\mathcal{G}_\varphi$ and $\mathcal{C}_\varphi$ may be regarded as the C*-algebra of all continuous functions of the right cosets. Now, let φ_1 and φ_2 be two extremal τ-invariant extensions of the state ω on $\mathfrak{A}$. For any $a,\ b \in \mathcal{A}$

$$A_n(a,b) \equiv \int (\gamma_g a^* \tau^n b) d\mu(g) \tag{V.3.83}$$

is gauge invariant and hence belongs to $\mathfrak{A}$. In (V.3.83) μ denotes the Haar measure on $\mathcal{G}$. So $\varphi_1(A_n(a,b)) = \varphi_2(A_n(a,b))$. Taking the limit $n \to \infty$ and using (V.3.79) one gets

$$\int \overline{f}_a^{(1)}(g) f_b^{(1)}(g) d\mu(g) = \int \overline{f}_a^{(2)}(g) f_b^{(2)}(g) d\mu(g), \tag{V.3.84}$$

where $f^{(j)}$ are defined as in (V.3.81) with φ_j replacing φ. Furthermore the map V from $\mathcal{C}_{\varphi_1}$ onto $\mathcal{C}_{\varphi_2}$ given by

$$V f_a^{(1)} = f_a^{(2)} \tag{V.3.85}$$

respects the product structure. It gives an isomorphism of the C*-algebras. Defining right and left translations of functions on $\mathcal{G}$ respectively by

$$(\varrho_g f)(g') = f(g'g); \quad (\lambda_g f)(g') = f(gg') \tag{V.3.86}$$

one sees that V commutes with all right translations. One can show that an isomorphism of two C*-subalgebras of the continuous functions on $\mathcal{G}$ which commutes with all right translations is given by a left translation (see appendix A in [Ara 77c]). Thus $V = \lambda(g)$ for some element $g \in \mathcal{G}$. This is precisely the claim of (V.3.80). It also follows that the stability subgroups of φ_1 and φ_2 are related through conjugation with this element g. In physical language the stability group is the *unbroken part of the gauge group* for the extended state. We see that it is determined by the state ω on $\mathfrak{A}$ uniquely up to a conjugation.

Theorem 3.4.2
Let ω be an α_t-invariant, extremal τ-invariant state on $\mathfrak{A}$ and φ an extremal τ-invariant extension to $\mathcal{A}$. Then there exists a continuous one-parameter subgroup of $\mathcal{G}$

$$t \in \mathbb{R} \rightarrow \varepsilon(t) \in \mathcal{G}$$

such that φ is invariant under the modified time translation

$$\alpha_t^{'} = \alpha_t \gamma_{\varepsilon(t)}. \tag{V.3.87}$$

This is a simple consequence of theorem 3.4.1. Since $\omega = \omega \circ \alpha_t$ the states φ and $\varphi \circ \alpha_t$ are two (possibly different) extensions of ω. Therefore they are related as in (V.3.80) where now, of course, g will depend on t.

The most interesting question concerns the extension of an extremal (α_t, β)-KMS-state ω over $\mathfrak{A}$. We shall not reproduce here the rather lengthy arguments establishing the properties of the extensions in full generality. They can be found in [Ara 77c]. Instead we shall comment on the main conclusions. The state ω is α_t-invariant and, since it is also primary it is an extremal α_t-invariant state. Under assumption 3.2.1. we can use the time translation group instead of the group τ^n. Theorem 3.4.1. holds also for this case and tells us that there is an extremal α_t-invariant extension φ to $\mathcal{A}$. It is unique up to gauge and has a stability group $\mathcal{G}_\varphi \subset \mathcal{G}$ (the unbroken part of the gauge symmetry). It may happen that $\mathcal{G}_\varphi$ has a subgroup $\mathcal{N}_\varphi$ called the *asymmetry group* with respect to which φ is a ground state or ceiling state. This means the following. Since φ is $\mathcal{G}_\varphi$-invariant we have in the GNS-representation $(\pi_\varphi(\mathcal{A}), \mathcal{H}_\varphi, |\Phi\rangle)$ an implementation of γ_g for $g \in \mathcal{G}_\varphi$ by unitaries leaving the state vector $|\Phi\rangle$ (the vector representative of φ) invariant. Their generators are self-adjoint operators on $\mathcal{H}_\varphi$, representing elements of the Lie algebra of $\mathcal{G}_\varphi$. The spectra of these generators are additive semigroups. This follows from the τ- (or α_t)-asymptotic Abelianness by essentially the same argument which establishes the additivity of the energy-momentum spectrum (Chapter II, theorem 5.4.1). So, if N is such a generator, its spectrum can either be one-sided (confined to non-negative or to non-positive reals) or it may extend from $-\infty$ to $+\infty$. The asymmetry group $\mathcal{N}_\varphi$ is that part of $\mathcal{G}_\varphi$ whose Lie algebra is represented by operators with a one-sided spectrum. In statistical mechanics this part does not play a rôle. It just means that the components (charges) associated with these directions in $\mathcal{G}$ are absent, their chemical potential is $-\infty$. One may eliminate the asymmetry group by considering instead of $\mathcal{A}$ the subalgebra $\mathcal{A}_1$ of invariant elements under $\mathcal{N}_\varphi$ (fixed point algebra for $\mathcal{N}_\varphi$) and instead of $\mathcal{G}_\varphi$ the quotient group $\mathcal{G}_\varphi/\mathcal{N}_\varphi$. We shall therefore formulate the conclusion only for the case where $\mathcal{N}_\varphi$ is absent.

Theorem 3.4.3
Let ω be an extremal (α_t, β)-KMS-state over $\mathfrak{A}$, φ an extremal α_t-invariant extension to $\mathcal{A}$ (unique up to gauge), $\mathcal{G}_\varphi$ the stabilizer (unbroken part of $\mathcal{G}$ in the state φ). If the asymmetry group $\mathcal{N}_\varphi$ is trivial then φ is an $(\alpha_t^{'}, \beta)$-KMS-state where

$$\alpha_t' = \alpha_t \gamma_{\varepsilon(t)} \tag{V.3.88}$$

and $\varepsilon(t)$ is a one-parameter subgroup *in the center of* $\mathcal{G}_\varphi$.

Note that if the gauge symmetry is completely broken or if the Lie algebra of $\mathcal{G}_\varphi$ has trivial center then $\alpha_t' = \alpha_t$ Note also that $\mathcal{G}_\varphi$ and the element in the center of the Lie algebra of $\mathcal{G}_\varphi$ defining $\varepsilon(t)$ are already essentially determined by ω. For two different extensions $\mathcal{G}_{\varphi_2} = g\mathcal{G}_{\varphi_1}g^{-1}$ where g is some fixed element of $\mathcal{G}$. The state φ need not necessarily be an extremal KMS-state. If not then its extremal KMS-components are also extensions of ω which are then, however, not α_t-invariant.

V.4 Modular Automorphisms of Local Algebras

Let us return to the consideration of the vacuum sector of a relativistic local theory. We use the concepts and notation of section III.4. The theorem of Reeh and Schlieder (subsection II.5.3) says that the state vector Ω of the vacuum is cyclic and separating for any algebra $\mathcal{R}(\mathcal{O})$ as long as the causal complement of $\mathcal{O}$ contains a non-void open set. Thus we can apply the Tomita-Takesaki theory. The vacuum defines a modular operator $\Delta(\mathcal{O})$, a modular conjugation $J(\mathcal{O})$ and a modular automorphism group $\sigma_\tau(\mathcal{O})$ for any such $\mathcal{R}(\mathcal{O})$. What is the meaning of these objects? We know that $\sigma_\tau(\mathcal{O})$ maps $\mathcal{R}(\mathcal{O})$ onto itself, $J(\mathcal{O})$ maps $\mathcal{R}(\mathcal{O})$ onto $\mathcal{R}(\mathcal{O})'$. But beyond this we have, in general, no geometric interpretation; the image under $\sigma_\tau(\mathcal{O})$ or $J(\mathcal{O})$ of $\mathcal{R}(\mathcal{O}_1)$ with $\mathcal{O}_1 \subset \mathcal{O}$ is not, in general, some local subalgebra. There are, however, interesting special cases where σ_τ and J do have a direct geometric meaning.

V.4.1 The Bisognano-Wichmann Theorem

Let W denote the wedge in Minkowski space

$$W = \{x \in \mathcal{M} : x^1 > |x^0|; \quad x^2, x^3 \text{ arbitrary}\}. \tag{V.4.1}$$

The 1-parameter subgroup $\Lambda(s)$ of special Lorentz transformations ("boosts" in the x^1-direction)

$$\Lambda(s) = \begin{pmatrix} \cosh s & \sinh s & 0 & 0, \\ \sinh s & \cosh s & 0 & 0, \\ 0 & 0 & 1 & 0 \\ 0 & 0 & 0 & 1 \end{pmatrix} \tag{V.4.2}$$

transforms W into itself. The transformation law of points may also be regarded as a relation in the Poincaré group. If $U(\Lambda(s))$, $U(x)$ are the unitary operators implementing respectively the boosts (V.4.2) and the space-time translations then

$$U(\Lambda(s))U(x)U(\Lambda(s))^{-1} = U(x(s)), \tag{V.4.3}$$

where

$$\begin{aligned} x^0(s) &= x^0 \cosh s + x^1 \sinh s, \\ x^1(s) &= x^0 \sinh s + x^1 \cosh s, \\ x^r(s) &= x^r \quad \text{for} \quad r = 2,\ 3. \end{aligned} \tag{V.4.4}$$

This relation may be extended to complex values of the parameter s. Setting

$$s = \lambda + i\mu \tag{V.4.5}$$

with λ, μ real we have

$$\begin{aligned} x^0(s) &= x^0(\lambda) \cos \mu + i \sin \mu\,(x^0 \sinh \lambda + x^1 \cosh \lambda), \\ x^1(s) &= x^1(\lambda) \cos \mu + i \sin \mu\,(x^0 \cosh \lambda + x^1 \sinh \lambda), \\ x^r(s) &= x^r \quad \text{for} \quad r = 2,\ 3. \end{aligned} \tag{V.4.6}$$

Introducing the generators, K for the boosts, P_μ for the translations

$$U(\Lambda(s)) = \exp iKs; \quad U(x) = \exp iP_\mu x^\mu \tag{V.4.7}$$

we get

$$\begin{aligned} &\exp iK(\lambda + i\mu)\, U(x) \exp -iK(\lambda + i\mu) \\ &= U(\cos \mu\, x(\lambda)) \exp\{-\sin\mu(P^0\eta^0 - P^1\eta^1)\}, \end{aligned} \tag{V.4.8}$$

with

$$\begin{aligned} \eta^0 &= x^0 \sinh \lambda + x^1 \cosh \lambda, \\ \eta^1 &= x^0 \cosh \lambda + x^1 \sinh \lambda, \\ \eta^r &= 0 \quad \text{for} \quad r = 2,\ 3. \end{aligned} \tag{V.4.9}$$

One notes that for $x \in W$ $\eta(\lambda)$ is a positive time-like vector for all λ. Therefore, due to the spectrum condition for energy-momentum in the vacuum sector, the second factor on the right hand side of (V.4.8) provides an exponential damping as long as

$$0 < \mu < \pi; \quad x \in W. \tag{V.4.10}$$

The first factor is unitary. We conclude that under the condition (V.4.10) the left hand side of (V.4.8) is a bounded operator and moreover an analytic function of the complex variable s in the strip $0 <$ Im s $< \pi$.

For $\mu = \pi$, $\lambda = 0$ we have

$$\Lambda(i\pi)x = x',$$

$$x'^\mu = \begin{cases} -x^\mu & \text{for} \quad \mu = 0,\ 1 \\ x^\mu & \text{for} \quad \mu = 2,\ 3. \end{cases} \tag{V.4.11}$$

We can invert the signs of x^r, by a spatial rotation through an angle of π around the 1-axis. Denoting this by $R_1(\pi)$ we get

$$R_1(\pi)\Lambda(i\pi)x = -x. \tag{V.4.12}$$

Using the CPT-operator Θ of Chapter II, theorem 5.1.4 we define the conjugate linear operator S by

$$S = \Theta U(R_1(\pi))U(\Lambda(i\pi)), \tag{V.4.13}$$

and obtain from (II.5.9)

$$SU(x)S^{-1} = U(x). \tag{V.4.14}$$

Let us evaluate the transformation law of a covariant field Φ under S. If we use spinorial notation and consider a field with n undotted, m dotted indices then a boost $\Lambda(\lambda)$ acts on each undotted index by the matrix $\exp(1/2\ \lambda\sigma_1)$, a rotation $R_1(\varphi)$ by $\exp(1/2\ i\varphi\sigma_1)$, where σ_1 denotes the first Pauli matrix. So the product $R_1(\pi)\Lambda(i\pi)$ brings a factor $\exp(i\pi\sigma_1) = -1$ for each undotted index. For dotted indices we have the complex conjugate transformation matrices (for real λ). The product $R_1(\pi)\Lambda(i\pi)$ gives $+1$. Altogether

$$U(R_1(\pi))U(\Lambda(i\pi))\Phi_r(0)U(\Lambda(i\pi))^{-1}U(R_1(\pi))^{-1} = (-1)^n\Phi_r(0), \tag{V.4.15}$$

where the polyindex r consists of n undotted, m dotted spinor indices. Together with (II.5.7) this yields for observable fields (where $n+m$ is even, $F=0$)

$$S\Phi_r(0)S^{-1} = \Phi_r^*(0). \tag{V.4.16}$$

Now we consider the (vector valued) distribution

$$\Phi(x_1)\Phi(x_2)\cdots\Phi(x_n)|\Omega\rangle \equiv U(x_1)\Phi(0)U(x_2-x_1)\Phi(0)\cdots U(x_n-x_{n-1})\Phi(0)|\Omega\rangle.$$

Applying (V.4.14), (V.4.16) we obtain formally

$$S\Phi(x_1)\Phi(x_2)\cdots\Phi(x_n)|\Omega\rangle = \Phi^*(x_1)\cdots\Phi^*(x_n)|\Omega\rangle.$$

We must remember, however, that on the way we used (V.4.8) and this gives an analytic continuation to complex values of s only if each of the arguments x_1, $(x_2-x_1),\ldots(x_n-x_{n-1})$ lies in W. For such configurations of points all x_i lie space-like to each other. So one can invert the order of factors on the right hand side and one must do so to approach the boundary of the analyticity domain in the right manner (see the corresponding discussion in the proof of the CPT-theorem in Chapter II). We obtain

$$S\Phi(x_1)\cdots\Phi(x_n)|\Omega\rangle = (\Phi(x_1)\cdots\Phi(x_n))^*\,|\Omega\rangle. \tag{V.4.17}$$

It remains to show that the special configurations of points used above suffice to conclude from the analyticity properties that (V.4.17) holds for all configurations of points in W. Then one has

$$SA|\Omega\rangle = A^*|\Omega\rangle \tag{V.4.18}$$

for all A in the polynomial algebra $\mathcal{P}(W)$, generated by fields smeared with test functions with support in W. One must show further that the domain $\mathcal{P}(W)|\Omega\rangle$ is a "core" for S i.e. that (V.4.18) for $A \in \mathcal{P}(W)$ uniquely determines S. For the technically quite tedious proof of these claims we refer to the original papers by Bisognano and Wichmann [Bis 75, 76]. Accepting this one knows that (V.4.18) holds also for $\mathcal{R}(W)$, the von Neumann algebra with which the fields in W are affiliated and one has finally

Theorem 4.1.1 (Bisognano and Wichmann)
If $\mathcal{R}(W)$ has a system of affiliated observable fields satisfying the axioms of Chapter II then the modular conjugation for the vacuum state is

$$J(W) = \Theta U(R_1(\pi)), \tag{V.4.19}$$

the modular operator is

$$\Delta(W) = e^{-2\pi K}, \tag{V.4.20}$$

the modular automorphisms σ_t act geometrically as the boosts (V.4.4) with $s = -2\pi t$.

Remarks. 1) Since the relations are purely geometrical one may wonder whether the theorem cannot be established by considering only the net $\mathcal{R}(\mathcal{O})$ without reference to a generating set of fields. This would demand to show that the operator $J(W)$, as determined from the Tomita-Takesaki theorem, has the geometric significance

$$J\mathcal{R}(\mathcal{O})J = \mathcal{R}(r\mathcal{O}), \tag{V.4.21}$$

where r is the reflection (V.4.11) This would also provide a proof of the CPT-theorem in the algebraic setting, defining Θ by (V.4.19).[1] Besides (V.4.21) one would have to prove the KMS-condition for the functions

$$F_{A,B}(t) = \omega_0(Be^{2\pi iKt}Ae^{-2\pi iKt}),$$

$$G_{A,B}(t) = \omega_0(e^{2\pi iKt}Ae^{-2\pi iKt}B), \quad A, B \in \mathcal{R}(W).$$

Looking at the argument above we see that the essential idea is to generate the algebra by translates $\alpha_x A$ with $x \in W$ and $A \in \mathcal{R}(\mathcal{O}_0)$ where $\mathcal{O}_0$ is a small neighborhood of the origin. The problem is then to control the growth of

$$e^{-K\mu}Ae^{K\mu}; \quad A \in \mathcal{R}(\mathcal{O}_0), \quad 0 < \mu < \pi. \tag{V.4.22}$$

If $\mathcal{O}_0$ shrinks to a point and A is replaced by a field $\Phi(0)$ transforming according to some finite dimensional representation of the Lorentz group then this is answered by (V.4.15). But we lack a study of the behaviour of (V.4.22) for small finite $\mathcal{O}_0$.

2) The trajectory of the point $x = (0,\ \varrho,\ 0,\ 0)$ under the boosts (V.4.2) can be interpreted as a uniformly accelerated motion with acceleration

$$b = \varrho^{-1}$$

For an observer moving on this trajectory and using his proper time $\tau = \varrho s$ as time coordinate the operator $H = \varrho^{-1}K$ generates time translations in his body

[1] Significant progress towards this goal has recently been made in [Borch 92, 95].

fixed coordinate system. According to theorem 4.1.1 the vacuum state looks to him like a thermal state with temperature

$$T = (2\pi\varrho)^{-1} = \frac{b}{2\pi}. \tag{V.4.23}$$

This very interesting result is closely related to the Hawking temperature of a black hole [Hawk 75]. In fact, the wedge W in Minkowski space, populated by observers which are constrained to stay always in W, provides the simplest example of a horizon (see Rindler [Rind 66]). If such an observer sends a signal to a region outside the wedge neither he nor his fellow travellers can receive an answer back. In this analogy between the wedge W, considered as the "Rindler universe", and a (permanent) black hole the Bisognano-Wichmann temperature (V.4.23) corresponds to the Hawking temperature. Motivated by Hawking's paper Unruh discussed the response of a uniformly accelerated detector to the vacuum state of a quantum field in Minkowski space [Unr 76]; see also [Dav 75], [de Witt 80]. Unruh idealized the detector as a collection of "molecules" (quantum mechanical systems with a discrete energy spectrum and negligeable spatial extension) coupled with the quantum field by an interaction Hamiltonian proportional to $\Phi(x)$ where x denotes the moving point on the trajectory. His result can be expressed in the following way: if the quantum field is in the vacuum state then the population of the energy levels of the molecules adjusts to the same distribution which the system at rest would have in a surrounding thermal bath at temperature (V.4.23). We shall return to the problem of quantum field theory in the presence of a gravitational background field (curved space-time) and the equivalence principle in quantum physics in Chapter VIII.

V.4.2 Conformal Invariance and the Theorem of Hislop and Longo

In general the wedge regions are the only ones for which the modular automorphisms (induced by the vacuum state) correspond to point transformations in Minkowski space. However, if the theory is conformally invariant there are wider classes of regions for which the modular automorphisms act geometrically. They include, as the most important case, the diamonds [Hisl 82].

The basic observation is most easily seen in the formalism described in subsection I.2.1 of where the conformal group is realized by pseudoorthogonal transformations in a 6-dimensional space with coordinates ξ^α. Consider rotations in the $\xi^1 - \xi^4$-plane

$$\begin{aligned} &T^{41}(\varphi)\xi \equiv \xi(\varphi),\\ &\xi^4(\varphi) = \xi^4 \cos\varphi + \xi^1 \sin\varphi,\\ &\xi^1(\varphi) = -\xi^4 \sin\varphi + \xi^1 \cos\varphi,\\ &\xi^\alpha(\varphi) = \xi^\alpha \quad \text{for} \quad \alpha \neq 1,\ 4. \end{aligned} \tag{V.4.24}$$

One verifies that if the original point $x(0)$

$$x^\mu(0) = \xi^\mu(\xi^4 + \xi^5)^{-1}, \quad \mu = 0,\ 1,\ 2,\ 3 \tag{V.4.25}$$

lies inside the wedge W then $x^\mu(\varphi)$ follows a continuous path, never reaching infinity for $0 \le \varphi \le \pi$. For $\varphi = \pi/2$ the wedge is mapped onto the diamond K_1 whose vertices are the points $x^0 = \pm 1$, $\mathbf{x} = 0$.

$$T^{41}\left(\frac{\pi}{2}\right) W = K_1, \tag{V.4.26}$$

$$K_1 = \{x : |x^0| + |\mathbf{x}| < 1\}. \tag{V.4.27}$$

To see this it is instructive, though somewhat tedious, to follow the orbits $x^\mu(\varphi)$ of the boundary points of W in detail. A quicker, though somewhat blind, way is to notice that

$$\left(\xi^4(\varphi) + \xi^5(\varphi)\right)\left(\xi^4 + \xi^5\right)^{-1} = \cos\varphi + x^1 \sin\varphi + \frac{1}{2}\left(1 - (x,x)\right)\left(1 - \cos\varphi\right)$$

cannot vanish for $0 < \varphi < \pi$ since, for $x \in W$, $x^1 > 0$ and $(x,x) < 0$. Thus the orbits cannot go to infinity. Then one checks that the inequalities characterizing the wedge

$$x^1 > |x^0|$$

go over for $\varphi = \pi/2$ into the inequalities characterizing the diamond K_1

$$\left|x^0\left(\frac{\pi}{2}\right)\right| + \left|\mathbf{x}\left(\frac{\pi}{2}\right)\right| < 1.$$

The modular transformations of W are the boosts (V.4.4) which are "pseudo-rotations" in the $\xi^1 - \xi^0$ plane.

$$\begin{aligned} &T^{10}(s)\xi \equiv \xi(s), \\ &\xi^0(s) = \xi^0 \cosh s + \xi^1 \sinh s, \\ &\xi^1(s) = \xi^0 \sinh s + \xi^1 \cosh s, \\ &\xi^\alpha(s) = \xi^\alpha \quad \text{for} \quad \alpha \neq 0,\ 1. \end{aligned} \tag{V.4.28}$$

This suggests that the modular group of K_1 corresponds to the conformal transformations

$$T(s) = T^{41}\left(\frac{\pi}{2}\right) T^{10}(s) T^{41}\left(\frac{\pi}{2}\right)^{-1} \equiv T^{04}(s). \tag{V.4.29}$$

The orbit of a point under these transformations is

$$\begin{aligned} &T^{04}(s)\xi \equiv \xi(s), \\ &\xi^0(s) = \xi^0 \cosh s + \xi^4 \sinh s, \\ &\xi^4(s) = \xi^0 \sinh s + \xi^4 \cosh s, \\ &\xi^\alpha(s) = \xi^\alpha \quad \text{for} \quad \alpha \neq 0,\ 4, \end{aligned} \tag{V.4.30}$$

or, in x-space

$$\begin{aligned} &x^0(s) = N(s)^{-1}\left(x^0 \cosh s + \tfrac{1}{2}\{1 + (x,x)\} \sinh s\right), \\ &x^i(s) = N(s)^{-1} x^i, \quad i = 1,\ 2,\ 3, \\ &N(s) = \left(x^0 \sinh s + \tfrac{1}{2}\{1 + (x,x)\} \cosh s + \tfrac{1}{2}\{1 - (x,x)\}\right). \end{aligned} \tag{V.4.31}$$

This can be simplified by introducing

$$x_+ = x^0 + |\mathbf{x}|, \quad x_- = x^0 - |\mathbf{x}|. \qquad \text{(V.4.32)}$$

The denominator $N(s)$ factors into

$$\frac{1}{4}e^s \left((1+x_+) + e^{-s}(1-x_+)\right)\left((1+x_-) + e^{-s}(1-x_-)\right)$$

and (V.4.31) becomes

$$x_\pm(s) = \left((1+x_\pm) - e^{-s}(1-x_\pm)\right)\left((1+x_\pm) + e^{-s}(1-x_\pm)\right)^{-1}. \qquad \text{(V.4.33)}$$

The modular conjugation for W is the inversion

$$x^1 \to -x^1; \quad x^0 \to -x^0; \quad x^r \to x^r; \quad \text{for} \quad r = 2,\ 3.$$

Conjugation with $T^{14}(\pi/2)$ leads to the inversion map for K_1

$$x^i \to -x^i(x,x)^{-1},\ i = 1,\ 2,\ 3; \quad x^0 \to x^0(x,x)^{-1}. \qquad \text{(V.4.34)}$$

Theorem 4.2.1 (Hislop and Longo)
In a free, massless field theory the modular automorphism group $\sigma_\tau(K_1)$ of the diamond region K_1 induced by the vacuum state has the geometric action (V.4.33) (with the parameter s replaced by $-2\pi\tau$). The modular conjugation $J(K_1)$ has the geometric action (V.4.34).

Proof. The above discussion shows that theorem 4.2.1. follows from the Bisognano-Wichman theorem if the theory is conformally invariant in the following sense

(i) algebraically: given any finitely extended region $\mathcal{O}$ there is a neighborhood of the identity $\mathcal{N}$ in the conformal group such that for $g \in \mathcal{N}$ one has an algebraic isomorphism α_g from $\mathcal{R}(\mathcal{O})$ to $\mathcal{R}(g\mathcal{O})$ respecting the net structure i.e.

$$\alpha_g \mathcal{R}(\mathcal{O}_1) = \mathcal{R}(g\mathcal{O}_1) \quad \text{for any} \quad \mathcal{O}_1 \subset \mathcal{O},\ g \in \mathcal{N}. \qquad \text{(V.4.35)}$$

In field theory this means that we must have a transformation law for fields under the conformal group conserving the algebraic relations (commutation relations and equation of motion) and that, for group elements sufficiently close to the identity, the argument of the field is transformed thereby according to the geometric action of the conformal group on Minkowski space as

$$x \to gx. \qquad \text{(V.4.36)}$$

It is clear that (V.4.36) cannot be required globally for all pairs g, x since conformal transformations do not give globally defined maps from $\mathcal{M}$ to $\mathcal{M}$. In the present context one can use the conformal invariance only for orbits of points which do not pass through infinity under the conformal transformations applied.

(ii) The vacuum state must be invariant under the conformal group

$$\omega_0(\alpha_g A) = \omega_0(A) \tag{V.4.37}$$

whenever $\alpha_g A$ is defined. In other words, the conformal invariance should not be spontaneously broken in the vacuum sector. Then α_g can be implemented by unitaries U_g in the usual fashion

$$U_g A|\Omega\rangle = \alpha_g A|\Omega\rangle. \tag{V.4.38}$$

For a massless free theory both items (i) and (ii) can be established. See e.g. Swieca and Völkel [Swiec 73], Schroer and Swieca [Schroer 74] where the unitaries U_g and the limitations in their geometric interpretation are studied in detail. In the case of a scalar field Φ the transformation law is

$$(\alpha_g \Phi)(x) = N(g,x)^{-1}\Phi(gx), \tag{V.4.39}$$

where N is the factor given in (I.2.46). □

Another interesting result in this context is

Theorem 4.2.2 [Buch 78]
In a free, massless theory the modular automorphism group $\sigma_\tau(V^+)$ induced by the vacuum state on the algebra of the forward light cone corresponds geometrically to the dilations

$$x^\mu(\tau) = e^{-2\pi\tau} x^\mu, \tag{V.4.40}$$

the modular conjugation to

$$x^\mu \to -x^\mu. \tag{V.4.41}$$

Proof. We start from the observation that rotations in the $\xi^0 - \xi^5$-plane

$$\begin{aligned} &T^{05}(\varphi)\xi \equiv \xi(\varphi), \\ &\xi^0(\varphi) = \xi^0 \cos\varphi + \xi^5 \sin\varphi, \\ &\xi^5(\varphi) = -\xi^0 \sin\varphi + \xi^5 \cos\varphi, \\ &\xi^\alpha(\varphi) = \xi^\alpha \quad \text{for} \quad \alpha \neq 0,\, 5 \end{aligned} \tag{V.4.42}$$

transform the diamond K_1 to the forward light cone V^+ for the angle $\varphi = \pi/2$ and that the orbit of points from K_1 is continuous in the interval $0 \le \varphi \le \pi$. Thus, if the theory is conformally invariant in the sense described in the proof of theorem 4.2.1, then the modular structure for V^+ results from that for K_1 by conjugation with $T^{05}(\pi/2)$ yielding instead of the orbits (V.4.30) the orbits

$$\begin{aligned} &\xi(s) = T^{54}(s)\xi, \\ &\xi^\mu(s) = \xi^\mu \quad \text{for} \quad \mu = 0,\, 1,\, 2,\, 3, \\ &\xi^4(s) + \xi^5(s) = e^{-s}(\xi^4 + \xi^5), \\ &\xi^4(s) - \xi^5(s) = e^{s}(\xi^4 - \xi^5). \end{aligned} \tag{V.4.43}$$

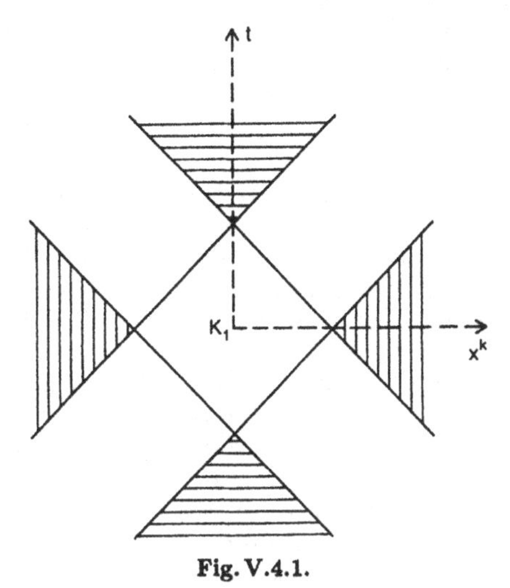

Fig. V.4.1.

This establishes (V.4.40) (putting $s = -2\pi\tau$ as in theorem 4.2.1.). In the same way one obtains (V.4.41) from (V.4.34). □

Remark. The inversion (V.4.34) maps K_1 onto K_1^r, the shaded region in Fig V.4.1.. The inversion (V.4.41) maps V^+ onto V^-. If in a theory the modular conjugations for $\mathcal{R}(K_1)$ and $\mathcal{R}(V^+)$ are as demanded by theorems 4.2.1. and 4.2.2. then

$$\mathcal{R}(K_1)' = \mathcal{R}(K_1^r), \tag{V.4.44}$$

$$\mathcal{R}(V^+)' = \mathcal{R}(V^-). \tag{V.4.45}$$

Since K_1^r and V^- are not the causal complements of K_1 and V^+, respectively, we do not have the relation $\mathcal{R}(\mathcal{O}'') = \mathcal{R}(\mathcal{O})$ (Chapter III.5) in such theories. This is due to the fact that observables commute here not only for space-like separation but also for time-like separation; the nonvanishing of commutators is restricted to light-like directions. There are indications that the theorems have a partial converse, namely that the vanishing of commutators in both space-like and time-like directions characterizes free massless theories. (See for instance Buchholz and Fredenhagen [Buch 77a, c]).

V.5 Phase Space, Nuclearity, Split Property, Local Equilibrium

V.5.1 Introduction

The principles formulated in Chapter III do not yet entail such basic features of experience as the existence of particles and a reasonable thermodynamic behaviour. The same applies, of course, to the axioms in Chapter II.1. We may recall that in Chapter II the occurrence of particles was put in "by hand" through the requirement that the mass operator $P_\mu P^\mu$ should have some discrete eigenvalues. If, furthermore, the masses are nonzero then this, in conjunction with the listed principles suffices to develop a (possibly incomplete) particle interpretation and determine the collision cross sections. Similarly, while we have defined what we mean by a thermodynamic equilibrium state, the existence of such states in the theory is not guaranteed by the principles stated so far.

Both questions have a common root. We must consider the analogue of classical phase space volumes i.e. the part of state space corresponding to a simultaneous limitation of energy and space volume. Loosely speaking, finite volumes in classical phase space should correspond to finite dimensional parts of state space in quantum physics. Starting from the vacuum representation this idea may be implemented in the following way.

Let K_r be the diamond whose base is a ball with radius r at time $t = 0$. We wish to consider the set of states which can be called "essentially localized" in K_r. While the localization of observables is a fundamental concept in our frame the localization of states is a less clear cut notion. It needs the vacuum as a reference state. [Knight 61] and [Licht 63] defined strictly localized states in a region $\mathcal{O}$ as those states which give the same expectation values as the vacuum for all measurements in the causal complement of $\mathcal{O}$. The corresponding state vectors are then given by

$$\Psi = W|\Omega\rangle \quad \text{with} \quad W^*W = \mathbb{1}, \quad W \in \mathcal{A}(\mathcal{O}), \tag{V.5.1}$$

i.e. states which are generated by an isometric operator from the field algebra of the region applied to the vacuum. Indeed, for such states

$$\langle\Psi|A|\Psi\rangle = \langle\Omega|A|\Omega\rangle \quad \text{for} \quad A \in \mathcal{R}(\mathcal{O}'). \tag{V.5.2}$$

A (superficial) paradox arises from the Reeh-Schlieder-theorem. Since $\mathcal{A}(\mathcal{O})|\Omega\rangle$ is dense in $\mathcal{H}$ the linear space spanned by the vectors of the form (V.5.1) is all of $\mathcal{H}$. So we can approximate any state vector by linear combinations of vectors describing states "strictly localized" in some region $\mathcal{O}$. This is due to the correlations between distant observables present in the vacuum state which, though decreasing fast with increasing distance, never vanish. By exploiting them judiciously one may, by an operation in K_r on the vacuum, create a localization center far away from K_r. But at what cost? It suffices for the following to restrict attention to the vacuum sector (uncharged states) and consider state vectors of the form

$$\Psi = A|\Omega\rangle \quad \text{with} \quad A \in \mathcal{R}(K_r). \tag{V.5.3}$$

The ratio of cost vs. effect may be measured by the quantity

$$c_A = \frac{\| A \|}{\| A|\Omega\rangle \|}, \tag{V.5.4}$$

the norm of the operator corresponding to the expenditure, the length of the resulting vector $A|\Omega\rangle$ to the achieved effect. It was shown by Haag and Swieca [Haag 65] that the states of the form (V.5.3) can indeed be interpreted as states which are "essentially localized" in K_r provide that $c_A \leq c$ where the choice of c determines the degree of tolerated delocalization which grows with c. The case (V.5.1) ("strict localization") corresponds to $c = 1$, the minimal value possible. With these considerations in mind we are now able to "measure" phase space volumes. The latter correspond to subsets of $\mathcal{H}$

$$\mathcal{M}_{E,r} = P_E \mathcal{R}^{(1)}(K_r)|\Omega\rangle, \tag{V.5.5}$$

where P_E is the spectral projector of the Hamiltonian to the interval $[0, E]$ and $\mathcal{R}^{(1)}$ denotes the unit ball in $\mathcal{R}$

$$\mathcal{R}^{(1)}(\mathcal{O}) = \{A \in \mathcal{R}(\mathcal{O}) : \| A \| \leq 1\}. \tag{V.5.6}$$

Again, the linear span of $\mathcal{M}_{E,r}$ has no memory of the localization region but the length of vectors in $\mathcal{M}_{E,r}$ decreases fast with increasing delocalization. The intuitive argument at the beginning of this section leads to the requirement that $\mathcal{M}_{E,r}$ should be "essentially finite dimensional". More precisely: there should be an ascending sequence of finite dimensional subspaces $\mathcal{H}_d \subset \mathcal{H}$ (d denoting the dimension) and for each $\varepsilon > 0$ a dimension $d(\varepsilon)$ such that the vectors in $\mathcal{M}_{E,r}$ orthogonal to $\mathcal{H}_{d(\varepsilon)}$ have length less than ε. This provided the motivation for the

Compactness requirement 5.1.1 [Haag-Swieca]
The set $\mathcal{M}_{E,r}$ is compact in the norm topology of $\mathcal{H}$.

It was recognized by Buchholz and Wichmann [Buch 86a] that the same intuitive picture leads to a much stronger requirement and that the estimates of the dependence of $d(\varepsilon)$ on E and r given in [Haag 65] can be considerably improved. Instead of the sharp cut-off in energy Buchholz and Wichmann use an exponential damping and replace the requirement 5.1.1 by the

Nuclearity requirement 5.1.2 [Buchholz-Wichmann]
The set

$$\mathcal{N}_{\beta,r} = e^{-\beta H} \mathcal{R}^{(1)}(K_r)|\Omega\rangle \tag{V.5.7}$$

is a nuclear set in $\mathcal{H}$ for any $\beta > 0$. The nuclearity index $\nu_{\beta,r}$ is bounded by

$$\nu_{\beta,r} < \exp\, c r^3 \beta^{-n} \tag{V.5.8}$$

(for large r, small β) with c, n some positive constants.

In the present context the nuclearity of $\mathcal{N}_{\beta,r}$ means that there is a positive trace class operator $T_{\beta,r}$ such that

$$\mathcal{N}_{\beta,r} \subset T_{\beta,r}\mathcal{H}^{(1)}; \quad \mathcal{H}^{(1)} \text{ the unit ball in } \mathcal{H}. \tag{V.5.9}$$

The nuclearity index is defined as

$$\nu_{\beta,r} = \inf \operatorname{tr} T_{\beta,r}, \tag{V.5.10}$$

the infimum being taken over all trace class operators satisfying (V.5.9).

The status of this requirement may be assessed from the following comments.

1) It has been tested in free field theories. There it holds if and only if the mass spectrum of particles satisfies

$$\sum f_i \exp -\beta m_i < \infty \quad \text{for all} \quad \beta > 0, \tag{V.5.11}$$

where f_i is the multiplicity of the mass value m_i including a factor $(2s+1)$ for spin orientations. It is worthwhile to note that it still holds if some fields are massless. We shall not reproduce these computations here but refer the critical reader to Buchholz and Junglas [Buch 86b], Buchholz and Jacobi [Buch 87b].

2) The requirement is necessary and sufficient to ensure "normal thermodynamic properties", namely the existence of KMS-states for all positive β for the infinite system and for (properly definable) finitely extended parts. For the latter the nuclearity index is directly the partition function [Buch 86a], [Jung 87], [Buch 89]. The r-dependence of $\nu_{\beta,r}$ in the estimate (V.5.8) expresses the proportionality of the free energy to the volume for large r. The dependence of ν on β gives the relation between energy and temperature in the high temperature region. It is determined by the density of energy levels in the asymptotic region. The possibility that the level density might increase stronger than exponentially, thereby giving rise to a maximal temperature [Hagedorn 67] can be accommodated by a modification of the estimate (V.5.8), replacing the power β^{-n} in the exponent by a function $\Phi(\beta)$ which becomes infinite at the Hagedorn temperature.

The definition of finitely extended "subsystems" is tied to a property of the net of local algebras first conjectured by Borchers and studied in detail by Doplicher and Longo [Dopl 84a], the "split property". The local algebras $\mathcal{R}(\mathcal{O})$ are not of type I i.e. not isomorphic to the algebra $\mathfrak{B}(\mathcal{H}_\mathcal{O})$ of some Hilbert space. This is the ultimate reason why it is not possible to define a (linear) subspace $\mathcal{H}_\mathcal{O} \subset \mathcal{H}$ which can be considered to correspond to the states "localized" in $\mathcal{O}$ and why we have to take recourse to constructions like (V.5.5) or (V.5.7) for this purpose. However, given two concentric (standard) diamonds with radii r_k where $r_2 > r_1$, the split property asserts that there exists a type-I-factor $\mathcal{N}$ such that

$$\mathcal{R}(K_1) \subset \mathcal{N} \subset \mathcal{R}(K_2). \tag{V.5.12}$$

It was shown in [Buch 86a] that this property follows from the nuclearity requirement 5.1.2 (at least if $r_2 - r_1$ is sufficiently large). Subsequently Buchholz, D'Antoni and Fredenhagen [Buch 87c] showed that it follows for any pair r_2, r_1 with $r_2 > r_1$ if the nuclearity index is bounded by the estimate (V.5.8). If the nuclearity index diverges at some finite temperature then the minimal separation $r_2 - r_1$ needed for the split is just the inverse Hagedorn temperature. In the next subsection we shall outline this argument and mention some consequences of the split property.

In a further development of this analysis Buchholz, D'Antoni and Longo showed that the nuclearity requirement can be formulated without using the Hamiltonian, working instead with the modular operator of $\mathcal{R}(K_r)$. Thereby the requirement becomes a purely local one [Buch 90a].

Another variant of phase space conditions stems from a remark due to Fredenhagen and Hertel (unpublished preprint 1979). Instead of starting from localized states and then damping the energy one may start from the set of linear forms over the quasilocal algebra $\mathfrak{A}$ which have energy below some bound E (or are exponentially damped) and consider their restrictions to a local algebra $\mathfrak{A}(K_r)$. Again one expects that the resulting set of linear forms over $\mathfrak{A}(K_r)$ must be nuclear. This approach is closer to the basic structure of the theory. One may note that an energy bound for states over $\mathfrak{A}$ can be expressed as a condition on their annihilator ideal (see the discussion around definition 3.2.3 of Chapter III). For a comparison of various phase space conditions see Buchholz and Porrmann [Buch 90c].

Technically it is somewhat more convenient to work with nuclear maps between Banach spaces rather than with nuclear sets in one space. The latter are then just the images of the unit ball under the former.

Definition 5.1.3
Let $\mathcal{E}$ and $\mathcal{F}$ be Banach spaces and Θ a bounded linear map from $\mathcal{E}$ into $\mathcal{F}$,

$$\| \Theta \| = \sup\{\| \Theta(E) \|, \ E \in \mathcal{E}, \ \| E \| = 1\}.$$

Generalizing the notion of a dimension function $d(\varepsilon)$, mentioned in the sequel of (V.5.6), one defines the "ε-content" $N(\varepsilon)$ of the map as the maximal number of elements $E_k \in \mathcal{E}^{(1)}$ (the unit ball of $\mathcal{E}$) such that the distance between any pair of their images exceeds ε

$$\| \Theta E_i - \Theta E_k \| > \varepsilon; \quad i \neq k, \quad \| E_k \| \leq 1.$$

Θ is called compact if $N(\varepsilon)$ is finite for every $\varepsilon > 0$.

If the growth of $N(\varepsilon)$ is bounded by

$$N(\varepsilon) < \exp(\varepsilon^{-q}) \quad \text{as} \quad \varepsilon \to 0 \tag{V.5.13}$$

then the smallest (positive) number q for which this holds is called the *order of the map* Θ.

Θ is called a nuclear map if it can be written as

$$\Theta E = \sum \varphi_i(E) F_i \tag{V.5.14}$$

with

$$\sum \| \varphi_i \| \| F_i \| < \infty, \tag{V.5.15}$$

where φ_i are bounded linear forms on $\mathcal{E}$ and $F_i \in \mathcal{F}$. We define the nuclearity index (or $\mathcal{L}^{(1)}$-norm) of Θ as

$$\| \Theta \|_1 = \inf \sum \| \varphi_i \| \| F_i \| \tag{V.5.16}$$

where the infimum is taken over all realizations of Θ in the form (V.5.14). Similarly one defines *p-nuclearity*, replacing (V.5.16) by

$$\| \Theta \|_p = \inf \left(\sum \| \varphi_i \|^p \| F_i \|^p \right)^{1/p} \tag{V.5.17}$$

where p is some positive number.

Comment. For the relation between the order of a map and nuclearity see [*Pietsch 1972*]. We only mention that a map of order $q < 1/2$ is certainly p-nuclear with $p \leq q/(1-2q)$.

V.5.2 Nuclearity and Split Property

The split property may be viewed as a sharpening of the locality principle. If $\mathcal{O}_1$ and $\mathcal{O}_2$ are space-like separated then the algebras $\mathcal{R}(\mathcal{O}_1)$ and $\mathcal{R}(\mathcal{O}_2)$ should not only commute but be "statistically independent". This means first that the partial states are uncoupled: if φ_k is an arbitrary pair of normal states over $\mathcal{R}(\mathcal{O}_k)$, $(k = 1, 2)$ then there exists an extension to a normal state φ over $\mathcal{R}(\mathcal{O}_1) \vee \mathcal{R}(\mathcal{O}_2)$. It means furthermore, that this extension can be so chosen that there are no correlations between the measuring results in $\mathcal{O}_1$ and $\mathcal{O}_2$ in the state φ i.e.

$$\varphi(A_1 A_2) = \varphi_1(A_1)\varphi_2(A_2) \quad \text{for} \quad A_k \in \mathcal{R}(\mathcal{O}_k), \tag{V.5.18}$$

and that for faithful states φ_k the extension remains faithful over

$$\mathcal{R}(\mathcal{O}_1) \vee \mathcal{R}(\mathcal{O}_2).$$

Irrespective of the physical interpretation it is a natural mathematical question to ask under what conditions two commuting von Neumann algebras $\mathcal{R}_1$, $\mathcal{R}_2$ acting on the same Hilbert space $\mathcal{H}$ are statistically independent. One prerequisite is "algebraic independence". This may be reduced to the condition that $A_1 A_2 = 0$ with $A_k \in \mathcal{R}_k$ implies that either $A_1 = 0$ or $A_2 = 0$. If $\mathcal{R}_1$ and $\mathcal{R}_2$ are factors then the algebraic independence is automatic (Chapter III, lemma 2.1.10). In the subsequent discussion we shall always assume that the $\mathcal{R}_k$ are factors. This covers the cases of principal physical interest and though this assumption could be avoided, it simplifies some formulations and arguments. Starting now from the algebraic (not yet topological) tensor product $\mathcal{C}$ of $\mathcal{R}_1$, $\mathcal{R}_2$ consisting of finite linear combinations of products A_1A_2 with $A_k \in \mathcal{R}_k$ we have the representation of $\mathcal{C}$ acting on $\mathcal{H}$ which we denote by π:

$$\pi(A_1A_2) = A_1A_2. \tag{V.5.19}$$

If $\mathcal{R}_1$ and $\mathcal{R}_2$ are algebraically independent then we have another representation, denoted by π_p, acting on $\mathcal{H} \otimes \mathcal{H}$, generated by

$$\pi_p(A_1A_2) = A_1 \otimes A_2, \tag{V.5.20}$$

(A_1 acting on the first factor $\mathcal{H}$ in the tensor product, A_2 on the second). One has

Proposition 5.2.1
Let $\mathcal{R}$, $\widehat{\mathcal{R}}$ be factors acting on $\mathcal{H}$ and $\mathcal{R} \subset \widehat{\mathcal{R}}$ Let $\Omega \in \mathcal{H}$ be a cyclic and separating vector for $\mathcal{R}$, $\widehat{\mathcal{R}}$ and $\widehat{\mathcal{R}} \wedge \mathcal{R}'$ In this situation the following conditions are equivalent

(i) $\mathcal{R}$ and $\widehat{\mathcal{R}}'$ are statistically independent,

(ii) there exists a vector $\eta \in \mathcal{H}$, cyclic and separating for $\mathcal{R} \vee \widehat{\mathcal{R}}'$ satisfying

$$\langle\eta|AB'|\eta\rangle = \langle\Omega|A|\Omega\rangle\langle\Omega|B'|\Omega\rangle \quad \text{for} \quad A \in \mathcal{R},\ B' \in \widehat{\mathcal{R}}', \tag{V.5.21}$$

(iii) there is a unitary operator W from $\mathcal{H}$ to $\mathcal{H} \otimes \mathcal{H}$ such that

$$W\pi(AB')W^* = \pi_P(AB'); \quad A \in \mathcal{R},\ B' \in \widehat{\mathcal{R}}', \tag{V.5.22}$$

(iv) there exists an intermediate type-I-factor $\mathcal{N}$:

$$\mathcal{R} \subset \mathcal{N} \subset \widehat{\mathcal{R}}. \tag{V.5.23}$$

If any of these conditions is satisfied the triple $(\mathcal{R}, \widehat{\mathcal{R}}, \Omega)$ defines a "standard split inclusion" in the terminology of [Dopl 84a]. Ω is called a standard vector for the split because it is cyclic and separating for $\mathcal{R}$, $\widehat{\mathcal{R}}'$ and $\mathcal{R} \vee \widehat{\mathcal{R}}'$. The existence of a product state vector η in the case of a free field has first been shown in [Buch 74].

Proof. (i)→(ii). The state ω induced by Ω on the various subalgebras under consideration is faithful on $\mathcal{R}$ and $\widehat{\mathcal{R}}'$ and thus, by assumption, there exists a faithful, normal product state ω_P on $\mathcal{R} \vee \widehat{\mathcal{R}}'$. Since Ω is cyclic and separating on $\mathcal{R} \vee \widehat{\mathcal{R}}'$ there is a vector representative $\eta \in \mathcal{H}$ of ω_P (theorem 2.2.2). The faithfulness of ω_P means that η is separating. We may choose η on the natural cone of Ω and then, by theorem 2.2.3 η is also cyclic.

(ii)→(iii). The map $W : \mathcal{H} \to \mathcal{H} \otimes \mathcal{H}$ defined by

$$WAB'|\eta\rangle = A|\Omega\rangle \otimes B'|\Omega\rangle \quad \text{for} \quad A \in \mathcal{R},\ B' \in \widehat{\mathcal{R}}' \tag{V.5.24}$$

is densely defined and has dense range. From (V.5.21) it follows that W is isometric. Hence it extends to a unitary map.

(iii)→(iv). The algebra $\mathfrak{B}(\mathcal{H}) \otimes \mathbb{1}$ is a type-I factor on $\mathcal{H} \otimes \mathcal{H}$. Its image

$$\mathcal{N} = W^* \mathfrak{B}(\mathcal{H}) \otimes \mathbb{1}\, W \tag{V.5.25}$$

is a type-I-factor on $\mathcal{H}$ which contains

$$\mathcal{R} = W^* \mathcal{R} \otimes \mathbb{1}\, W \tag{V.5.26}$$

and commutes with

$$\widehat{\mathcal{R}}' = W^* \mathbb{1} \otimes \widehat{\mathcal{R}}'\, W \tag{V.5.27}$$

and hence is contained in $\widehat{\mathcal{R}}$.

(iv)→(i). An infinite type-I-factor with infinite commutant is unitarily equivalent to $\mathfrak{B}(\mathcal{H}) \otimes \mathbb{1}$ on $\mathcal{H} \otimes \mathcal{H}$. Thus we have (V.5.25). If we have furthermore a standard vector Ω and the inclusion relations (V.5.23) then the identifications (V.5.26), (V.5.27) and

$$\eta = W^* |\Omega\rangle \otimes |\Omega\rangle, \quad \omega_P(C) = \langle \eta | C | \eta \rangle, \quad C \in \mathcal{R} \vee \widehat{\mathcal{R}}' \tag{V.5.28}$$

show that $\mathcal{R}$ and $\widehat{\mathcal{R}}'$ are statistically independent. □

Our main objective in this subsection is to show that the split property follows from the nuclearity requirement 5.1.2 We start from the following observation. Let K_1, K_2 be concentric standard diamonds with radii r_1, r_2 respectively, where $r_2 - r_1 = \delta > 0$, $A \in \mathcal{R}(K_1)$, $B' \in \mathcal{R}(K_2)'$. Then

$$[\alpha_t A, B'] = 0 \quad \text{for} \quad |t| < \delta. \tag{V.5.29}$$

Lemma 5.2.2 (Buchholz, D'Antoni, Fredenhagen)
If A and B' satisfy (V.5.29) then one can construct a continuous function f of a real variable such that

$$\langle \Omega | A B' | \Omega \rangle = \langle \Omega | A f(H) B' | \Omega \rangle + \langle \Omega | B' f(H) A | \Omega \rangle \tag{V.5.30}$$

and which furthermore has "almost exponential decrease":

$$\lim_{|E| \to \infty} (\exp |E|^{\kappa}) \, |f(E)| = 0 \quad \text{for} \quad 0 \leq \kappa < 1. \tag{V.5.31}$$

Proof. Consider the following two functions

$$h_+(t) = \omega\left((\alpha_{-t} A) B'\right) = \langle \Omega | A e^{iHt} B' | \Omega \rangle, \tag{V.5.32}$$

$$h_-(t) = \omega\left(B' \alpha_{-t} A\right) = \langle \Omega | B' e^{-iHt} A | \Omega \rangle. \tag{V.5.33}$$

Due to the positivity of the spectrum of the Hamiltonian h_+ is the boundary value of an analytic function in the upper half plane, h_- is the boundary value of an analytic function in the lower half plane and, due to (V.5.29) the boundary values coincide in the interval $-\delta < t < \delta$. So there is a function $h(z)$, analytic in the cut complex t-plane, the cuts running along the real axis from $-\infty$ to $-\delta$ and from δ to $+\infty$ such that h has the boundary values h_+ respectively h_- as the cuts are approached from above or below. We have

$$h(0) = \langle \Omega | AB' | \Omega \rangle. \tag{V.5.34}$$

By the transformation

$$w \to z = \frac{2\tau w}{w^2+1} \tag{V.5.35}$$

the interior of the unit circle in the complex w-plane is mapped onto the complex z-plane with cuts along the real axis from $-\infty$ to $-\tau$ and from $+\tau$ to $+\infty$, the upper half plane corresponding to the upper half disc $|w| < 1$, $0 < \varphi \equiv \arg w < \pi$, the lower half plane to the lower half disc. Evaluation of $h(0)$ by a Cauchy integral around the unit circle in the w-plane

$$h(0) = \frac{1}{2\pi} \int_0^{2\pi} h\left((2\tau e^{i\varphi})(1+e^{2i\varphi})^{-1} \right) d\varphi$$

yields in terms of vacuum expectation values

$$2\pi \langle \Omega | AB' | \Omega \rangle$$

$$= \int_0^{\pi} d\varphi \left(\langle \Omega | A \exp\left(iH\tau/\cos\varphi\right) B' | \Omega \rangle + \langle \Omega | B' \exp\left(iH\tau/\cos\varphi\right) A | \Omega \rangle \right). \tag{V.5.36}$$

This relation holds for all τ with $0 < \tau < \delta$. If we multiply it with a smooth function $g(\tau)$ whose support is in the interval $[0, \delta]$ and integrate over τ we get

$$\langle \Omega | AB' | \Omega \rangle = \langle \Omega | A f(H) B' | \Omega \rangle + \langle \Omega | B' f(H) A | \Omega \rangle$$

with

$$2\pi f(E) = \tilde{g}(0)^{-1} \int_0^{\pi} d\varphi \tilde{g}\left(\frac{E}{\cos\varphi}\right) \quad \text{where} \quad \tilde{g}(E) = \int g(\tau) e^{iE\tau} d\tau. \tag{V.5.37}$$

Now, as Jaffe has shown [Jaffe 67], there exist smooth functions g with the indicated support whose Fourier transforms decrease almost exponentially (in the sense of (V.5.31) and do not vanish at $E = 0$. Picking such a function g one sees that f is continuous and satisfies the claims of the lemma. □

Now, the required nuclearity of the map

$$\Theta_\beta : \mathcal{R}(K_r) \to \mathcal{H}; \quad \Theta_\beta A = e^{-\beta H} A | \Omega \rangle; \quad A \in \mathcal{R}(K_r) \tag{V.5.38}$$

with a nuclearity index bounded by

$$\nu_\beta < \exp\left(\frac{\beta_0}{\beta}\right)^n \tag{V.5.39}$$

implies the nuclearity of the map

$$\Theta_f : \mathcal{R}(K_r) \to \mathcal{H}; \quad \Theta_f A = f(H)A|\Omega\rangle; \quad A \in \mathcal{R}(K_r) \tag{V.5.40}$$

when f satisfies the conditions of lemma 5.2.2. To see this write

$$\Theta_f = \sum \Theta_k; \quad \Theta_k = P_k f(H) e^{\beta H} \Theta_\beta, \tag{V.5.41}$$

with P_k the spectral projector of H for the interval $[k-1, k]$. Θ_k is obviously nuclear and its index satisfies

$$\| \Theta_k \|_1 < f_k \exp\left(\beta k + \left(\frac{\beta_0}{\beta}\right)^n\right); \quad f_k = \sup |f(E)| \quad \text{for} \quad k-1 \leq E \leq k. \tag{V.5.42}$$

Since this holds for any positive value of β we pick β for each k so that the exponent in (V.5.42) becomes minimal. This happens for $\beta = \beta_0 (n/\beta_0 k)^{1/n+1}$, yielding

$$\| \Theta_k \|_1 < f_k \exp c_n (\beta_0 k)^{n/n+1},$$

where the constant $c_n \leq 2$ for all $n > 0$. Thus

$$\| \Theta_f \|_1 \leq \sum \| \Theta_k \|_1 < \infty.$$

From the nuclearity of Θ_f it follows that we have

$$f(H)A|\Omega\rangle \equiv \Theta_f A = \sum \varphi_i(A)|\Psi_i\rangle; \quad A \in \mathcal{R}(K_r) \tag{V.5.43}$$

where $\Psi_i \in \mathcal{H}$, φ_i is a linear form on $\mathcal{R}(K_r)$ and $\sum \| \varphi_i \| \| \Psi_i \| < \infty$. In fact, since Θ_f gives a map from a von Neumann algebra to normal states over it, the linear forms φ_i can be taken in $\mathcal{R}_*$, the predual of $\mathcal{R}$, i.e. the φ_i are normal linear functionals over $\mathcal{R}(K_r)$. Then

$$\langle \Omega \mid B' f(H) A|\Omega\rangle = \sum \varphi_i(A)\langle \Omega|B'|\Psi_i\rangle$$

with an analogous expression for $\langle \Omega|Af(H)B'|\Omega\rangle$. Putting $r = r_1$ we see by (V.5.30) that ω is an absolutely convergent sum of factorizing normal linear forms over the algebraic tensor product $\mathcal{C}$ of $\mathcal{R}(K_{r_1})$ with $\mathcal{R}(K_{r_2})'$ and therefore ω is a normal state in the representation π_P. This shows that the representations π and π_P are not disjoint. Actually one finds that π and π_P are unitarily equivalent. We shall not prove this here but refer the sceptical reader to [Buch 87c, 91b] for this part of the argument.

Summing up we have

Theorem 5.2.3
If Θ_β is nuclear with index bounded by

$$\nu < \exp c\beta^{-n}$$

(for some positive constants c and n) then we have the split property and Ω is a standard vector.

V.5.3 Open Subsystems

The split property provides a tool for constructing some analogue of finitely extended subsystems. Instead of one box we have two diamonds K_k with $K_2 \supset K_1$; instead of the ground state in the box we have a product vector η satisfying (V.5.21). The three objects K_1, K_2, η define the split. They determine W by (V.5.24) and $\mathcal{N}$ by (V.5.25) Let us use the single symbol Λ for a split. There is a (linear) subspace $\mathcal{H}_\Lambda \subset \mathcal{H}$

$$\mathcal{H}_\Lambda = \mathcal{N}|\eta\rangle = W^*(\mathcal{H} \otimes \Omega)W, \tag{V.5.44}$$

whose vectors correspond to strictly localized states in K_2 in the sense of (V.5.2) (though they do not exhaust them). The projector on $\mathcal{H}_\Lambda$ is

$$P_\Lambda = W^*(\mathbb{1} \otimes P_\Omega)W \in \mathcal{N}', \tag{V.5.45}$$

with P_Ω the 1-dimensional projector on Ω in $\mathcal{H}$. One notes that there is an isometric operator $V \in \mathcal{R}(K_2)$ such that

$$V|\Omega\rangle = |\eta\rangle. \tag{V.5.46}$$

Indeed, since Ω is cyclic and separating for $\mathcal{R}(K_2)'$, setting

$$VB'|\Omega\rangle = B'|\eta\rangle \quad \text{for} \quad B' \in \mathcal{R}(K_2)' \tag{V.5.47}$$

defines V consistently on a dense set of vectors. Relation (V.5.21) implies that V is isometric on its domain of definition and hence extends to an isometric operator on $\mathcal{H}$. From (V.5.47) it follows directly that V commutes with any $C' \in \mathcal{R}(K_2)'$ and thus $V \in \mathcal{R}(K_2)$.

This implies that every unit vector in $\mathcal{H}_\Lambda$ can be obtained by applying an operator with unit norm from $\mathcal{R}(K_2)$ to the vacuum. Therefore, by the nuclearity condition 5.1.2, $\exp -\beta H$ maps the unit ball of $\mathcal{H}_\Lambda$ onto a nuclear set in $\mathcal{H}$ i.e. $\exp -\beta H\, P_\Lambda$ is a trace class operator and one may define the partition function for Λ (to inverse temperature β) as

$$Z_{\beta,\Lambda} = \operatorname{tr} P_\Lambda e^{-\beta H} P_\Lambda. \tag{V.5.48}$$

[Jungl 87]. Then, with r_2 the radius of K_2

$$Z_{\beta,\Lambda} \leq \nu_{\beta,r_2} \tag{V.5.49}$$

(compare (V.5.8), (V.5.10)). For thermodynamic arguments (large r_2 and fixed $r_2 - r_1$) one may identify the nuclearity index $\nu_{\beta,r}$ itself with the partition function as originally suggested in [Buch 86a].

To carry the analogy with the treatment of finite systems in non-relativistic quantum statistical mechanics further one would like to define a "box Hamiltonian", charge operators for the box etc. (the box being replaced by a split Λ). A first step towards this is the following observation by Buchholz, Doplicher and Longo [Buch 86c]:

Observation 5.3.1
Let $\mathcal{G}$ be an unbroken symmetry group of the theory. Thus we have automorphisms α_g, $g \in \mathcal{G}$ with geometric action[1]

$$\alpha_g \mathcal{R}(\mathcal{O}) = \mathcal{R}(g\mathcal{O})$$

which are implemented by unitaries $U(g) \in \mathfrak{B}(\mathcal{H})$ leaving the vacuum invariant

$$U(g)|\Omega\rangle = |\Omega\rangle.$$

Then there is also a "local representation" $U_\Lambda(g)$ of $\mathcal{G}$, given by

$$U_\Lambda(g) = W^*(U(g) \otimes \mathbb{1})W. \tag{V.5.50}$$

Besides being a unitary representation of $\mathcal{G}$ the U_Λ satisfy

$$U_\Lambda(g) \in \mathcal{N}; \quad U_\Lambda(g)|\eta\rangle = |\eta\rangle, \tag{V.5.51}$$

and, for $A \in \mathcal{R}(K_r)$ with $r < r_1$ and g sufficiently close to the identity

$$U_\Lambda(g) A U_\Lambda(g)^{-1} = \alpha_g A. \tag{V.5.52}$$

(V.5.52) holds if there is a path from the identity to g such that $g' K_r$ stays in K_1 for all g' along this path.

Since $P_\Lambda \in \mathcal{N}'$ we may equally well take the restriction of U_Λ to $\mathcal{H}_\Lambda$

$$U_\Lambda^{\mathrm{rest}} \equiv U_\Lambda P_\Lambda = P_\Lambda U_\Lambda P_\Lambda, \tag{V.5.53}$$

which also gives a representation of $\mathcal{G}$.

Remark. In field theory the generators of symmetries can be written as integrals of densities over 3-space at a given time. These densities, e.g. components of the energy momentum tensor or conserved currents are Wightman fields. Doplicher suggested that the split property may be used to construct these fields from the local algebras [Dopl 82]. An attempt to realize this may start from observation 5.3.1 [Buch 86c]. There remain, however, some unresolved difficulties. One results from the fact that the subspace $\mathcal{H}_\Lambda$ contains infinitely many orthogonal states which with respect to observations in K_1 look almost like the vacuum (with respect to K_2' they look exactly like the vacuum). Typical examples of these are state vectors $W^*(T|\Omega\rangle \otimes |\Omega\rangle)$ with T a unitary localized far away from K_2. Such states may be regarded as excitations on the surface of Λ. There are too many of them with arbitrarily high energy. Therefore $H_\Lambda^{\mathrm{rest}}$ can not yet be regarded as the box Hamiltonian.

[1] In the case of gauge transformations and charges we would have to start here from the algebra of operations, the field algebra $\mathcal{A}$ instead of the observable algebra.

V.5.4 Modular Nuclearity

The nuclearity condition really concerns the local structure of the theory. This is not evident from the formulation 5.1.2 which uses the global Hamiltonian and global vacuum state. The work of Buchholz, D'Antoni and Longo [Bu 90a] shows how 5.1.2 can be replaced by a local condition. Let K_1, K_2 be concentric diamonds as above with $r_2 - r_1 = \delta > 0$, ω a faithful state on $\mathcal{R}(K_2)$, not necessarily the vacuum. Take the GNS-representation of $\mathcal{R}(K_2)$ generated from ω. It gives a Hilbert space $\mathcal{H}$, a cyclic, separating vector $\Omega \in \mathcal{H}$ and the associated Tomita-Takesaki operators Δ and J. Put

$$R = \left(1 + \Delta^{-1/2}\right)^{-1}. \qquad \text{(V.5.54)}$$

Then we have

Modular Nuclearity Condition 5.4.1
The map $\Xi : \mathcal{R}(K_1) \to \mathcal{H}$ defined by

$$\Xi A = RA|\Omega\rangle, \quad A \in \mathcal{R}(K_1) \qquad \text{(V.5.55)}$$

is nuclear of some order q.

The map Ξ is closely related to the maps Ξ_λ defined as

$$\Xi_\lambda A = \Delta^\lambda A|\Omega\rangle, \quad A \in \mathcal{R}(K_1), \quad 0 < \lambda < \frac{1}{2}. \qquad \text{(V.5.56)}$$

One notes that the order q_λ of Ξ_λ is symmetric around $\lambda = 1/4$ since

$$\Delta^\lambda A|\Omega\rangle = J\Delta^{1/2-\lambda}A^*|\Omega\rangle.$$

We shall be interested in Ξ_λ for $\lambda \le 1/4$ and write $\Xi_\#$, $q_\#$ for the case $\lambda = 1/4$. One has

Proposition 5.4.2

$$q_\lambda = (4\lambda)^{-1} q_\#; \quad \frac{1}{2} q_\# \le q \le q_\#, \quad 0 < \lambda \le \frac{1}{4}. \qquad \text{(V.5.57)}$$

Comment. From the discussion of simple models it appears that the orders q, q_λ may be assumed to be zero i.e. the ε-content of the maps increases less than the exponential of any power of ε^{-1} for $\varepsilon \to 0$.

Proof. The comparison of the orders q_λ for different values of λ follows from the simple lemma

Lemma 5.4.3
If $\Theta_\kappa = T^\kappa \Theta_1$ is a family of maps from the Banach space $\mathcal{E}$ into a Hilbert space

where T is a positive operator whose domain contains the range of Θ_1 then for $0 < \kappa \leq 1$ the order q_κ of Θ_κ satisfies $q_\kappa = \kappa^{-1} q_1$.

The estimate $q < q_\#$ in (V.5.57) follows from $R = \left(\Delta^{1/4} + \Delta^{-1/4}\right)^{-1} \Delta^{1/4}$ and the fact that $\left(\Delta^{1/4} + \Delta^{-1/4}\right)^{-1}$ is bounded. To establish the upper bound for $q_\#$ one considers the map Ξ_* from $\mathcal{R}(K_1)$ to the predual of $\mathcal{R}(K_2)'$ (the set of normal linear forms on $\mathcal{R}(K_2)'$):

$$(\Xi_* A)(B') = \langle \Omega | B' A | \Omega \rangle; \quad A \in \mathcal{R}(K_1), \quad B' \in \mathcal{R}(K_2)'. \tag{V.5.58}$$

From the identity

$$\| \Delta^{1/4} A | \Omega \rangle \|^2 = \langle \Omega | j(A) A | \Omega \rangle = (\Xi_* A)(jA) \leq \| A \| \| \Xi_* A \|$$

it follows that the ε-contents $N_\#(\varepsilon)$, $N_*(\varepsilon)$ of $\Xi_\#$ and Ξ_* satisfy $N_\#(\varepsilon) \leq N_*(\varepsilon^2/2)$ and therefore $q_\# < 2q_*$. On the other hand

$$(\Xi_* A)(B') = \langle \Omega | B' A | \Omega \rangle = \langle \left(1 + \Delta^{-1/2}\right) B'^* \Omega | \left(1 + \Delta^{-1/2}\right)^{-1} A \Omega \rangle$$

$$= \langle B'^* + j(B') \Omega | \Xi A \rangle \leq 2 \| B' \| \| \Xi A \| .$$

So $q_* \leq q$. □

The modular nuclearity condition for the vacuum state and the energy nuclearity 5.1.2 are closely related. One has

Proposition 5.4.4
Let Ω be the vacuum state vector, $\Theta_\beta : \mathcal{R}(K_1) \to \mathcal{H}$

$$\Theta_\beta A = e^{-\beta H} A | \Omega \rangle, \quad A \in \mathcal{R}(K_1), \tag{V.5.59}$$

q_β the order of the map Θ_β. Then it follows from requirement 5.1.2 that $q_\# \leq 4 q_\beta$ for $\beta = \delta / c$ where $\delta = r_2 - r_1$ and c is a numerical constant < 5.

Proof. With $A \in \mathcal{R}(K_1)$, $S = J \Delta^{1/2}$ the modular operators referring to K_2 we have

$$h_+(t) \equiv \langle \Phi | U(t) S A | \Omega \rangle = \langle \Phi | U(t) A^* | \Omega \rangle.$$

For Φ in the domain of S^* (or, equivalently, in the domain of $\Delta^{-1/2}$)

$$h_-(t) \equiv \langle \Phi | S U(t) A | \Omega \rangle = \langle U(t) A \Omega | S^* \Phi \rangle = \langle \Omega | A^* U(-t) | S^* \Phi \rangle.$$

Due to the positivity of the Hamiltonian $h_+(t)$ is the boundary value on the real axis of an analytic function in the upper half complex t-plane, $h_-(t)$ similarly the boundary value of an analytic function in the lower half plane. For $|t| < \delta = r_2 - r_1$ we have $\alpha_t A \in \mathcal{R}(K_2)$ and therefore

$$h_-(t) = \langle \Phi | \alpha_t A^* | \Omega \rangle = h_+(t) \quad \text{for} \quad |t| < \delta.$$

We can then apply the technique of lemma 5.2.2. and obtain

$$\langle \Phi | SA | \Omega \rangle = \langle \Phi | f(H) SA | \Omega \rangle + \langle \Phi | S f(H) A | \Omega \rangle.$$

Since $\Delta^{-1/2} R$ is a bounded operator, putting $\Phi = R\Psi$ ensures that Φ is in the domain of S^* for any $\Psi \in \mathcal{H}$. Thus we obtain

$$RSA|\Omega\rangle = Rf(H)SA|\Omega\rangle + RSf(H)A|\Omega\rangle,$$

and, since $RS = JR$ and $SA|\Omega\rangle = A^*|\Omega\rangle$,

$$RA|\Omega\rangle = Rf(H)A|\Omega\rangle + JRf(H)A^*|\Omega\rangle; \quad A \in \mathcal{R}(K_1).$$

Here f is any function constructed as in (V.5.37) and it has the properties described in lemma 5.2.2. The nuclearity of Θ_f discussed in the sequel of (V.5.40) implies then the nuclearity of Ξ as defined in (V.5.55). The comparison of the orders of the maps Ξ and Θ_β is achieved by means of the following lemma.

Lemma 5.4.5
There is a numerical constant $c > 0$ such that for $\beta \leq (r_2 - r_1)/c$ and any $\Psi \in \mathcal{H}$

$$\inf \parallel f(H)|\Psi\rangle \parallel \leq k^{-1} \parallel e^{-\beta H}|\Psi\rangle \parallel^k \cdot \parallel |\Psi\rangle \parallel^{1-k},$$

where the infimum is taken over all choices of f satisfying the required conditions and $k = (1 + c\beta/(r_2 - r_1))^{-1}$.

This leads to the upper bound for $q_{\#}$ stated in proposition 5.4.4. We refe for this computation to the original paper [Buch 90a]. □

V.6 The Universal Type of Local Algebras

We claim now that the algebra $\mathcal{R}(K)$ of a diamond is (as a W*-algebra) isomorphic to a unique mathematical object: the hyperfinite factor of type III_1. This means that physical information distinguishing different theories or different sizes of K is not contained in the algebraic structure or topology of an individual algebra $\mathcal{R}(K)$. The information comes from the relation between the algebras of different regions, from the net. The universality of $\mathcal{R}(K)$ may be seen as analogous to the situation in quantum *mechanics* where we can associate to each system or subsystem an algebra of type I, i.e. an algebra isomorphic to the set of all bounded operators on a Hilbert space. The change from the materially defined systems in mechanics to "open subsystems" corresponding to sharply defined regions in space-time in a relativistic local theory forces the change in the nature of the algebras from type I to type III_1. The fact that there is only one hyperfinite factor of type III_1 up to W*-isomorphy has been proved by Haagerup [Haager 87] based on work by Connes.

A von Neumann algebra $\mathcal{R}$ is called *hyperfinite* if it is the weak closure of an ascending sequence of finite dimensional algebras, in short if it is the W*-inductive limit of finite dimensional algebras. One should be aware of the fact that a subalgebra of a hyperfinite algebra is not necessarily again hyperfinite. But the W*-inductive limit of an ascending sequence of hyperfinite algebras is hyperfinite. Clearly the factor I_∞, the set of all bounded operators on a separable Hilbert space, is hyperfinite. It is the inductive limit of the matrix algebras of finite dimensional subspaces. The hyperfiniteness of $\mathcal{R}(K)$ follows then from the split property provided that this property holds for any pair of concentric diamonds with radii $r_2 > r_1$ for arbitrarily small $r_2 - r_1$ since $\mathcal{R}(K_2)$ is the inductive limit of the intermediate type I factors as r_1 tends to r_2. We have seen that this in turn is a consequence of the nuclearity assumption if the growth of the nuclearity index with the temperature is bounded by (V.5.8) i.e. if the level density does not increase too fast with the energy (no Hagedorn temperature).

Reasons for demanding that $\mathcal{R}(K)$ should be a factor have been discussed earlier (see e.g. section 5 of Chapter III). Perhaps one should keep in mind that a more thorough study of this question is desirable. At present we see no indication suggesting that local W*-algebras might have a non trivial center. However, if one wants to envisage this possibility then the claim at the beginning of this section has to be replaced by the more cautious statement: in the central decomposition of $\mathcal{R}(K)$ only factors of type III_1 can appear.

The conviction that $\mathcal{R}(K)$ must be of type III dates from the work of Araki [Ara 64c] who established this in a theory of free fields by explicit computation. It was recognized by Driessler [Driess 77] that in a theory which is invariant under dilatations there is a very simple argument showing that the relative dimension function of projectors in $\mathcal{R}(K)$ can only take the values 0 and ∞. We recall that this was von Neumann's original definition of type III (see subsection 2.1 of Chapter III). Using theorem 2.4.1 of the present Chapter, one can sharpen this and say that the type must be III_1 in this case. A dilatation by a scale factor λ maps $\mathcal{R}(K)$ onto $\mathcal{R}(\lambda K)$. For $\lambda \to 0$ $\mathcal{R}(K)$ is asymptotically Abelian with respect to this action since the region $\mathcal{O}_\lambda^c = (\lambda K)' \cap K$, the intersection of the causal complement of λK with K grows with decreasing λ and the W*-inductive limit of $\mathcal{R}(\mathcal{O}_\lambda^c)$ for $\lambda \to 0$ is $\mathcal{R}(K)$.

Of course we cannot have (unbroken) dilatation invariance if there are non-vanishing masses. But there are good reasons to believe that a realistic theory should become conformally invariant in the short distance limit. Among them we may count the dominant rôle of chirality in the standard model of elementary particle physics. The existence of an "ultraviolet fixed point", of a scaling limit, also plays a crucial rôle for quantum field theory in curved space-time (see Chapter VIII, section 3). In our context this relates to the question of the geometric significance of the modular automorphisms of $\mathcal{R}(K)$. If the theory has a scaling limit then the action should be approximately geometric in a well defined sense. This question was studied by Fredenhagen [Fred 85a]. We describe the main results but refer for the proofs to the original paper.

First, let $\mathcal{R}_k$ $(k = 1,\ 2)$ be von Neumann algebras with a common cyclic and separating vector Ω, S_k the Tomita operators defined in (V.2.1), Δ_k the modular operators and $\sigma_t^{(k)}$ the modular automorphisms. If $\mathcal{R}_2 \supset \mathcal{R}_1$, the domain of S_2 contains the domain of S_1 and in restriction to the domain of S_1 the two operators S_2 and S_1 are equal. This implies some relations between the modular operators. In particular one has

Proposition 6.1
Let $\mathcal{R}_2 \supset \mathcal{R}_1$ and $A \in \mathcal{R}_1$ such that $\sigma_t^{(2)} A \in \mathcal{R}_1$ for $|t| \leq \tau$. Then

$$\| \left\{ (1+\Delta_2)^{-1} - (1+\Delta_1)^{-1} \right\} A\Omega \| \\ \leq \frac{2}{\pi} \tanh^{-1}\left(e^{-\pi\tau}\right) \left\{ \| A\Omega \|^2 + \| A^*\Omega \|^2 \right\}^{1/2}. \qquad \text{(V.6.1)}$$

One applies this estimate to the case where $\mathcal{R}_2$ is the algebra of the wedge region W (see (V.4.1)) in the vacuum sector, $\mathcal{R}_1$ is the algebra of a diamond K tangent to the wedge at the origin, Ω is the vacuum state vector. As λ decreases λK will contract to a small subdiamond of the same nature, ultimately to the origin. Since we know the geometric significance of the modular automorphisms of the wedge algebra we can estimate the parameter range of t for which an element of $\mathcal{R}(\lambda K)$ will stay in $\mathcal{R}(K)$ under the action of $\sigma_t^{(W)}$. Then proposition 6.1. leads to

Proposition 6.2
For each $f \in \mathcal{L}^{(1)}(\mathbb{R})$ and $0 < \lambda < 1$ there exists a constant $c_f(\lambda)$ such that

$$\| \int dt\, f(t) \left\{ \Delta_K^{it} - \Delta_W^{it} \right\} A\Omega \|^2 \leq c_f(\lambda) \left\{ \| A\Omega \|^2 + \| A^*\Omega \|^2 \right\} \qquad \text{(V.6.2)}$$

for all $A \in \mathcal{R}(\lambda K)$ and one has

$$c_f(\lambda) \to 0 \quad \text{as} \quad \lambda \to 0. \qquad \text{(V.6.3)}$$

The constant c_f neither depends on the details of the theory nor on the size of K.

This shows that the action of $\sigma_t^{(K)}$ on the elements of $\mathcal{R}(\lambda K)$ becomes almost geometric for small λ; it approaches the action of $\sigma_t^{(W)}$. This can be used to show that if the theory has a scaling limit then the spectra of Δ_K and Δ_W coincide. To determine the Connes invariant of $\mathcal{R}(K)$ we need information on the spectra of modular operators for all normal states, not only for the vacuum. To obtain this Fredenhagen proves the following variant of the theorem 2.4.1.

Proposition 6.3
Let ω be a faithful normal state on a von Neumann algebra $\mathcal{R}$ and Δ_ω its modular operator. Then a necessary and sufficient condition for the number $k \in \mathbb{R}^+$

to be in the spectrum of Δ_ω is that for each $\varepsilon > 0$ there is some $A \in \mathcal{R}$ with $\omega(A^*A) = 1$ such that for all $B \in \mathcal{R}$

$$|\omega(AB) - k\,\omega(BA)| \leq \varepsilon\,\{\omega(B^*B) + k\,\omega(BB^*)\}^{1/2}. \qquad \text{(V.6.4)}$$

He then proceeds to show that in a theory which has a scaling limit this criterion is satisfied for all values of $k \in \mathbb{R}^+$. So the Connes invariant is maximal, the type is III_1.

VI. Particles. Completeness of the Particle Picture

VI.1 Detectors, Coincidence Arrangements, Cross Sections

VI.1.1 Generalities

In section 1 of Chapter III we claimed that the theory is fully characterized once the abstract net of observable algebras $\mathfrak{A}(\mathcal{O})$ is known. This means that all physical consequences must be derivable from this net. For comparison with experiments in high energy physics the most interesting ones concern the types of particles which occur and their collision cross sections. What is a particle? In section 3 of Chapter I we argued that the state vectors in an irreducible representation of $\overline{\mathfrak{P}}$ describe a single particle, a particle alone in the world. This was taken as the starting point for the discussion of the particle aspects of the theory in Chapter II, sections 3 and 4. Within the Hilbert space setting provided by the Wightman axioms the single particle subspace $\mathcal{H}^{(1)}$ was defined as the part belonging to the discrete spectrum of the mass operator $M = (P_\mu P^\mu)^{1/2}$. From $\mathcal{H}^{(1)}$ we constructed the states of asymptotic particle configurations by introducing a product composition $\otimes^t$ between state vectors which becomes unambiguous and has an obvious physical interpretation when it is applied to states which at time t are localized in far separated regions. It was then shown that this condition is satisfied at asymptotic times for any subset of single particle states and that $\otimes^t$ converges for $t \to \pm\infty$ to a composition $\otimes^{\text{out}}$, $\otimes^{\text{in}}$, respectively of the state vectors of $\mathcal{H}^{(1)}$. The proof depended on the assumption of a fast decrease of correlation functions with increasing space-like separation. This assumption in turn was shown to be a consequence of locality and positivity of energy provided that the vacuum was clearly separated from all other states by a mass gap. When compared to the claim at the beginning of this section the results and methods[1] of Chapter II fall short on several counts.

(i) They make essential use of unobservable (charge carrying) fields. Alternatively put, the Hilbert space used must contain all superselection sectors. Though this is no objection in principle to the claim that the net of observables

[1] With the exception of the approach by Araki and Haag [Ara 67] which was briefly described in II.4.3. This approach will be carried further below.

suffices for the construction of collision states because the unobservable fields may be constructed with the methods of Chapter IV, it indicates that a more direct approach, closer to experimental procedure, is warranted.

(ii) The existence of a discrete part of the mass spectrum and the completeness of the particle picture ($\mathcal{H}^{\mathrm{in}} = \mathcal{H}^{\mathrm{out}} = \mathcal{H}$) was put in by assumption, unrelated to the properties of the local net.

(iii) The method fails if there is no mass gap. It may be argued that in the regime of high energy collisions the additional complications connected with the vanishing of the photon mass (slow decrease of the correlations in the vacuum state, infrared photon clouds) are of peripheral interest and that their effect can be dealt with once collision theory in a purely massive model is understood. Yet it is unsatisfactory to base the approach on a simplification which we know to be not true in reality. We should take this also as an indication for the need to find a more natural approach.

Taking up the suggestion from item (i) we note that experimentally all information comes from the use of detectors and coincidence arrangements of detectors. The essential features used are that a detector is a macroscopically well localized positive observable which gives no signal in the vacuum state. In the mathematical set up of the theory the two requirements cannot be strictly reconciled due to the Reeh-Schlieder theorem; the algebra $\mathcal{R}(\mathcal{O})$ of a strictly finite region does not contain positive operators with vanishing vacuum expectation value. However we may represent a detector centered at the origin by a positive, almost local element C of the algebra of observables, sufficiently well approximated in norm by an element $C_r \in \mathcal{R}(K_r)$. Specifically

$$C \in \mathfrak{A}^{+}; \quad \| C \| = 1; \quad \omega_0(C) = 0, \tag{VI.1.1}$$

and, for any prescribed (arbitrarily small) inaccuracy $\varepsilon > 0$ there is a radius r such that

$$\| C - C_r \| < \varepsilon \quad \text{for some} \quad C_r \in \mathcal{R}^{+}(K_r). \tag{VI.1.2}$$

We may consider r for given ε as the essential size of the detector. A detector centered at the point x is then, of course, represented by

$$C(x) = \alpha_x C, \tag{VI.1.3}$$

and a coincidence arrangement of n detectors C_k at time t is described by the product

$$D_n = C_1(x_1) \ldots C_n(x_n); \quad x_k = (t, \mathbf{x}_k) \tag{VI.1.4}$$

with $| \mathbf{x}_i - \mathbf{x}_k | > R$ where R is large compared to the essential extensions r_k of the detectors C_k. If in a state ω we have $\omega(C) = p$ with p large compared to ε we know that the state deviates noticeably from the vacuum in the region K_r, it has a local excitation. If $\omega(D_n)$ differs significantly from zero we know that ω has at least n coexisting localization centers at time t.

Before proceeding further we should consider a question which will be undoubtedly on the mind of one who strives for a precise correspondence between experimental procedure and mathematical representation. Suppose an experimentalist shows you an instrument saying that this is a very good muon detector. He gives you the engineering manual which describes how it is constructed. Then he says with an innocent smile: can you work out the algebraic element which corresponds to this manual in the mathematical description? Conversely, given C satisfying (VI.1.1), (VI.1.2), how can one construct a corresponding instrument? The answer is: we do not want to bother and we need not know in such detail. On the experimental side the development of an apparatus optimally suited for a particular purpose (highly selective sensitivity) is a lengthy process involving trial and error. The same situation prevails on the theoretical side. The conditions (VI.1.1) and (VI.1.2) are so weak that they allow infinitely many algebraic elements. Most of them will correspond to poor detectors if one has a specific task in view. But theoretical guidance for choosing suitable ones exists. We may choose C of the form

$$C = L^*L, \tag{VI.1.5}$$

where $L \in \mathfrak{A}_S$ has energy-momentum transfer limited to a prescribed region Δ in p-space in the complement of the closed forward cone i.e.

$$\int L(x)f(x)\,d^4x = 0 \quad \text{for} \quad \text{supp}\tilde{f} \subset \Delta^c; \quad \Delta \cap \overline{V}^+ = \emptyset, \tag{VI.1.6}$$

where $\tilde{f}$ denotes the Fourier transform of f and $L(x) = \alpha_x L$. In addition L shall be essentially contained in $\mathcal{R}(K_r)$ in the sense of (VI.1.2). Of course, if q is the diameter of Δ this is possible only for

$$rq \gg 1.$$

One can define the effective volume of the detector as

$$\mathcal{V}_C \equiv \frac{4\pi}{3} r_C^3 = \frac{\int \| \, [L(x), L^*] \, \| \; d^3x}{\| \, L \, \|^2}; \quad x^0 = 0. \tag{VI.1.7}$$

One should note that we can choose L in the observable algebra even if we want to register particles which carry some charge and have a sharp mass. For this purpose we have to choose Δ space-like. Then L can have nonvanishing matrix elements between state vectors in the charged sector on a mass hyperboloid. We shall furthermore choose Δ to be sufficiently separated from V^+. Then not only the vacuum but all states with energy below some value δ will be annihilated by L and thus C registers only excitations with energy above δ in the volume $\mathcal{V}_C$. δ fixes a boundary between "hard excitations" which are registered and "soft excitations" which are not registered. Without loss of generality we may restrict attention to states which carry a total energy below some value E. Then the compactness requirement in the form suggested by Fredenhagen and Hertel (see Chapter V, subsection 5.1) suggests that a finite number of different C_k will suffice to determine the partial state in the detector region (i.e. the restriction of ω to $\mathcal{R}(K_r)$ up to any desired accuracy in the norm topology of $\mathcal{R}(K_r)_*$). The sensitivity of a detector to specific states has to be tested by monitoring. Once the single particle states of various types are known (see below) it is a finite problem to choose detectors, selectively sensitive to a specific type. Furthermore, the supports Δ_k of the chosen L_k will then also give information about the energy-momentum content. We can distinguish local excitations which carry different momenta. In this way we can ultimately deduce all information from the probabilities for response of geometric arrangements of detectors, knowing a priori not more than that each of them responds to some local excitation carrying some approximate momentum.

Let us reconsider now our understanding of the particle picture. We call a state ω *at most n-fold localized at time t* if it cannot trigger any $(n+1)$-fold coincidence arrangement at this time. We shall express this as the demand that[2]

$$\int_{|\mathbf{x}_i-\mathbf{x}_k|>R} \omega\{C_1(t,\mathbf{x}_1)\ldots C_{n+1}(t,\mathbf{x}_{n+1})\}d^3x_1\ldots d^3x_{n+1} < \varepsilon_D, \qquad \text{(VI.1.8)}$$

for any choice of the detectors conforming with the mentioned requirements. Here we may first fix the tolerated background probability ε_D, choose the sizes r_k of the detectors and the separation distance R which has to be large compared to every r_k. If we use preexisting information about the energy bounds of the state ω then a finite number of different choices of the detectors C_k will suffice in the test (VI.1.8). Clearly the choice of r_k and, more significantly, of R determines what we regard as a single localization center. We are interested in the regime of "small particle physics" where we expect to encounter matter concentrations of small intrinsic extension, not rocks or other large chunks of cohesive matter. This will already be ensured by the energy bounds on the state since we know from experience that with growing extension of cohesive matter also the mass will increase. So for each energy bound E there is a radius R_E which bounds the intrinsic size of possible chunks of cohesive matter. This empirical knowledge must, of course, also follow from the theory. We have included it in the property (VI.1.13) below which implies that, given the total maximal energy, we can choose R sufficiently large so that a signal from an n-fold coincidence implies that there are at least n distinct coexisting localization centers at the respective time.

If ω satisfies (VI.1.8) and has no component with less than n localization centers i.e.

$$\omega \neq \lambda\omega_{n-1} + (1-\lambda)\omega'; \quad \text{for} \quad \lambda \neq 0,$$

where ω_{n-1} is a state satisfying (VI.1.8) with n replaced by $n-1$, then we say that ω is exactly n-fold localized at the time.

Remarks. (i) Since we are integrating in (VI.1.8) over infinitely extended space one might worry that the left hand side could become infinite and the condition empty. Indeed $\int C(x)d^3x$ is not an element of the quasilocal algebra and it is

[2] Instead of the integral one might, at this stage, equally well take the supremum of the integrand for $|\mathbf{x}_i - \mathbf{x}_k| > R$ as the relevant quantity. We shall see, however, that (VI.1.8) is the most convenient starting point for the analysis. See remark (i) below.

unbounded. However, within the subset of states with total energy below E there is a uniform bound

$$\omega\left(\int C(x)d^3x\right) < \left(\frac{\mathcal{V}_C}{\delta}\right) E. \tag{VI.1.9}$$

This key lemma has been proved by Buchholz (lemma 2.2. in [Buch 90b]). The only feature needed besides locality is the restriction that we are dealing with states in a sector in which the translations can be implemented by unitaries satisfying the spectrum condition. An intuitive argument runs as follows. The response of the detector indicates that there is an energy larger than δ in the volume $\mathcal{V}_C$ around the point x. So the expectation value of the energy density at time t at $\mathbf{x}$ is larger than $\delta/\mathcal{V}_C\ \omega(C(x))$ and the total energy of the state larger than $\delta/\mathcal{V}_C \int \omega(C(x)d^3x)$.

(ii) Instead of $C(x_1)\ldots C(x_{n+1})$ in (VI.1.8) we may equally well use

$$\int_{|\mathbf{x}_i-\mathbf{x}_k|>R} \omega\left(L^*(x_1)\ldots L^*(x_{n+1})L(x_{n+1})\ldots L(x_1)\right) d^3x_1\ldots d^3x_{n+1} < \varepsilon_D' \tag{VI.1.10}$$

since the commutation of a factor $L(x)$ with $L(x+y)$ or $L^*(x+y)$ introduces only a small change when $|\ \mathbf{y}\ | > R$, a change which remains small even when integrated over $\mathbf{y}$. This shows that for states with energy below E *the maximal number of localization centers stays uniformly bounded for all times by*

$$n_{\max} = \frac{E}{\delta}. \tag{VI.1.11}$$

We may generalize (VI.1.9) to

$$0 \leq \int \omega\left(L^*(x_1)\ldots L^*(x_{n+1})L(x_{n+1})\ldots L(x_1)\right) \prod d^3x_k$$
$$< \mathcal{V}_C^{n+1} n_{\max}(n_{\max}-1)\ldots\times(n_{\max}-n). \tag{VI.1.12}$$

VI.1.2 Asymptotic Particle Configurations

A state describing a single stable particle, alone in the world, can be characterized as a state which is singly localized at all times; it cannot trigger a two-fold coincidence at any time. Actually, since we consider detectors which have a threshold δ as the minimal energy needed for a response, we are talking here about a single "hard" particle. It may be accompanied by an unregistered background whose energy density exceeds nowhere the threshold level $\delta/\mathcal{V}_C$. We note, however, that "hard" does not mean massive. Photons of energy beyond δ are included.

For an intuitive picture it is helpful to introduce an (unbounded, non local) observable $N(t)$, the *localization number at time* t which assigns the value n to a state wich is n-fold localized at time t. Mathematically it is a linear form over a domain in $\mathfrak{A}_*$ which contains the states of bounded energy. More intuitively it may be considered as an unbounded operator with non negative integer eigenvalues in a Hilbert space. Of course $N(t)$ depends on the choice of R, ε_D and of the r_k. But we expect that this becomes irrelevant for large times. We expect that with increasing time the configuration expands, its diameter growing linearly with time. The probability densities $\omega\left(C(t,\mathbf{x}_1)\ldots C(t,\mathbf{x}_{n+1})\right)$ then must decrease correspondingly because the integral over the $\mathbf{x}_k$ stays bounded. In other words the theory must have the property that for sufficiently large times $t > T$ (depending on ω)

$$\int_{R_E<|\mathbf{x}_1-\mathbf{x}_2|<R(t)} \omega\left(C(t,\mathbf{x}_1)C(t,\mathbf{x}_2)\right) d^3x_1 d^3x_2 < \varepsilon \tag{VI.1.13}$$

if we take for any $0 < \kappa < 1$

$$R(t) < \text{ const. } |\, t\,|^{\kappa}\,. \tag{VI.1.14}$$

The essential part of the coincidence integrals (VI.1.8) for large times is expected to come from separations $R > R(t)$ so that we may let R grow with $|\, t\,|$ in the definition of the localization number $N(t)$ and, correspondingly let ε_D tend to zero. This means that the only parameter which remains relevant for the definition of $N(t)$ at large times is δ. Note that (VI.1.13) contains also the (expected) property that the intrinsic extension of cohesive matter is bounded by R_E independently of the time.

If (VI.1.13) is satisfied we can expect that for sufficiently large times the individual localization centers in an n-fold localized state will no longer influence each other. Since the maximal localization number is uniformly bounded by (VI.1.11) this means that each asymptotic localization center will be a stable, hard particle (possibly accompanied by an unregistered background with energy density below $\delta/\mathcal{V}_C$). In the limit $N(t)$ will become the asymptotic particle number. More precisely, we expect that

$$\varphi_t(A) = \int \omega(\alpha_x L^* A L)\, d^3x, \quad x = (t,\mathbf{x}), \quad A \in \mathcal{R}(\mathcal{O}) \tag{VI.1.15}$$

should converge for $|\, t\,| \rightarrow \infty$ and, if

$$\lim \|\, \varphi_t \,\| = \lim \int \omega(\alpha_x L^* L) d^3x \neq 0 \tag{VI.1.16}$$

we anticipate that the limit of (VI.1.15) gives (apart from normalization) the expectation value of A in a state which is permanently singly localized i.e. the limit describes a mixture of stable single particle states.

Starting from the qualitative picture described above Buchholz drew up the following strategy for the analysis of the particle content of the theory [Buch 87a].

1) One wants to show the convergence (VI.1.15) and the existence of a nonvanishing limit for some choices of L. The set of nonvanishing limits gives the

particle content. We may note that closely related to the question of convergence is the property (VI.1.13).

2) In order to obtain a mathematically well defined object which can replace the "single particle subspace" $\mathcal{H}^{(1)}$ of Chapter II one must remove the arbitrary threshold δ. If the theory has no mass gap then this step may demand a generalization of the notion of state for a sharply defined particle. This is no tragedy. Instead of states we must consider *weights* on the algebra $\mathfrak{A}$. This is mathematically well defined. In our context a weight may be regarded as a positive linear form on the subalgebra $\mathcal{C}$ of detectors. A precise definition will be given later. $\mathcal{C}$ does not contain the unit element and therefore a weight cannot be normalized in the standard way. The single particle weights are the limit elements for $x^0 \to \infty$ of

$$\int \omega(\alpha_x C)\, d^3x; \quad C \in \mathcal{C} \tag{VI.1.17}$$

for states of bounded energy (considered as positive linear forms over $\mathcal{C}$).

3) One studies the decomposition of single particle weights into pure components. This replaces the familiar decomposition of $\mathcal{H}^{(1)}$ into irreducible parts. It turns out that the decomposition leads to pure single particle weights which have sharp momentum, mass and spin and allow the distinction of different particle types occurring in the theory.

4) Using a pure single particle weight (instead of a state) in the GNS-construction one obtains a representation of $\mathcal{C}$. One shows then that under the standard assumptions for the theory this representation can be extended to a representation of $\mathfrak{A}$ on a separable Hilbert space which is locally normal i.e. equivalent to the vacuum representation when restricted to the subalgebra of a finite region. The pure weight itself appears as an improper state vector, like a plane wave in quantum mechanics, which we denote by $\mid p, \alpha\rangle$. Here p is the energy-momentum of the weight, α combines the remaining classification parameters i.e. spin and particle type. One would like to show that for fixed p the index α can run only through a finite number of values and also that the set of possible mass values $m = (p^\mu p_\mu)^{1/2}$ is a discrete set.

5) The functions

$$C \to \Gamma_{\alpha\beta}(C; p) = \langle p, \beta \mid C \mid p, \alpha\rangle \tag{VI.1.18}$$

give the sensitivity of the detector C. If there is only a finite number of particle types in a finite mass interval one can then construct special elements C^γ which are sensitive only for a particular value of the index γ:

$$\Gamma_{\alpha\beta}(C^\gamma; p) = \delta_{\alpha\gamma}\delta_{\beta\gamma}C^\gamma(p). \tag{VI.1.19}$$

With the help of these one can find the momentum-space densities $\varrho_\gamma^{\text{out}}(\mathbf{p})$, $\varrho_\gamma^{\text{in}}(\mathbf{p})$ at asymptotic times $\pm\infty$ for any state ω by the formula

$$\int C^{(\gamma)}(\mathbf{p})\omega\left(\varrho_\gamma^{\pm}(\mathbf{p})\right) h\left(\frac{\mathbf{p}}{p^0}\right) d^3p = \lim_{t\to\pm\infty} \int \omega\left(C^{(\gamma)}(t,\mathbf{x})\right) h\left(\frac{\mathbf{x}}{t}\right) d^3x. \tag{VI.1.20}$$

Here $\pm$ stands for *out, in*, respectively and h is any smooth function. Letting h tend towards a δ-function in velocity space one gets the probability density for finding a particle of type γ with the corresponding asymptotic momentum in the state ω. The formula (VI.1.20) was derived in [Ara 67] under much more restrictive assumptions. As outlined there, it allows the determination of the collision cross sections. In the present context (VI.1.20) follows once one can identify the asymptotic velocity of a particle with $\mathbf{p}/p^0$. To obtain a useful algorithm for the computation of the cross sections one has to devise an efficient way of filtering out states with prescribed incoming configurations. One way of doing this, using only the vacuum expectation values of observables, has been devised by Buchholz and Stein [Stein 89], [Buch 91a]. It is necessarily more tedious than the methods described in Chapter II and it gives the cross sections, not the S-matrix. But it is clear that a price has to be paid if one does not restrict attention to theories with a mass gap and wants to avoid using charge carrying fields. The price is remarkably low compared to the difficulties incurred if one really wants a precise formula for hadron cross sections in QCD in terms of Green's functions of quark fields.

6) There remains the question of completeness of the particle picture. This is the question whether a state ω is completely specified by the knowledge of its asymptotic particle content. This may again be reduced to the convergence of an expression like (VI.1.17). If there is an energy-momentum tensor so that the total energy can be expressed as a space integral over a local density then one can take instead of C in (VI.1.17) the (suitably smeared out) energy density. The convergence of this implies that the total energy can be accounted for by the contributions from the single particle weights in the asymptotic configuration.

How much of this program has been achieved to date? If one focuses on the aim of relating all the features mentioned to simple structural properties of the theory which, apart from locality, spectrum condition and Poincarè symmetry will have to include nuclearity and some aspects of the dynamical law (e.g. asymptotic Abelianness of $\mathfrak{A}$ with respect to α_t) then there remain at present some gaps. There are plausibility arguments indicating that these gaps can be filled. The precise conditions which are necessary and sufficient to ensure the convergence of (VI.1.15) are not yet known. One knows that (VI.1.15) has weak limit points and that any one of these is a single particle weight. The first problem appears therefore to find lower bounds for (VI.1.15) which exclude the vanishing of all limits. If one has an energy-momentum tensor then the argument indicated under item 6) of the strategy can serve this purpose. The decomposition theory of single particle weights and the analysis of the properties of pure single particle weights has been carried through. We shall describe this in the next section. Still it is an essential and not yet accomplished task to show the convergence of (VI.1.15) because, if there are several limit points, one cannot derive the crucial formula (VI.1.20) on which the determination of cross sections depends.

VI.2 The Particle Content

VI.2.1 Particles and Infraparticles

In the last section a single particle state was defined as a state which is permanently singly localized. How does this geometric characterization relate to the more common one which associates single particle states with the discrete part of the mass spectrum? For the case of a theory with a minimal nonvanishing mass this has been discussed by Enss. He finds that in this case the two characterizations are equivalent if the theory satisfies a compactness requirement [Enss 75]. We indicate the argument.

In a purely massive theory we can implement the Poincaré symmetry by unitary operators in the Hilbert space containing all sectors and we can split off the center of mass motion in the subspace orthogonal to the vacuum as described in Chapter I, subsection 3.4 Thus we can write

$$\mathcal{H} = \mathcal{H}_C \otimes \mathfrak{h}, \tag{VI.2.1}$$

where $\mathcal{H}_C = \mathcal{L}^{(2)}(\mathbb{R}^3)$ describes the center of mass motion and $\mathfrak{h}$ takes care of all the remaining degrees of freedom. A general state vector Ψ in $\mathcal{H}$ can be written as a function on $\mathbb{R}^3$ with values in $\mathfrak{h}$

$$\mathbf{p} \in \mathbb{R}^3 \to \Psi(\mathbf{p}) \in \mathfrak{h}. \tag{VI.2.2}$$

Here $\mathbf{p}$ is the center of mass momentum and

$$\langle \Psi \mid \Psi \rangle = \int \langle \Psi(\mathbf{p}) \mid \Psi(\mathbf{p}) \rangle \, d^3p. \tag{VI.2.3}$$

If Ψ has total energy below E and is localized at some time, say at $t = 0$, in some given region then the compactness requirement says that there is, for each $\mathbf{p}$ a finite dimensional subspace $\mathfrak{h}_N(\mathbf{p}) \subset \mathfrak{h}$ such that $\Psi(\mathbf{p})$ can be approximated sufficiently well by a vector in $\mathfrak{h}_N(\mathbf{p})$ (N denoting the dimension). Moreover, if we shift the localization region by a spatial translation the subspace $\mathfrak{h}_N(\mathbf{p})$ does not change since such a translation multiplies $\Psi(\mathbf{p})$ only by a phase factor. So one concludes that the components of singly localized states (for a fixed choice of R) at time $t = 0$ lie essentially in $\mathfrak{h}_N(\mathbf{p})$. For each time t we have such subspaces $\mathfrak{h}_N(\mathbf{p}, t) \subset \mathfrak{h}$ characterizing the states which are singly localized at the respective time and they result from $\mathfrak{h}_N(\mathbf{p})$ by application of e^{iHt} since the Hamiltonian H commutes with the space translations. The $\mathbf{p}$-component of a permanently singly localized state must lie in the intersection of all these subspaces. If there are permanently singly localized states this intersection cannot be empty. So for each $\mathbf{p}$ there must be a finite dimensional subspace of $\mathfrak{h}$ which is almost stable under the mass operator. The mass operator M must have a discrete part in its spectrum. The converse, namely that a subspace of $\mathcal{H}$, belonging to a discrete eigenvalue of M, contains only states which are permanently singly localized follows from the discussion in Chapter II, section 4 in theories with mass gap.

If the theory has no mass gap the above argument fails on two counts. For mass zero states we cannot separate off the center of mass motion; in addition the Poincarè symmetry may not be implementable in all sectors. It may then be no longer true that the existence of a particle can be recognized by the appearance of a discrete eigenvalue of M. In quantum electrodynamics the 1-electron states do not belong to an eigenspace of M; the electron mass is the lower bound of the mass spectrum in the sector of charge 1 but the hyperboloid $p^2 = m^2$ carries zero weight in the spectral decomposition. This is one aspect of the "infrared problem" in QED. It was worked out by Schroer in a simplified model and he coined the term *infraparticle* for a particle like the electron which is not associated to a discrete eigenvalue of M [Schroer 63].

In the standard field theoretic approach this aspect concerns the nature of the singularity of the Feynman amplitudes at $p^2 = m^2$. Unfortunately even the nature of the singularity depends on gauge conventions. So it is difficult to extract the information about the spectrum of M in the physical Hilbert space. See [Kibble 68 a, b], Faddeev and Kulish [Fadd 71]. Focusing attention on the observables the infraparticle aspect and the spontaneous breaking of the Lorentz symmetry has been taken up again by Fröhlich, Morchio and Strocchi [Fröh 79a, b] and by Buchholz who showed that the absence of a discrete eigenvalue of M for states with an electric charge is a direct consequence of Gauss' law [Buch 86d]. We sketch this argument.

Let $F^{\mu\nu}(x)$ denote the electromagnetic field, considered as an operator valued distribution acting in the Hilbert space of a primary representation of the observable algebra in which the translations are implementable with P^μ-spectrum in $\overline{V}^+$. Pick a test function f with compact support in a region space-like to the origin and scale it:

$$f_R(x) = R^{-2} f(R^{-1}x). \tag{VI.2.4}$$

Then, for large R the (unbounded) observable $F^{\mu\nu}(f_R)$ may be regarded as some weighted average of the flux through large spheres around the origin. Specifically, if we take in polar coordinates

$$f(x) = f_1(t,r) f_2(\vartheta, \varphi)$$

then

$$F^{\mu\nu}(f_R) = \int \Phi^{\mu\nu}(Rt, Rr) f_1(t,r)\,dt\,dr, \tag{VI.2.5}$$

where

$$\Phi^{\mu\nu}(t', r') = \int F^{\mu\nu}(t', r', \vartheta, \varphi) f_2(\vartheta, \varphi) r'^2 \sin\vartheta \; d\vartheta\, d\varphi \tag{VI.2.6}$$

is the flux through a sphere of radius r' at time t', averaged over the angles with the weight function f_2. The essential input for the subsequent argument is the claim that for all states of interest in elementary particle physics the limit of the expectation value

$$\lim_{R\to\infty} \omega\left(F^{\mu\nu}(f_R)\right) \equiv f^{\mu\nu}(f) \tag{VI.2.7}$$

exists, does not vanish for all f in charged states and that the fluctuation stays bounded as $R \to \infty$

$$\omega\left(F^{\mu\nu}(f_R)^2\right) < \infty. \tag{VI.2.8}$$

The justification of these claims relies on Gauss' law, due to which we should be able to measure the charge of a state by the (space-like asymptotic) flux of the electric field. In charged states the expectation value of $F^{\mu\nu}(x)$ at large distances should deviate sufficiently from the vacuum expectation value to yield a nonvanishing limit as demanded in (VI.2.7) but the fluctuations should remain bounded. The fluctuations (VI.2.8) in the vacuum state can be estimated using standard properties of Wightman functions. They involve the correlation between $F^{\mu\nu}(x)$ and $F^{\mu\nu}(y)$ for very large space-like x and y. One does not expect that this estimate is significantly affected by the presence of charges in finite regions.

Granted (VI.2.7), (VI.2.8) one notes that in a primary folium $F^{\mu\nu}(f_R)$ converges weakly to a multiple of the identity (on a dense domain) since the limit is affiliated with the center (compare theorem 3.2.2 in Chapter III). Thus $f^{\mu\nu}(f)$ is independent of the state ω in the folium. Now one considers the commutator of the mass operator with $F^{\mu\nu}$. One has

$$\left[M^2, F^{\mu\nu}(f_R)\right] = iR^{-1}\left(P^{\varrho}F^{\mu\nu}((f_{,\varrho})_R) + F^{\mu\nu}((f_{,\varrho})_R)P^{\varrho}\right), \tag{VI.2.9}$$

where $f_{,\varrho} = \partial_\varrho f$. If ω is a state with sharp mass the expectation value of the left hand side vanishes and, if it is a state with bounded energy, we can use (VI.2.7) to evaluate the limit of the bracket on the right hand side. So we obtain for such states

$$\omega(P^{\varrho})c_{\varrho} = 0 \quad \text{for} \quad c_{\varrho} = f^{\mu\nu}(f_{,\varrho}). \tag{VI.2.10}$$

Since c_ϱ cannot vanish for all choices of f and of indices μ, ν the momentum spectrum of states with sharp mass is restricted to a subset of the mass hyperboloid which has lower dimension i.e. the directions of the spatial momentum would have to be restricted for such states. This is impossible for normalizable states in the folia under consideration.

So one can conclude that at least all electrically charged particles are infraparticles. They do not correspond to a discrete eigenvalue of M; moreover one finds that the Lorentz symmetry is not implementable in a sector of states with nonvanishing electric charge. The Lorentz symmetry is spontaneaously broken. Nevertheless one can attribute a sharp mass to a charged particle and one has a well defined discrete set of mass values for the particle types occurring in the theory. The point is only that there remains no (normalizable) charged state which is permanently singly localized if we let the threshold of the allowed detectors tend to zero. There remain "improper single particle states", *weights* σ_P corresponding to an electron with sharp energy-momentum p; the values of p which occur fill the mass hyperboloid $p^2 = m^2$ but the representations of $\mathfrak{A}$ which are induced by weights with different p are inequivalent. One may say that the velocity of the electron gives a superselection rule. This has a simple physical reason. An electrically charged particle moving with constant velocity is accompanied by an electromagnetic field (e.g. the Lorentz transformed Coulomb field of a particle at rest). For different velocities the flux of this field through some segment of a sphere with arbitrarily large radius will be different (and nonvanishing). The asymptotic flux cannot be changed by the action of any element of the quasilocal algebra.

As a consequence of this superselection rule no coherent superpositions of weights σ_P with different values of p are possible, no wave packets corresponding to normalizable strict 1-electron states can be formed. Yet electron interference is a salient fact which is explained in quantum mechanics by applying the superposition principle to 1-electron wave packets. Obviously this quantum mechanical idealization is good enough for the discussion of electron interference experiments in spite of the fact that QED tells us that, strictly speaking, there are no such coherent wave packets. The seeming paradox may serve as a warning against overrating the significance of idealizations in the mathematical description of a physical situation. The reader is encouraged to work out how the quantum mechanical description of an electron interference experiment can be justified within the field theoretic setting. Here we only remark that such an experiment concerns the partial state in a finite space-time region and that the initial information we have about it is only up to some background with energy density below some threshold. The phenomenon studied must be insensitive to this ignorance. Thus also the soft electromagnetic radiation which is necessarily generated by the interaction of the primary electron beam with a diffracting crystal and external electromagnetic fields changing the electron velocity may be ignored. It causes an uncertainty of the quantum mechanical wave function which remains irrelevant for the phenomenon. In cosmological applications of quantum field theory one should, however, be careful to take the infraparticle aspect into account.

VI.2.2 Single Particle Weights and Their Decomposition

The following analysis, relying on (partly unpublished) arguments and results of Buchholz and Porrmann [Buch 91a, 94] is based on the geometric definition of the particle concept and applies therefore equally to particles and infraparticles. It covers essentially the items 2) - 4) of the strategy outlined at the end of subsection 1.2. We state here the main results and give some comments and sketches of proofs.

A representation of $\mathfrak{A}_S$ in which the translation group is implementable with P^μ-spectrum contained in $\overline{V}^+$ will be called, for short, a *positive energy representation.* In the following $\mathfrak{S}$ denotes the set of all physically allowed states which generate via the GNS-construction a positive energy representation and $\mathfrak{S}_E$ the subset of states with energy below E. The unitaries implementing the translations in a positive energy representation can be fixed uniquely by the following theorem.

Theorem 2.2.1
Let π be a positive energy representation of $\mathfrak{A}_S$ and $\pi(\mathfrak{A}_S)'' = \mathcal{R}$. Then:

a) There is a choice of the unitary group $U(x)$ implementing the translation automorphisms α_x

$$\pi(\alpha_x A) = U(x)\pi(A)U(x)^*$$

such that

$$U(x) \in \mathcal{R}. \tag{VI.2.11}$$

This fixes $U(x)$ up to unitaries in the center $\mathcal{R} \cap \mathcal{R}'$.

b) There is a choice of $U(x) \in \mathcal{R}$ such that the P^μ-spectrum has a Lorentz invariant lower boundary in each subspace of $\mathcal{H}$ which reduces $\mathcal{R}$ i.e. in each $P\mathcal{H}$ when P is a projector from $\mathcal{R}'$.

Properties a) and b) fix $U(x)$ uniquely. We call the representation $x \to U(x)$ having these properties the *canonical implementation* of α_x.

Part a) of the theorem is due to Borchers [Borch 66], part b) to Borchers and Buchholz [Borch 85]. We shall not give the proof.

Let $\mathcal{L}_0 \subset \mathfrak{A}_S$ be the set of almost local annihilators and $\mathcal{L}$ the left ideal in $\mathfrak{A}_S$, algebraically generated from $\mathcal{L}_0$. Its elements are finite linear combinations of elements of the form AL_0 with $A \in \mathfrak{A}_S$, $L_0 \in \mathcal{L}_0$. Thus $\mathcal{L}$ is a subset of the Doplicher ideal (III.3.21) and differs from it only by the additional requirement that L_0 shall be almost local and $\mathcal{L}$ is not completed in the norm topology. Let $\mathcal{L}^*$ be the set of adjoints of $\mathcal{L}$ and

$$\mathcal{C} = \mathcal{L}^*\mathcal{L} \tag{VI.2.12}$$

the set of finite linear combinations of elements $L_1^*L_2$ with $L_k \in \mathcal{L}$. $\mathcal{C}$ is a *-algebra, again, of course, not norm closed.

Proposition 2.2.2

(i) For $C \in \mathcal{C}$ one has

$$q_E(C) \equiv \sup \int |\, \omega(C(\mathbf{x}))\,|\, d^3x < \infty, \tag{VI.2.13}$$

where the supremum is taken over all $\omega \in \mathfrak{S}_E$. q_E is a seminorm on $\mathcal{C}$ and it is invariant under translations in space and time

$$q_E(\alpha_x C) = q_E(C). \tag{VI.2.14}$$

(ii) For $A \in \mathfrak{A}_S$ and $L_k \in \mathcal{L}$

$$q_E(L_1^* A L_2) \leq \|\, A \,\|\, (q_E(L_1^*L_1) q_E(L_2^*L_2))^{1/2}. \tag{VI.2.15}$$

The proposition is a consequence of (VI.1.9). Let us consider now, with $\omega \in \mathfrak{S}_E$, the sequence of positive linear forms on $\mathcal{C}$

$$\overline{\omega}_t(C) = b(t)^{-1} \int_t^{t+b(t)} \omega(C(t', \mathbf{x}))dt' d^3x, \qquad \text{(VI.2.16)}$$

where $b(t)$ increases monotonously to infinity as $t \to \infty$. The time averaging may be unnecessary since, due to the intuitive reasoning of section 1, we hope that $\int \omega(C(t,\mathbf{x}))d^3x$ itself converges for $t \to \infty$. But it helps in the subsequent argument. Irrespective of the question of convergence the sequence $\{\overline{\omega}_t(C)\}$ has limit points as $t \to \infty$ since it is uniformly bounded by $q_E(C)$ due to proposition 2.2.2. In fact the sequence of forms $\{\overline{\omega}_t\}$ has weak limit points in the space $\overline{\mathcal{C}}^*$, the topological dual of the space $\overline{\mathcal{C}}$.[1] These weak limit points are positive linear forms over $\mathcal{C}$ and one can study their properties. The worst possibility is that all limit points of (VI.2.16) are zero for any $\omega \in \mathfrak{S}_E$. If this happens then the theory has no particle content. Every state which is localized at some finite time will then dissolve ultimately and the total probability of any detector signal at very late times goes to zero even if all of space is paved with detectors. As mentioned above, given a proper local formulation of the conservation laws (e.g. existence of an energy-momentum tensor as a Wightman field associated with the local algebras) one can eliminate this possibility.

Let $\mathcal{W}_E$ denote the set of nonvanishing limit points in $\overline{\mathcal{C}}^*$ of (VI.2.16) as ω ranges through $\mathfrak{S}_E$ and put $\mathcal{W} = \cup \mathcal{W}_E$. One has

Proposition 2.2.3
Let $\sigma \in \mathcal{W}$. Then

(i) σ is a positive linear form on $\mathcal{C}$, a weight on $\mathfrak{A}_S$ and

$$\sigma(C^*C) \le q_E(C^*C) \quad \text{for} \quad \sigma \in \mathcal{W}_E. \qquad \text{(VI.2.17)}$$

(ii) σ is translation invariant

$$\sigma(\alpha_x C) = \sigma(C). \qquad \text{(VI.2.18)}$$

(iii) For g running through the Poincaré group $\sigma_g(C) \equiv \sigma(\alpha_g C)$ and $\sigma(C_1 \alpha_g C_2)$ are continuous functions of g.

(iv) For $\sigma \in \mathcal{W}_E$ the Fourier transform of $\sigma(C_1 \alpha_x C_2)$ has support in $\overline{V}^+ - \Delta_E$ where $\Delta_E = \{p \in \overline{V}^+ : p^0 < E\}$.

(v)

$$\int | \sigma(C_1 \alpha_{\mathbf{x}} C_2) | \, d^3x < \infty. \qquad \text{(VI.2.19)}$$

[1] This follows from a slight generalization of theorem 2.2.11 of Chapter III (Alaoglu's theorem). The natural topology on $\mathcal{C}$ is given by the family of seminorms $\{q_E\}$. $\overline{\mathcal{C}}$ denotes the closure of $\mathcal{C}$ in this topology. Since E is arbitrary the fact that q_E is only a seminorm is not relevant.

Comments. The claims (i) and (ii) are evident from the definition of σ. For the invariance of σ under time translations the averaging over time in the definition (VI.2.16) is used. For the proof of the continuity properties we refer to [Buch 91b]. Concerning (iv) one observes that $\int e^{-ipx}\alpha_x C d^4x$ transfers an energy-momentum p and therefore, if $\omega \in \mathfrak{S}_E$,

$$\int e^{-ipx}\omega(C_1\alpha_x C_2)d^4x = 0 \quad \text{unless} \quad p + \Delta_E \subset \overline{V}^+. \tag{VI.2.20}$$

One has to show then that (VI.2.20) survives the passage from ω to σ.

Most significant is (v). The estimate (VI.2.19) relates to the intuitively expected property (VI.1.13). It means that σ is singly localized and, since it is also stationary, it is permanently singly localized. Thus it may rightfully be called a "*single particle weight*" (an improper single particle state). The proof of (v) follows from a judicious application of proposition 2.2.2 and locality.

The next step is to use σ for the GNS-construction of a representation of $\mathfrak{A}$. Since σ is a positive linear form over $\mathcal{C}$ one may follow the standard GNS-procedure to obtain a Hilbert space $\mathcal{H}_\sigma$, regarding $\mathcal{C}$ as a linear space equipped with the scalar product

$$\langle C_2 \mid C_1\rangle = \sigma(C_2^* C_1), \tag{VI.2.21}$$

dividing out the Gelfand ideal and completing in the topology provided by the norm $\| \, |C\rangle \, \| = (\sigma(C^*C))^{1/2}$. The facts that σ is not a state but a weight and that $\mathcal{C}$ is not closed in the norm topology of $\mathfrak{A}$ do not affect this construction. Furthermore, since $\mathcal{C}$ is a left modul of $\mathfrak{A}$ we can obtain a representation of $\mathfrak{A}$ acting in $\mathcal{H}_\sigma$ by

$$\pi_\sigma(A) \mid C\rangle = \mid AC\rangle. \tag{VI.2.22}$$

One checks that π_σ is consistently defined by (VI.2.22) and that

$$\| \pi_\sigma(A) \| \leq \| A \| . \tag{VI.2.23}$$

The representation π_σ should not be considered as exotic. If $\mathfrak{S}_E$ satisfies the compactness criterion in the sense of Fredenhagen and Hertel then π_σ is locally normal i.e. in restriction to the algebra of a finite region it is quasiequivalent to π_ω where ω is the state from which σ is obtained by the limit of (VI.2.16). We may also assume that the Hilbert space is separable. Only there is no vector representative of σ in $\mathcal{H}_\sigma$, since σ is not a state.

The translation invariance of σ implies that one can also implement the translations by unitary operators $U_\sigma^{(0)}(x)$ defined by

$$U_\sigma^{(0)}(x) \mid C\rangle = \mid \alpha_x C\rangle, \tag{VI.2.24}$$

and find, due to item (iv) of proposition 2.2.3 that for $\sigma \in \mathcal{W}_E$ the spectrum of (the generators of) $U_\sigma^{(0)}(x)$ is contained in $\overline{V}^+ - \Delta_E$. We have added the upper index (0) to U_σ to indicate that this is not the "canonical" implementation of the translations in π_σ. We can shift the spectrum by multiplying $U_\sigma^{(0)}(x)$ with a factor e^{iqx} and, due to the known lower bound, achieve that the spectrum is moved inside $\overline{V}^+$ choosing a sufficiently large positive time-like q. Thus we can apply theorem 2.2.1 and obtain a unique canonical implementation, denoted by $U_\sigma(x)$, with the properties listed in the theorem.

We look now at the decomposition of the representation π_σ or, alternatively put, at the decomposition of weights in $\mathcal{W}$. $\mathcal{W}$ is a convex set of translationally invariant weights, and σ will, in general be in the interior, i.e. it will be a mixture. Any orthogonal family of projectors in the commutant $\pi_\sigma(\mathfrak{A})'$ will decompose the Hilbert space into subspaces invariant under $\pi_\sigma(\mathfrak{A})$ and $U_\sigma(x)$ because the latter are in $\pi_\sigma(\mathfrak{A})''$ due to theorem 2.2.1. Using the projectors in a maximal Abelian subalgebra of the commutant we get a decomposition of π_σ into irreducible representations and, correspondingly, a decomposition of σ into *pure single particle weights* σ_z

$$\sigma = \int \sigma_z \, d\mu(z). \tag{VI.2.25}$$

We have used z as the label distinguishing the different pure single particle weights arising from a finest decomposition of limit points of (VI.2.16).

Proposition 2.2.4
Let σ_z be a pure single particle weight. Then

(i) the generated representation π_z of $\mathfrak{A}$ is an irreducible, positive energy, locally normal representation;

(ii) the canonical implementation of the translations is given by

$$U_z(x) \mid C\rangle_z = e^{ipx} \mid \alpha_x C\rangle_z, \tag{VI.2.26}$$

where the 4-momentum $p \in \overline{V}^+$ is uniquely determined by z.

Comments. Claim (i) is rather evident from the above discussion. For claim (ii) we note that, since σ_z retains the translation invariance (VI.2.18) of σ we obtain an implementation $U_z^{(0)}(x)$ just as in (VI.2.24). Furthermore $U_z(x)U_z^{(0)}(x)^{-1}$ must commute with $\pi_z(\mathfrak{A})$ since both choices of unitaries implement the translations. Since π_z is irreducible this quotient must be a 1-dimensional representation of the translation group i.e. it is e^{ipx} for some 4-vector p which lies in the closed forward cone due to the positive energy property.

One may note that for $\sigma \in \mathcal{W}$ and $C \in \mathcal{C}$

$$\omega_{\sigma,C}(A) \equiv \langle C \mid \pi_\sigma(A) \mid C\rangle \tag{VI.2.27}$$

is a (proper) state over $\mathfrak{A}$ and that the weight σ corresponds to the improper vector $|\mathbb{1}\rangle$, (an unbounded linear form on $\mathcal{H}$). The momentum p occurring in (VI.2.26) is, in physical terms, the sharp momentum of the pure single particle weight σ_z which one may regard as an idealized eigenstate of the momentum space operators P^μ to spectral value p just like a plane wave in quantum mechanics.

From proposition 2.2.3, item (iii), one conludes that if σ is a pure single particle weight to momentum p then σ_Λ is one to momentum Λp. Two possibilities exist then. Either the representations π_σ and π_{σ_Λ} are equivalent or they are disjoint. In the first case one has an ordinary particle and can form wave packets by coherent superposition $\int \varphi(p) \mid p\rangle d\mu(p)$ where $d\mu(p)$ is the Lorentz invariant measure concentrated on the mass hyperboloid, φ is a wave function and $\mid p\rangle$ denotes the improper GNS-vector $\mid \mathbb{1}\rangle$. This wave packet is then a normalizable state vector in $\mathcal{H}$. In the second case we have an infraparticle. The velocity $\mathbf{p}/p^0$ is a superselection quantity. No normalizable single particle state of this type exists. The Lorentz symmetry is spontaneously broken.

Special consideration must be given to the Lorentz transformations which leave p unchanged, the "little group" or "stability group" $\mathcal{G}_p$ of p. For $p^2 > 0$ this group is isomorphic to the 3-dimensional rotation group. Then there are (almost local) elements $C \in \mathcal{C}$ which are invariant under $\mathcal{G}_p$. Under reasonable assumptions (see below) it follows that this symmetry is not spontaneously broken i.e. there is a projective unitary representation implementing $\mathcal{G}_p$ in $\mathcal{H}_z$. This allows an adaptation of Wigner's arguments described in Chapter I, section 3, showing that a pure, massive single particle weight may be chosen to have sharp integer or half integer value of the spin. The same argument may be applied in the case of mass zero to the 1-parametric subgroup of $\mathcal{G}_p$ given by the rotations around the direction of the spatial momentum. This leads to a sharp helicity of such pure, massless single particle weights. However it can no longer be inferred that this helicity is restricted to integer or half integer values. Summing up: The particle content of the theory is given by $\mathcal{W}$, the non vanishing limit points of (VI.2.16) as $t \to \infty$ for all states of bounded energy. To each extremal element of $\mathcal{W}$ (pure single particle weight) there is a sharp momentum $p \in \overline{V}^+$ and we may restrict attention to those having a definite spin (resp. helicity). We may consider the Lorentz invariant set $\{\sigma_\Lambda\}$ arising from a *pure* single particle weight σ by Lorentz transformations as corresponding to a particle type. The Lorentz symmetry may be spontaneously broken in π_σ. Then the above set contains the incoherent pure weights of an infraparticle.

One would like to show, using again some version of the compactness or nuclearity assumption, that there is only a finite number of distinct particle types for fixed mass. This is the "reasonable assumption" mentioned above. A proof of this does not yet exist.

VI.2.3 Further Remarks on the Particle Picture and Its Completeness

The strategy described at the end of subsection 1.2 and the part of it carried through to date (section 2.2) give the most clear cut definition of what we mean by the particle content of the theory from the point of view of observables, taking into account all complications arising from superselection rules, infrared problems, long range correlations. It will still require some hard work to close the remaining gaps in this program but there is little doubt that this can be achieved and will result in a good understanding of how various structural properties of the net $\{\mathfrak{A}(\mathcal{O})\}$ relate to different aspects of the particle picture.

Leaving aside, for the moment, the question of convergence of (VI.1.15) or (VI.2.16) the simple and most important qualitative conclusion is the following. If one considers the partial state in a region $K_R(x)$ (the diamond with radius R and center x) then, for any state $\omega \in \mathfrak{S}_E$, as x moves to time-like infinity $x = (t, \mathbf{v}t)$ with $| \mathbf{v} | \leq 1$ the state will look with overwhelming probability like the vacuum state in this region, $\| (\omega - \omega_0) |_{(K_R(x))} \| \to 0$ as $| t | \to \infty$ Furthermore the deviation from the vacuum, registered by the elements $C(x)$ with $C \in \mathcal{C}$, will, at least for almost all $\mathbf{v}$ with $| \mathbf{v} | < 1$, decrease like $| t |^{-3}$. For large times $| t |^3 \omega(C(x))$ will look like (the restriction of) a single particle weight in the region.

The convergence of

$$t^3 \int \omega \left(C(t, \mathbf{v}t)\right) h(\mathbf{v}) d^3v \qquad \text{(VI.2.28)}$$

when the support of h is in the interior of the ball $| \mathbf{v} | \leq 1$ concerns the rescattering problem: can a bounded number of particles in unbounded space rescatter infinitely often or will there emerge ultimately a configuration where no further encounter is possible? This problem has been studied extensively in non-relativistic quantum mechanics (see. e.g. [Enss 83], [Mour 79], [*Sigal 1983*], [Graf 90] and references given there). In purely massive quantum field theory satisfying the Wightman axioms Bros and Iagolnitzer [Bros 88], [Iag 87] have studied its relation to the analytic structure of n-point functions.

Let us try to describe the main intuitive ingredient of this problem. If we have a state which at time t describes n far separated particles of definite mass localized in respective neighborhoods of the points $\mathbf{x}_k$ then the probability for a collision at time t' around a space point $\mathbf{y}$ will be determined by the momentum space probabilities in each of the single particle states around the values $\mathbf{p}_k$ which are calculated classically from the needed velocities $\mathbf{v}_k = (\mathbf{y} - \mathbf{x}_k)/(t' - t)$ and the masses of the particles. Only if there is a significant probability of these momenta for at least two particles we can expect a collision at $(t', \mathbf{y})$. Of course this argument neglects the interparticle forces but, if $| \mathbf{x}_k - \mathbf{x}_l |$ and $t' - t$ are large and the forces not of too long range then the modification is not significant. In a collision process with center $(t', \mathbf{y})$ one has conservation of energy-momentum. This limits the possible spatial configurations of the reaction products at subsequent times. Extrapolating from this argument, once we know that a state ω of bounded energy is n-fold localized at time t with large distances between the individual localization centers then we can hope to decompose ω approximately into a convex combination (mixture) of components ω_λ where each λ corresponds to a subsequent *history* (or fate) during a (large) time interval $[t, t + \tau]$, each history being characterized by a finite set of space-time points y_i marking the centers of events (collision centers during the interval) together with causal ties between events (a tie corresponding to a particle with momentum geometrically determined by the space-time points it connects and its mass). In other words, a history in the time interval can be pictured like a Feynman graph with the vertices marking points in space-time, the lines momenta, with momentum conservation at every vertex but with the further restriction that the mass value of each momentum must be that of a stable physical particle. This reduces the problem of convergence to a classical one (a billard ball problem in which the balls may fracture or recombine) and to which the method of Hunziker [Hunz 68] can be applied, showing that even as $\tau \to \infty$ only a finite number of events are possible. In a purely massive quantum field theory this picture is supported by Symanzik's structure analysis mentioned in section 2.5 of Chapter II. Within the scope of renormalized perturbation theory in quantum electrodynamics the problem of "asymptotic completeness" (completeness of the particle picture) has recently been studied by Steinmann [Steinm 91] but the comparison of his results with the geometric picture remains to be done.

VI.3 The Physical State Space of Quantum Electrodynamics

One consequence of the possible appearance of infrared clouds is that in quantum electrodynamics there are innumerably many sectors satisfying the Borchers selection criterion even for states with the same charge. It is then convenient to make a coarser distinction and combine all sectors which coincide in their information about the content of outgoing (or incoming) *massive* particles in their respective folia of states into one class. Since electrically charged particles are massive such a class fixes in particular the electric charge and for this reason it was called a *charge class* in [Buch 82b]. We shall keep this terminology here though these classes still give a much finer distinction than that provided by the charge quantum numbers.

As in Chapter IV we consider only states with vanishing energy density at space-like infinity. This is guaranteed by restricting attention to "positive energy representations" in the sense of subsection 2.2. Let $(V^+ + a)$ denote the open positive light cone with apex a in position space. A massive particle will ultimately (as $t \to \infty$) enter inside this cone, no matter how the apex a is chosen since it moves slower than light. Similarly, an incoming massive particle can be detected in any cone $V^- + a$. Thus, if π is an arbitrary positive energy representation of the net of observable algebras, all information about outgoing massive particles in its folium of states can be obtained from the part $\pi(\mathfrak{A}(V^+ + a))''$ and the information about incoming massive particles from $\pi(\mathfrak{A}(V^- + a))''$ for any a. Let us denote the centers of these algebras by $\mathfrak{Z}_a^+$, $\mathfrak{Z}_a^-$, respectively and the center of the total algebra by $\mathfrak{Z}$, Thus

$$\mathfrak{Z}_a^{\pm} = \pi(\mathfrak{A}(V^{\pm} + a))'' \cap \pi(\mathfrak{A}(V^{\pm} + a))'; \quad \mathfrak{Z} = \pi(\mathfrak{A})'' \cap \pi(\mathfrak{A})'. \qquad \text{(VI.3.1)}$$

One finds

Theorem 3.1 [Buch 82b]

(i) Every element $Z \in \mathfrak{Z}_a^+$ is invariant under translations

$$Z(x) = U(x)ZU(x)^{-1} = Z. \tag{VI.3.2}$$

(ii) $\mathfrak{Z}_a^+$ is independent of a. We write

$$\mathfrak{Z}_a^+ = \mathfrak{Z}^+. \tag{VI.3.3}$$

(iii)

$$\mathfrak{Z}^+ \subset \mathfrak{Z}. \tag{VI.3.4}$$

The same statements hold for the elements of $\mathfrak{Z}_a^-$.

Proof. Part (ii) follows immediately from (i), part (iii) almost immediately: For $A \in \mathfrak{A}(\mathcal{O})$, $\mathcal{O} \subset V^+ + a$ and $Z \in \mathfrak{Z}_a^+$ we get from (i)

$$[Z(x), \pi(A)] = [Z, \pi(A)] = 0.$$

Thus, for any x, also $[Z, \pi(\mathfrak{A}(\mathcal{O} + x))] = 0$ and, since the algebras $\pi(\mathfrak{A}(\mathcal{O} + x))$ generate $\pi(\mathfrak{A})$, Z commutes with $\pi(\mathfrak{A})$. So $Z \in \mathfrak{Z}$.

It remains to prove part (i). Let n be any positive time-like 4-vector. We show that $U(tn)\mathfrak{Z}_a^+U(tn)^{-1}$ commutes with $\mathfrak{Z}_a^+$ for any real t. For t positive $(V^+ + a + tn) \subset (V^+ + a)$. For negative times $(V^+ + a + tn) \supset (V^+ + a)$ which implies $U(tn)\pi(\mathfrak{A}(V^+ + a))'U(tn)^{-1} \subset \pi(\mathfrak{A}(V^+ + a))'$. So by the definition of $\mathfrak{Z}_a^+$ we have the claimed commutativity for both positive and negative t. Replacing a by $a + t'n$ we see that the algebra

$$\mathfrak{C} = \vee_t\, U(tn)\mathfrak{Z}_a^+U(tn)^{-1}$$

is Abelian. $\mathfrak{C}$ is also stable under the 1-parametric translation group $U(tn)$ which has a generator with positive spectrum. So theorem 2.2.1 can be applied to $\mathfrak{C}$, telling us that apart from $U(tn)$ there is another implementation of these translations by unitaries $V(tn)$ which belong to $\mathfrak{C}''$. Since $\mathfrak{C}$ is Abelian each element of this algebra is invariant under translations by tn and the same holds a forteriori for the elements of $\mathfrak{Z}_a^+$. By choosing different directions of n we can generate the whole translation group and thus verify the claim (i).

The same argument applies, of course, if we replace the future directed cones V^+ by past directed cones V^-. So the theorem holds also if $\mathfrak{Z}^+$ is replaced by $\mathfrak{Z}^-$. □

Given a general positive energy representation π we may decompose it with respect to the part of the center $\mathfrak{Z}^+ \vee \mathfrak{Z}^-$. This leads to representations which are primary on each $\mathfrak{A}(V^+ + a)$ and on each $\mathfrak{A}(V^- + a)$ and in which the translations are still implementable (with positive energy). Since we expect that $\mathfrak{Z}$ is larger than $\mathfrak{Z}^+ \vee \mathfrak{Z}^-$ such a representation may still be decomposed into primary representations of the total algebra. However, under reasonable assumptions this further decomposition will lead to representations whose restrictions to $\mathfrak{A}(V^+ + a)$ and to $\mathfrak{A}(V^- + a)$ are equivalent and therefore do not differ in any superselection quantum number relating to massive particles. For this reasons the term "charge class" was introduced for the set of representations contained in a (positive energy) representation in which $\mathfrak{Z}^+ \vee \mathfrak{Z}^-$ is trivial.

Asymptotic Photon Fields. What distinguishes different irreducible representations of the total algebra within one charge class? There are the different infrared clouds of incoming or outgoing photons. To describe them one would like to define asymptotic creation and annihilation operators for photons. They have been constructed at least for certain classes of representations by a suitable modification of the LSZ-procedure described in subsection 3.3 of Chapter II. For the representation generated from the vacuum state this is treated in [Buch 77b]. The photon is no infraparticle; therefore one will have a subspace $\mathcal{H}^1$ belonging to the sharp mass zero describing the states of a single photon in the vacuum sector. Picking a smooth, but otherwise arbitrary element A of the observable algebra $\mathfrak{A}$ one considers angular averages of translates of A in positive light-like directions

$$A_t = -2t \int (\partial_0 A)(t, t\mathbf{e}) \sin \vartheta \, d\vartheta d\varphi \qquad \text{(VI.3.5)}$$

where $\mathbf{e}$ is a unit vector in 3-space with polar angles ϑ, φ. Applying this to the vacuum state vector Ω one gets

$$\pi_0(A_t) \mid \Omega\rangle = \mid \mathbf{P} \mid^{-1} \left(e^{it(H-|\mathbf{P}|)} - e^{it(H+|\mathbf{P}|)}\right) HA \mid \Omega\rangle. \qquad \text{(VI.3.6)}$$

Here $\mathbf{P}$ denotes the 3-momentum, H the Hamiltonian. Averaging A_t over a time interval $\beta(t)$ which grows slowly to infinity as $t \to \infty$ (such that $t^{-1}\beta(t) \to 0$) we put

$$\overline{A}_t = \beta(t)^{-1} \int_t^{t+\beta(t)} A_{t'} dt', \qquad \text{(VI.3.7)}$$

and find that $\pi_0(\overline{A}_t) \mid \Omega\rangle$ converges strongly

$$\pi_0(\overline{A}_t) \mid \Omega\rangle \to P_1 \pi_0(A) \mid \Omega\rangle, \qquad \text{(VI.3.8)}$$

where P_1 is the projector on the 1-photon subspace $\mathcal{H}^1$. From the space-time support properties ob $\overline{A}_t$ and the Reeh-Schlieder theorem it follows then that $\pi_0(\overline{A}_t)$ converges strongly on a dense domain in $\mathcal{H}$. It defines an operator

$$A^{\text{out}} = \lim_{t\to\infty} \pi_0(\overline{A}_t). \qquad \text{(VI.3.9)}$$

The energy-momentum transfer of A^{out} is confined to light-like vectors. One can use these operators to construct a Wightman field $F^{\text{out}}_{\mu\nu}$ which behaves in every respect like a free Maxwell field. This field defines a net $\mathfrak{F}^{\text{out}}(\mathcal{O})$ of local algebras. Of course, though $F^{\text{out}}_{\mu\nu}(x)$ is formally a local field it should not be interpreted as an (idealized) local observable at the point x. It has this significance only for asymptotic observations. (See fig. II.3.1 in Chapter II). Still there remain some geometric relationships between the net $\mathcal{F}^{\text{out}}$ and the net of local observables. By construction we have

$$\mathcal{F}^{\text{out}}(V^+) \subset \pi_0(\mathfrak{A}(V^+))''. \qquad \text{(VI.3.10)}$$

Furthermore, if $x \in V^-$ and n is a positive light-like vector then $x+tn$ becomes space-like to any point in V^+ for sufficiently large t. Since $F^{\rm out}_{\mu\nu}(x)$ propagates only in light-like directions ("Huygen's principle") we have

$$\mathcal{F}^{\rm out}(V^-) \subset \pi_0(\mathfrak{A}(V^+))'. \tag{VI.3.11}$$

The latter relation can be strengthened to

$$\mathcal{F}^{\rm out}(V^-) = \pi_0(\mathfrak{A}(V^+))'. \tag{VI.3.12}$$

So one has

$$\left(\pi_0(\mathfrak{A}(V^+)) \vee \mathcal{F}^{\rm out}(V^-)\right)'' = \mathfrak{B}(\mathcal{H}). \tag{VI.3.13}$$

The missing information in $\pi_0(\mathfrak{A}(V^+))''$ is supplied by the radiation field $\mathcal{F}^{\rm out}$. For the detailed arguments and proofs we refer to [Buch 82b] where it is also shown how to construct different representations of $\mathfrak{A}$ in each charge class, called "infrared minimal" because their folia contain a state which coincides with the vacuum in restriction to a chosen space-like cone (compare section 3 of Chapter IV). For these one can construct radiation fields $F^{\rm in}$ and $F^{\rm out}$. It should, however, not be inferred that these infrared minimal charge classes exhaust all possibilities or even that they contain all physical states which might be of interest. Kraus, Polley and Reents [Kraus 77] have described other states, obtained from an "infravacuum", which have some advantages for the treatment of the infrared problem arising in collision theory of charged particles. They contain a sufficiently chaotic background radiation field so that the additional infrared photons, produced in the bremsstrahlung can be considered as a small perturbation which does not change the equivalence class of the representation.

Let us now consider the asymptotic observables for charged particles. According to the discussion in the last section they should reduce to the algebra generated by the momenta and spins of these particles. If we disregard spin they form an Abelian algebra in contrast to the asymptotic observables for uncharged particles which are generated by the non-Abelian algebra of free fields $\Phi^{\rm out}$ associated to the various particle types (compare [Fröh 79b]). The values of the particle momenta are not fixed by the label ξ of the class, but a change in asymptotic momentum must be accompanied by a change in the radiation field.

Finally, let us look once more at the implementation $U(x)$ of the translations in a representation of the observable algebra in which $\mathcal{F}^{\rm out}$ can be defined. (e.g. an infrared minimal representation). Since $\mathcal{F}^{\rm out}$ is stable under translations there exist, by theorem 2.2.1, unitary operators $V^{\rm out}(x) \in \mathcal{F}^{\rm out\,''}$ implementing the translation group on $\mathcal{F}^{\rm out}$. We may split $U(x)$ into

$$U(x) = V^{\rm out}(x) U^{\rm out}_M(x), \tag{VI.3.14}$$

with $U^{\rm out}_M \in \mathcal{F}^{\rm out\,'}$ and interpret $U^{\rm out}_M$ as the translations acting only on the massive outgoing particles. Then one finds that in the spectrum of $U^{\rm out}_M$ the mass values of all massive particles (charged or uncharged) appear as discrete eigenvalues (see [Fröh 79b], [Buch 82b]) whereas in the spectrum of $U(x)$ the charged particle masses appear only as lower boundaries.

VII. Principles and Lessons of Quantum Physics. A Review of Interpretations, Mathematical Formalism, and Perspectives

Introduction

In the previous chapters we used the standard language of quantum theory. We have extensively used the terms "observable" in the sense of Dirac's book of 1930 and "state" in the sense of von Neumann's book of 1932. This language implies an essential resignation as compared to the "classical" ideal of the scope of physics prevailing before the advent of quantum mechanics. Eminent physicists, among them Planck, Einstein, and Schrödinger did not accept this resignation as unavoidable. Over 60 years have passed since the great debates. We have witnessed fundamental new discoveries, an enormous growth of knowledge, and the development of the theory from non-relativistic quantum mechanics to relativistic quantum field theories with applications ranging from laboratory experiments in high energy physics to cosmology. Yet, with a few minor modifications and changes in emphasis the standard language remained viable and no alternative of comparable usefulness has been created. Nevertheless some measure of dissatisfaction or at least an uneasiness that something is missing in the standard picture persisted throughout the years and is even more acutely felt today. The large amount of literature devoted to the interpretation of quantum mechanics bears witness to this. It therefore seems useful to review the main arguments advanced during this long-lasting dispute, trying to sharpen the points at issue, and to assess the strengths and weaknesses of various points. This will be attempted in section VII.1. Another matter is the specific mathematical structure of quantum physics. Much work has been devoted to understanding the rôle of Hilbert space and complex algebras from some deeper-lying operational principles. A brief sketch of such endeavors will be given in section VII.2. Finally, in section VII.3, I want to present a point of view which I call the "evolutionary picture". It replaces "measurement result" by the more general notion of "event" and may be regarded either as a different idealization from the one implied by the Bohr–Heisenberg cut or, more ambitiously, as part of the conceptual structure of a wider theory to be developed.

The road to disharmony is paved by efforts to clarify misunderstandings

VII.1 The Copenhagen Spirit. Criticisms, Elaborations

We must bear in mind that the mathematical formalism of quantum mechanics and its first remarkable successes in physical applications existed prior to the philosophical discussions concerning the interpretation and proper understanding of the lessons. The point of view resulting from the intense struggle with this problem in the circle of Niels Bohr was called by Heisenberg the "Copenhagen Spirit", presumably to contrast it with the position held by Einstein. The first and most obvious fundamental lesson appeared to be the *indeterminacy of the laws of nature* (originally called acausal behavior of atomic objects). This was seen as an inescapable consequence of phenomena like radioactive decay, quantum jumps between stationary states of an atom, and diffraction experiments. In the last of these an individual electron or photon produces a sharply localized effect (e.g., a dot on a photographic plate) whereas the distribution of these effects shows an interference pattern that depends on the experimental arrangement in a way which cannot be reconciled with the assumption that each particle follows a specific trajectory in space, but which is easily described in a wave picture. In fact, it was this contrast between localization of effects and spreading of waves which led Max Born to the probability interpretation of the Schroedinger wave function. Let us define *indeterminacy* as the claim that the optimal attainable knowledge of the past does not enable us to predict future behavior with certainty. This leaves open the question of whether God knows the future, a question that lies outside the scope of physics. After the formulation of Heisenberg's uncertainty relations, the closer analysis of indeterminacy by the discussion of many thought experiments led to a much deeper negative statement: it is not possible to assume that an electron has, at a particular instant of time, any position in space; in other words, the concept of position at a given time is not a meaningful attribute of the electron. Rather, "position" is an attribute of the interaction between the electron and a suitable detection device. More generally, a phenomenon which we observe does not reveal a property of the "atomic object". The phenomenon is created in the act of detection. This fundamental point is somewhat veiled in the standard language. If we say that some "observable" of a quantum object is measured this suggests that there is some corresponding property of the object which may have different numerical values in the individual case and the purpose of the experiment is to determine this value. By contrast Bohr stresses that the measurement result is a property of the compound system of measuring device plus object and that the full description of the experimental arrangement is an essential part of the definition of the phenomenon. Thus, Max Born's probability for the "position of a particle" within some region of space should be understood as the *probability of an effect* if a detector is placed in this region.

Niels Bohr's Epistemological Considerations. "We must be able to tell our friends what we have done and what we have learned". This key sentence of Bohr alludes to several facts. First, our knowledge about the physical world derives from observation; in physics observations usually involve planned experiments. The description of such an experiment must be given in unequivocal terms; the stated conditions and the results must be reproducible by others. Bohr concludes that, no matter how sophisticated and abstract the technical language of physics may become, we may not transcend the language of classical physics in the *description* of the experiment. Why "classical"? Planck's quantum of action introduces a discrete element into physics which implies some discontinuity. In particular, if we want to control or verify the specified conditions this will involve some interference which will change the conditions slightly in an uncontrollable way due to the indeterminacy expressed in the uncertainty relations. Therefore Bohr argues that the acquisition of knowledge about an atomic object demands a dichotomy. On the one hand there is the experimental arrangement which must be described in classical terms. On the other, there is the atomic object. The observed phenomena depend on both. They give only indirect information about the object and we cannot speak about the behavior of the object independently of the means of observation.

Generalizing this dichotomy Heisenberg argued that we cannot avoid introducing a (notional) cut between the "physical system" we want to study and the means of observation. The cut may be shifted but not eliminated. This precludes in particular the consideration of the whole universe as a physical system in the sense of the formalism of quantum theory (see the remarks on page 3 of this book). Bohr asks us to remember "that in the drama of nature we are both actors and spectators". This fitted well with Bohr's deepest conviction, the principle of *complementarity*: there are many aspects of nature. We may study one by a particular experimental arrangement. This precludes the use of another arrangement which would be needed to study another aspect. The more precisely we want to grasp one aspect the more the other escapes us. Therefore it is folly to try to catch reality by one complete and precise mathematical counterfeit. I remember a conversation in 1953 in which Bohr expressed his dismay at my juvenile criticism saying: "Of course you can change the mathematics. But this will not affect the essential lessons we learned". Still, [illegible]e accept the general lesson there remains the question of how to proceed from these qualitative considerations to a quantitative prediction for a specific experiment. How do we translate the description of an experimental arrangement into mathematical symbols? For this Bohr relied on the *correspondence principle* which establishes a close connection between models in classical mechanics and quantum mechanics. With the enormous change in the scope of problems to which quantum physics addresses itself today, this question must be seriously reconsidered. Perhaps the word "classical" has been overemphasized both in the epistemological analysis and in the reliance on the correspondence principle. We shall return to this below.

Criticisms and Elaborations. One of the points of Einstein's criticism (not the most relevant one) concerned indeterminacy. He was deeply convinced that the fundamental laws of nature cannot be probabilistic ("God does not play at dice"). One cannot argue about convictions or the ways of God. The opposite conviction, namely that strictly deterministic laws, reducing the world to clockwork, would be a nightmare and cannot possibly be the last word, appears to me more attractive. Of course, metaphysical beliefs – one way or another – do not provide a basis for discussions of physical theory. In judging the merits of attempts to overcome the apparent indeterminacy within the realm of physics by the assumption of hidden degrees of freedom and quantum forces, the criteria must be either the ability of the proposal to suggest finer experiments or its convincing naturalness, explaining a wide range of phenomena with few assumptions. The material addressed must at least include spin correlations, the exclusion principle, particle transmutations, and, ultimately, also the more subtle effects treated in quantum field theory. The existing proposals seem to be far from meeting these demands and do not look promising to me. Therefore I shall not discuss this criticism further but accept indeterminacy as a feature of the laws of nature.

The deeper problem which disturbed Einstein concerns the question of "reality". In past centuries physicists proceeded from the assumption of a real outside world which exists separate from our consciousness; the task of physics was precisely to describe this world, to discover the "laws of nature" which governed it. Quite a different picture was drawn by philosophers for whom the mind–body problem was a central issue. There was the recognition of the primacy of consciousness with conclusions ranging from the extreme idealism of Berkeley to Kant's distinction between "things as such" about which we can know nothing and "things of appearance" which are partly shaped by a priori given faculties of our mind.

Prior to the advent of quantum mechanics, the ideas of positivism, demanding the elimination of all concepts in science not directly related to observations, had exerted some influence on the attitude of physicists. It was natural that positivists should consider the Copenhagen Spirit as a corroboration of their doctrine. Niels Bohr did not share this view. On the one hand the abstract mathematics of quantum mechanics involves concepts far removed from direct perception. But one should also note an interesting aspect of Bohr's view on complementarity: he saw it as a generalization of the relativity principle, in the sense that different observers can catch different aspects of reality, depending now on their choice rather than their position; the totality of these aspects which may be "symbolically" united by the mathematical formalism constitutes a full picture (of nature). An analogy might be the use of charts in the description of a manifold.

The discussions on the interpretation of quantum mechanics raised doubts about the separability of the laws of physics from processes in the human mind and led to a broad spectrum of opinions.

It should be understood that our discussion here cannot concern the merits of different philosophical positions but only the question of whether the discoveries in the atomic and subatomic regime *force* us to reject the assumption that physics is dealing with a real outside world whose laws can be stated without reference to human consciousness. In other words, does our inability in "ascribing conventional physical attributes to atomic objects" and the "impossibility of any sharp separation between the behavior of atomic objects and the interaction with the measuring instruments which serve to define the conditions under which the phenomena appear" [Bohr 49] imply that we have to include into physical considerations mental aspects like consciousness, intelligence in the planning of an experiment, and the free will in deciding on its execution? I think that this can be answered by a clear no.[1] The raw material of quantum physics which the theory tries to order consists of facts which can be documented, such as dots on a photographic plate. The (classically described) experimental arrangement is also a "physical system", not endowed with consciousness; the total experiment can be automatically registered and documented as a computer print-out. Of course it is logically possible to say that only the last step in the observation procedure, namely the studying of the print-out by a human observer with subsequent consciousness establishes a fact. But since the quantum mechanical uncertainties are practically negligible in this step such a position cannot help in the solution of the riddles with which the fathers of quantum mechanics were confronted. It may be objected that in the formalism of quantum mechanics the notion of "fact" does not appear and is replaced by "measurement result". Indeed one may claim that there are no precisely definable facts, no absolutely reliable documents. But this is not remedied by replacing the word "fact" by "measurement result" and shifting the burden to the judgment of consciousness. The unreliability of the memory of impressions far exceeds that of a registered document. Rather, to the extent to which the objection is relevant, it may indicate that a physical theory can never reach an absolutely precise picture of nature and that at each step of the

[1] In his "Remarks on the mind–body question" [Wig 63] Wigner writes: "If one formulates the laws of quantum mechanics in terms of probabilities of impressions, these are *ipso facto* the primary concepts with which one deals" and: "The principal argument is that thought processes and consciousness are the primary concepts, that our knowledge of the external world is the content of our consciousness and that consciousness therefore cannot be denied". To avoid misunderstanding my above "no" I should stress that I fully agree with the second sentence and do not dispute the logic of the first. What I want to assert is that in resolving the problems implied by the inabilities and impossibilities emphasized by Bohr it is neither necessary nor even helpful to include consciousness in the description of quantum physical laws.

development certain idealizations or, if one wishes, asymptotic notions have to be used.

Stepping down from lofty generalities to a practical example, let us consider a typical experiment in high energy physics. It yields a computer record of many individual runs, which is subsequently studied by various teams of experts. Hopefully they will agree that one such document may be described in the words: A positron and an electron in the storage ring collided and produced a multitude of particles, concentrated mainly in two narrow jets (event number 1). These jets contained a $B, \bar{B}$-pair, identified by subsequent decays $B \rightarrow D \rightarrow K \rightarrow \ldots$ (events $2, 3, \ldots$) and corresponding decays of $\bar{B}$. If we trust the experts we must accept, as raw material for the theory, that not only the computer records produced by electric discharges in macroscopic detectors but also the mentioned subatomic events are facts. The account is told in a way which can be understood by our friends. The appearance of words unknown some decades ago and not in the vocabulary of classical physics does not affect the essence of Bohr's analysis but may be a warning against taking some formulations too literally. In conclusion I believe that the phenomena to which quantum theory applies may be appropriately described in an "*as if*" realism, where facts, whether observed or not, are assumed to exist and constitute the cornerstones of the theory. It should not be taken for granted that the notion of "fact" is synonymous with "macroscopic change". The definition of what constitutes a fact may involve idealizations depending on the regime considered and the precision demanded, and it may change with the development of physical knowledge. But I do not see any essential conflict between quantum physics and that part of Einstein's creed which he poetically expressed in the words: "Out yonder there was this huge world, which exists independently of us human beings and which stands before us like a great eternal riddle, at least partially accessible to our inspection."

We must now come back to the important issue of our inability to "ascribe conventional physical attributes to atomic objects". First we must ask what we mean by an atomic object or, more generally, by a "physical system". The relevance of this question is brought into sharp focus in quantum field theory. From previous chapters of this book it is evidently not obvious how to achieve a division of the world into parts to which one can assign individuality. We have moved far away from Maxwell's ode on atoms: "Though in the course of ages catastrophes have occurred ... the foundation stones of the material universe remain unbroken and unworn. They continue this day as they were created – perfect in number and measure and weight."[2] Instead we used a division according to regions in space-time. This leads in general to open systems. Under special circumstances we can come from there to the materially defined systems of quantum mechanics, claiming for instance that in some large region of space-time we have precisely an electron and a proton whose ties to the rest of the world may be neglected. One of the essential elements in singling out such a material system and assigning to it an individual, independent (at least temporary)

[2]Quoted from the book "Subtle is the Lord" by A. Pais.

existence is its *isolation*, i.e., the requirement that in a large neighborhood we have a vacuum-like situation. However this is not always enough. Suppose that in the course of time the electron and the proton, originally close together, become widely separated, still remaining isolated. Then we might want to consider them as separate individuals and wish to assign to each of them some notion of "state" (some attribute). The formalism of quantum mechanics, supported by experiments, tells us that this is not always possible. There may be persistent correlations of a non-classical character (in the experiments concerned with the EPR-phenomenon they are spin-correlations). It should be noted that persistent correlations occur also in classical mechanics; there they do not put the assumption of individual existence of the partners in doubt. The conservation of angular momentum in the classical mechanics of a two-particle system implies that if we know the 3-vector of total angular momentum at one time then the measurement of the angular momentum of one particle much later tells us exactly what the angular momentum of the other particle must be, no matter how far apart the particles are then. In quantum theory, however, the angular momentum is not a numerical 3-vector. The two-particle wave function belonging to angular momentum zero is not a product of single particle wave functions but a linear combination of such – which, moreover, can be chosen in different ways. Thus even if we envisage "unconventional" attributes for each particle, such as wave functions, this will not suffice to describe the situation. In other words, in spite of the large separation of the particles they cannot be regarded as individuals in any realistic sense as long as there remain correlations of this kind, conserving a common property due to some past encounter. This indicates that in the endeavor to divide the world into parts, the "facts" (e.g., events marking approximately some space-time point) are more basic than "quantum objects". The latter notion corresponds to causal links between the events. We shall return to this in section VII.3.

Can quantum theory stand on its own feet or does it need some props involving a knowledge of classical theory? There are several facets to this question. For instance: given the formalism of quantum electrodynamics (supplemented by nuclear physics) together with the interpretation of the basic symbols in terms of local observables and the general rules relating to states, observables, and probabilities, can we derive the existence of states describing the behavior of billiard balls, electric power generators, etc.? Can we retrieve the body of knowledge collected in the text books of classical physics and show that it results as an approximation from the given quantum theory? This is a tough task but the strategy is clear. One has to go over to a "coarse-grained description" defining suitable "collective coordinates" which can be related to classical quantities. Then one must show that for these quantities one obtains an adequate approximation to the previously known deterministic laws. As an instructive example in quantum mechanics Hepp and Lieb [Hepp 73] have treated the case of a pendulum regarded as an essentially rigid body consisting of many atoms. There is one collective coordinate, the angle θ between the pendulum and the vertical. If one starts from a quantum state in which this angle and its angular velocity are prescribed (which is approximately possible because of the large mass) and the internal state of the solid is the ground state, then during the motion there will be a transfer of energy from the macroscopic energy of the collective motion to the internal degrees of freedom and it may be shown that, due to the large number of atoms, the resulting time dependence of θ follows the deterministic equations of motion of a damped harmonic oscillator. Moreover, if one starts with a state which is the coherent superposition of states corresponding to different initial values of θ, then the phase relation is very soon lost by the interaction with the internal degrees of freedom. More generally, for "almost" any initial state the so-called reduced density matrix for the motion of θ (i.e., the description disregarding the situation of individual atoms) becomes diagonal in the θ-representation after a very short time. So, with respect to θ any ensemble can be considered practically as a classical distribution of harmonic oscillators with different initial conditions. There remain no interference effects. This *decoherence* of collective degrees of freedom appears to be a rather general feature. It is essential in establishing that the quantum theory allows us to recover approximately classical behavior under suitable circumstances. A more elaborate discussion of the derivation of classical properties from quantum theory and of decoherence may be found in the book by Omnès [*Omnès 1994*] where references are also given to other work.

What can one learn from this concerning the problem of measurement in quantum theory? A measuring instrument for "quantum objects" is a very special system. Typically it has several stable or metastable states, close together in energy but differing strongly in their macroscopic appearance. Treating the measurement as a collision process between the object and such an instrument we may distinguish three different types of degrees of freedom: firstly, those relating to the object, secondly, the macroscopic (collective) variables of the instrument and, finally its internal degrees of freedom. One may assume that initially one has a product state $\varrho_1 \otimes \varrho_2 \otimes \varrho_3$ where ϱ_3 could be the ground state or equilibrium state at some temperature of the internal degrees of freedom, ϱ_2 is adequately characterized by some numerical values of the macroscopic variables which we denote summarily by α_0. Believing in the effectiveness of decoherence the final state will be a mixture of states, each of which belongs to some value α of the macroscopic variables (the measurement results). Thus it defines a probability $p(\alpha; \varrho_1)$ for the occurrence of the result α which depends, as indicated, on the initial state of the object. If the object survives we shall furthermore have a correlation between α and the final state of the object, which, if we are not interested in the final state of the internal motion, can be described by a density matrix ϱ_1' depending in general on the measurement result α and the initial state of the object ϱ_1. Up to here we have only used the arguments described above. Now, for the ideal maximal measurements axiomatized by Dirac and von Neumann we should find that ϱ_1' does not depend on ϱ_1. In its dependence on α it should assign to each α a pure state from some family determined by the instrument. This will be the case if there exists a special orthogonal family of initial states $|\psi_n\rangle$ for which the interaction process with the instrument leads to a strict (deterministic) correlation between n and the measurement result α. In the mathematical formalism this family corresponds to the eigenstates of the observable A representing the instrument. We refer here to the ideal case where the measuring device is assumed to define a self-adjoint operator acting in the Hilbert space of the pure states of the object. Denoting by α_n the result attached with certainty to the initial state $|\psi_n\rangle$ one obtains for a general pure initial state $\psi = \sum c_n\psi_n$ the probability $p(\alpha_n;\psi) = |c_n|^2$ and also the prediction that the final state of the object which is coupled to the result α_n is just ψ_n. For a realistic instrument we cannot expect such a sharp correspondence between α and a specific initial and final state of the object. But there are good instruments for which this is at least approximately true. For them the final state of the object is sharply fixed by the measuring result α whereas the information from the initial state is largely forgotten in the subensemble associated with a fixed value of α. It enters only in the fraction of systems in this subensemble which is given by the probability $p(\alpha;\psi)$. Focusing future attention on one such subensemble, say on the one characterized by α_n, and ignoring the rest, we have an apparently discontinuous change of the state of the system from ψ to ψ_n which is often called the "reduction" or "collapse" of the wave function. It is not due to any physical effect but to our method of book keeping, in going over from the original ensemble to the conditional probabilities in a subensemble associated with a specific measurement result. Of course the realization of a measurement result as a *fact* and the possibility of using it for the characterization of a subensemble of objects for subsequent study is not fully reconcilable with the quantum theoretic description of the dynamics of the total system of object plus apparatus. In the above discussion it is attributed to the decoherence effect and to the additional assumption that in each individual case one value of α must be realized, i.e., it becomes a fact.

We may ask whether there is a generic reason for the special feature of "good instruments". Looking at typical detectors like a photographic plate or a bubble chamber we note that the macroscopic change consists in a local change in some small part of it. The distinction between measurement results is given by the position at which this change occurs. Then we can understand the existence of a preferred family of initial states of the object. A wave packet which is sharply localized at the time of arrival can produce an effect only in the neighborhood of the place where it meets the detector; the final state of the object after detection will again be a reasonably sharply localized wave packet emanating from the region where the effect was produced. Perhaps we may take this as an indication that in the measurement problem we should distinguish between the primary phenomenon due to some irreversible event to which we can assign a reasonably well-defined region in space-time and the accessories in the experimental arrangement (like external fields, shields ...) which serve to define the possibility for such events.

This has a bearing on several other questions. First it suggests that not every self-adjoint operator can be realized by a measuring instrument. Secondly, though position and momentum variables appear on an equal footing in the formalism of quantum mechanics, there is a hierarchy with respect to measurement procedure. A momentum measurement ultimately also demands the detection of localized effects, irrespective of whether we use the relation of momentum to velocity, or to wave number, or appeal to its conservation laws. So the primary phenomena are again localized events. The precision depends on the control of the situation in a large region of space-time.

Finally we must come back to the correspondence principle. If we mean by "correspondence" merely that in the realm of electrodynamics and mechanics we can retrieve the laws of the classical theory as an approximation valid for states with high quantum numbers then we do not leave the topic of the previous discussion. There is, however, also the amazing parallelism of the mathematical structure of the theories (Poisson brackets corresponding to commutators, conservation of the form of equations of motion). This has led to the notion of "quantization" of a classical model as a well-defined procedure, regarded as forming in some way an indispensable background of the quantum theory. Related to this there is the problem mentioned in connection with the epistemological considerations of Bohr: How do we achieve the translation between the description of an experimental arrangement and the mathematical symbols of quantum theory? Is it necessary for this purpose to know a "corresponding" classical model theory? Looking at modern models developed in high energy physics such as quantum chromodynamics, it is true that they do have a corresponding classical counterpart, namely the classical field theory with the same form of the Lagrangean. But this is not of much help for the translation between the description of experimental hardware and the mathematical symbols of the theory and it is questionable whether such a formal correspondence is a relevant aspect in a fully developed quantum theory. However, even in this regime there remains one classical feature which is used in an essential way: the assumption of a given space-time continuum. In the preceding chapters we tried to demonstrate how this last classical relic can be used in developing a detailed interpretation of phenomena. The situation is not essentially changed if we go over to curved space-time, as long as the gravitational field can be treated as a "classical" background field. In quantum gravity, however, this last relic is lost and this leads to conceptual problems which are unsolved today.

Summarizing the previous discussion, it appears that the quantum theory of a specific regime in physics (general formalism, specification of the degrees of freedom, their commutation relations, equations of motion and their relation to space-time) constitutes a physical theory which is complete and satisfactory in almost all respects. There remains only one feature which needs additional careful consideration. The statistical predictions are verified by the study of many individual cases. Each of them yields a measurement result, regarded as the transition from a possibility to a fact. If we do not want to attribute this transition to a change in the consciousness of an observer, then we must say that the realization of individual facts out of a large number of possibilities is a basic feature in nature which is not explained by decoherence but has to be either separately postulated or explained in a picture of nature using new concepts (see section VII.3). In this we should, however, keep in mind that quantum theory is not only concerned with statistical predictions but also with precise statements about structural properties, for instance the internal wave function of an atom in its ground state which, in contrast to the wave function of the center of mass motion, may be directly regarded as an element of reality.

The Time Reflection Asymmetry of the Statistical Conclusions in Quantum Theory. The standard formulas for probabilities in quantum theory, as given for instance in the books by Dirac or von Neumann, refer to a given "arrow of time". They are not symmetric with respect to time reversal. Let us illustrate this by a simple argument. Suppose in some ensemble we measure successively two maximal observables with discrete spectra A_1 and A_2. The times of measurement shall be t_1 and t_2 with $t_2 > t_1$. We use the Heisenberg picture and denote the orthonormal systems of eigenstates respectively by ψ_k, φ_α. The number of cases in which the results k *and* α are registered shall be $N_{k\alpha}$. The ensemble consists of N individuals. We can look at the conditional probabilities in the sample

$$p_{k\to\alpha} \equiv N_{k\alpha}/N_k\,; \quad N_k = \sum_\alpha N_{k\alpha}\,, \tag{VII.1.1}$$

and

$$p_{\alpha\to k} \equiv N_{k\alpha}/N_\alpha\,; \quad N_\alpha = \sum_k N_{k\alpha}\,. \tag{VII.1.2}$$

Equation (VII.1.1) is the relative frequency (probability) that, given the result k in the first measurement, we observe α in the second. Equation (VII.1.2) gives the probability that, if α is observed in the second measurement, it followed result k in the first. Now the standard formula in quantum mechanics says that

$$p_{k\to\alpha} = |\langle\varphi_\alpha \mid \psi_k\rangle|^2\,, \tag{VII.1.3}$$

irrespective of the ensemble studied. If this is true then

$$p_{\alpha\to k} = p_{k\to\alpha} N_k/N_\alpha$$

depends on the ensemble. In other words, we need more information if we want to make probability statements about a past situation.

In using the term "ensemble" we must, of course, recognize that this does not mean an arbitrary collection of individuals. The common feature which is shared by all the individuals of which it is composed, the criterion used in admitting an individual as a member of it, must be so clearly stated that it enables "our friends" to check any statistical regularities we claim to have found in it. Also we shall not speak of an ensemble unless there exist such regularities. In other words, given the criteria defining the ensemble there should be probabilities for specific occurrences which can be verified. The time asymmetry in the basic probability formula results from the claim that it is possible to define an ensemble entirely by specifications preceding the first measurement. In an indeterministic theory it is then impossible to define an ensemble entirely by "post-selective" specifications. In Bohr's discussion the time asymmetry appears as obvious. For instance: "The irreversible amplification effects on which the registration of the existence of atomic objects depends reminds us of the essential irreversibility inherent in the very concept of observation" [Bohr 58]. We might also speak about *the essential irreversibility inherent in the notion of fact.* The above-mentioned asymmetry is not due to the "psychological arrow of time" nor to any interdiction of statistical statements about the past. The possibility of registration of sequences of events in a document whose analysis is not believed to produce any changes allows us to include some "post-selection criteria" in the definition of an ensemble as discussed by Aharonov, Bergmann and Lebowitz [Ahar 64]. We can enhance the situation considered at the beginning by adding measurements A at time $t < t_1$ and B at time $t' > t_2$, study the statistics of the A_1, A_2 measurements in an ensemble selected by having yielded the value "a" in the A-measurement and "b" in the B-measurement. Since the first specification already defines an ensemble, the additional requirement of b just selects a subensemble. The probability of obtaining the results k, α given a and b can be written

$$p(k, \alpha; a, b) = N^{-1} w(a, k, \alpha, b)\,; \quad N = \sum_{k', \alpha'} w(a, k', \alpha', b)\,, \qquad \text{(VII.1.4a)}$$

$$w(a, k, \alpha, b) = tr\, P_b P_\alpha P_k P_a P_k P_\alpha P_b\,, \qquad \text{(VII.1.4b)}$$

where the P's are 1-dimensional projectors corresponding to the indicated properties and the sum over k', a' runs over a complete orthonormal family of eigenstates of A_1, A_2. Due to the invariance of the trace under cyclic permutations we have

$$w(a, k, \alpha, b) = w(b, \alpha, k, a)\,; \quad p(k, \alpha; a, b) = p(\alpha, k; b, a)\,. \qquad \text{(VII.1.5)}$$

Anyone who believes in the dogma that all basic laws of nature must be invariant under time reversal may rejoice here. They should, however, recall that this formula applies only to very specially defined ensembles excluding most applications in quantum theory and that it is of rather modest value for the purpose of obtaining retrodictive probability statements for quantum events. An intriguing view of "quantum past" has been described by Dyson [Dy 92].

VII.2 The Mathematical Formalism

In the first chapters of his book (1930 edition) Dirac erects an edifice which, purely on the basis of architectural beauty, could rank among the great masterpieces of human creations. No traces of the dirt in the previous discussions of measurements remain visible. It may be appropriate to quote some lines: "We introduce some symbols which we say denote physical things such as states of a system or dynamic variables. These symbols we shall use in algebraic analysis in accordance with certain axioms which will be laid down. To complete the theory we require laws by which any physical conditions may be expressed by equations between the symbols and by which, conversely, physical results may be inferred from equations between the symbols." The central place among the axioms is held by the superposition principle, abstracted from interference phenomena but carefully shifted to the realm of symbols for states and deprived of direct operational interpretation. It leads to the consideration of complex linear spaces, ultimately Hilbert spaces. The symbols for dynamic variables, considered by themselves, obey the axioms of abstract *-algebras. In relation to the state symbols they appear as linear operators acting on Hilbert space. Many authors in later years have tried to analyse whether these axioms can be replaced by principles with more direct operational significance, and how they are embedded in a wider setting, etc. Such endeavors are sometimes classified as "uninteresting for physics". But, apart from their obvious relation to many interesting mathematical questions, we must also consider the possibility that even the basic conceptual-mathematical structure of quantum physics as we know it may be superseded by a more general theory. It is then relevant to understand the cross connections between various elements of the theory. Of course it is quite impossible to present here any comprehensive review. We can only indicate a few lines of approach, mention some results, and add a few remarks.

Focusing on the statistical aspect we may start from the general notion of "state" as representing an ensemble of individual systems, characterized by some selection criterion which suffices to guarantee statistical regularities in subsequent observations. Within the set of such "states of a system" we have the operationally defined process of mixing which implies that this set is a convex body. As described in section I.1, this may be used to define a real linear space V with a distinguished positive cone $V^{(+)}$ and a distinguished linear form e (an element of the dual space) which defines a norm on the elements of $V^{(+)}$ by

$$\|\omega\| = (e, \omega)\,; \quad \omega \in V^{(+)}. \tag{VII.2.1}$$

Since $V^{(+)}$ is a pointed cone, every element of V can be written (in many ways) as

$$\varphi = \omega_1 - \omega_2\,; \quad \varphi \in V,\ \omega_i \in V^{(+)}. \tag{VII.2.2}$$

One can then define the norm of φ by $\|\varphi\| = \inf(\|\omega_1\| + \|\omega_2\|)$ where the infimum is taken over all possible decompositions. Thereby V becomes a real Banach space. Due to its special construction it is called a "base norm space".

A convex cone is characterized by its facial structure. A face $F \subset V^{(+)}$ is a convex subcone stable under purification, i.e., if a normalized state $\omega \in F$ is a mixture as in (I.1.11) then ω_1 and ω_2 also shall belong to F. A pure state is an extremal point of the convex body, i.e., one for which no non-trivial decomposition of the form (I.1.11) is possible. The ray defined by a pure state is a 1-dimensional face.

Assumption 1. There exist pure states (extremals with respect to the convex structure). Any state can be obtained (in many ways) as a mixture of pure states:

$$\omega = \sum \lambda_i \alpha_i ; \quad \lambda_i > 0, \ \sum \lambda_i = 1 . \tag{VII.2.3}$$

We use symbols α, β generically to denote pure, normalized states. In (VII.2.3) ω is a normalized general state.

Among the measurements which can be made on a state (on each individual of the ensemble) we consider the "propositions" (in Mackey's terminology "questions") which can only have two alternative outcomes "yes" or "no". Applying a proposition E to a state ω we get a non-negative number $p(E;\omega)$, the probability of the yes-answer. To proceed with the analysis we need

Assumption 2. Any proposition E can be applied to any state ω.

This allows us to consider a proposition E as a linear form over V, an element of the dual space V^* (in fact of the positive cone $V^{*(+)}$) and write

$$p(E;\omega) = (E, \omega) . \tag{VII.2.4}$$

Comment. There may be good reasons to believe that this assumption cannot hold in nature (see section VIII.1). One should be aware that a serious restriction here will demand a departure from the standard mathematics of quantum theory.

In the set of propositions we have a natural partial ordering. We may call $E_2 \leq E_1$ if $(E_2, \omega) \leq (E_1, \omega)$ for all $\omega \in V^{(+)}$. The special element $e \in V^{(+)}$ may be regarded as the trivial proposition for which every state gives the yes-answer. It is the largest element in the mentioned partial ordering. Next we make

Assumption 3. There exist finest propositions ("atoms" which are minimal with respect to the mentioned partial ordering). Every non-minimal proposition contains some finest proposition below it.

Comment. Assumptions 1 and 3 correspond to the "type I" situation (compare sections III.2 and V.2). This is adequate for quantum mechanics; finer topological considerations need not concern us here.

While the previous assumptions (though by no means harmless) are of a fairly general nature, the next one, establishing a close relation between V and V^*, is one of the very specific pillars of the architecture.

Assumption 4. There is a one-to-one correspondence between pure states and finest propositions. Denoting the proposition corresponding to the pure state α by $g(\alpha)$ one has

$$(g(\alpha), \alpha) = 1 , \tag{VII.2.5}$$

i.e., in the ensemble α the finest proposition $g(\alpha)$ yields with certainty the answer yes. Further, the probability function is symmetric

$$(g(\alpha), \beta) = (g(\beta), \alpha). \qquad \text{(VII.2.6)}$$

As a natural generalization we add:
There is a one-to-one correspondence between the faces of $V^{(+)}$ and propositions. For the general proposition E the corresponding face F_E consists of all states which give with certainty the yes-answer to the proposition E. This implies that to every face F there is a complementary face F' corresponding to proposition $e - E$.

Consequences.

1) Due to (VII.2.2) and assumption 1 one can extend the map $\alpha \to g(\alpha)$ to a linear map from V into V^*. This, together with (VII.2.6), implies that we have a real-valued scalar product in V defined by

$$\langle \varphi_1 \mid \varphi_2 \rangle = (g(\varphi_1), \varphi_2) = \langle \varphi_2 \mid \varphi_1 \rangle ; \quad \varphi_i \in V . \qquad \text{(VII.2.7)}$$

This scalar product is non-negative between states and vanishes identically if the states belong to complementary faces. Thus one may say that $V^{(+)}$ is a self-polar cone in the real linear space V, equipped with the scalar product (VII.2.7).

2) The set of faces is an orthocomplemented lattice. Since any subset of states generates a face by convex combinations and purification, we can define the face $F_1 \vee F_2$ as the face generated by the union of the states contained in F_1 and F_2 and $F_1 \wedge F_2$ by the intersection. $F \to F'$ gives an orthocomplementation:

$$(F_1 \vee F_2)' = F_1' \wedge F_2' . \qquad \text{(VII.2.8)}$$

Thus by assumption 4 the set of propositions is also an orthocomplemented lattice.

Following Birkhoff and von Neumann [Birk 36] the operations $\wedge, \vee, '$ are interpreted as corresponding to the logical operations of "and", "or", and negation. This structure in the set of propositions has accordingly been called a quantum logic. The question is then which additional "axioms" are needed to show that the lattice of propositions is isomorphic to the set of orthogonal projectors of a Hilbert space. The required properties of the lattice are "semi-modularity" and "orthomodularity". Various choices of postulates leading to this have been given. See, e.g., [Pir 64], [Jauch 69]. The relation of orthocomplemented modular lattices to projective geometry and quantum mechanics is elaborated in [*Varadarajan 1968*]. A generalization of the standard algebraic structure, proposed in [Jord 34] led to "Jordan algebras", a topic which attracted considerable mathematical interest. For a review see, e.g., [*Braun and Koecher 1966*].

Comments. The relation (VII.2.6) expresses a symmetry between "state preparing instruments" and "analyzing instruments" and is thus related to time-reversal invariance. In Dirac's notation it is visible in the use of two different types of symbols for pure states, the "bras" and the "kets". Of course (VII.2.6) is the absolute square of the scalar product in Hilbert space.

Instead of looking for additional axioms concerning the lattice of propositions, one may classify the possible structures of self-polar cones. This has been a subject of considerable interest in pure mathematics (for a review see, e.g., [*Iochum 1984*]). To translate the problem into algebraic language one considers the set of linear transformations of V which transform $V^{(+)}$ into itself. They have a physical counterpart, namely the operations by physical instruments which do not destroy all systems in the ensemble but leave a final ensemble after the interaction of the instrument with the members of the initial ensemble. In short, they produce a change of state. The reason why such operations are represented by linear transformations of V is that they will conserve the weights in convex combinations: $T\sum \lambda_i\omega_i = \sum \lambda_i T\omega_i$ if $\lambda_i > 0$, $\sum \lambda_i = 1$, $\|\omega_i\| = 1$, which may be naturally extended to non-normalized states by $Ta\omega = aT\omega$. Parallel to the decomposition theory of von Neumann algebras one has here a "central decomposition" of V into primary components. We shall not address here the fine topological distinctions which arise in the case of infinite-dimensional V but focus only on the algebraic situation for finite-dimensional subspaces V_n of V whose intersection with $V^{(+)}$ gives a face $V_n^{(+)}$ generated by a finite number n of pure states. Correspondingly we restrict attention to the operations transforming $V_n^{(+)}$ into itself. Then the essential property needed is the "homogeneity", i.e., the feature that any interior point of $V_n^{(+)}$ can be reached from any other such point by an operation. For homogeneous, self-polar, finite-dimensional cones there is a representation theorem (see, e.g., [Vin 65]) stating that its faces are in one-to-one correspondence with the subspaces of a Hilbert space over the field of either real numbers or complex numbers or quaternions; in particular the 1-dimensional subspaces (the rays) in this Hilbert space correspond to the pure states. The difference between the three alternatives shows up in the faces generated by two pure states. In the complex case V_2 is 4-dimensional. The scalar product (VII.2.7) in it is the Lorentz scalar product, so V_2 is isomorphic to Minkowski space and $V_2^{(+)}$ to the forward light cone. The description in terms of a 2-dimensional complex Hilbert space for the pure states is given by the spinorial representation of the Lorentz group. This surprising parallel between the simplest non-trivial faces in quantum mechanics and relativistic space-time has induced speculations about a deeper connection, such as von Weizsäcker's notion of "Ur" (see, e.g., [*von Weizsäcker 1988*] and the ideas of Finkelstein on space-time code [Fink68]. Unfortunately this has so far not led to a viable theory. For the quaternionic case see, e.g., [Fink 62]. For an argument in favor of the complex case see [Ara 80].

Among the vast literature devoted to the questions discussed in this section let me mention [*Ludwig 1987*], [*Alfsen 1971*], [*Alfsen and Shultz 1976*], [Alf

78, 80]; [Miel 69, 74] (theory of filters); [Foul 60], [Pool 68a, 68b] (Baer *-semigroups), [Conn74], [Belli 78] (derivations defined by faces).

Final Comments. It may be worthwhile to draw attention to two aspects. First, the appearance of the *linear* space V is very general and essentially only due to the statistical setting. The possibility of considering mixed ensembles leads to the convex structure of state space which is extended to a linear structure in an embedding space. Thus this type of linearity is not due to any approximation to a nonlinear structure and has no deep physical basis (apart from the existence of ensembles with statistical regularities). The structure of V and V^* arising from the assumptions 1–4 is obviously of the nature of an idealization, distinguished by demands for simplicity and symmetry and justified by its success in a wide field of applications. In spite of its striking beauty, emphasized by Dirac, we cannot expect that this specific structure will remain the framework of physical theories for ever.

VII.3 The Evolutionary Picture

We maintained above that the findings of physics (including quantum physics) can and should be stated in the picture of an "as if" realism, that physics by its very method does not transcend the dividing line between an assumed real outside world and the world of the mind with its impressions, emotions, ... The surprising behavior of "atomic objects" shows that the "elements of reality" cannot be those in which classical physics believed. What can they be? Let us start with quantum *mechanics* where we assume that it is possible to define a single electron as a "physical system". One says that we can – if we want – measure the position of the electron at a chosen time. Then one adds that it would lead to a host of difficulties to believe that the electron has a definite position when it is not measured. If we look without bias at the experiments alluded to, we must recognize that the "element of reality" in the position measurement is a dot on a photographic plate or a flash from a scintillation screen. These are indeed not properties of the electron but properties of the interaction process of the electron with another system. They are "events" characterized by a reasonably sharp position in space *and in time.* Other experiments showing interference effects, etc., indicate that we should not believe that the electron by itself has any position at a given time. The natural conclusion is that "position" is not an attribute of the electron whereas space-time position is an attribute of a certain class of events. What about the electron itself? We experience it as a causal tie or link between two events, its "birth" in the electron source and its death (or transmutation) in the interaction with the detector. We can distinguish different types of causal links, the simplest ones being stable particles. Avoiding the poorly defined distinction between elementary particles and composite ones we use the term "particle" for any stable, essentially rigid, structure (nucleus, atom, crystal, ...). It corresponds to the "permanently singly localized states" of Chapter VI. A causal link also has attributes which we can regard as real. They are structural properties. In the case of the electron they are for us today only its mass, electric charge, magnitude of spin, and magnetic moment. For the next generation of high energy physicists they may include some internal structure. We know empirically that stable particles can be isolated and that each type has a specific internal structure which can be regarded as a true attribute, an "element of reality". In fact it is one of the central achievements of quantum theory that it can determine the possible structures of stable bodies. The ground state wave function in the center of mass system provides this information in quantum mechanics. On the other hand, the wave function for the center of mass *motion* should not be considered as an attribute of an individual nor as an element of reality. It refers to probability predictions for the occurrence of future events.

Let us anticipate some objections. First, why focus on stable particles? This relates to the "division problem" mentioned earlier. To isolate some part of the universe which can be considered as independent of the rest we must resort to idealizations. In order to recognize simple laws we begin with the simplest situations: a system of a few particles isolated in a large volume for a long time. We know that such an idealized situation can be realized to a good approximation though never perfectly (we cannot reach absolute zero temperature or provide perfect shielding, etc.). One may, of course, also consider an "open system". But then one has to specify the boundary conditions defining its interaction with the rest of the world and this is obviously a much less simple task. Let us look at the idealized situation where we have initially just two clearly separated particles, an electron and a crystal. Experience tells us that after a possible collision we find a final situation which can be described in each individual case as a system of several particles, say n_1 photons, n_2 electrons, and a residual crystal, possibly electrically charged. Furthermore, we can attribute to each such particle an approximate momentum. The theory reproduces these findings in the following way. It cannot describe individual processes but treats a statistical ensemble of them (irrespective of whether we claim to know a pure state wave function or only a statistical matrix). If we start with a pure state then the final state will again be described by a wave function. It is a sum of terms of which each corresponds to one of the final situations described above. This wave function contains in addition the phase relations between the various terms in the sum. But we also know (from the theory) that this becomes unobservable in principle in the idealization in which "final" refers to the limit $t \to \infty$. Thus, effectively, the final ensemble is a mixture of subensembles, each corresponding to one of the situations encountered experimentally in an individual case. The theory suggests, however, that for finite times after the collision there should be possible experiments which determine the phase relation between different "outgoing channels" or different momentum configurations. One may argue then that the individual process cannot be considered as completed at finite times. But then one must look seriously at the question of how the measurement of the phase relation in the outgoing wave function, say one microsecond after the collision, can be realized. Except in the case of existence of metastable particles with comparable lifetime, the interference between different final channels has never been considered and we may wait patiently for the proposal of such an experiment. In the case of long-lived metastable states, the processes of formation and decay should be regarded as distinct events. To exhibit interference between different momentum configurations in one channel one needs equipment which is very extended in space with precisely controlled external fields. The presence of such equipment conflicts (almost by definition) with the assumed isolation of the originally defined system in a sufficiently large, prescribed region of space-time. This reminds us of Bohr's demand for a full description of the experimental arrangement before we speak of phenomena. However, there is a change of emphasis. The relevant feature needed is the idealization involved in separating an individual system from the rest of the world. In the regime to which local quantum physics applies and limitations due to global circumstances beyond our control may be ignored, there is no limit in principle to the precision with which the definition of such an idealized system can be achieved and it has a precise sense to say that under such circumstances a certain set of possible sequences of real events exists from which precisely one is realized in the individual case.[3]

Another frequently voiced opinion is that macroscopic amplification is essential for the concept of a real event. Indeed, this would be so if we wanted to equate the notion of event with that of a document which we can show to our friends. One might argue that there are no facts unless they can be documented. But this is an unduly rigid position rarely respected by the practising physicist in quest of understanding and expansion of knowledge. What is important is that the facts postulated in a theory fit naturally to documented facts. In that respect the above discussion shows that in the presently available theory the notion of event or fact indeed needs some idealization but that this does not mean complex systems or macroscopic amplification but rather isolation in a sufficiently large space-time volume. Then a collision process between particles (elementary or composite), leading from an initial configuration to a final configuration and fixing some approximate position in space-time, may be regarded as a closed event, a fact. The event can be assumed to exist whether we observe it or not and obviously only few events will be documented. The sharpness of the mentioned space-time position depends, of course, on the nature of the event. For low-energy reactions it will be very fuzzy, occupying a large region; in high-energy reactions we can approach quite sharp localization. We shall return to this aspect later.

So far the only deviation from the orthodox view has been to replace the notion of "measurement result" by the more general notion of "event", which is

[3]This language has some (superficial) similarity with the "consistent history approach" [Griff 84], [*Omnès 1994*], [Gell 90]. But when we talk here about real unregistered events we do not mean an arbitrary choice of sequences of projection operators satisfying the "consistency conditions" with respect to the recorded events, but a very specific set, determined by the properly idealized physical situation.

considered as a fact independent of the presence of an observer. This, however, has important consequences. An event is irreversible. It is the transition from a possibility to a fact. We are raised in the belief that the fundamental laws of physics are invariant under time reversal, that they do not stipulate an arrow of time. What is the evidence which led to this belief, which by now is so firmly entrenched? Could it possibly be a prejudice? There is the manifest invariance under time reflection of the basic dynamic equations in classical physics. This symmetry is somewhat clouded by the question about the appropriate boundary conditions. In usual applications we use the retarded solution of the Maxwell equations ("Sommerfeld's Ausstrahlungsbedingung", radiation damping). This asymmetry arises because we assume that we know the past and not the future. So it may be considered as "anthropomorphic" or, if one wishes, a consequence of the "psychological arrow of time". Another basic piece of experience is the apparent irreversibility of all natural processes, leading to the second law of thermodynamics. It is interesting to note that Planck, as a firm believer in the exact validity of the entropy principle under all circumstances, was motivated in his studies of black-body radiation by the wish to put the irreversibility of radiation processes into evidence. Instead of showing this he discovered the quantum of action and it is slightly ironical that he had to use Boltzmann's probabilistic interpretation of entropy in his arguments. It was thus the success of statistical mechanics in explaining the thermodynamic laws starting from a time-reversal-invariant microscopic theory which convinced most physicists that the apparent irreversibility was a feature of the macroscopic world resulting from more basic laws which did not prefer a direction of time.

Let us recall the essential point in the arguments of Boltzmann and Gibbs leading from reversible microscopic laws to irreversibility in a coarse-grained description. Different macroscopic states have extremely different statistical weights or probabilities, roughly defined as the number of microscopic configurations leading to the same coarse-grained appearance. Among the macrostates – which can be reached from a particular one according to the dynamic law – the equilibrium state has the highest weight. Therefore, starting from a state with low weight (far away from equilibrium) there is an overwhelming probability that it develops into a state of higher weight, i.e., moves closer to equilibrium. So far the argument does not imply any preference for one time direction. It would also be overwhelmingly probable that the earlier state was one of higher weight, i.e., closer to equilibrium. In a laboratory experiment the asymmetry is again introduced by the psychological arrow of time. There is an initial time at which the experimenter creates a state far away from equilibrium because he wants to find something interesting. The various arrows of time are discussed in [*Zeh 1984*]. The "biological arrow" and with it the "psychological arrow" is dominated by the existence of the solar system. We may ask next why the astronomical situation is as it is, why in astrophysics the "thermodynamic arrow" at any place is obtained by continuous transport of the arrow from another place. In the approach of statistical mechanics to entropy production the irreversible Boltzmann equation is derived from reversible microscopic laws by adding some natural assumptions about the initial state. The time-reversed situation, where we apply the corresponding conditions to the final state and calculate backwards in time, is excluded as obviously absurd. As a reason for this asymmetry one can appeal to the cosmological conditions, i.e., the expansion of the universe. One may accept this latter as an unexplained fact and speculate that there might be a phase in which the universe contracts and the thermodynamic arrow turns around. But such questions are clearly beyond the scope of our discussion here.

Our question is rather: why should there be any difference between quantum physics and classical physics with respect to the status of irreversibility? A short answer is that quantum physics introduces an element of discreteness manifested in the existence of stable structures and the "indivisibility of a quantum process". This is closely tied to indeterminacy. The future is open, not precisely determined by the past. Though some remnant of time-reversal symmetry persists in quantum mechanics and quantum field theory, there is the asymmetry of the basic statements discussed at the end of section VII.1. In Bohr's view this "reminds us of the essential irreversibility inherent in the very concept of observation". In other words it is tied to the psychological arrow of time. But if we do not want to place the concept of observation into the center of physics we must ask ourselves: what would be the natural picture if we claim that there are discrete, real events, i.e., random, irreversible choices in nature?

Starting from this question we come almost unavoidably to an evolutionary picture of physics. There is an evolving pattern of events. At any stage the past consists of the part which has been realized, the future is open and allows possibilities for new events. Altogether we have a growing graph or, using another mathematical language, a growing category whose objects are the events and whose (directed) arrows are the causal links. We shall assume further that the relation to space-time is provided by the events. Each event marks roughly a region in space-time, the extension and sharpness of which depends on the nature of the event. No independent localization properties of the links are assumed. This corresponds to the earlier remark that the simplest type of causal link is a particle and that "position" is not a real attribute of a particle. In fact it is essential for the understanding of many apparent paradoxes in quantum physics (including the EPR-effect) that a causal link becomes real only after both the source and the target events have been realized. Before this it remains a "potential link", analogous to a free valence bond searching for a partner. For example, after the source event of emission of a particle we represent it by a roughly spherical wave function. This should not be interpreted as relating to the probability for the changing position of a point-like particle but rather to the probability for the space-time location of the collision center in a subsequent event. Only after the realization of this target event may we (retrospectively) assign an approximate world line and incoming momentum to the particle. Let us assume here for simplicity that customary space-time, in which patterns of events and links can be embedded, has been independently defined. A pattern of events and links prior to a given time is a history.

The quantum laws concern two aspects. On the one hand they must determine the intrinsic structure of links and events (for instance the internal wave functions or structure functions of particles). On the other hand they must give probability laws for the formation of specific patterns, including the positions of collision centers. The formulation of such laws in the evolutionary picture is an unaccomplished task. The existing theory provides some guidance but it is only in the simplest situations (the case of extremely low density of matter) that a reasonably clean definition of events and links is visible and a comparison with the standard formulation has been attempted. Let us sketch a strongly simplified version of this which shows some essential aspects. To each type of link α (here a type of particle) we have an associated Hilbert space $\mathcal{H}_\alpha$ and we may consider all the subsequently mentioned spaces as subspaces of the Fock space generated from the $\mathcal{H}_\alpha$ of all types. Consider for simplicity "maximal" events (corresponding to the strongest possible decisions). They specify their backward links completely. If the event has two backward links of types α and β then it selects a specific product vector $\varphi_\alpha \otimes \varphi_\beta \in \mathcal{H}_\alpha \otimes \mathcal{H}_\beta$ and transforms it to a vector in the tensor product space $\mathcal{H}_\gamma \otimes \mathcal{H}_\delta \otimes \ldots$ corresponding to the outgoing channel[4]. This vector, however, is not a product vector but a linear combination of such. Its expansion into a sum of product vectors depends on a choice of bases in the factor spaces. The selection of a particular product vector is realized only by the subsequent events, since links become established only after both source and target event are realized. A space-like surface not passing through any event defines a "subjective past" consisting of the pattern of all earlier events. Among these events there are saturated ones for which all forward links are absorbed by some other event inside this subjective past and there are others still having free valence links for the formation of future events. To such a subjective past we associate a state which summarizes the probability predictions for possible extensions of the pattern to the future. In our simplified picture the state depends only on the subpattern of the unsaturated past events. As the space-like surface is shifted to the future, the associated state changes as new events appear. This change, analogous to the "reduction of the wave packet", corresponds to the transition from a possibility to a fact. Let us illustrate this in the example of Fig. VII.3.1 in which the wavy line indicates the chosen space-like surface. We are interested in the extension of the past history by the pattern of events 4 and 5 and the newly established links. The temporal order of $1, 2, 3$ is irrelevant but it is assumed that no other events of the past can play a rôle for the events linked to 3. Events $1, 2, 3$ fix unit vectors (not products)

$$\Phi_1 \in \mathcal{H}_\gamma \otimes \mathcal{H}'_1; \quad \Phi_2 \in \mathcal{H}_\delta \otimes \mathcal{H}'_2; \quad \Phi_3 \in \mathcal{H}_\alpha \otimes \mathcal{H}_\beta. \tag{VII.3.1}$$

Events 4 and 5 are represented by the rank-1 operators (in Dirac notation)

$$c\,|\Phi_4\rangle\,\langle\varphi_\alpha \otimes \varphi_\gamma|, \quad c'\,|\Phi_5\rangle\,\langle\varphi_\beta \otimes \varphi_\delta|, \tag{VII.3.2}$$

[4]We have made the further simplifying assumption that the choice of a specific outgoing channel is included in the characteristics of the event.

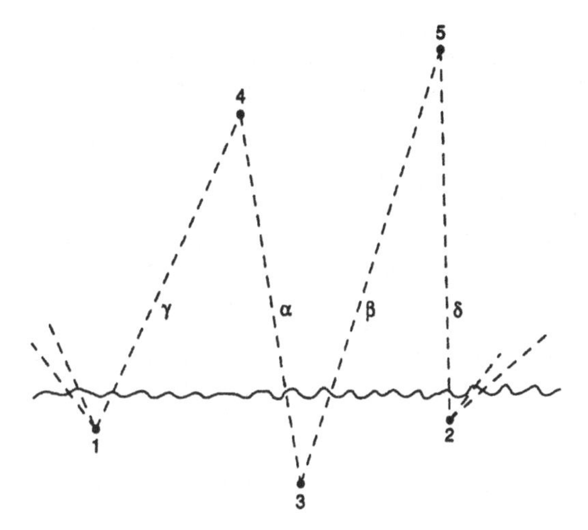

Fig. VII.3.1.

where the φ_λ are specific unit vectors in the subspaces $\mathcal{H}_\lambda$ $(\lambda = \alpha, \beta, \gamma, \delta)$ and Φ_4, Φ_5 unit vectors in the tensor product spaces of the new outgoing channels. The constants c, c', together with the selection of the backward ties, i.e., the vectors $\langle \ldots |$, determine the probability of a single event. Thus the probability of event 4 is obtained by applying the first operator of (VII.3.2) to $\Phi_1 \otimes \Phi_3$. This yields a vector whose squared length gives this probability. To obtain the joint probability of events 4 and 5 we have to apply both operators of (VII.3.2) to $\Phi_1 \otimes \Phi_2 \otimes \Phi_3$ and square the length of the resulting vector. This joint probability shows correlations even though these events may lie space-like to each other. They are due to the fact that the two events have backward causal links to a common source (event 3). Moreover the vector Φ_3, determined by event 3, does not specify a product vector $\varphi_\alpha \otimes \varphi_\beta$ before both events are realized and thus it is not possible to assign individual "states" to the not yet established links. It is this feature which distinguishes the joint probability of events from the case of classical correlations which would result if we could assume an individual state for each link (possibly unknown) and then consider correlations between these states of the links. A prime example is the EPR-phenomenon.

We consider the decay of a spin-zero particle into two spin-$\frac{1}{2}$ particles followed by a measurement of the spin orientation of the two particles with respect to respective directions $\mathbf{e}_1, \mathbf{e}_2$ prescribed by Stern–Gerlach magnets. This may be idealized as the situation pictured in Fig. VII.3.1 where events 1 and 2 correspond to the setting of the Stern–Gerlach magnets, event 3 to the decay process, and events 4 and 5 to the interaction between the decay particles and the two Stern–Gerlach arrangements, each allowing only a binary decision whose results are denoted by $+$ or $-$. Since the events 1 and 2 concern the setting of classical apparatus, the links γ and δ are already fixed by these events and may be characterized by the directions $\mathbf{e}_1, \mathbf{e}_2$. Disregarding the motion in space and focusing only on the spin, the vector Φ_3 is the unique singlet state in the Hilbert space of 2-particle spin states. If 4 is the event with outcome + then φ_α is realized as the single particle state $\varphi_+(\mathbf{e}_1)$ (spin oriented in the $+\mathbf{e}_1$ direction). Since the arrangement is such that we are sure that one of the results + or − must happen, the constants c and c' are equal to 1. The prescription described above for finding the joint probabilities thus leads to the well-known quantum-mechanical expressions. By contrast, suppose we assume that after the event 3 each particle has a definite state, characterized by some variables ξ so that the ensemble may be described by a distribution function $\varrho(\xi_1, \xi_2)$ and there is a probability $p_+(\mathbf{e}, \xi)$ that in the state ξ we get the answer + if the orientation of the magnet is in direction $\mathbf{e}$, then the joint probability of the result ++ in the settings $\mathbf{e}_1, \mathbf{e}_2$ will be

$$w_{++}(\mathbf{e}_1, \mathbf{e}_2) = \int p_+(\mathbf{e}_1, \xi_1) p_+(\mathbf{e}_2, \xi_2) \varrho(\xi_1, \xi_2) d\mu(\xi_1, \xi_2) , \qquad \text{(VII.3.3)}$$

with corresponding expressions if we replace some of the + signs by − signs. We know that the $p_\pm$ and ϱ are non-negative, that $p_+(\mathbf{e}, \xi) + p_-(\mathbf{e}, \xi) = 1$ and, due to the experimentally observed conservation law in each individual case, $w_{++}(\mathbf{e}, \mathbf{e}) = w_{--}(\mathbf{e}, \mathbf{e}) = 0$. Therefore, if ξ, ξ' is a pair for which the distribution function ϱ does not vanish, then

$$p_+(\mathbf{e}, \xi) p_+(\mathbf{e}, \xi') = p_-(\mathbf{e}, \xi) p_-(\mathbf{e}, \xi') = 0 .$$

If neither $p_+(\mathbf{e}, \xi)$ nor $p_-(\mathbf{e}, \xi)$ vanishes, then both p_+ and p_- must vanish for the setting $\mathbf{e}$ at ξ'. This is impossible because their sum must be 1. So we conclude that contributions to (VII.3.3) can come only from states where $p_+(\mathbf{e}_1, \xi_1)$ equals 0 or 1, i.e., where we have a deterministic coupling between the state ξ and the result of the measurement in any direction. This is the case in which ξ can be regarded as a "hidden variable" restituting determinism. It is the case for which Bell derived his famous inequality [Bell 64]. For a simple derivation see [Wig 78]. This inequality disagrees with the quantum-mechanical prediction and, according to common belief, also with the experimental results. Our discussion shows that not only is it impossible to assume hidden variables but that even the assumption of individual states of the two particles after event 3 and before events 4 and 5 is not tenable.

The decision that one possible pattern of events should be realized may be regarded as a free choice of nature, limited only by the probability assignment. The amount of freedom thus accorded to nature is larger than in the standard view where the experimenter forces nature to decide only on the answer to a proposed question. It must be stressed, however, that also in the standard use of quantum-theoretical formalism the element of free choice by nature cannot be eliminated. It is merely pushed into the background by focusing on ensembles instead of individual cases. Thus one may derive from the dynamic law governing the time development of "states" (representing ensembles) that, in the case of complex systems, the density matrix very rapidly becomes effectively diagonal in suitably chosen collective coordinates whatever the initial state may have been. "Effectively" means that in no realistic experiment will the off-diagonal terms play a rôle ("decoherence"). One concludes then that this final ensemble *may be thought of* as a mixture of subensembles in each of which the collective coordinates have specific values. This is perfectly correct as far as statistical predictions for subsequent measurements are concerned. However, it does not explain the fact that in each individual case nature has decided on one specific set of values (e.g., the position of a dot on a photographic plate), a decision not controlled by the experimenter and not described by the time development of the density matrix. A striking example of the ambiguities involved in the step from the statistics of an ensemble to conclusions about individual cases will be discussed below. It is interesting to note that Dirac advanced the idea of a free choice of nature in 1927 at the 5th Solvay Congress. Bohr was not happy with this formulation at that time but used it in his later writings with some reservations.

Comparison with Standard Procedure. Consider the interaction of a cosmic ray μ-meson with a body of superheated water. The phenomenon observed will be a string of bubbles which we attribute to "elementary" quantum processes

$$\mu + \text{atom} \rightarrow \mu + \text{ion} + e + \gamma\,, \tag{VII.3.4}$$

each such process creating a localized disturbance which acts as a germ for vaporization. This verbal description corresponds to the event picture in which we idealize (VII.3.4) as a closed process which can be separated to a good approximation from the subsequent amplification. We would further like to assume that the localization of each bubble is not primarily due to the amplification process but that it is an attribute of the event (VII.3.4). In fact it should be much more sharply definable than the extension of a bubble. What can we learn about this from the conventional treatment of an isolated event of this sort?

We have a Fock space of incoming particles. The initial state is described as a tensor product of two single-particle wave functions of the respective center of mass motion (we treat the atom as a single particle). The final state is described as a vector in Fock space resulting from the application of the S-matrix to this tensor product. It is a sum of terms describing the different channels. We write as usual $S = \mathbb{1} + iT$ and, for a particular final channel (suppressing spin indices)

$$\psi^{\text{out}}(p_1' \dots p_n') = \int \tau(p_1' \dots p_n'; p_1, p_2)\psi_1^{\text{in}}(p_1)\psi_2^{\text{in}}(p_2)\delta^4(\textstyle\sum p_k' - \sum p_k)d\mu(p_1)d\mu(p_2)\,, \tag{VII.3.5}$$

with $d\mu(p_i) = \delta(p_i^2 - m_i^2)\Theta(p_i^0)d^4p$.

Using

$$\delta^4(q) = (2\pi)^{-4}\int e^{ixq}d^4x$$

and noting that $e^{ix\sum p_k'}$ represents in any channel just the space-time translation by x in the Fock space of outgoing particles (similarly $e^{ix\sum p_k}$ for the incoming particles) we may write (VII.3.5) in vector notation as

$$\Psi^{\text{out}} = \int \mathfrak{T}_x \Psi^{\text{in}} d^4x \,,$$

regarding this as a mapping in Fock space where

$$\mathfrak{T}_x = U(x)\mathfrak{T}U^{-1}(x)$$

is the translation by the 4-vector x of an operator $\mathfrak{T}$ whose matrix elements are the functions $\tau(p'_i; p_k)$. The latter are needed only on the subspace of momentum conservation and their extension away from there is arbitrary. They are smooth functions of the momenta except at thresholds. We can choose the extensions so that these functions are smooth (except at thresholds). Then $\mathfrak{T}_x$ is essentially a local operator centered around x. The localization of $\mathfrak{T}_x$ will be poor in the case of long range forces or "weak processes" like soft photon emission and, to a lesser degree, also because of the threshold singularities. Let us leave aside these problems and focus on hard inelastic events. The characteristics of an event include the nature of the backward links, i.e., the charges, mass and spin values, and internal structure of the incoming particles and, although they should not include detailed information about forward links since these are fixed only in subsequent events, we may include in our case the choice of a specific final channel and even some rough specification of the momenta of outgoing particles since this concerns mutually exclusive possibilities, provided the isolation is adequate. We demanded that we should be able to attribute a sharply defined space-time region to the event. This is not yet provided by the sharpness of localization of $\mathfrak{T}_x$ (which corresponds roughly to the range of the interaction) but requires that if we make a cell division in x-space, writing

$$\int \mathfrak{T}_x dx = \sum \mathfrak{T}_k \,; \quad \mathfrak{T}_k = \int \mathfrak{T}_x g_k(x) dx \,; \quad \sum g_k(x) = 1 \,; \quad \Psi_k = \mathfrak{T}_k \Psi^{\text{in}} \,,$$

with the function g_k having support in the cell k, then, for appropriately chosen cell division, the individual terms Ψ_k may be considered as describing (incoherent) alternatives, one of which is selected by nature in the individual case. By contrast, believing in the absolute validity of the quantum theoretical formalism, one concludes that the phase relation of different Ψ_k can be made evident or, in other words, that the needed size of the cells depends strongly on far away circumstances surrounding the process, not only on the event itself (i.e., on the presence of instruments which are far away at the time of the event). To assess the significance of this difference we have to study the statistics of an ensemble of such processes followed by subsequent measurements on the final state. The relevant test experiment is a very precise control of the energy–momentum of all initial and final particles. The assumption of an extension a_ν of the event in the ν-direction implies a limitation in the control of the momentum balance ΔP_ν of order h/a_ν. This raises the question of how precisely the relevant part of past history can be controlled in all samples of the ensemble. Here the following consideration may be instructive. If the overlap region of the wave functions of incoming particles were sufficiently narrow, then only a single term, Ψ_k, would occur. But this is usually not the case. Consider the opposite extreme where we take the initial state of the atom in (VII.3.4) as an equilibrium state in a large vessel so that its position is almost unknown. If β is the inverse temperature the state can be described by a density matrix diagonal in the momentum representation, given (non-relativistically) by

$$\langle \mathbf{p}'|\varrho|\mathbf{p}\rangle = \delta^3(\mathbf{p}'-\mathbf{p})e^{-\beta\mathbf{p}^2/2m}.$$

(we have disregarded the normalization which involves the size of the volume). Now we note that precisely the same density matrix also arises as a mixture of Gaussian wave packets, minimal at some time t, with width

$$\lambda = h(\beta/2m)^{1/2},$$

distributed with uniform density in space and time. Numerically, taking m as the proton mass, this gives at a temperature of 1 K a value $\lambda = 2\times 10^{-7}$ cm. Thus it does not make any difference for the statistics of any subsequent experiment whether we assume that the initial state is built up from plane waves or from localized packets of size λ. The origin of this ambiguity is, of course, the non-uniqueness of the decomposition of an impure state and we see here that we cannot confine attention to decompositions into mutually orthogonal states because we considered mixtures of packets which overlap and are minimal at different times. We are reminded again of the fact that the study of statistics in an ensemble allows widely different pictures for the individual case.

Still, there is no known law of nature which would prevent the control of the momenta of incoming particles and the measurement of the momenta of outgoing particles with arbitrary precision. Such an overall high precision experiment would be, in the standard language, the one complementary to the well-known high-energy experiments where we see by inspection in the individual case the existence of a collision center from which the tracks of particles emerge. The precision in the definition of this collision center may not be great, but it is much sharper than the controlled localization of the incoming particles. Thus here is a question indicating a difference between the evolutionary picture and standard formalism. It should be settled by experiments but this appears at present to be beyond feasibility because we would need precision measurements on all initial and final particles and the precision attainable will presumably be relative to the total momenta involved and therefore not effective for the high energy reactions in which we expect the localization to be sharp.

Let us turn now to patterns of events and links in the low density situation. A link, corresponding to a particle, is mathematically described by an irreducible representation of the total symmetry group which is the direct product of a global gauge group with the Poincaré group. The vectors in this representation space give the charge quantum numbers and a wave function for the center of mass motion and spin orientation. The event is described by a *reducible* representation resulting from the tensor product of the irreducible representations associated with the backward links, followed by "quasiprojection" by an operator $\mathfrak{T}_k$. After the event this representation is decomposed again into a sum of tensor products of irreducible representations, each term corresponding to a specific channel of outgoing particles which furnish possible links to subsequent events. A new event is realized by the fusion (tensor product) of such links originating from different past events. We have been careful so far to speak of representations, not of vectors in the representation spaces. The reason is that, in contrast to the simplified picture described before, $\mathfrak{T}_k$ is not a rank 1 operator and we can only include so much information about backward links as corresponds to the characteristics we can attribute to events. These include the approximate momenta determined retrospectively from the location of the source event. The τ-functions in (VII.3.5) do not factor in the variables of outgoing particles. This means that we cannot attribute a specific single-particle state to a free valence link and this implies in turn that we cannot treat the probabilities of the formation of subsequent patterns as a classical stochastic process. While this complication is not very relevant for position patterns in the case where the mean free path is very large compared to the unsharpness of localization of events so that all momenta can be taken as quite well-defined though unknown, there is no corresponding mechanism providing a specification of the state of spin orientation of the individual particle. This is demonstrated by the experiments concerning the EPR-effect for spin.

Concluding Remarks. The conceptual structure proposed above incorporates the essential message of quantum physics and does not seem to be at odds with known experimental findings. At the present stage it is not clear whether this structure should be regarded only as an idealization suitable in a certain regime of phenomena or whether a fundamental theory based on this picture can be developed. This would demand a more general definition of events and links, in other words a deeper understanding of the "division problem". It might demand a finer division of "decisions of nature", related to the quantum of action rather than to collision processes between stable structures. The relation of events to space-time must be clarified. It is here that some differences from the standard formalism will be manifested. One of the factors in favor of the picture presented is precisely this point. It seems ultimately unsatisfactory to accept space-time as a given arena in which physics has to play. This feature persists even in general relativity where a 4-dimensional space-time continuum is a priori assumed and only its metric structure depends on the physical situation. In particular, in the absence of all matter and all events there would still remain this continuum, void of significance. This aspect was one of the factors that motivated the author to introduce the notion of "event" as a basic concept with the ultimate aim of understanding space-time geometry as the relations between events [Haag 90a]. The other motivation was, of course, the desire to separate the laws of quantum physics from the presence of an observer [Haag 90b]. In this respect it appears that theorists discussing quantum processes inside a star or in the early universe necessarily transcend Bohr's epistemology.

Usually the orthodox interpretation is then silently ignored but there are some efforts to build a rational bridge from the standard formalism to such areas of physical theory, most prominently the work by Gell-Mann and Hartle [Gell 90, 94]. It uses the concept of "consistent histories" introduced by Griffiths [Griff 84] and extended by Omnès [*Omnès 1994*]. One criticism of this concept is that consistent histories embodying some established facts are highly non-unique. This led Omnès to the distinction between "reliable properties" and truth.

Still another motivation comes from the following consideration. The general mathematical structure of standard quantum theory is extremely flexible. Its connection to physical phenomena depends on our ability to translate the description of circumstances (e.g., experimental apparatus) to a specification of operators in Hilbert space. Apart from the case of very simple systems, the success in this endeavor is due to the fact that for most purposes no precise mathematical specification is needed. Thus, for the treatment of collision processes in quantum field theory it suffices to give a division of "all" observables into subsets which relate to specified space-time regions. However, in addition to this classification of observables one uses the postulate of strict relativistic causality. Some consequences of this postulate have been verified by the check of dispersion relations to regions with an extension far below 10^{-13} cm. On the other hand it seems highly unlikely that the construction of an instrument of intrinsic size of, say, 10^{-15} cm and the control of its placement to such an accuracy could be possible even in principle, i.e., that we may assume the existence of such *observables*. But it is not unlikely that we can attribute to high energy *events* a localization of this order of magnitude though we have no means of verifying this in the individual case. Thus the indirect check by means of dispersion relations could be explained by the existence of sharply localized events rather than sharply localized observables.

The realization of a specific result in each individual measurement has been recognized by many authors as a challenge to the theory of measurement which cannot be explained using only the dynamic law of quantum theory applied to the interaction of a quantum system with a macroscopic device but needs an additional postulate. In the words of Omnès this is "a law of nature unlike any other". In a series of papers Blanchard and Jadczyk suggested a formalism in which irreversibility is introduced in the dynamics of the coupling of a quantum system with a classical one and thereby obtained a (phenomenological) description of this aspect of measurements (see, e.g., [Blanch 93, 95]).

The evolutionary understanding of reality was proposed many years ago by A.N. Whitehead [*Whitehead 1929*]. His writings have influenced philosophers and theologians, but few physicists. A notable exception are the papers by H.P. Stapp in which he outlines a theory of events having many features in common with the evolutionary picture described above [Stapp 77, 79]. It is a pity that these seminal papers did not receive the attention they deserve and unfortunate that I became aware of them too late to incorporate an adequate discussion of this work. The first two postulates in [Stapp 77] are identical with those underlying the evolutionary picture. Differences in views concern his postulate 3 (momentum conservation) and the meaning of causal independence. Especially the discussion of the EPR-effect in [Stapp 79] differs from the treatment above and leads to a different assessment of the lessons. In physics D. Finkelstein suggested an approach to the space-time problem based on similar concepts [Fink 68]. C.F. von Weizsäcker tried for many years to draw attention to the fundamental difference between facts as related to the past and possibilities as related to the future and argued that for this reason the statistical statements in physics must always be future directed [Weiz 73].

Whatever the ultimate fate of these ideas, we should recognize that the standard formalism of quantum physics is not sacrosanct and will probably be modified in future theories. With regard to the interpretation there is no basic disagreement with the epistemological analysis of Niels Bohr but an appeal to accept that physical theory always transcends the realm of experience, introducing concepts which can never be directly verified by experience though they must be compatible with it.

VIII. Retrospective and Outlook

VIII.1 Algebraic Approach vs. Euclidean Quantum Field Theory

The virtues of the algebraic approach are

1) Naturalness of the conceptual structure; directness and unambiguity of the physical interpretation.

2) The sometimes amazing harmony between physical questions and mathematical structures. The theory of von Neumann algebras and C*-algebras offers an adequate language for the concise formulation of the principle of locality in special relativistic quantum physics and tools for further development of the physical theory. But more striking are cases where the discussion of physical questions and the development of a mathematical structure proceed parallel, ignorant of each other and motivated by completely disjoint objectives, till at some time the close ties between them are noticed and a mutually fruitful interaction between physicists and mathematicians sets in. One example in our context is the formulation of equilibrium quantum statistical mechanics in the algebraic setting and the inception of the Tomita-Takesaki theory of modular automorphisms. Another one is the correspondence between the statistics parameter and the Jones index.

3) The fact that a few principles stake out a vast territory in a qualitatively correct manner. The essential principles are locality in the sense of special relativity, stability (positivity of the energy), nuclearity. The consequences range from the emergence of the particle picture with all its ramifications (statistics, charge structure, collision theory) to a natural characterization of equilibrium states and the second law of thermodynamics.

4) A rather detailed understanding of the anatomy of the theory with the recognition of those points where further structural elements are needed and where modifications are demanded if one aspires to a synthesis with the principles of general relativity.

The weakness of the algebraic approach is that it has not enabled us yet to construct or even characterize a specific theory. It has given a frame and a language not a theory. The many qualitative consequences are all in agreement with experience but there are few quantitative consequences. Some steps towards a distinction of specific theories within the general frame, involving the notions of "germs" and "scaling" will be mentioned at the end of this section.

Quite another distribution of strengths and weaknesses is met in the approach via the Feynman path integral in the Euclidean domain. This has been by far the most widely used approach in the past two decades. Here one starts from an explicit expression for the Lagrangean in terms of fields. It defines the action

$$W = \int \mathcal{L}(x) d^4x \tag{VIII.1.1}$$

(taken over all space-time) as a functional of classical field distributions, denoted collectively by Φ. One argues that the generating functional of the hierarchy of Schwinger functions is given by the functional Fourier transform of e^{-W} when the time coordinate is taken as purely imaginary.[1] From the Schwinger functions physical information is obtained using the Osterwalder-Schrader theorem and the interpretation of τ-functions in the setting of general quantum field theory as described in Chapter II.

The path integral thus provides a method of quantization of a classical model. It replaces the canonical formalism and is of much greater elegance in relativistic field theory. While in quantum mechanics, where Feynman first presented the path integral as a method of solving the Schrödinger equation, the scheme is equivalent to canonical quantization, this equivalence is not so clear in relativistic gauge field theories. The path integral bypasses Hilbert space and local algebras. To establish the connection to the latter one needs the Osterwalder-Schrader reflection positivity and growth estimates. This has been established in simple models but it is not clear whether one really wants it in its full strength. The path integral is more directly related to τ-functions than to Wightman functions. One gets the expectations of products of local quantities in temporal order in a distinguished state (the vacuum). It is conceivable that the time ordered product plays a more basic rôle in the foundation of the theory than the non commutative, associative product of our algebras. The following remarks may give some indications in this direction. If we look at the efforts to relate the Hilbert space formalism of quantum mechanics to natural operational principles it appears from the discussion at the end of VII.2 that one central assumption is that any observable can be measured in any state. Since, clearly, we cannot make a measurement before the state is prepared this assumption is reasonable only if we distinguish between observables and observation procedures and if we can choose for each observable an observation procedure at an arbitrary time. This means that one needs a dynamical law which identifies procedures at different times as the same observable, and this law should not dependent on the state. General relativity indicates that the dynamical law must involve the state to some extent. The relevance of this within the context of special relativistic quantum physics is open, but we may take it as an indication that the dynamical law cannot be regarded as an algebraic relation in $\mathfrak{A}_S$ (see the remarks in subsection III.3.3) but arises on the level of von Neumann algebras

[1] Here Fermi fields (classical spinorial fields) and Bose fields have to be treated differently. The integration over the Fermi fields (Berezin integration) is the simpler one and can be carried through in closed analytic form if the Fermi fields occur only bilinearly in the Lagrangean.

and therefore needs at least the weak topology induces by states. If we have local normality then, in restriction to finite diamonds, all allowed states lead to the same primary representation and we can get algebraic relations between observables at different times as implied by (III.3.44).

In the quantization of a classical field theory by the path integral one meets again problems familiar from canonical quantization. The devil dwells, as usual, in details. The counterpart of the representation problem mentioned in section II.1 is the flexibility in the definition of the functional integral. This is either handled by splitting e^{-W} into two parts, the first corresponding to a known reference measure, the second to a deviation from it. If the reference measure is taken as a Gaussian measure, correponding to a free field theory, then the path integral yields the "magic formula" of Gell-Mann and Low. There remains the renormalization problem and, in gauge theories, the handling of constraints due to redundant degrees of freedom. An alternative method, developed in constructive quantum field theory and in the lattice approach to gauge theories, starts from a problem with a finite number of degrees of freedom by introducing an ultraviolet cut-off (replacing the continuum of space or space-time by a discrete lattice) and confining the system to a finite box. Then one can (in principle) evaluate the functional integrals and study the dependence of the Schwinger functions on the ultraviolet cut-off, the box volume and on the adjustable parameters of the model. Renormalization is then replaced by showing that the adjustable parameters can be chosen as functions of the cut-offs in such a way that one obtains a well defined limit for the Schwinger functions when the lattice spacing tends to zero and the box volume to infinity.

The primary question concerns, of course, the choice of the Lagrangean. In finding out which fields are needed there is some guidance from phenomenology of high energy physics. However this is not enough as exemplified by the meson theories of strong interactions and the Fermi type theories of weak interactions which dominated the scene through some decades. The most valuable theoretical clue in the search for a specific theory has been the local gauge principle. It is the generalization from the Maxwell-Dirac system (considered as a classical field theory) to the Yang-Mills theories with non Abelian local gauge groups. This principle is well understood and natural on the classical level and, given the group, it leaves little freedom in the choice of the Lagrangean. Since we do not know yet how to formulate this principle directly in quantum physics the characterization of the theory by a classical model has remained indispensable for heuristic guidance in the search for a good theory. It may be considered as the central challenge to the algebraic approach to incorporate the local gauge principle into its conceptual frame.

In conclusion, it seems to me unwise to limit attention to one of the two approaches. The Lagrangean and the Feynman path integral are at present indispensable tools in the characterization and study of a specific theory. Together with the local gauge principle they pose questions which in the algebraic approach are not understood and should be tackled. On the other hand, basing the theory on a Lagrangean and the path integral only, is certainly too narrow.

Many aspects discussed in the previous chapters are not easily accessible there or not visible at all. The conceptual basis is too weak and needs insights from other sources even to arrive at the physical consequences of a specific model. The division of the task of characterizing the theory into two steps: the formulation of a classical model and its subsequent quantization cannot be ultimately satisfactory. We need a synthesis of the knowledge gained in the different approaches.

Germs. We may take it as the central message of Quantum Field Theory that all information characterizing the theory is strictly local i.e. expressed in the structure of the theory in an arbitrarily small neighborhood of a point. For instance in the traditional approach the theory is characterized by the Lagrangean density. Since the quantities associated with a point are very singular objects it is advisable to consider neighborhoods. This means that instead of a fiber bundle one has to work with a sheaf. The needed information consists then of two parts: first the description of the germs (of a presheaf), secondly the rules for joining the germs to obtain the theory in a finite region. We shall only address the first part here.

As J.E. Roberts first pointed out[2] the notions of a presheaf in state space and its germs are naturally related to the net structure of the algebras. Consider for each algebra $\mathcal{R}(\mathcal{O})$ the set of (normal) linear forms on it. It is a Banach space $\Sigma(\mathcal{O})$. The subset of positive, normalized forms are the states. The algebra $\mathcal{R}(\mathcal{O})$, considered as a Banach space, is the dual space of $\Sigma(\mathcal{O})$. To simplify the language let us denote by $\mathcal{R}$ the algebra of the largest region considered and by Σ the set of linear forms on it. In Σ we have a natural restriction map. If $\mathcal{O}_1 \subset \mathcal{O}$ then $\mathcal{R}(\mathcal{O}_1)$ is a subalgebra of $\mathcal{R}(\mathcal{O})$; hence a form $\varphi \in \Sigma(\mathcal{O})$ has a restriction to $\mathcal{R}(\mathcal{O}_1)$ thus yielding an element of $\Sigma(\mathcal{O}_1)$. We may also regard an element of $\Sigma(\mathcal{O})$ as an equivalence class of elements in Σ, namely the class of all forms on Σ which coincide on $\mathcal{R}(\mathcal{O})$. In this way we may pass to the limit of a point, defining two forms to be equivalent with respect to the point x

$$\varphi_1 \underset{x}{\sim} \varphi_2 \tag{VIII.1.2}$$

if there exists *any* neighborhood of x on which the restrictions of φ_1 and φ_2 coincide. Such an equivalence class will be called a germ at x and the set of all germs at x is a linear space Σ_x. It is, however, no longer a Banach space. Let us choose a 1-parameter family of neighborhoods $\mathcal{O}_r$ of x, e.g. standard double cones with base radius r centered at x. Denote the norm of the restriction of a form φ to $\mathcal{R}(\mathcal{O}_r)$ by $\|\varphi\|(r)$. This is a bounded function of r, non-increasing as $r \to 0$. The equivalence class $[\varphi]_x$ of forms with respect to the point x (the germ of φ) has a characteristic, common to all its elements, the germ of the function $\|\varphi\|(r)$.

The next question is how to obtain a tractable description of Σ_x by which the distinction between different theories shows up. Following Haag and Ojima

[2] Private communication around 1985

[Haag 96] one may appeal to the compactness or nuclearity properties (see V.5). Roughly speaking this property means that finite parts of phase space should correspond to finite dimensional subspaces in $\mathcal{H}$. Here phase space is understood as arising from a simultaneous restriction of the total energy and the space-time volume considered. Let Σ_E denote the subspace of forms which arise from matrix elements of the observables between state vectors in $\mathcal{H}_E$, the subspace of $\mathcal{H}$ with energy below E. Correspondingly we have $\Sigma_E(\mathcal{O})$, the restriction of these forms to $\mathcal{R}(\mathcal{O})$ and $\Sigma_E(x)$ their germs at x. The essential substance of the compactness requirement is that for any chosen accuracy ε there is a finite-dimensional subspace $\Sigma_E^{(n)}(\mathcal{O})$, with its dimensionality n increasing with E and r, such that for any $\varphi \in \Sigma_E(\mathcal{O})$ there is an approximate $\hat{\varphi} \in \Sigma_E^{(n)}(\mathcal{O})$ satisfying

$$\|\varphi - \hat{\varphi}\| \leq \varepsilon \|\varphi\|. \tag{VIII.1.3}$$

As ε is decreased n will increase. Since n is an integer, however, this will happen in discrete steps. For our purposes we shall adopt the following version of the compactness requirement whose relation to other formulations will not be discussed.

For fixed E there is a sequence of positive, smooth functions $\varepsilon_k(r)$, $k = 0, 1, 2\ldots$ with $\varepsilon_0 = 1$ and $\varepsilon_{k+1}/\varepsilon_k$ vanishing at $r = 0$. For each k there is a subspace $\mathcal{N}_k$ of $\Sigma_E(x)$ consisting of those germs for which $\|\varphi\|(r)$ vanishes stronger than ε_k at $r = 0$. There is a natural number n_k giving the dimension of the quotient space $\Sigma_E(x)/\mathcal{N}_k$. The n_k and the germs of the functions ε_k are characteristic of the theory.

We take the ε_k to be optimally chosen so that $\|\varphi\|(r) = \varepsilon_k$ is attained for some $\varphi \in \mathcal{N}_{k-1}$. To coordinatize the germs in Σ_E we choose an ordered basis $A_{k,l}(r) \in \mathcal{R}(\mathcal{O}_r)$ with $\|A_{k,l}\| = 1$ such that $|\varphi(A_{k,l}(r)| = \varepsilon_k(r)\|\varphi\|$ for $\varphi \in \Sigma_E/\mathcal{N}_k$, $l = 1, \ldots (n_k - n_{k-1})$. It follows that $\lim_{r\to 0} A_{k,l}(r)\varepsilon_k^{-1}(r)$ exists as a bounded sesquilinear form on $\mathcal{H}_E$ which might be called a point-like field. However, since we know so far nothing about the dependence of the structure on the energy bound E this object could change with increasing E.

In the case of a dilatation invariant theory the situation is much simpler. On the one hand there are energy independent "scaling orbits" $A_{k,l}(r)$ provided by the dilatations. Further, the ε_k (leading to the same numbers n_k for all energies) are powers

$$\varepsilon_k(E, r) = (Er)^{\gamma_k}. \tag{VIII.1.4}$$

Therefore point-like fields

$$\Phi_{k,l} = \lim A_{k,l}(r) r^{-\gamma_k} \tag{VIII.1.5}$$

can be defined as sesquilinear forms on any $\mathcal{H}_E$ Their bound in $\mathcal{H}_E$ increases in proportion to E^{γ_k}. So the analysis of Fredenhagen and Hertel [Fred 81b] applies which shows that the $\Phi_{k,l}$ are Wightman fields which furthermore have specific dimensions γ_k. The set of these is the Borchers class of observable fields in the theory and γ_k gives an ordering in this set. Of course, with Φ also the derivatives

$\partial_\mu \Phi$ will appear in this set and one can also define a product of such fields as (the dual of) the germs of

$$r^{-(\gamma_1+\gamma_2)}\varphi(A_1(r)A_2(r)), \tag{VIII.1.6}$$

where the indices 1, 2 stand for k_1, l_1 and k_2, l_2. Since $A_1A_2 \in \mathcal{R}(\mathcal{O}_r)$ and has norm bounded by 1 only the values of φ on the basis elements $A_{k,l}$ for low k will be important and give a contribution proportional to $r^{\gamma_k-\gamma_1-\gamma_2}$. Symbolically one can express this as an operator product expansion

$$\Phi_1\Phi_2 = \sum c_{k,l}(r)\Phi_k \tag{VIII.1.7}$$

where the coefficient functions $c_{k,l}$ behave like $r^{\gamma_k-\gamma_1-\gamma_2}$.

In the physically interesting case of *asymptotic* dilatation invariance it appears reasonable to expect that the essential aspects of this structure remain valid, at least for low k-values. Recently Buchholz and Verch [Buch 95] have formulated ideas of a renormalization group analysis in the algebraic language. Perhaps their approach can provide a starting point for a rigorous discussion of the germs in a theory with asymptotic dilatation invariance.

The accuracy functions ε_k and corresponding dimensionalities give some information about the theory. If, in addition, one can define pointlike fields and an operator product decomposition (which one might hope for in the case of an asymptotically dilatation invariant theory) then one might also obtain some relations between the $\Phi_{k,l}$, corresponding to field equations. One can, however, not expect that this gives complete information about the dynamics. There remains the problem of joining the germs. One idea concerning this has been suggested in [Fred 87]. There the additional problem of unobservable fields and gauge invariance has not been considered. Perhaps the analogue of the gauge connections of the classical theory will enter in the prescription for joining germs of neighboring points, involving degrees of freedom not contained in the individual germ (at least not in low order).

Scaling. Scaling considerations have been a valuable tool in many areas, among them the analysis of the short distance behavior in Quantum Field Theory (see e.g. [Sy 72]). In the field theoretic setting one may start from the transformations

$$\Phi_\lambda(x) = N(\lambda)\Phi(\lambda x), \quad 0 < \lambda < 1, \tag{VIII.1.8}$$

where Φ denotes a basic field and N is an appropriately chosen numerical function depending on Φ. In the algebraic setting one has the problem thet (VIII.1.8) does not define a mapping from $\mathcal{A}(\mathcal{O})$ onto $\mathcal{A}(\lambda\mathcal{O})$ unless the theory is dilatation invariant because, due to the field equations, an algebraic element can be expressed in many different ways as a functional of the basic fields and each such way gives a different image under the transformation since the field equations change with λ. To deal with this Buchholz and Verch [Buch 95] introduce an algebra $\hat{\mathcal{A}}$ whose elements are orbits $A(\lambda)$ in the quasilocal algebra $\mathcal{A}$ which are restricted by two conditions

(i) if $A(1) \in \mathcal{A}(\mathcal{O})$ then $A(\lambda) \in \mathcal{A}(\lambda\mathcal{O})$.

(ii) if the momentum transfer of $A(1)$ is contained in a region Δ of p-space then the momentum transfer of $A(\lambda)$ shall be contained in the region $\lambda^{-1}\Delta$.

The authors show that these two restrictions suffice to translate the field theoretic discussion based on (VIII.1.8) to the algebraic approach. In particular one can characterize the different possibilities for the short distance behavior. Of special interest among these are the cases in which there exists a scaling limit for $\lambda \to 0$, leading to a dilatation invariant limit theory, in particular the case of "asymptotic freedom" in which the scaling limit for the observables reduces to a free massless theory. One may regard the scaling limit as the theory in tangent space (compare section VIII.2). Since the scaling limit has higher symmetry than the original theory one may expect that the superselection analysis for the tangent space theory along the lines of Chapter IV leads to a larger gauge group. Buchholz suggested that in this way the significance of a local gauge group in the theory should be understood. For example in QCD the color group is not visible in the global structure due to confinement but appears in the tangent space theory [Buch 94, 96].

VIII.2 Supersymmetry

This prominent topic of the past fifteen years has not been mentioned in the previous chapters. Though its relevance for the description of physical phenomena is not clear at this stage there are several features indicating that it may contain essential clues for the future development of the theory.

One of them relates to the desire for a unification of geometric and internal symmetries. In the setting of III.3.2 the continuous symmetries form a Lie group $\mathcal{G}$ which is a semidirect product of the geometric symmetry group $\mathcal{G}_{\text{geo}}$ which must be a subgroup of the conformal group and contains the Poincaré group and an internal symmetry group $\mathcal{G}_{\text{int}}$ which transforms each $\mathcal{A}(\mathcal{O})$ into itself.[1] The fact that each $g \in \mathcal{G}$ has a uniquely determined geometric part (i.e. that there is a projection from $\mathcal{G}$ to $\mathcal{G}_{\text{geo}}$) implies that $\mathcal{G}_{\text{int}}$ is an invariant subgroup of $\mathcal{G}$ and we can write

$$g = (h,\ k) \quad \text{with} \quad h \in \mathcal{G}_{\text{int}};\ k \in \mathcal{G}_{\text{geo}}, \tag{VIII.2.1}$$

with multiplication law

$$(h,k)(h',k') = (h\alpha_k h', kk'), \tag{VIII.2.2}$$

where α_k is some action of $\mathcal{G}_{\text{geo}}$ on $\mathcal{G}_{\text{int}}$. The Lie algebra of $\mathcal{G}_{\text{int}}$ must therefore be a representation space of $\mathcal{G}_{\text{geo}}$, in particular of the Poincaré group. In the

[1] We consider symmetries of the field algebra $\mathcal{A}$, remembering that by the remarks in IV.1.1 and IV.1.3 we should not expect any exact internal symmetries of the observable algebra $\mathfrak{A}$.

mid sixties the question of whether there can be a non trivial action of $\mathcal{G}_{\text{geo}}$ on $\mathcal{G}_{\text{int}}$ was studied and, under some additional physically reasonable assumptions, it was answered by "no" (see e.g. [O' Raif 65], [Mich 65], [Cole 67], [Orz 70], [Garb 78a,b]). There can be no generators of $\mathcal{G}_{\text{int}}$ which do not commute with the translations or which have any vectorial or tensorial transformation character under the Lorentz group. The conclusion was that $\mathcal{G}$ can only be a direct product of $\mathcal{G}_{\text{geo}}$ and $\mathcal{G}_{\text{int}}$; there is no interplay between the geometric and the internal part. The aspect changes when one considers a "super Lie algebra" or, in mathematical terms, a "$\mathbb{Z}_2$-graded Lie algebra". Just as an ordinary Lie algebra it is a linear space equipped with a bracket operation but it has two distinct subspaces: the even elements (grade 0) and the odd elements (grade 1). Denoting the grade of element a by $s(a)$ the bracket has the property

$$[a,b] = (-1)^{s(a)s(b)+1}[b,a] \tag{VIII.2.3}$$

and satisfies the generalized Jacobi identity

$$[a,[b,c]]+(-1)^{s(a)s(b)+s(a)s(c)}[b,[c,a]]+(-1)^{s(c)s(a)+s(c)s(b)}[c,[a,b]] = 0. \tag{VIII.2.4}$$

The sign factors by which (VIII.2.4) differs from the usual Jacobi identity can be read off from (VIII.2.3). The cyclic permutations of the elements a, b, c can be obtained by two transpositions and each transposition must be accompanied by the sign factor in (VIII.2.3). Thus, from the order a, b, c to the order b, c, a we have the transposition $a \leftrightarrow b$ followed by $a \leftrightarrow c$ which together yield a sign $(-1)^{s(a)s(b)+s(a)s(c)}$. In the physical context the odd elements correspond to fermionic, the even elements to bosonic operators and the bracket is the anticommutator or the commutator, respectively. Such a fusion of commutation and anticommutation relations into a unified structure was proposed by Berezin and Kac [Bere 70]. The fascinating aspects of this structure became apparent when Wess and Zumino introduced a super Lie algebra containing the generators of the Poincaré group in which the energy momentum operators were given by the bracket of fermionic generators, [Wess 74]. The "baby model" of this algebra involved only the components Q_r $(r = 1,2)$ of a "spinorial charge", its Hermitean conjugate Q_r^*, the generators P^μ and $M_{\mu\nu}$ of translations and Lorentz transformations, respectively. Here we used the van der Waerden notation for 2-component spinors described in subsection I.2.1. The commutation relations of $M_{\mu\nu}$ with the other quantities are fixed by the requirement that the Q_r and Q_r^* transform as undotted and dotted spinors, respectively. Writing the 4-vector P^μ as a rank 2 spinor $P_{r\dot{s}}$ the remaining commutation relations are

$$[Q_r, Q_s^*]_+ = P_{r\dot{s}}; \quad [Q_r, Q_s]_+ = 0, \tag{VIII.2.5}$$

$$[Q_r, P^\mu] = 0, \tag{VIII.2.6}$$

together with those resulting by Hermitean conjugation. The suffix + indicating the anticommutator has been written only for emphasis. In the graded Lie algebra $[\ ,\ ]_+$ is just the bracket between odd elements.

One remarkable feature is the appearance of spinorial charges which extend the geometric symmetries in a non trivial way. Applying relations (VIII.2.5), (VIII.2.6) to single particle states with mass $m \neq 0$ we obtain a multiplet of particles with equal mass but different spin. The spin content of such a multiplet is easily evaluated looking at the restriction of the commutation relations to the "little Hilbert space" (degeneracy space in the center of mass system where $P^\mu = (m, 0)$ and thus $P_{r\dot{s}}$ is the numerical matrix $m\mathbb{1}$). There (VIII.2.5) just says that $m^{-1/2}Q_r$, $m^{-1/2}Q_r^*$ satisfy the canonical anticommutations relations. We have a system of two pairs of fermionic creation-annihilation operators in the degeneracy space. Their relation to the stability subgroup of the Lorentz group, the 3-dimensional rotation group, is

$$[M_k, Q_r] = -\frac{1}{2}\sum_s (\sigma_k)_{rs} Q_s, \quad k = 1,\ 2,\ 3;\ r,\ s = 1,\ 2. \tag{VIII.2.7}$$

We have used the usual notation for the angular momentum operators i.e. M_3 instead of M_{12}. Q_1^* and Q_2 raise the 3-rd component of the spin by 1/2, Q_2^* and Q_1 lower it by 1/2. The degeneracy space is spanned by four vectors: a bottom state $|b\rangle$, annihilated by the Q_r and belonging to spin 0, the doublet $Q_r^*|b\rangle$ belonging to spin 1/2, and a top state $|t\rangle = Q_1^* Q_2^* |b\rangle$, again with spin 0.

The most significant aspect of supersymmetry is, however, the expression for the energy operator given by (VIII.2.5)

$$P^0 = Q_1^* Q_1 + Q_2^* Q_2. \tag{VIII.2.8}$$

The right hand side is a manifestly positive operator. Thus the positivity of the energy, one of the central pillars of the theory, is automatically included in the supersymmetry relations. This is reminiscent of Dirac's original idea to draw the square root of the relativistic Hamiltonian. The incredible fruitfulness of this idea is well known to all of us.

The baby model can be embellished by taking several pairs of spinorial generators Q_r^L $(L = 1, \cdots N)$ which are transformed into each other by an internal symmetry group $\mathcal{G}_{\text{int}}$. The possible commutation relations are severely restricted. Adapting the argument of [Cole 67] to the case at hand one finds that in a massive theory the unbroken part of the super Lie algebra can only contain the generators of $\mathfrak{P}$ and of $\mathcal{G}_{\text{int}}$ besides the Q^L, Q^{L*} and that $\mathfrak{P}$ and $\mathcal{G}_{\text{int}}$ still commute with each other. The graded sub-(Lie)-algebra generated from the Q^L, Q^{L*} contains only the P^μ and possibly "central charges" of $\mathcal{G}_{\text{int}}$ which commute with all other elements in the algebra. In particular the Lorentz group and the non central internal symmetries are not generated from the spinorial charges. The relations (VIII.2.5) are replaced by

$$[Q_r^L, Q_s^{M*}]_+ = \delta^{LM} P_{r\dot{s}}. \tag{VIII.2.9}$$

The spinorial generators commute with the translations. The spin content of supermultiplets in this case is discussed in detail in [*Lopuszanski 1991*.]

A much more interesting structure emerges in the massless case with conformal invariance. Then we may have two types of spinorial charges, denoted by

Q^L and Q'^L. Together with their Hermitean conjugates they generate a bosonic symmetry group which is the direct product of the conformal group and an internal symmetry group U(N). The explicit form of these relations and their derivation was given by Haag, Lopuszanski and Sohnius [Haag 75].

So far the discussion concerned the global charges. In a local theory the elements of the super Lie algebra should arise as integrals of local densities over a space-like surface, in the case of the spinorial charges as integrals of local, rank 3 spinor fields. The commutation relations of these fields are the analogue of the current algebra in the bosonic case. (VIII.2.5) suggests that the bracket of the rank 3 spinor fields must lead to the energy momentum tensor which is related to diffeomorphisms of space-time in an analogous way as the electric current is related to local gauge transformations. Thus, in conjunction with a local gauge principle one may hope to be led to an approach to a quantum theory of gravity. It is beyond the scope of this book to discuss the very extensive work which has been done in this direction, called *supergravity*. A survey is given in [*Wess and Bagger 1983*]. See also the reprint volumes [*Salam and Sezgin 1989*], [*Ferrara 1987*].

VIII.3 The Challenge from General Relativity

VIII.3.1 Introduction

The synthesis between the tenets of quantum physics and those of general relativity has remained an unsolved problem for over sixty years. The early discussions between Einstein and Bohr on physical reality may well be seen as an omen forshadowing the difficulties encountered. Local quantum physics as described in the previous chapters accepts space-time, including its causal structure, as a given arena in which physics is staged. One may say that the space-time coordinates, tied to a reference system established by an observer, are classical quantities belonging to the observer side of the Heisenberg cut. The global structure of space-time is needed for the commutation relations between observables, in particular for the causal commutativity at (arbitrarily large) space-like separations. The local metric structure is needed in the formulation of dynamical laws in quantum field theory. Translation invariance is necessary for the definition of energy-momentum which, in turn, is central for the formulation of stability and nuclearity. Now, if we wish to include gravity and understand it as a modification of the metric field depending on the matter distribution and hence on the prevailing physical state, we can distinguish three levels.

On the first we still retain space-time equipped with a classical metric field, attached to the observer side of the Heisenberg cut, but abandon the specialization to Minkowski space and allow metric fields with curvature (see subsection I.2.3). At this level much of the previously described structure can be taken over. We can regard the theory still to be defined in terms of a net $\mathcal{A}(\mathcal{O})$ of *-algebras associated with open regions of the space-time manifold M and satisfying causal commutativity. This level is appropriately called quantum field theory (resp. local quantum physics) in curved space-time. The main problem which has to be resolved there is caused by the absence of translation invariance (possibly of all geometric symmetries). Thus we do not have axiom **A** of the Wightman frame and must find another formulation of stability. This can be done and there are some interesting consequences which will be discussed in the next subsections.

On the second level one may still retain a 4-dimensional space-time continuum as an ordering structure for the net of observables but assume no longer that it is equipped with a (classical) metric field. This level may be called (semi-conventional) quantum gravity. The problem here is to generalize manifold theory to build an object which we might call a manifold with an event structure over it from its germs. The macroscopic metric, being classical, must then be understood in analogy to spontaneous symmetry breaking. It is determined by the prevailing physical state which breaks the diffeomorphism invariance and refers to collective degrees of freedom in regions large compared to the Planck length. Much work has been devoted to the program of quantizing Einstein's theory of gravity either with or without inclusion of matter fields and gauge fields. Important progress has been made in this approach by formulating the theory in terms of spinorial variables replacing the metric tensor $g_{\mu\nu}$ [Asht 87]. For further developments along this line and references see [*Ashtekar 1991*].

On the third level, aiming at the famous TOE (theory of everything), the notion of space-time itself is abandoned as a primary concept. To provide the mathematics which might be needed in this venture several ideas have been put forward in the past decade: non commutative geometry, quantum spaces and quantum groups, superstring theory. Conceptually one has to overcome the Heisenberg cut and develop a theory of real event structures. We cannot enter this level in this book and shall only offer a few remarks in subsection 3.4.

VIII.3.2 Quantum Field Theory in Curved Space-Time

Assume that a classical gravitational background field is given. The large masses from which it originated shall not concern us. Mathematically we have a given 4-dimensional manifold M with a given pseudo-Riemannian (hyperbolic) metric. In a chart (coordinate system within some region) the latter is described by a classical metric field $g_{\mu\nu}(x)$. We want to define the net $\mathcal{A}(\mathcal{O})$ on the space-time manifold M. To focus on a tractable example we consider the case where the net arises from a Wightman field obeying a linear, generally covariant field equation, resulting from the field equations of the Minkowski space theory by replacing the derivatives ∂_μ by the covariant derivatives D_μ. In the case of a scalar field the Klein-Gordon equation is replaced by

$$(\Box_g + m^2)\Phi = 0 \qquad \text{(VIII.3.1)}$$

where

$$\Box_g = |g|^{-1/2}\partial^\mu |g|^{1/2}\partial_\mu \qquad \text{(VIII.3.2)}$$

and $|g|$ denotes the absolute value of the determinant of the metric tensor $g_{\mu\nu}$. The commutation relations are

$$[\Phi(x), \Phi(y)] = i\Delta(x, y)\mathbb{1}, \tag{VIII.3.3}$$

where Δ is the difference between the retarded and the advanced Green's function of the Klein-Gordon equation. It is a uniquely defined distribution on $M \times M$ provided M is globally hyperbolic.

Let us start from a more general setting. We consider first a larger net of algebras which contains only the kinematical information that we are dealing with a scalar field. We shall call it the net of Borchers-Uhlmann algebras and denote it by $\mathcal{A}_U$ since it was first mentioned in [Borch 62], [Uhl 62]. For a region $\mathcal{O} \subset M$ $\mathcal{A}_U(\mathcal{O})$ is the tensor algebra of C^∞-test functions with support in $\mathcal{O}$. It is the direct sum of monomial parts

$$\mathcal{A}_U = \oplus_{n=0}^{\infty} \mathcal{A}_U^n \tag{VIII.3.4}$$

where $\mathcal{A}_U^n$ consists of functions of n points. The unit element is the number 1 in the zero grade part. The general element is a hierarchy of functions

$$F = \begin{pmatrix} f^{(0)} \\ f^{(1)}(x) \\ f^{(2)}(x_1, x_2) \\ \vdots \end{pmatrix} \tag{VIII.3.5}$$

of which only a finite number are non vanishing. Addition is grade-wise. The algebraic product is the (not symmetrized) tensor product. With $F_1 \in \mathcal{A}_U^n$, $F_2 \in \mathcal{A}_U^m$, $F_1 F_2 \in \mathcal{A}_U^{n+m}$:

$$(F_1 F_2)(x_1 \ldots x_{n+m}) = f_1^{(n)}(x_1 \ldots x_n) f_2^{(m)}(x_{n+1} \ldots x_{n+m}). \tag{VIII.3.6}$$

An involution is defined by reversing the order of arguments plus complex conjugation. $F \to F^*$ corresponds to $f^{(n)}(x_1 \ldots x_n) \to \overline{f}^{(n)}(x_n \ldots x_1)$. Thus $\mathcal{A}_U$ is a *-algebra. Moreover there is a natural action of the local diffeomorphism group of M on $\mathcal{A}_U$. If g is a diffeomorphism mapping $\mathcal{O}_1$ on $\mathcal{O}_2$ then α_g is given by[1]

$$\left(\alpha_g f^{(n)}\right)(x_1 \ldots x_n) = f^{(n)}(g^{-1}x_1 \ldots g^{-1}x_n); \quad x_k \in \mathcal{O}_2. \tag{VIII.3.7}$$

It maps $f^{(n)} \in \mathcal{A}_U^n(\mathcal{O}_1)$ to $\alpha_g f^{(n)} \in \mathcal{A}_U^n(\mathcal{O}_2)$. A state ω is a normalized positive linear form on $\mathcal{A}_U$. It may be characterized by a hierarchy of distributions $\{w^{(n)}\}$ where $w^{(n)} \in \mathcal{D}'(M^n)$ and

$$\omega(F) = \sum w^{(n)}(f^{(n)}). \tag{VIII.3.8}$$

[1]One could consider a different transformation law, putting on the right hand side of (VIII.3.7) a factor $\prod J^p(x_k)$ where J is the Jacobian of the transformation g. The value of p determines whether Φ is a scalar or a scalar weight of some order. In fact, if we omit this factor and want Φ to be a scalar then we would have to understand $\Phi(f^{(1)})$ to mean formally $\int \Phi(x) f^{(1)}(x)|g|^{1/2} d^4x$. But this will not be relevant for the discussion in this subsection.

The hierarchy must satisfy the positivity condition

$$\omega(F^* F) \geq 0 \quad \text{for all} \quad F \qquad \text{(VIII.3.9)}$$

and the normalization

$$w^{(0)} = 1. \qquad \text{(VIII.3.10)}$$

Of course, $\mathcal{A}_U$ does not carry much information. To inject the dynamical law and the commutation relations we have to divide $\mathcal{A}_U$ by an ideal $\mathfrak{J}$, the elements which are zero due to (VIII.3.1), (VIII.3.3). The dynamical ideal, for instance, is generated from the elements in the grade-1 part of the form

$$(\Box_g + m^2)^* f^{(1)},$$

resulting from (VIII.3.1) by averaging with a test function. Still, $\mathcal{A} = \mathcal{A}_U/\mathfrak{J}$ does not determine the germ of the theory. There remain uncountably many inequivalent representations of $\mathcal{A}_U(\mathcal{O})$ for arbitrarily small $\mathcal{O}$. This follows, as in the case of the Minkowski space theory, from the arguments in II.1 since the field equations help only to reduce the algebra to the algebra of Φ and its time derivative on a space-like surface and then we remain with the canonical commutation relations. In Minkowski space this problem was handled by referring to a distinguished state, the vacuum, and subsequently assuming local normality of all physically allowed representations with respect to the vacuum representation. The characterization of the vacuum involves, however, global aspects and, in the case of curved space-time it is not evident how to select a distinguished state [Full 73]. Insisting on the principle of local definiteness we have to characterize a unique folium of physically allowed states for sufficiently small $\mathcal{O}$. For the free theory the first proposal in this direction was to require that the 2-point function should have the *Hadamard form* [Adl 77, 78], [Wald 77, 78], [Full 81a, b]

$$w^{(2)}(x_1, x_2) = u\sigma^{-1} + v \ln|\sigma| + w, \qquad \text{(VIII.3.11)}$$

where $\sigma(x_1, x_2)$ is the square of the geodesic distance between x_1 and x_2 and u, v, w are smooth functions of x_1 and x_2. This was required to hold in a sufficiently small neighborhood of an arbitrary point x. Due to the field equations and commutation relations u and v are uniquely determined by the geometry in the neighborhood of x i.e. by the $g_{\mu\nu}$ and their derivatives. Only the last term, the function w, can depend on the individual state in the folium. Important progress in understanding the significance of the Hadamard form for $w^{(2)}$ in a free theory is due to Radzikowski [Radz 92], who related it to Hörmander's concept of wave front sets and "microlocal analysis". This leads also to a definition of Wick ploynomials. We shall return to this below. In [Full 81a, b] it was shown that if the metric is globally hyperbolic and regular then the Hadamard form persists at later times if it is satisfied in the vicinity of some space-like surface.

Another approach which is not confined to free theories and sheds some light on the relation between the short distance behaviour of states and the dynamical law assumes that the partial state in the neighborhood of any point has a scaling limit. This allows the reduction of the theory to the tangent space of a point. One finds that the tangent space theory has translation invariance and a distinguished translation invariant state which results as the scaling limit of an arbitrary physical state of the full theory [Haag 84], [Fred 87]. To describe the procedure we use a coordinate system in the neighborhood $\mathcal{O}$ of the point x to which we want to contract but the conclusions are intrinsic. Consider a vector field X which vanishes at x and is of the form

$$X^{\mu}(y) = (y^{\mu} - x^{\mu}) + O(y - x)^2. \qquad \text{(VIII.3.12)}$$

Note that this is a coordinate independent restriction. The form of $X^{\mu}(y)$ will differ in different charts only by terms which are at least quadratic in $(y - x)$. Consider next the orbits of points in $\mathcal{O}$ under the motion

$$\frac{dy^{\mu}(\lambda)}{d\lambda} = \lambda^{-1} X^{\mu}(y(\lambda)); \quad y^{\mu}(1) = y \in \mathcal{O}; \quad \lambda \in [0,\ 1]. \qquad \text{(VIII.3.13)}$$

$y(\lambda)$ will move to x as $\lambda \to 0$; so (VIII.3.13) contracts the neighborhood to the point x. The solutions of (VIII.3.13) define a semigroup of local diffeomorphisms $g_X(\lambda)$

$$g_X(\lambda)y = y(\lambda), \qquad \text{(VIII.3.14)}$$

and, at $\lambda = 0$, the parametrized curve $y(\lambda)$ has a well defined tangent vector which we denote by $\eta_X y$, considering η_X as a diffeomorphism from $\mathcal{O}$ into the tangent space at x. One has

$$\eta_X g_X(\lambda) = \lambda \eta_X. \qquad \text{(VIII.3.15)}$$

Corresponding to the diffeomorphism $g_X(\lambda)$ we have a morphism $\alpha(g_X(\lambda))$ from $\mathcal{A}_U(\mathcal{O})$ to $\mathcal{A}_U(g_X(\lambda)\mathcal{O})$, defined by (VIII.3.7) with g replaced by $g_X(\lambda)$. Similarly we have a morphism $\alpha(\eta_X)$ from $\mathcal{A}_U(\mathcal{O})$ to the tensor algebra $\widehat{\mathcal{A}}_U$ of test functions on the tangent space $\mathcal{T}_x$ with support in $\widehat{\mathcal{O}} = \eta_X \mathcal{O}$

$$\left(\alpha(\eta_X) f^{(n)}\right)(z_1 \ldots z_n) = f^{(n)}(y_1 \ldots y_n) \quad \text{with} \quad z_k = \eta_X y_k. \qquad \text{(VIII.3.16)}$$

We have denoted the coordinates in $\mathcal{T}_x$ by z and shall denote objects relating to tangent space by a roof $\widehat{\ }$. By (VIII.3.15) one has

$$\alpha(\eta_X)\alpha\left(g_X(\lambda)\right)\alpha(\eta_X^{-1}) \equiv \widehat{\alpha}(\lambda) = \alpha(\widehat{D}(\lambda)) \qquad \text{(VIII.3.17)}$$

where $\widehat{D}(\lambda)$ is the dilation in tangent space:

$$\left(\widehat{\alpha}(\lambda)\widehat{f}^{(n)}\right)(z_1 \ldots z_n) = \widehat{f}(\lambda^{-1} z_1 \ldots \lambda^{-1} z_n). \qquad \text{(VIII.3.18)}$$

Definition 3.2.1
A state ω on $\mathcal{A}_U(\mathcal{O})$ is said to have a scaling limit with respect to the contracting vector field X if there exists a scaling function $N(\lambda)$ (positive, monotone) such that for all natural numbers n and all $f^{(n)} \in \mathcal{D}(\mathcal{O}^n)$ $\lim_{\lambda\to 0} N(\lambda)^n \omega(\alpha(g_X(\lambda))f^{(n)})$ exists and is nonvanishing for some $f^{(n)}$.

Since $\alpha(\eta_X)$ is a positive linear map we may regard the limit as a state on $\widehat{\mathcal{A}}_U$, the tensor algebra of test functions in tangent space, and write

$$\widehat{\omega}_X(\widehat{f}^{(n)}) = \lim N(\lambda)^n \omega\left(\alpha(g_X(\lambda))\alpha(\eta_X^{-1})\widehat{f}^{(n)}\right) = \lim N(\lambda)^n \omega\left(\alpha(\eta_X^{-1})\widehat{\alpha}(\lambda)\widehat{f}^{(n)}\right). \tag{VIII.3.19}$$

The last form follows from (VIII.3.17), (VIII.3.18) and has the advantage that it is defined on all of $\widehat{\mathcal{A}}_U$ since the support of $\widehat{\alpha}(\lambda)\widehat{f}^{(n)}$ will move inside $\widehat{\mathcal{O}}$ for sufficiently small λ whenever $\widehat{f}^{(n)}$ has compact support. The map on tangent space, introduced by the factor $\alpha(\eta_X^{-1})$ is essential for the first item in the following theorem which lists the basic properties of the scaling limit.

Theorem 3.2.2
Assume that ω has a scaling limit at x with respect to the contracting vector field X. Then:

(i) The limit state (VIII.3.19) is independent of the choice of the vector field X within the class (VIII.3.12). It depends only on the base point x and we shall therefore denote it by $\widehat{\omega}_x$.

(ii) The scaling function is "almost a power" i.e.

$$\lim_{\lambda'\to 0} \frac{N(\lambda'\lambda)}{N(\lambda')} = \lambda^{\alpha}. \tag{VIII.3.20}$$

We shall call $d = 4 + \alpha$ the canonical dimension of the field Φ.[2]

(iii) The limit state has the scaling behaviour

$$\widehat{\omega}_x(\widehat{\alpha}(\lambda)\widehat{f}^{(n)}) = \lambda^{-n\alpha}\widehat{\omega}_x(\widehat{f}^{(n)}). \tag{VIII.3.21}$$

(iv) If ω is primary[3] then the scaling limit exists for a dense set of states in the folium of ω and it leads to the same limit state $\widehat{\omega}_x$ for all of them.

(v) If the approach to the limit is uniform for x varying in some neighborhood (see (VIII.3.25) below) then the limit state is invariant under translations in tangent space i.e.

$$\widehat{\omega}_x(\alpha(T_a)\widehat{F}) = \widehat{\omega}_x(\widehat{F}) \quad \text{where} \quad (T_a\widehat{f}^{(n)})(z_1\ldots z_n) = \widehat{f}^{(n)}(z_1 - a\ldots z_n - a). \tag{VIII.3.22}$$

[2]Negative values of d would mean that the $w^{(n)}$ have no singularity at coincident points.
[3]This refers to the von Neumann algebra determined by the GNS-representation π_ω of $\mathcal{A}_U(\mathcal{O})$. The transition from the unbounded operators $\pi_\omega(F)$ to bounded operators is afforded by polar decomposition as indicated in section III.1.

The proof of the theorem is given in [Fred 87]. Here we shall only describe the additional assumption which is needed to establish the property (v) which is crucial for the following. The basic relation from which the argument starts is most easily seen if we choose the bases in the tangent spaces as corresponding to the coordinate axes and choose the contracting vector field to a base point x so that the term of order $(y-x)^2$ is absent. Then

$$g_x(\lambda)y = x + \lambda(y-x); \quad \eta_x y = y - x, \tag{VIII.3.23}$$

and thus

$$g_x(\lambda)\eta_x^{-1}T(a) = g_{x+\lambda a}(\lambda)\eta_{x+\lambda a}^{-1} = g_x(\lambda)\eta_x^{-1} + O(\lambda). \tag{VIII.3.24}$$

The validity of (v) depends on the possibility of neglecting the term of order λ on the right hand side of (VIII.3.24) in the limit $\lambda \to 0$. The precise condition for this is the existence of a bound, uniform in x within some neighborhood, for the difference between the limit and the scaling value at finite small λ

$$|\omega_x(\lambda; F) - \lim_{\lambda\to 0} \omega_x(\lambda; F)| < R(\lambda; F) \quad \text{for} \quad x \in \mathcal{O}, \tag{VIII.3.25}$$

where $\omega_x(\lambda; F)$ denotes the expression in definition 3.2.1.

From $\widehat{\omega}_x$ we get by the GNS-construction a Hilbert space representation of the tangent space algebra $\widehat{\mathcal{A}}_U$ in which $\widehat{\omega}_x$ is represented by a state vector $|\Omega\rangle$. The translation invariance of the limit state implies that we also have a representation of the translation group by unitary operators $U(a)$ acting in this Hilbert space with $|\Omega\rangle$ as an invariant vector. This allows the definition of energy-momentum operators in the tangent space theory and therefore the formulation of a principle of local stability.

Postulate 3.2.3 (Local Stability)
The energy-momentum spectrum in the tangent space theory at x is confined to the positive cone $\overline{V}^+$ (defined by the metric $g_{\mu\nu}(x)$ at the point x).

The causal commutation relations of the theory on M translate to causal commutation relations in the tangent space theory. It seems probable that $\widehat{\omega}_x$ must also be Lorentz invariant, though the necessary conditions have not been investigated. Thus the tangent space theory is expected to satisfy the axioms **A-E** of a quantum field theory in Minkowski space. Moreover, if the dimension of Φ is $d = 1$ (which in the case of the free field follows from the canonical commutation relations (VIII.3.3) and remains to be true if the theory is asymptotically free in the short distance limit) then the 2-point function in the tangent space theory is determined to be that of a massless free field theory in Minkowski space and then it follows from standard arguments (using the positivity (VIII.3.9) and causality) that the tangent space theory reduces to the theory of a massless free field in Minkowski space.

The dependence of $\widehat{\omega}_x$ on the base point x in a fixed coordinatization of M and of the tangent bundle is given by the affine connection:

$$\nabla_\mu \widehat{\omega}_x = 0, \tag{VIII.3.26}$$

$$\nabla_\mu = \frac{\partial}{\partial x^\mu} - \Gamma^\alpha_{\mu\beta} K_\alpha^\beta, \tag{VIII.3.27}$$

$$K_\alpha^\beta \widehat{\omega}_x(\widehat{f}) = \widehat{\omega}_x \left(\sum z_k^\beta \frac{\partial}{\partial z_k^\alpha} \widehat{f} \right), \tag{VIII.3.28}$$

where Γ are the Christoffel symbols.

For the 2-point Wightman function of the original state ω one obtains the short distance behaviour

$$w^{(2)}(x_1, x_2) = \sigma^{-1} + \Delta w^{(2)} \tag{VIII.3.29}$$

where $\Delta w^{(2)}$ must be less singular than of order 2 in the coordinates. Thus the leading singularity agrees with that of the Hadamard form though (VIII.3.29) would still allow a singularity in $\Delta w^{(2)}$ stronger than logarithmic for general physically allowed states. Here it must be remarked that (VIII.3.29) is not sufficient to guarantee the principle of local definiteness whereas the Hadamard form suffices for this purpose [Verch 94] (see also [Lüd 90] and the literature quoted there).

Let us return now to Radzikowski's proposal of characterizing the class of allowed states in terms of the wave front sets of Wightman distributions. This contains more information than the scaling limit; it relates to the germs, not only to the tangent space theory. Moreover it allows a stronger formulation of the spectrum condition, replacing Wightman's axiom A at least for free fields. An additional bonus is that one can define Wick polynomials of free fields as in the Minkowski space theory.

The wave front set of a distribution u, denoted by WFu, is a refinement of the notion of singular support, lifting it from base space to the cotangent bundle. It is a local notion in the sense that only the behavior of u in an arbitrarily small neighborhood of a point in base space is relevant. Note, however, the for $u = w^{(n)}$ the base space is $4n$-dimensional i.e. we consider points $X = x_1 \ldots x_n$ in configuration space. Introducing a chart we can first "localize" the distribution u by multiplying it with a smooth function h with support in some neighborhood $\mathcal{U}$ of X and then take the ordinary Fourier transform of uh, yielding a smooth function $\widehat{uh}(\xi)$ in "momentum space". The pair X, ξ is called a *regular directed point* if there exists some neighborhood $\mathcal{U}$ of X and some *conic* neighborhood C_ξ of ξ such that for every smooth function h with support in $\mathcal{U}$ the function $(1 + |\xi|)^N \widehat{uh}(\xi)$ stays bounded in C_ξ for all $N > 0$. *Conic* means that with $\xi' \in C_\xi$ also $\lambda \xi'$ is contained in C_ξ for all $\lambda > 0$. For the distinction between regular directed points and others only arbitrarily small neighborhoods of X are relevant. Therefore the definition does not depend on the chart and the pairs X, ξ may be regarded as points in the cotangent bundle.

Definition 3.2.4

The wave front set WFu of the distribution u in M is the complement of the set of all regular directed points in the cotangent bundle T^*M, excluding the trivial point $\xi = 0$.

One has

Theorem 3.2.5 [Radz 92]
Let Φ be a free field on the globally hyperbolic manifold M, satisfying (VIII.3.1), (VIII.3.3) and ω a quasifree state. The the following conditions are equivalent

(i) the 2-point function $w^{(2)}$ has the wave front set

$$WFw^{(2)} = \{x_1, k_1;\ x_2, -k_2\} \quad \text{with } x_1, k_1 \sim x_2, k_2;\ k_1^0 > 0.$$

where $x, k \sim x, k$ means that there is a lightlike geodesic γ from x_1 to x_2 with cotangent vectors k_1 and k_2 respectively.

(ii) $w^{(2)}$ has the Hadamard form.

Since for a quasifree state all truncated functions for $n \neq 2$ vanish the 2-point function suffices to define the state. The support properties of the $w^{(n)}$ in momentum space are such that pointwise products of these distributions are well defined. As in Minkowski space Wick polynomials of the free field can be defined [Brun 95]. A construction of Hadamard states has been given by Junkers [Junk 95].

The notion of wave front sets may be applied equally well to the $w^{(n)}$ of an interacting theory. Thus it is natural to try to formulate the general spectrum condition in terms of wave front sets. In an interacting theory it is, however, not clear whether the singular support is confined to light-like or even time-like separation of the points. Brunetti, Fredenhagen and Köhler [Brun 95] suggested the following prescription. Consider a graph $\mathcal{G}_n$ whose vertices represent points in the set $\{x_1, \ldots, x_n\}$ and whose edges represent connections between (a subset of) pairs x_i, x_j by smooth paths from x_i to x_j. To every edge e one assigns a covariantly constant *time-like* covector $k(e)$ which is future directed if $i < j$ but not related to the tangent vector of the path. In $\mathcal{G}_n$ there appears with every edge e also the inverse e^{-1} which carries the momentum $k(e^{-1}) = -k(e)$. The proposed condition is then

Proposed Condition 3.2.6 ("Microlocal spectrum condition")
The wave front set of $w^{(n)}$ is contained in the set $\{x_1, k_1, \ldots, x_n, k_n\}$ for which there exists a graph $\mathcal{G}_n$ as described above with $k_i = \sum k(e_\lambda)$ where the sum runs over all edges which have the point x_i as their source. (The trivial momentum configuration $k_1 = \ldots k_n = 0$ is excluded).

Probably the condition is not optimal but the fact that the k_i are all time-like and the distinction between future- and past direction is determined by the sequence of the index of x_i implies that pointwise multiplication of the distribution is defined.

Summing up: To make quantum field theory in curved space-time well defined (to the same extent as a corresponding theory in Minkowski space) one needs a specification of the set of physically allowed states as demanded by the principle of local definiteness. This is partially achieved if one requires that the states have a scaling limit. This gives, for every point in M, a reduced theory in the tangent space which allows a formulation of local stability. The tangent space theories are typically expected to be isomorphic to a free, massless, local theory in Minkowski space. The transport of the tangent space theories from one base point to another is governed by (VIII.3.26) to (VIII.3.28). The scaling limit is, however, not yet sufficient to characterize the germ of the theory. The requirements on the wave front set are stronger and yield this information. It should be borne in mind that these restrictive conditions do not characterize a specific state but a folium where they apply to a dense set of states. There is, in general, no distinguished state corresponding to the vacuum.

VIII.3.3 Hawking Temperature and Hawking Radiation

Classsical general relativity leads to the conclusion that very massive stars ultimately end by gravitational collapse, leading at some stage to the formation of a black hole from whose inside no signal can reach an outside observer. Furthermore, for the outside world the black hole has some aspects of a thermodynamic system in equilibrium ("no hair theorems", entropy) [*Hawking and Ellis 1980*], [Bek 73]. In a seminal paper Hawking argued that a black hole has a surface temperature

$$T = \hbar(8\pi MG)^{-1} \tag{VIII.3.30}$$

which can be understood by considering quantum field theory in the curved space-time due to the gravitational field of the collapsing star. Here M is the mass of the star, G the gravitational constant [Hawk 75]. We have written Planck's constant explicitly since it is one of the surprising aspects that here for the first time one sees a nontrivial interplay between quantum physics and general relativity.

Hawking's first computation related the (partial) state of the quantum field at late time far away from the black hole to the state at an early time before the star had begun to callapse. Up to this time the metric was practically static. For a globally static metric one has a well defined ground state of the quantum field and one may assume that at very early times the field is in the ground state of the static metric. For later times the state is then determined by the field equations. Hawking used a geometric optics approximation to solve this initial value problem in the case of the free field equations (VIII.3.1) in the metric provided by a classical model of stellar collapse. He then found that at late times the state describes an outgoing thermal radiation corresponding to the temperature (VIII.3.30). Though there is now little doubt that this computation yields the essential features of the state at late times there remained some uneasiness concerning the corrections and the wish for a more direct understanding of the claimed universality of the effect.

Several subsequent papers addressed themselves to the simpler situation of a (spherically symmetric) eternal black hole whose outside region can be described as Schwarzschild space. For a review see e.g. [Kay 87]. In standard spherical coordinates r, ϑ, φ the black hole radius is

$$r_0 = 2MG, \tag{VIII.3.31}$$

and the metric for $r > r_0$

$$ds^2 = \left(1 - \frac{r_0}{r}\right) dt^2 - \left(1 - \frac{r_0}{r}\right)^{-1} dr^2 - r^2(d\vartheta^2 + \sin^2\vartheta \, d\varphi^2). \tag{VIII.3.32}$$

This metric is static in the Schwarzschild time t but the fact that at $r = r_0$ the metric components $g_{rr} = \infty$, $g_{tt} = 0$ means that $r = r_0$ is a horizon and t is not a good coordinate there. There are other coordinate systems which extend Schwarzschild space to the interior of the black hole. If t is expressed in such coordinates it becomes ambiguous on the horizon. There is an analogy of Schwarzschild space with the Rindler space W which may be described as the wedge in Minkowski space

$$x^1 > |x^0|; \quad x^0, x^2, x^3 \quad \text{arbitrary}. \tag{VIII.3.33}$$

There the boundary $x^1 = |x^0|$ is a horizon. An observer in W, sending a signal to the Minkowski region $x^1 < |x^0|$ can never get an echo back if he is confined to move within W. The orbit

$$\begin{aligned} x^0(t) &= \varrho \sinh t, \\ x^1(t) &= \varrho \cosh t \end{aligned} \tag{VIII.3.34}$$

is a uniformly accelerated motion in Minkowski space and ϱt is the proper time on this orbit. One can coordinatize W by ϱ, t and one recognizes that the "time translation" $t \to t + a$ in these coordinates is a Lorentz transformation in Minkowski space, hence a symmetry. Therefore the metric in W is static with respect to t. This Rindler time t is analogous to the Schwarzschild time. Remembering the Bisognano-Wichmann theorem V.4.1.1 we know that the Minkowski vacuum is a thermal state in W with respect to Rindler time translations. The temperature $(2\pi)^{-1}$ is analogous to the Hawking temperature and agrees with it if the accelerations are properly compared. It is amusing to recall that the papers [Bis 76] and [Hawk 75] resulted from completely disjoint motivations and the striking parallelism was first noticed in [Sew 80]. Earlier Unruh had presented arguments showing that a linearly accelerated detector in Minkowski space on the orbit (VIII.3.34) should respond to the vacuum state similar as one at rest in a medium filled with Planck radiation at the temperature $(2\pi\varrho)^{-1}$, [Unr 76], compare also [Dav 75]. This is the Bisognano-Wichmann temperature scaled to the proper time of the detector. Although these analogies are suggestive they do not suffice to understand the case of an eternal spherical black hole because in the Rindler case we have more information. Besides the Rindler time translation there is (on the extension of W to Minkowski space) another time-like Killing vector field, our usual time translation, and this is used in defining the vacuum as a reference state. Only in the neighborhood of the horizon the analogy between Rindler and Schwarzschild is good. This suggests that the discussion should focus on the local situation of the state in the neighborhood of the horizon. Furthermore the formation of the black hole by stellar collapse should not be ignored since a permanent black hole is a physically unrealistic mental construct.

So we shall return to the case of a spherically symmetric collapse, paying special attention to the state near the horizon after the black hole formation. Outside of the star and the black hole one has the Schwarzschild metric (VIII.3.32). This follows from Birkhoff's theorem which says that a spherically symmetric metric in a part of space without matter (and hence vanishing Ricci tensor) is always of the form (VIII.3.32); it has a time-like Killing vector field defining the Schwarzschild time even if the metric is not static in the total space. In fig. VIII.3.1. we have used a time coordinate τ which remains meaningful on the horizon. The collapsing star is indicated by the horizontally shaded region, the black hole by diagonal shading. At $\tau = 0$ the radius of the star crosses the Schwarzschild radius r_0, the horizon begins. Also indicated are outgoing light rays from the neighborhood of the crossing point and the surface $t = 0$. In the region

$$\tau > 0; \quad r > r_0, \tag{VIII.3.35}$$

it will be convenient sometimes to use the "tortoise coordinate"

$$r^* = r + r_0 \ln \left(\frac{r}{r_0} - 1\right) \tag{VIII.3.36}$$

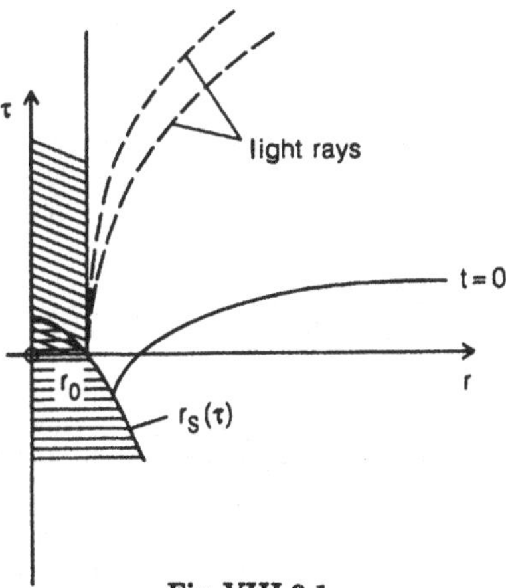

Fig. VIII.3.1.

instead of r. The horizon is then moved to $r^* \to -\infty$. For large r the difference between r^* and r is not significant. The time coordinate τ used in the figure was chosen as

$$\tau = t + r^* - r, \tag{VIII.3.37}$$

noting that $t + r^*$ stays finite on the horizon. We are interested in the partial state in a region $\mathcal{O}$ which is far away from the black hole at a very late time, centered near a point $r = R,\ t = T$ so that

$$T \gg R; \quad R \gg r_0. \tag{VIII.3.38}$$

Specifically we are interested in the response of a detector placed in this region. We may represent it by the observable

$$C = Q^* Q; \quad Q = \Phi(h), \tag{VIII.3.39}$$

with appropriately chosen test function h having support in $\mathcal{O}$. The computation of the expectation value $\omega(C)$ is rendered possible by the following remarks [Fred 90].

1) If f is a c-number solution of the covariant Klein-Gordon equation (VIII.3.1) and Φ satisfies (VIII.3.1) then the integral over a space-like surface Σ

$$\int_\Sigma \left(\Phi(x) \frac{\partial f(x)}{\partial \sigma^\mu} - \frac{\partial \Phi(x)}{\partial \sigma^\mu} f(x) \right) d\sigma^\mu \equiv \Phi_f \tag{VIII.3.40}$$

is independent of Σ. Here $d\sigma^\mu$ is the surface element. Q can be written in the form (VIII.3.40) with

$$f(t, \mathbf{x}) = \int f_{t'}(t, \mathbf{x}) dt' \tag{VIII.3.41}$$

where $f_{t'}$ is the solution of (VIII.3.1) with the Cauchy data

$$f_{t'}(t, \mathbf{x}) \Big|_{t=t'} = 0; \quad \frac{\partial f_{t'}(t, \mathbf{x})}{\partial t} \Big|_{t=t'} = h(t', \mathbf{x}). \tag{VIII.3.42}$$

So $f_{t'}$ differs from zero only if t' is in the t-support of h. To verify this note that in the neighborhood of $\mathcal{O}$, where r and t are very large r_0/r is negligible and for a surface $t =$ const. the surface element $d\sigma^\mu$ is adequately given in Schwarzschild coordinates by $d\sigma^\mu = (dv, 0, 0, 0)$ where dv is the spatial volume element. Then Φ_f becomes

$$\int dt' \left(\Phi(t, \mathbf{x}) \frac{\partial f_{t'}(t, \mathbf{x})}{\partial t} - \frac{\partial \Phi(t, \mathbf{x})}{\partial t} f_{t'}(t, \mathbf{x}) \right) dv.$$

Since $f_{t'}$ is a solution of (VIII.3.1) we can use again the independence of the integration surface and put $t = t'$. Then by (VIII.3.42) the above expression reduces to $\int \Phi(t, \mathbf{x}) h(t, \mathbf{x}) d^4x = \Phi(h)$. If we now want to apply (VIII.3.40) choosing for Σ the surface $\tau = 0$ we have to solve the Cauchy problem for the solution $f_{t'}$ of (VIII.3.1) with initial data given by (VIII.3.42) down to the surface $\tau = 0$. Given this we can express $\omega(C)$ in terms of the 2-point function $\omega\left(\Phi(x_1)\Phi(x_2)\right)$ near the plane $\tau = 0$.

2) The solution of (VIII.3.1) with such initial data has, for $\tau \geq 0$, its support outside the horizon and the star. So we need only the Schwarzschild metric. Equation (VIII.3.1) gives then, after separating off the angular part, using r^* instead of r, putting $f_{t'} = r^{-1}Y_{l,m}(\vartheta,\varphi)\psi_l(r^*)$

$$\left(\frac{\partial^2}{\partial t^2} - \frac{\partial^2}{\partial r^{*2}} + V_l(r^*)\right)\psi_l = 0. \tag{VIII.3.43}$$

The shape of the "barrier" V is indicated in fig. VIII.3.2. The height is of order $((l+1/2)/r_0)^2$, the width of order r_0. A wave packet moving according to (VIII.3.43), having its support at $t = T$ around $r = R$ with the relations (VIII.3.38), will at $\tau = 0$ be split into two clearly distinct parts: the first, having penetrated the barrier, is centered at very large negative values of r^* i.e. very close to the horizon; the other will be at very large positive values of r (or r^*). This is intuitively understood by looking at the analogous barrier penetration problem in Schrödinger wave mechanics. It is discussed by Dimock and Kay [Dim 87]. Refined estimates [Fred 90] show that around $\tau = 0$ we have

$$\psi_l(r^*) = \psi_+ + \psi_- + \Delta$$

with $\operatorname{supp}\psi_+ = [a(T),\infty]$, $\operatorname{supp}\psi_- = [-\infty,-a(T)]$ and Δ tending to zero uniformly with all its derivatives as $T \to \infty$ and $a(T) \to \infty$ for $T \to \infty$.

3) For very large T the counting rate $\omega(C)$ is thus related to the 2-point function around $\tau = 0$ by a term involving only ψ_-, one involving only ψ_+ and a cross term which will, however, tend to zero since $w^{(2)}$ decreases fast for large space-like separation of the points. The term with ψ_+ relates to signals received by outward directed detectors (looking away from the star); this will be not much affected by the presence of the black hole and we shall not discuss it. The term with ψ_- involves $w^{(2)}$ only in a short distance neighborhood of the crossing point $\tau = 0$, $r = r_0$. In fact, it gives a scaling limit of ω at this point. So, if we believe that all physically allowed states have the same scaling limit and that this is obtained by putting $\omega^{(2)} = \sigma^{-1}$, we can evaluate $\omega(C)$ (in the limit $T \to \infty$). The result is (for $T \to \infty$)

$$\omega(C) = \sum_{l,m}\int_{-\infty}^{\infty} |D_l(\varepsilon)|^2\varepsilon^{-1}\left(e^{\beta\varepsilon}-1\right)^{-1}|\tilde{h}_{l,m}(\varepsilon)|^2 d\varepsilon. \tag{VIII.3.44}$$

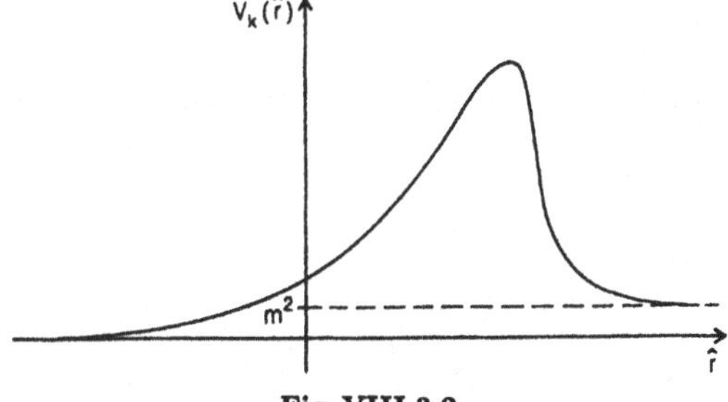

Fig. VIII.3.2.

Here $|D_l|^2$ is the barrier penetration factor,

$$\tilde{h}_{l,m}(\varepsilon) = \text{const.} \int h(x) Y_{l,m}(\vartheta, \varphi) e^{i\varepsilon(t-r^*)} d^4x, \qquad \text{(VIII.3.45)}$$

$$\beta = 4\pi r_0. \qquad \text{(VIII.3.46)}$$

To ensure that C represents a detector h has to be chosen so that $\tilde{h}(\varepsilon)$ has support only for positive ε (see Chapter VI). Then $|\tilde{h}_{l,m}(\varepsilon)|^2$ is the sensitivity of the detector to quanta of energy ε, angular momentum l, m. (VIII.3.46) shows that the asymptotic counting rate is the one produced by an outgoing radiation of temperature $(4\pi r_0)^{-1}$, modified by the barrier penetration effect.

The result (VIII.3.44) to (VIII.3.46) is just a corroberation of Hawking's original prediction. The derivation presented above shows that the effect is independent of the details of past history up to the formation of the black hole and relates to the short distance limit of the state at the surface of the black hole which is – by the assumption of local definiteness – the same for all physically allowed states. In fact, in the idealization used, the radiation originates from the surface of the horizon at the *moment of its formation* ($\tau = 0$) and at large times there is a steady flow of outgoing radiation, not decreasing as $T \to \infty$. This conflicts with energy conservation. The energy radiated away must be compensated by a loss of mass of the star inside the black hole ("black hole evaporation"). This in turn leads to a shrinking of the Schwarzschild radius with increasing τ and thereby the causal structure is changed so that the radiation arriving at (T, R) relates to points on the horizon at times $\tau(T)$, increasing with T; thus the black hole surface at all times (not only at $\tau = 0$) is responsible. To turn this qualitative argument into a quantitative one we must determine the back reaction on the metric due to the energy distribution in the state of the quantum field. The natural approach to this problem would be to use the Einstein equations (I.2.94), replacing in the outside region $T_{\mu\nu}(x)$ by $\omega\left(T_{\mu\nu}(\Phi; x)\right)$, the expectation value of the energy-momentum density of the field Φ in the prevailing state ω. This has led to an extensive literature attempting to define a "renormalized energy-momentum tensor", starting from the classical expression for $T_{\mu\nu}(\Phi; x)$ and the Hadamard form of the 2-point function of the state. It has been shown that $T_{\mu\nu}$ can be defined as an operator valued distribution in the Wightman sense *uniquely up to an unknown c-number function* but it is precisely this c-number function which we need here. For a survey see [*Wald 1984*]. The possibility of defining a Wick product may help in resolving this ambiguity.

Let us add a last remark concerning the universality (model independence). While it appears that the Bekenstein entropy and the Hawking temperature are completely model independent these quantities do not have direct observational significance since they concern the black hole surface in relation to the vector field of Schwarzschild time translation which should properly be regarded as an asymptotic symmetry. In the transition region between the surface and the observer the dynamics of the quantum field plays a rôle. In the free field model it produces the barrier penetration factor. If we envisage instead a realistic quantum field theory, say the standard model of elementary particles, some of the above arguments can be adapted. But the determination of the change of the state in the transition region is a highly non trivial problem in quantum field theory which cannot be circumvented by thermodynamical arguments.

VIII.3.4 A Few Remarks on Quantum Gravity

The gravitational constant G, together with $\hbar$ and c, determines a mass, the Planck mass

$$M_P = \left(\frac{\hbar c}{G}\right)^{1/2} = 10^{19} m_{\text{proton}}. \qquad \text{(VIII.3.47)}$$

It is the mass for which the Compton wave length and (half of) the Schwarzschild radius become equal. The corresponding length, the *Planck length* is

$$l_P = \frac{\hbar}{M_P c} = \left(\frac{G\hbar}{c^3}\right)^{1/2} = 10^{-33}\,\text{cm}. \qquad \text{(VIII.3.48)}$$

It appears clear that at distances of the order of l_P a classical metric has no place and probably the picture of space-time as a 4-dimensional continuum becomes unreasonable. On the other hand it seems that classical space-time, equipped with a (classical) causal and metric structure is well established at distances above 10^{-16} cm [4] and that in this regime the principles of local quantum physics, described in the earlier chapters apply well. The modifications at larger scales, related to space-time curvature, can be treated in a semiclassical manner along the lines indicated in subsections 3.2, 3.3. In this treatment there remains still the problem of a self-consistent description of the back reaction, hinging on the definition of the expectation value of the energy-momentum tensor in a given state of the quantum fields.

The crux of the interface of quantum physics and gravitation is, however, the short distance regime. There, below $10^{10} l_P$ we have no direct guidance from experiment and cannot expect any. We can speculate and try to produce a scheme whose merits can be tested by establishing contact with extrapolations from high energy physics and ultimately by an understanding of the relation of mass scales in particle physics to the Planck mass. Much work and ingenuity has been devoted to this quest. We mentioned some lines of approach at the end of subsection 3.1. One recent proposal [Dopl 95] is to define a non commutative geometry of x-space by stipulating commutation relations between the coordinates involving the Planck length and restricting the simultaneous precision of the measurability in different directions. Space-time regions are then replaced by states over the algebra of the x_μ.

In a minimal adaptation of the algebraic approach and the locality principle one could keep the idea of a net of algebras which, however, should be labelled now by elements of a partially ordered set $\mathcal{L}$ (instead of regions in $\mathbb{R}^4$). $\mathcal{L}$

[4] This bound is not meant to have physical significance. It just reflects ignorance and might be pushed down by several orders of magnitude.

could be atomic, with minimal elements (atoms) replacing microcells in space-time. If one dislikes the somewhat artificial picture of arranging these atoms in a physical lattice with lattice distance l_P in $\mathbb{R}^4$ then a notion of neighborhood can be introduced, for instance, by requiring that to any atom a there exist two other atoms a_1 and a_2 (neighbors of a) such that $b \in \mathcal{L}$ with $b > a_1$, $b > a_2$ implies $b > a$. If $\mathcal{L}$ is a lattice in the mathematical sense (equipped with operations $\vee$ and $\wedge$) then this gives the relation $(a_1 \vee a_2) \wedge a = a$, which implies a non-Boolean structure. In its coarse grained structure $\mathcal{L}$ should go over into the Boolean lattice of regions in $\mathbb{R}^4$. The algebras associated to atoms should naturally be chosen to be all isomorphic to a finite dimensional algebra. Since no work based on such a picture has yet been done the preceding remarks should be understood as "an artist's impression of a scenario". The judgment of its worth must be left to the future.

Bibliography

Alfsen, E. M. *Compact convex sets and boundary integrals.* Ergebnisse Math. **57**, Berlin: Springer 1971

Alfsen, E. M. and Shultz, F. W. *Non commutative spectral theory for affine function spaces on convex sets.* Memoirs of the A. M. S. **172**, 1976

Arnold, V. I. *Mathematical methods of classical mechanics.* New York: Springer 1978

Baumgärtel, H. *Operatoralgebraic Methods in Quantum Field Theory.* Berlin: Akademie Verl. 1995

Baumgärtel, H. und Wollenberg, M. *Causal Nets of Operator Algebras. – Mathematical Aspects of Algebraic Quantum Field Theory.* Berlin: Akademie Verl. 1992

Bell, J. S. *Collected papers in quantum mechanics. Speakable and unspeakable in quantum mechnics.* Cambridge Univ. Press 1987

Bergmann, P. G. *Introduction to the theory of relativity.* Dover: Prentice-Hall 1976

Birrell, N. D. and Davies, P. C. W. *Quantum fields in curved space.* Cambridge Univ. Press 1982

Bjorken, J. D. and Drell, S. D. *Relativistic quantum fields.* New York: McGraw-Hill 1965

Bogolubov, N. N. and Shirkov, D. V. *Introduction to the theory of quantized fields.* London: Interscience 1958

Bogolubov, N. N., Logunov, A. A. and Todorov, I. T. *Introduction to axiomatic quantum field theory.* Reading, Mass.: Benjamin 1975

Bogolubov, N. N., Logunov, A. A., Oksak, A. I. and Todorov, I. T. *General principles of quantum field theory.* Dordrecht: Kluwer 1990

Bohr, N. *Atomphysik und menschliche Erkenntnis.* Braunschweig: Vieweg Verl. 1985

Bratteli, O. and Robinson, D. W. *Operator algebras and quantum statistical mechanics I. C^*- and W^*-algebras, symmetry groups, decomposition of states.* New York: Springer 1979

Bratteli, O. and Robinson, D. W. *Operator algebras and quantum statistical mechanics II. Equilibrium states models in quantum statistical mechanics.* New York: Springer 1981

Braun, H. and Koecher, M. *Jordan algebras.* Berlin: Springer 1966

Dixmier, J. *Von Neumann algebras.* Amsterdam: North-Holland 1981

Dixmier, J. *C^*-algebras.* Amsterdam: North-Holland 1982

Dunford, N. and Schwartz, J. T. *Linear operators. Part I: General theory.* New York: Interscience Publishers 1958

Emch, G. E. *Algebraic methods in statistical mechanics and quantum field theory.* New York: Wiley 1972

d'Espagnat, B. *Reality and the physicist.* Cambridge: Cambridge Univ. Press 1989

Ferrara, S. (ed.) *Supersymmetry. Vol 1 and 2.* Amsterdam: North Holland, World Scientific 1987

Feynman, R.P. and Hibbs, A. R. *Quantum mechanics and path integrals.* New York: McGraw-Hill 1965

Friedrichs, K. O. *Mathematical aspects of the quantum theory of fields.* New York: Intersciences Publishers 1953

Gelfand, I. M. and Shilov, G. E. *Generalized functions. Vol.I–III.* New York: Academic Press 1964

Gelfand, I. M. and Vilenkin, N. *Generalized functions. Vol.IV.* New York: Academic Press 1964

Glimm, J. and Jaffe, A. *Quantum physics. A functional integral point of view.* Berlin: Springer 1987

Hamermesh, M. *Group theory and its application to physical problems.* Reading: Addison-Wesley 1964

Hawking, S. W. and Ellis, G. F. R. *The large scale structure of space–time.* Cambridge Univ. Press 1980

Heitler, W. *The quantum theory of radiation.* London: Oxford University Press 1936

Hepp, H. *Théorie de la renormalisation.* Berlin: Springer 1968

Horuzhy, S. S. *Introduction to algebraic quantum field theory.* Dordrecht: Riedel 1989

Iochum, B. *Cônes autopolaires et algèbres de Jordan.* Lecture Notes in Mathematics, Heidelberg: Springer 1984

Israel, R. B. *Convexity in the theory of lattice gases.* Princeton: Univ. Press 1979

Itzykson, C. and Zuber, J.-B. *Quantum field theory.* New York: McGraw-Hill 1980

Jauch, J. M. and Rohrlich, F. T. *The theory of photons and electrons. The relativistic quantum field theory of charged particles with spin one-half.* 2nd, expanded ed. Berlin: Springer 1976

Jost, R. *The general theory of quantized fields.* Prividence, R. I.: American Math. Soc. 1965

Kadanoff, L. P. and Baym, G. *Quantum statistical mechanics.* New York: Benjamin 1962

Kadison, R. V. and Ringrose, J. R. *Fundamentals of the theory of operator algebras. Vol I: Elementary theory.* New York: Academic Press 1983

Kadison, R. V. and Ringrose, J. R. *Fundamentals of the theory of operator algebras. Vol II: Advanced theory.* Orlando: Academic Press 1986

Kastler, D. (ed.) *The algebraic theory of superselection sectors.* Singapore, World Scientific 1990

Ludwig, G. *An axiomatic basis for quantum mechanics.* Heidelberg: Springer 1987

Lopuszanski, J. *An introduction to symmetry and supersymmetry in quantum field theory.* Singapore: World Scientfic 1991

Moser, J. *Dynamical systems. Theory and applications.* Springer: New York 1975

Naimark, M. A. *Linear representations of the Lorentz group.* Oxford: Pergamon Press 1964

Naimark, M. A. *Normed rings.* Groningen: P. Noordhoff 1972

Ohnuki, Y. *Unitary representations of the Poincare group and relativistic wave equations.* Singapore: World Scientific 1988

Omnès, R. *The Interpretation of Quantum Mechanics.* Princeton: Princeton University Press 1994

Pedersen, G. K. *C^*-algebras and their automorphism groups*. London: Academic Press 1979

Rohrlich F. T. *Classical charged particles*. Reading, Mass.: Addison–Wesley 1965

Ruelle, D. *Statistical physics*. New York: Benjamin 1969

Salam, A. and Sezgin, E. (eds.) *Supergravities in diverse dimensions. Vol. 1 and 2*. Amsterdam: North Holland, World Scientific 1989

Sakai, S. *C^*-algebras and W^*-algebras*. Berlin: Springer 1971

Schilpp, P. A. (ed.) *Albert Einstein: Philosopher-Scientist*. Evanston: The Library of Living Philosophers 1949

Schwartz, L. *Théorie des Distributions*. Paris: Hermann 1957

Sewell, G. L. *Quantum theory of collective phenomena*. Oxford Science Publ., Clarendon Press 1989

Sigal, I. M. *Scattering theory for many–body quantum mechanical systems*. Lecture Notes in Mathematics Vol. 1011, Springer: Heidelberg 1983

Steinmann, O. *Perturbative expansions in axiomatic field theory*. Berlin: Springer 1971

Stratila, S. *Modular theory in operator algebras*. Ed. Acad. Bukarest 1981

Stratila, S. and Zsido, L. *Lectures on von Neumann algebras*. Turnbridge Wells: Abacus Press 1979

Streater, R. F. and Wightman, A. S. *PCT, Spin and statistics and all that*. New York: Benjamin 1964

Takesaki, M. *Tomita's theory of modular Hilbert algebras and its application*. Lecture Notes in Mathematics. Berlin: Springer 1970

Takesaki, M. *Theory of Operator Algebras*. Berlin: Springer 1979

Thirring, W. *A course in mathematical physics*, Vol. 1–4. Heidelberg: Springer 1981

Varadarajan, V. S. *Geometry of quantum theory*. New York: Van Nostrand 1968

Van der Waerden, B. *Die gruppentheoretische Methode in der Quantenmechanik*. Berlin: Springer 1932

Wald, R. M. *General relativity*. Chicago: Univ. of Chicago Press 1984

Warner, F. *Foundations of differentiable manifolds and Lie groups.* Glenview, Ill.: Scott, Foresman and Co. 1971

Weinberg, S. *Gravitation and cosmology: principles and application of the general theory of relativity.* New York: Wiley 1972

v. Weizsäcker, C. F. *Aufbau der Physik.* München: Hanser Verl. 1985

Wentzel, G. *Quantum theory of fields.* New York: Intersciences Publishers 1949

Wess, J. and Bagger, J. *Supersymmetry and supergravity.* Princeton: Princeton Univ. Press 1983

Wheeler, J. A. and Zurek, W. H. (eds.) *Quantum Theory and Measurement.* Princeton: Princeton University Press 1983

Whitehead, A. N. *Process and Reality,* original edition Cambridge 1929

Wigner, E. P. *Group theory and its application to the quantum mechanics of atomic spectra.* New York: Academic Press 1959

Wigner, E. P. *Symmetries and Reflections.* Bloomington: Indiana Univ. Press 1967

Zeh, H. D. *Die Physik der Zeitrechnung.* Heidelberg: Springer 1984

Author Index and References

Adler, S.

[Adl 77] – Liebermann, J. and Ng, Y. J. *Regularisation of the stress–energy tensor for massless vector particles propagating in a general background metric.* Ann. of Phys. **106**, 279 (1977) VIII.3.2

[Adl 78] – Liebermann, J. and Ng, Y. J. *Trace anomaly of the stress–energy tensor for massless vector particles propagating in a general background metric.* Ann. of Phys. **113**, 294 (1978) VIII.3.2

Aharonov, Y.

[Ahar 59] – and Bohm, D. *Significance of electromagnetic potentials in the quantum theory.* Phys. Rev. **115**, 485 (1959) I.5.4

[Ahar 64] – Bergmann, P. G. and Lebowitz, J. L. *Time symmetry in the quantum process of measurements.* Phys. Rev. **134 B**, 1410 (1964) VII.1

Alfsen, E. M.

See bibliography

[Alf 78] – and Shultz, F. W. *State space of Jordan-Algebras.* Acta Math. **140**, 155 (1978) VII.2

[Alf 80] – Hanche-Olsen, H. and Shultz, F. W. *State space of C^*-Algebras.* Acta Math. **144**, 267 (1980) VII.2

Araki, H.

[Ara 61] *Connection of spin with commutation rules.* J. Math. Phys. **2**, 267 (1961) II.5.1, IV.4

[Ara 62a] *Einführung in die axiomatische Quantenfeldtheorie.* Lecture Notes ETH Zürich, Hepp, K. (ed.) (1962) III.1

[Ara 62b] – Hepp, K. and Ruelle, D. *On the asymptotic behavior of Wightman functions in space-like directions.* Helv. Phys. Acta. **35**, 164 (1962) II.4.4

[Ara 63a] – and Woods, E. J. *Representations of the canonical commutation relations describing a nonrelativistic infinite free Bose gas.* J. Math. Phys. **4**, 637 (1963) II.1.1

[Ara 63b] *A lattice of von Neumann algebras associated with the quantum theory of a free Bose field.* J. Math. Phys. **4**, 1343 (1963) III.4.2
[Ara 64a] *Von Neumann algebras of local observables for free scalar field.* J. Math. Phys.. **5**, 1 (1964) III.4.2
[Ara 64b] *On the algebra of all local observables.* Prog. Theoret. Phys. **32**, 844 (1964) III.3.2
[Ara 64c] *Type of von Neumann algebras associated to the free field.* Prog. Theoret. Phys. **32**, 956 (1964)
[Ara 67] – and Haag, R. *Collision cross sections in terms of local observables.* Commun. Math. Phys. **4**, 77 (1967) II.4.1, II.4.4, VI.1.1
[Ara 72] *Remarks on spectra of modular operators of von Neumann algebras.* Commun. Math. Phys. **28**, 267 (1972) V.2.4
[Ara 77a] – and Sewell, G. L. *KMS conditions and local thermodynamic stability of quantum lattics systems.* Commun. Math. Phys. **52**, 103 (1977) V.1.6
[Ara 77b] – and Kishimoto, A. *Symmetry and equilibrium states.* Commun. Math. Phys. **52**, 211 (1977) V.3.4
[Ara 77c] – Haag, R., Kastler, D. and Takesaki, M. *Extension of KMS states and chemical potential.* Commun. Math. Phys. **53**, 97 (1977) V.3.4
[Ara 77d] *Relative entropy of states of von Neumann algebras.* Publ. RIMS, Kyoto Univ. **13**, 173 (1977)
[Ara 80] *On a characterization of the state space of quantum mechanics.* Commun. Math. Phys. **75**, 1 (1980) VII.2
[Ara 82] – and Yamagami, S. *On quasi-equivalence of quasi-free states of the canonical commutation relations.* Publ. RIMS, Kyoto Univ. **18**, 238 (1972)

Arnold, V. I.
See bibliography

Ashtekar, A.
See bibliography
[Asht 87] *New Hamiltonian formulation of general relativity.* Phys. Rev. **D36**, 1587 (1987) VIII.3.1

Bagger, J.
See bibliography: Wess

Bakamjian, B.
[Bakam 53] – and Thomas, L. H. *Relativistic particle dynamics II.* Phys. Rev. **92**, 1300 (1953) I.3.4

Bannier, U.
[Ban 88] *Allgemein kovariante algebraische Quantenfeldtheorie und Rekonstruktion von Raum-Zeit.* Ph. D. Thesis, Hamburg University (1988) III.3.1

Bargmann, V.
[Barg 47] *Irreducible unitary representations of the Lorentz group.* Ann. Math. **48**, 568 (1947) I.3.2
[Barg 54] *On unitary ray representations of continous groups.* Ann. Math. **59**, 1 (1954) I.3.1
Baumann, K.
See [Schmidt 56]
Baumgärtel, H.
[Baumg 89] *Quasilocal algebras over index sets with a minimal condition.* Operator theory: Adv. and appl. **41**, 43 (1989)
Baym, G.
See bibliography: Kadanoff
Bekenstein, J. D.
[Bek 73] *Black holes and entropy.* Phys. Rev. **D7**, 2333 (1973) VIII.3.3
Bell, J. S.
See bibliography
[Bell 64] *On the Einstein-Podolsky-Rosen paradox.* Physics **1**, 195 (1964) III.1, VII.3
Béllissard, J.
[Belli 78] – and Iochum, B. *Homogeneous self dual cones versus Jordan algebras, the theory revisited.* Ann. Inst. Fourier, Grenoble **28**, 27 (1978) VII.2
Berezin, F.
[Bere 70] and Kac, G. Mat. Sbornik **82**, 331 (1970) (in russian) VIII.2
Bergmann, P. G.
See bibliography
See [Ahar 84]
Birkhoff, G.
[Birk 36] and von Neumann, J. *The logic of quantum mechanics.* Ann. Math. **37**, 823 (1936) VII.2
Birrell, N. D.
See bibliography
Bisognano, J. J.
[Bis 75] – and Wichmann, E.H. *On the duality condition for a Hermitean scalar field.* J. Math. Phys. **16**, 985 (1975) III.4.2, V.4.1
[Bis 76] – and Wichmann, E.H. *On the duality condition for quantum fields.* J. Math. Phys. **17**, 303 (1976) III.4.2, V.4.1
Bjorken, J. D.
See bibliography
Blanchard, Ph.
[Blanch 93] – and Jadzyk, A. *On the interaction between classical and quantum systems.* Phys. Lett. A **175**, 157 (1993) VII.3

[Blanch 95] – and Jadzyk, A. *Event enhanced quantum theory and piecewise deterministic dynamics.* Ann. der Physik **4**, 583 (1995) VII.3

Bleuler, K.

[Bleu 50] *Eine neue Methode zur Behandlung der longitudinalen und skalaren Photonen.* Helv. Phys. Acta **23**, 567 (1950) I.5.3

Bogolubov, N. N.

See bibliography

[Bog 57] – and Parasiuk, O. S. *Über die Multiplikation der Kausalfunktionen in der Quantentheorie der Felder.* Acta Math. **97**, 227 (1957) II.2.4

[Bog 58] *A new method in the theory of superconductivity I.* Sov hys. JETP **7**, 41 (1958) II.1.1

Bohm, D.

See [Ahar 59]

Bohr, N.

See bibliography

[Bohr 49] *Discussions with Einstein.* In: Schilpp: Albert Einstein: Philosopher-Scientist. The Library of Living Philosophers 1949. VII.1

Bongaarts, P. J. M.

[Bon 70] *The electron–position field coupled to external electromognetic potentials as an elementary C^*–algebra theory.* Ann. Phys. **56**, 108 (1970)

[Bon 77] *Maxwell's equations in axiomatic quantum field theory. I. Field tensor and potentials.* J. Math. Phys. **18**, 1510 (1977)

[Bon 82] *Maxwell's equations in axiomatic quantum field theory. II. Covariant and noncovariant gauges.* J. Math. Phys. **23**, 1818 (1982)

Borchers, H. J.

[Borch 60] *Über die Mannigfaltigkeit der interpolierenden Felder zu einer kausalen S-Matrix.* Nuovo Cimento **15**, 784 (1960) II.5.5

[Borch 62] *On the structure of the algebra of the field operators.* Nuovo Cimento **24**, 214 (1962) VIII.3.2

[Borch 63] – Haag, R. and Schroer, B. *The vacuum state in quantum field theory.* Nuovo Cimento **29**, 148 (1963)

[Borch 64] – and Zimmermann, W. *On the self-adjointness of field operators.* Nuovo Cimento **31**, 1074 (1964) III.1

[Borch 65a] *On The vacuum state in quantum field theory II.* Commun. Math. Phys. **1**, 57 (1965) IV.1

[Borch 65b] *Local rings and the connection of spin with statistics.* Commun. Math. Phys. **1**, 291 (1965) IV.1

[Borch 66] *Energy and momentum as observables in quantum field theory.* Commun. Math. Phys. **2**, 49 (1966) VI.2.2
[Borch 67] *A remark on a theorem of B. Misra.* Commun. Math. Phys. **4**, 315 (1967)
[Borch 69a] *On the implementability of automorphism groups.* Commun. Math. Phys. **14**, 305 (1969)
[Borch 69b] *On groups of automorphisms with semibounded spectrum.* In: Systèmes à un nombre infini de degrés de liberté. CNRS, Paris (1969)
[Borch 72] – and Hegerfeldt, G. C. *The structure of space–time transformations.* Commun. Math. Phys. **28**, 259 (1972)
[Borch 84] *Translation group and spectrum condition.* Commun. Math. Phys. **96**, 1 (1984)
[Borch 85] – and Buchholz, D. *The energy–momentum spectrum in local field theories with broken Lorentz symmetry.* Commun. Math. Phys. **79**, 169 (1985) VI.2.2
[Borch 85a] *Locality and covariance of the spectrum.* Fizika **17**, 289 (1985)
[Borch 90] – and Yngvason, J. *Positivity of Wightman functionals and the existence of local nets.* Commun. Math. Phys. **127**, 607 (1990) III.1
[Borch 92] *CTP theorem in two–dimensional theories of local observables.* Commun. Math. Phys. **143**, 315 (1992) V.4.1
[Borch 95] *On the use of modular groups in quantum field theory.* Ann. Inst. Henri Poincaré **63**, 331 (1995) V.4.1

Borek, R.
[Borek 85] *Representations of the current algebra of a charged massless Dirac field.* J. Math. Phys. **26**, 339 (1985)

Bratteli, O.
See bibliography

Braun, H.
See bibliography

Brenig, W.
[Bren 59] – and Haag, R. *Allgemeine Quantentheorie der Stossprozesse.* Fortschr. Phys. **7**, 183 (1959) II.3.1

Bros J.
[Bros 76] – Buchholz, D. and Glaser, V. *Constants of motion in local quantum field theory.* Commun. Math. Phys. **50**, 11 (1976)
[Bros 88] – and Iagolnitzer, D. *Two particle asymptotic completeness in weakly coupled quantum field theories.* Commun. Math. Phys. **119**, 331 (1988) VI.2.3
[Bros 94] – and Buchholz, D. *Towards a relativistic KMS-condition.* Nucl. Phys. **B 429**, 291 (1994) V.1.5

Brunetti, R.
[Brun 95] – Fredenhagen, K. and Köhler, M. *The microlocal spectrum condition and Wick polynomials of free fields on curved spacetimes.* Commun. Math. Phys. to appear 1996 VIII.3.2
Buchholz, D.
See [Bros 76], [Bor 85]
[Buch 74] *Product states for local algebras.* Commun. Math. Phys. **36**, 287 (1974) V.5.2
[Buch 77a] – and Fredenhagen, K. *Dilations and interaction.* J. Math. Phys. **18**, 1107 (1977) V.4.2
[Buch 77b] *Collision theory for massless bosons.* Commun. Math. Phys. **52**, 147 (1977) VI.3
[Buch 77c] – and Fredenhagen, K. *A note on the inverse scattering problem in quantum field theory.* Commun. Math. Phys. **56**, 91 (1977) V.4.2
[Buch 82a] – and Fredenhagen, K. *Locality and the structure of particle states.* Commun. Math. Phys. **84**, 1 (1982) IV.1, IV.1, IV.3
[Buch 82b] *The physical state space of quantum electrodynamics.* Commun. Math. Phys. **85**, 49 (1982) VI.3
[Buch 85] – and Epstein, H. *Spin and statistics of quantum topological charges.* Fysica **17**, 329 (1985) IV.3.4
[Buch 86a] – and Wichmann, E. *Causal independence and the energy-level density of states in local quantum field theory.* Commun. Math. Phys. **106**, 321 (1986) V.5
[Buch 86b] – and Junglas, P. *Local properties of equilibrium states and the particle spectrum in quantum field theory.* Lett. Math. Phys. **11**, 51 (1986) V.5.1
[Buch 86c] – Doplicher, S. and Longo, R. *On Noether's theorem in quantum field theory.* Ann. Phys. **170**, 1 (1986) V.5.3
[Buch 86d] *Gauss' law and the infraparticle problem.* Phys. Lett. **B174**, 331 (1986) VI.2.1
[Buch 87a] *On particles, infraparticles and the problem of asymoptotic completeness.* Proc. IAMP Conf. Marseille, World Scientific 1987 VI.1.1, VI.1.2
[Buch 87b] – and Jacobi, P. *On the nuclearity condition for massless fields.* Lett. Math. Phys. **13**, 313 (1987) V.5.1
[Buch 87c] – D'Antoni, C. and Fredenhagen, K. *The universal structure of local algebras.* Commun. Math. Phys. **111**, 123 (1987) V.5.1, V.5.2
[Buch 88] – Mack, G. and Todorov, I. *The current algebra on the circle as a germ of local field theories.* Nucl. Phys. **B** (Proc. Suppl. 5B), 20 (1988) IV.5
[Buch 89] – and Junglas, P. *On the existence of equilibrium states in local quantum field theory.* Commun. Math. Phys. **121**, 255 (1989) V.5.1

[Buch 90a] – D'Antoni, C. and Longo, R. *Nuclear maps and modular structure II.* Commun. Math. Phys. **129**, 115 (1990) V.5.1, V.5.4

[Buch 90b] *Harmonic analysis of local operators.* Commun. Math. Phys. **129**, 631 (1990) VI.1.1

[Buch 90c] – and Porrmann, M. *How small is the phase space in quantum field theory?* Ann. Inst. H. Poincare **52**, 237 (1990) V.5.1

[Buch 91a] – Porrmann, M. and Stein, U. *Dirac versus Wigner: Towards a universal particle concept in local quantum field theory.* Phys. Lett. **267 B**, 377 (1991) VI.1.2

[Buch 91b] – and Yngvason, J. *Generalized nuclearity conditions and the split property in quantum field theory.* Lett. Math. Phys. **23**, 159 (1991) V.5.2

[Buch 92] – Doplicher, S., Longo, R. and Roberts, J. E. *A new look at Goldstone's theorem.* Rev. Math. Phys. Special Issue, 49 (1992) III.3.2

[Buch 94] *On the manifestation of paricles.* In: Proc. Beer Sheva Conf. 1993 Math. Phys. towards the 21st century, A. N. Sen and A. Gersten (eds.), Ben Gurion of the Negev Press 1994 VI.2.2, VIII.1

[Buch 95] – and Verch, R. *Scaling algebras and renormalization group in algebraic quantum field theory.* Rev. Math. Phys. **7**, 1195 (1995) VIII.1

Carey, A. L.

[Car 77] – Gaffney, J. M. and Hurst, C. A. *A C^*-algebraic formulation of the quantization of the electromagnetic field.* J. Math. Phys. **18**, 629 (1977)

[Car 78] – Gaffney, J. M. and Hurst, C. A. *A C^*-algebraic formulation of gauge transformations of the second kind for the electromagnetic field.* Rep. Math. Phys. **13**, 419 (1978)

Coester, F.

[Coest 60] – and Haag, R. *Representation of states on a field theory with canonical variables.* Phys. Rev. **117**, 1137 (1960) II.1.1

[Coest 65] *Scattering theory for relativistic particles.* Helv. Phys. Acta. **38**, 7 (1965) I.3.4

Colemann, S.

[Col 67] – and Mandula, J. *All possible symmetries of the S-Matrix.* Phys. Rev. **159**, 1251 (1967) III.3.2, VII.2

Connes, A.

[Conn 73] *Une classification des facteurs de type III.* Ann. Sci. Ecole Norm. Sup. **6**, 133 (1973) V.2.4

[Conn 74] *Caractérisation des espaces vectoriels ordonnés sous-jacents aux algèbres de von Neumann.* Ann. Inst. Fourier, Grenoble **24(4)**, 121 (1974) VII.2, V.2.4

Cuntz, J.

[Cuntz 77] *Simple C^*-algebras generated by isometries.* Commun. Math. Phys. **57**, 173 (1977) IV.4

D'Antoni, C.

See [Buch 87c, 90a]

Davies, E. B.

[Dav 70] – and Lewis, J. T. *An operational approach to quantum propability.* Commun. Math. Phys. **17**, 239 (1970)

Davies, P. C. W.

See bibliography: Birrell

[Dav 75] *Scalar particle production in Schwarzschild and Rindler metrics.* J. Phys. **A8**, 609 (1975) V.4.1, VII.3.3

Dell'Antonio, C.

[Dell'Ant 61] *On the connection of spin with statistics.* Ann. Phys. **16**, 153 (1961) II.5.1

Dimmock, J.

[Dim 87] – and Kay, B. S. *Classical and quantum scattering theory for linear scalar on the Schwarzschild metric I.* Ann. of Phys. **175**, 366 (1987) VIII.3.3

Dirac, P. A. M.

[Dir 38] *Classical theory of radiating electrons.* Proc. Roy. Soc. (London) **A 167**, 148 (1938) I.2.2

Dixmier, J.

See bibliography

Doplicher, S.

See [Buch 86c, 92]

[Dopl 65] *An algebraic spectrum condition.* Commun. Math. Phys. **1**, 1 (1965) III.3.2

[Dopl 66] – Kastler, D. and Robinson, D. W. *Covariance algebras in field theory and statistical mechanics.* Commun. Math. Phys. **3**, 1 (1966)

[Dopl 67] – Kadison, R. V., Kastler, D. and Robinson, D. W. *Asymptotically Abelian systems.* Commun. Math. Phys. **6**, 101 (1967) V.3.4

[Dopl 68a] – and Kastler, D. *Ergodic states in a non commutative ergodic theory.* Commun. Math. Phys. **7**, 1 (1968)

[Dopl 68b] – Regge, T. and Singer I. M. *A geometrical model showing the independence of locality and positivity of the energy.* Commun. Math. Phys. **7**, 51 (1968) II.5.4

[Dopl 69a] – Haag, R. and Roberts, J. E. *Fields, observables and gauge transformations I.* Commun. Math. Phys. **13**, 1 (1969) IV.1

[Dopl 69b] – Haag, R. and Roberts, J. E. *Fields, observables and gauge transformations II.* Commun. Math. Phys. **15**, 173 (1969) IV.1, IV.2, IV.4

[Dopl 71] – Haag, R. and Roberts, J. E. *Local observables and particle statistics I.* Commun. Math. Phys. **23**, 199 (1971) IV.1, IV.2

[Dopl 74] – Haag, R. and Roberts, J. E. *Local observables and particle statistics II.* Commun. Math. Phys. **35**, 49 (1974) IV.1, IV.2

[Dopl 82] *Local aspects of superselection rules.* Commun. Math. Phys. **84**, 105 (1982) V.5.3

[Dopl 83] – and Longo, R. *Local aspects of superselection rules II.* Commun. Math. Phys. **88**, 399 (1983) V.5.3

[Dopl 84a] – and Longo, R. *Standard and split inclusions of von Neumann algebras.* Invent. Math. **73**, 493 (1984) V.5.1, V.5.2

[Dopl 84b] – and Roberts, J. E. *Compact Lie groups associated with C^*-Algebras.* Bull. Am. Math. Soc. **11**, 333 (1984)

[Dopl 88] – and Roberts, J. E. *Compact group actions on C^*-algebras.* J. Operator Theory **19**, 283 (1988) IV.4

[Dopl 89a] – and Roberts, J. E. *Endomorphisms of C^*-algebras, cross products and duality for compact groups.* Ann. of Math. **130**, 75 (1989) IV.4

[Dopl 89b] – and Roberts, J. E. *Monoidal C^*-categories and a new duality theory for compact groups.* Invent. Math. **98**, 157 (1989) IV.4

[Dopl 90] – and Roberts, J. E. *Why there is a field algebra with a compact gauge group describing the superselection structure in particle physics.* Commun. Math. Phys. **131**, 51 (1990) IV.3.3, IV.4

[Dopl 95] – Fredenhagen, K. and Roberts, J. E. *The structure of spacetime at the Planck scale and quantum fields.* Commun. Math. Phys. **172**, 187 (1995) VIII.3.4

Drell, S. D.

See bibliography: Bjorken

Driessler, W.

[Driess 74] *On the type of local algebras in quantum field theory.* Commun. Math. Phys. **53**, 295 (1974) V.6

[Driess 79] *Duality and the absence of locally generated superselection sectors for CCR-type algebras.* Commun. Math. Phys. **70**, 213 (1979)

[Driess 86] – Summers, S. J. and Wichmann, E. H. *On the connection between quantum fields and von Neumann algebras of local operators.* Commun. Math. Phys. **105**, 49 (1986) III.1

Drinfeld, V. G.
[Drin 87] *Quantum Groups.* In Proc. ICM, Berkeley 1986, Academic Press 1987 IV.5

Drühl, K.
[Drühl 70] – Haag, R. and Roberts, J. E. *On parastatistics.* Commun. Math. Phys. **18**, 204 (1970) IV.1

Dunford, N.
See bibliography

Dyson, F. J.
[Dy 49] *The S-Matrix in quantum electrodynamics.* Phys. Rev. **75**, 1736 (1949) I.5.5, II.2.4
[Dy 67] – and Lenard, A. *Stability of matter I.* J. Math. Phys. **8**, 423 (1967) V.1.4
[Dy 68] – and Lenard, A. *Stability of matter II.* J. Math. Phys. **9**, 698 (1968) V.1.4
[Dy 92] *Quantum Past: The Limitations of Quantum Theory.* Schrödinger Lecture May 92, Imperial College London. unpublished VII.1

Eckmann, J. P.
[Eck 73] – and Osterwalder, K. *An application of Tomita's theory of modular Hilbert algebras: duality for free Bose algebras.* Funct. An. **13**, 1 (1973)

Edwards, C. M.
[Edw 70] *The operational approach to algebraic quantum field theory.* Commun. Math. Phys. **16**, 207 (1970)

Ekstein, H.
[Ek 56a] *Theory of time dependent scattering for multichannel processes.* Phys. Rev. **101**, 880 (1956) II.3.1
[Ek 56b] *Scattering in field theory.* Nuovo Cimento **4**, 1017 (1956) II.3.1
[Ek 69] *Presymmetry II.* Phys. Rev. **184**, 1315 (1969) I.1, V.3.3, VIII.1

Ellis, G. F. R.
See bibliography: Hawking

Emch, G. E.
See bibliography

Enss, V.
[Enss 75] *Characterisation of particles by means of local observables.* Commun. Math. Phys. **45**, 35 (1975) VI.2.1
[Enss 83] *Asymptotic observables on scattering states.* Commun. Math. Phys. **89**, 245 (1983) VI.2.3

Epstein, H.
See [Buch 85]
[Ep 63] *On the Borchers class of a free field.* Nuovo Cimento **27**, 886 (1963) II.5.5

[Ep 67] *CTP-Invariance of the S-Matrix in a theory of local observables.* J. Math. Phys. **8**, 750 (1967) IV.2.4

[Ep 69] - Glaser, V. and Martin, A. *Polynomial behaviour of scattering amplitudes at fixed momentum transfer for theories with local observables.* Commun. Math. Phys. **13**, 257 (1969) II.5.2

[Ep 71] - and Glaser, V. *Le role de la localité dans la renormalisation pertubative en théorie quantique des champs.* In *Statistical Mechanics and Quantum Field Theory.* Stora, R. and DeWitt, C. (eds.), London: Gordon and Breach (1971) II.2.3, II.2.4

d'Espagnat, B.
See bibliography

Evans, D. E.

[Ev 80] *A review of semigroups of completely positive maps.* In: *Mathematical problems in theoretical physics.* Osterwalder, K. (ed.), Berlin: Springer 1980

Ezawa, H.

[Ez 67] - and Swieca, J. A. *Spontaneous breakdown of symmetries and zero-mass states.* Commun. Math. Phys. **5**, 330 (1967) III.3.2

Faddeev, L.

[Fadd 71] - and Kulish, P. P. *Asymptotic conditions and infrared divergences in quantum electrodynamics.* Theor. Math. Phys. **4**, 745 (1971) VI.2.1

[Fadd 90] *Integrable models for quantum groups.* In: *Fields and particles.* Proc. Schladming 1990, Mitter, H., Schweiger, W. (eds.), Springer: Berlin 1990 IV.5

Fannes, M.

[Fann 77a] - and Verbeure, A. *Correlation inequalities and equilibrium states.* Commun. Math. Phys. **55**, 125 (1977) V.1.6

[Fann 77b] - and Verbeure, A. *Correlation inequalities and equilibrium states II.* Commun. Math. Phys. **57**, 165 (1977) V.1.6

Federbush, P.

[Fed 68] *Local operator algebras in the presence of superselection rules.* J. Math. Phys. **9**, 1718 (1968)

Fell, J. M. G.

[Fell 60] *The dual spaces of C^*-algebras.* Trans. Amer. Math. Soc. **94**, 365 (1960) III.2.2

Ferrara, S.
See bibliography

Feynman, R.
See bibliography

[Feyn 48] *Space-time approach to non-relativistic quantum mechanics.* Rev. Mod. Phys. **20**, 367 (1948) I.4, VII.1

Finkelstein, D.

[Fink 62] – Jauch, J.M., Schiminovich, S. and Speiser, D. *Foundations of quaternion quantum mechanics.* J. Math. Phys. **3**, 207 (1962) VII.2

[Fink 68] *Space–time code.* Phys. Rev. **184**, 1261 (1968) VII.3

Foulis, D. J.

[Foul 60] *Baer * semigroups.* Proc. Amer. Math. Soc. **11**, 648 (1960) VII.2

Fredenhagen, K.

See [Brun 96], [Buch 77a, 77c, 82a, 87c], [Dopl 95]

[Fred 81a] *On the existence of antiparticles.* Commun. Math. Phys. **79**, 141 (1981) IV.3.4

[Fred 81b] – and Hertel, J. *Local algebras of observables and pointlike localized fields.* Commun. Math. Phys. **80**, 555 (1981) III.1

[Fred 85a] *On the modular structure of local algebras of observables.* Commun. Math. Phys. **97**, 79 (1985) V.6

[Fred 85b] *A remark on the cluster theorem.* Commun. Math. Phys. **97**, 461 (1985) II.4.4

[Fred 87] – and Haag, R. *Generally covariant quantum field theory and scaling limits.* Commun. Math. Phys. **108**, 91 (1987) III.3.1, VIII.3.2

[Fred 89a] – Rehren, K. H. and Schroer, B. *Superselection sectors with braid group statistics and exchange algebras I.* Commun. Math. Phys. **125**, 201 (1989) IV.5

[Fred 89b] *Structure of superselection sectors in low dimensional quantum field theory.* Proceedings Lake Tahoe Conf. 1989. Plenum Publ. Corp. 1990 IV.5

[Fred 90] – and Haag, R. *On the derivation of Hawking radiation associated with the formation of a black hole.* Commun. Math. Phys. **127**, 273 (1990) VIII.3.3

[Fred 92] – Rehren, H. and Schroer, B. *Superselection sectors with braid group statistics and exchange algebras II. Geometric aspects and conformal covariance.* Rev. Math. Phys. Special issue 1992, 113 (1992) IV.5

Friedrichs, K. O.

See bibliography

Fröhlich, J.

[Fröh 79a] – Morchio, G. and Strocchi, F. *Infrared problem and spontaneous breaking of the Lorentz group in QED.* Phys. Lett. **89B**, 61 (1979) VI.2.1

[Fröh 79b] – Morchio, G. and Strocchi, F. *Charged sectors and scattering states in quantum electrodynamics.* Ann. of Phys. **119**, 241 (1979) VI.2.1, VI.3

[Fröh 79c] *The charged sectors of quantum electrodynamics in a framework of local observables.* Commun. Math. Phys. **66**, 223 (1979) VI.3

[Fröh 88] *Statistics of fields. The Yang-Baxter equation and the theory of knots and links.* In: *Nonperturbative quantum field theory.* G. 't Hooft, A. Jaffe, G. Mack, P. K. Mitter and R. Stora (eds.), Plenum Publishing Corporation(1988) IV.5

[Fröh 89] – and Marchetti, P. A. *Quantum field theories of vortices and anyons.* Commun. Math. Phys. **121**, 177 (1989) IV.5

[Fröh 90] – and Gabbiani, F. *Braid statistics in local quantum theory.* Rev. Math. Phys. **2**, 251 (1990) IV.5

[Fröh 91] – and Kerler, T. Zürich, preprint 1991 IV.5

Fulling, S. A.

[Full 73] *Non uniqueness of canonical field quantization in Riemannian space-time.* Phys. Rev. **D7**, 2850 (1973) VIII.3.2

[Full 81a] – Sweeny, M. and Wald, R. M. *Singularity structure of the two-point function in quantum field theory in curved space time.* Commun. Math. Phys. **63**, 257 (1981) VIII.3.2

[Full 81b] – Narcowich, F. J. and Wald, R. M. *Singularity structure of the two-point function in quantum field theory in curved space time II.* Ann. of Phys. **136**, 243 (1981) VIII.3.2

Gabbiani, F.

See [Fröh 90]

Gaffney, J. M.

See [Car 77, 78]

Garber, W. D.

[Garb 78a] – and Reeh, H. *Nontranslationally covariant currents and associated symmetry generators.* J. Math. Phys. **19**, 597 (1978) VIII.2

[Garb 78b] – and Reeh, H. *Nontranslationally covariant currents and symmetries of the S-Matrix.* J. Math. Phys. **19**, 985 (1978) VIII.2

Garding, L.

[Gard 54a] – and Wightman, A.S. *Representations of the anticommutation relations.* Proc. Nat. Acad. Sci. **40**, 617 (1954) II.1.1

[Gard 54b] – and Wightman, A.S. *Representations of the commutation relations.* Proc. Nat. Acad. Sci. **40**, 622 (1954) II.1.1

Gelfand, I. M.

See bibliography

Gell-Mann, M.
[Gell 51] – and Low, F. *Bound states in quantum field theory.* Phys. Rev. **84**, 350 (1951) II.2.4
[Gell 90] – and Hartle, J. B. In: *Complexity, Entropy and the Physics of Information.* W. Zurek (ed.). Reading: Addison Wesley 1990, and *Proc. of the 25th International Conf. on High Energy Physics, Singapore 1990.* K. K. Phua and Y. Yamaguchi (eds.), Singapore: World Scientific 1990
[Gell 94] – and Hartle, J. B. *Proc of the NATO-Workshop "On the physical origins of time asymmetry".* J. Halliwell, J. Perez-Mercader and W. Zurek (eds.), Cambridge Univ. Press 1994 VII.3
Glaser, V.
See [Ep 69, 71], [Bros 76]
[Glas 57] – Lehmann, H. and Zimmermann, W. *Field operators and retarded functions.* Nuovo Cimento **6**, 1122 (1957) II.2.6
Glimm, J.
See bibliography
[Glimm 61] *Type I C^*-algebras.* Ann. Math. **73**, 572 (1961) II.1.1
Goldstone, J.
[Gold 61] *Field theories with "superconductor" solutions.* Nuovo Cimento **19**, (1961) III.3.2
Graf, G. M.
[Graf 90] *Asymptotic completeness for N–body short range quantum systems: a new proof.* Commun. Math. Phys. **132**, 73 (1990) VI.2.3
Green, H. S.
[Green 53] *A generalized method of field quantization.* Phys. Rev. **90**, 270 (1953) I.3.4, IV.1
Greenberg, O. W.
[Greenb 64] – and Messiah, A. M. L. *Symmetrization postulate and its experimental foundation.* Phys. Rev. **136B**, 248 (1964) IV.1
[Greenb 65] – and Messiah, A. M. L. *Selection rules for parafields and the absence of para particles in nature.* Phys. Rev. **138B**, 1155 (1965) IV.1
Grifiths, R. B.
[Griff 84] *Consistent histories.* J. Stat. Phys. **36**, 219 (1984) VII.3
Guido, D.
[Guido 95a] – and Longo, R. *An algebraic statistics theorem.* Commun. Math. Phys. **172**, 517 (1995) IV.5
[Guido 95b] – and Longo, R. *The conformal spin and statistics theorem.* Preprint Dipart. Mat. Univ. Rome Tor Vergata 1995 IV.5

Gunson, J.
[Gun 67] *On the algebraic structure of quantum mechnanics.* Commun. Math. Phys. **6**, 262 (1967)

Gupta, S. N.
[Gup 50] *Theory of longitudinal photons in quantum electrodynamics.* Proc. Phys. Soc. (London) **A 63**, 681 (1950) I.5.3

Haag, R.
See [Bren 59], [Coest 60], [Borch 63], [Ara 67], [Dopl 69a, 69b], [Drühl 70], [Dopl 71, 74], [Ara 77c], [Fred 87, 90]
[Haag 54] Lecture Notes Copenhagen, CERN T/RH1 53/54 II.1.1, II.1.2
[Haag 55a] *On quantum field theories.* Dan. Mat. Fys. Medd. **29** no. 12 (1955) II.1.1, II.1.2
[Haag 55b] *Die Selbstwechselwirkung des Elektrons.* Z. Naturforsch. **10a**, 752 (1955) I.2.2
[Haag 57] *Discussion des "axiomes" et des propriétés asymptotiques d'une théorie des champs locale avec particules composées.* In: *Les problèmes mathématique de la théorie quantique des champs* (Lille 1957). Paris: Centre National de la Recherche Scientifique 1959 II.4.1, III.1
[Haag 58] *Quantum field theories with composite particles and asymptotic conditions.* Phys. Rev. **112**, 669 (1958) II.4.2, II.4.4
[Haag 62a] – and Schroer, B. *Postulates of quantum field theory.* J. Math. Phys. **3**, 248 (1962) III.1, III.4.2
[Haag 62b] *The mathematical structure of the Bardeen–Cooper–Schrieffer model.* Nuovo Cimento **25**, 287 (1962) III.3.2
[Haag 63] *Bemerkungen zum Nahwirkungsprinzip in der Quantenphysik.* Ann. d. Physik **11**, 29 (1963) III.4.2
[Haag 64] – and Kastler, D. *An algebraic approach to quantum field theory.* J. Math. Phys. **5**, 848 (1964) I.1, III.1
[Haag 65] – and Swieca, J. A. *When does a quantum field theory describe particles?* Commun. Math. Phys. **1**, 308 (1965) V.5.1
[Haag 67] – Hugenholtz, N. M. and Winnink, M. *On the equilibrium states in quantum statistical mechanics.* Commun. Math. Phys. **5**, 215 (1967) V.1.1
[Haag 70] – Kadison, R. V. and Kastler, D. *Nets of C^*-algebras and classification of states.* Commun. Math. Phys. **16**, 81 (1970)
[Haag 73] – Kadison, R. V. and Kastler, D. *Asymptotic orbit of states of a free Fermi gas.* Commun. Math. Phys. **33**, 1 (1973)
[Haag 74] – Kastler, D. and Trych–Pohlmeyer, E. B. *Stability and equilibrium states.* Commun. Math. Phys. **38**, 173 (1974) V.3.2

[Hepp 65] *On the connection between the LSZ and Wightman quantum field theory.* Commun. Math. Phys. **1**, 95 (1965) II.4.4

[Hepp 72] *Quantum theory of measurement and macroscopic observables.* Helv. Phys. Acta **45**, 237 (1972)

[Hepp 73] - and Lieb, E. H. Helv. Phys. Acta **46**, 573 (1973) VII.1

Hertel, J.

See [Fred 81]

Hertel, P.

[Hert 72] - Narnhofer, H. and Thirring, W. *Thermodynamic functions for Fermions with gravostatic and electrostatic interactions.* Commun. Math. Phys. **28**, 159 (1972) V.1.4

Hibbs, A. R.

See bibliography: Feynman

Hislop, P. D.

[Hisl 82] - and Longo, R. *Modular structure of the local algebras associated with the free massless scalar field theory.* Commun. Math. Phys. **84**, 71 (1982) V.4.2

Horuzhy, S. S.

See [Poli 73], bibliography

van Hove, L.

[Hove 52] *Les difficultés de divergences pour un modèle particulier de champs quantifié.* Physica **18**, 159 (1952) II.1.1

Hugenholtz, N. M.

See [Haag 67]

[Hug 67] *On the factor type of equilibrium states in quantum statistical mechanics.* Commun. Math. Phys. **6**, 189 (1967)

[Hug 75] - and Kadison, R. V. *Automorphisms and quasifree states of the CAR-algebra.* Commun. Math. Phys. **43**, 181 (1975)

Hunziker, W.

[Hunz 68] *S-matrix in classical mechanics.* Commun. Math. Phys. **8**, 282 (1968) VI.2.3

Hurst, C. A.

See [Car 77, 78]

Iagolnitzer, D.

See [Bros 88]

[Iag 87] *Asymptotic completeness and multiparticle analysis in quantum field theories.* Commun. Math. Phys. **110**, 51 (1987) VI.2.3

Iochum, B.

See [Belli 78], bibliography

Israel, R. B.

See bibliography

Itzykson, C.

See bibliography

Jacobi, P.
See [Buch 87b]
Jadczyk, A. Z.
See [Blanch 93, 95]
[Jad 69] *On spectrum of internal symmetries of the algebraic quantum field theory.* Commun. Math. Phys. **12**, 58 (1969)
Jaffe, A.
See bibliography: Glimm
[Jaffe 67] *High energy behavior in quantum field theory I. Strictly localizable fields.* Rev. Phys. **158**, 1454 (1967) V.5.2
Jauch, J. M.
See [Fink 62], bibliography
[Jauch 69] – and Piron, C. *On the structure of quantal proposition systems.* Helv. Phys. Acta **42**, 842 (1969) VII.2
Jones, V. F. R.
[Jones 83] *Index for subfactors.* Invent. Math. **72**, 1 (1983) IV.5
Jordan, P.
[Jord 34] – von Neumann, J. and Wigner, E. P. *On an algebraic generalization of the quantum mechanical formulation.* Ann. Math. **35**, 29 (1934) VII.2
Jost, R.
See bibliography
[Jost 57] *Eine Bemerkung zum CTP-Theorem.* Helv. Phys. Acta. **30**, 409 (1957) II.5.1
Junglas, P.
See [Buch 86b, 89]
[Jung 87] *Thermodynamisches Gleichgewicht und Energiespektrum in der Quantenfeldtheorie.* Thesis, University Hamburg 1987 V.5.1
Junker, W.
[Junk 95] *Adiabatic vacua and Hadamard states for scalar quantum fields on curved spacetime.* Thesis Hamburg 1995 VIII.3.2
Kac, G.
See [Bere 70]
Kadanoff, L. P.
See bibliography
Kadison, R. V.
See [Dopl 67], [Haag 70, 73], [Hug 75], bibliography
[Kad 65] *Transformation of states in operator theory and dynamics.* Topology **3**, Suppl. 2, 177 (1965)
[Kad 67] *The energy-momentum spectrum of quantum fields.* Commun. Math. Phys. **4**, 258 (1967)
Kahn, B.
[Kahn 38] and Uhlenbeck, G. E. *On the theory of condensation.* Physica **5**, 399 (1938) II.2.2

Kamefuchi, S.
See [Ohnu 68], [Ohnu 69]
Kastler, D.
See [Haag 64], [Dopl 66, 67, 68a], [Haag 70, 73, 74], [Ara 77c], bibliography
[Kast 66] – Robinson, D.W. and Swieca, J. A. *Conserved currents and associate symmetries; Goldstone's theorem.* Commun. Math. Phys. **2**, 108 (1966) III.3.2
[Kast 66] – and Robinson, D.W. *Invariant states in statistical mechanics.* Commun. Math. Phys. **3**, 151 (1966)
[Kast 72] – Loupias, G., Mebkhout, M. , and Michel, L. *Central decomposition of invariant states.* Commun. Math. Phys. **27**, 192 (1972)
[Kast 76] *Equilibrium states of matter and operator algebras.* Sympos. Math. **20**, 49 (1976) V.3.4
[Kast 79] – and Takesaki, M. *Group duality and KMS-states.* Commun. Math. Phys. **70**, 193 (1979)
[Kast 82] – and Takesaki, M. *Group duality and KMS-states II.* Commun. Math. Phys. **85**, 155 (1982)
Kay, B.
See [Dim 87]
[Kay 87] – and Wald, R. M. *Some recent developments related to the Hawking effect.* In: Proceedings Int. Conf. on Differential Geom. Methods in Theoret. Phys., Doebner, H. D. and Hennig, J. D. (ed.)., Singapore: World Scientific 1987 VIII.3.2, VIII.3.3
Kerler, T.
See [Fröh 91]
Kibble, T. W. B.
[Kibble 68a] *Mass shell singularities of Green's functions.* Phys. Rev. **173**, 1527 (1968) VI.2.1
[Kibble 68b] *Asymptotic states.* Phys. Rev. **174**, 1882 (1968) VI.2.1
Kishimoto, A.
See [Ara 77b]
Knight, J. M.
[Knight 61] *Strict localisation in quantum field theory.* J. Math. Phys. **2**, 459 (1961) V.5.1
Koecher, M.
See bibliography: Braun
Köhler, M.
[Köhl 95] *New examples for Wightman fields on a manifold.* Class. Quantum Grav. **12**, 1413 (1995) VIII.3.2
Kosaki, H.
[Kos 86] *Extension of Jones' theory on index to arbitrary factors.* J. Func. Anal. **66**, 123 (1986) IV.5

Kraus, K.
[Kraus 77] – Polley, L. and Reents, G. *Models for infrared dynamics I. Classical currents.* Ann. Inst. Henri Poincare **26**, 109 (1977) VI.3
Kubo, R.
[Kubo 57] *Statistical mechanical theory of irreversible processes I.* J. Math. Soc. Japan **12**, 570 (1957) V.1.1
Kulish, P. P.
See [Fadd 71]
Landau, L. J.
[Land 69] *A note on extended locality.* Commun. Math. Phys. **135**, 246 (1969) III.4.2
[Land 87] *On the violation of Bell's inequality in quantum theory.* Phys. Lett. **120**, 54 (1987)
Landshoff, P. V.
[Lands 67] – and Stapp, H. P. *Parastatistics and a unified theory of identical particles.* Ann. Phys. (N.Y.) **45**, 72 (1967) IV.1
Laporte, O.
[Lap 31] – and Uhlenbeck, G. E. *Application of spinor analysis fo the Maxwell and Dirac equations.* Phys. Rev. **37**, 1380 (1931) I.2.1
Lebowitz, J. L.
See [Ahar 64]
Lehmann, H.
See [Glas 57]
[Leh 54] *Über Eigenschaften von Ausbreitungsfunktionen und Renormierungskonstanten quantisierter Felder.* Nuovo Cimento **11**, 342 (1954) II.3.3
[Leh 55] – Symanzik, K. and Zimmermann, W. *Zur Formulierung quantisierter Feldtheorien.* Nuovo Cimento **1**, 425 (1955) II.1.2, II.3.3
[Leh 57] – Symanzik, K. and Zimmermann, W. *On the formulation of quantized field theories II.* Nuovo Cimento **6**, 320 (1957) II.2.6
Lenard, A.
See [Dy 67, 68]
Lewis, J. T.
See [Car 77, 78]
Licht, A. L.
[Licht 63] *Strict localisation.* J. Math. Phys. **4**,1443 (1963) V.5.1
Lieb, E. H.
See [Hepp 73]
[Lieb 75a] – and Thirring, W. *Bound for the kinetic energy of fermions wich proves the stability of matter.* Phys. Rev.

Lett. **35**, 687 (1975) and *erratum* Phys. Rev. Lett. **35**, 1116 (1975) V.1.4
Liebermann, J.
See [Adl 77, 78]
Logunov, A. A.
See bibliography: Bogolubov
Longo, R.
See [Hisl 82], [Dopl 83, 84], [Buch 86c, 90a, 92], [Guido 95a, 95b]
[Longo 89] *Index of subfactors and statistics of quantum fields.* Commun. Math. Phys. **126**, 217 (1989) IV.5
[Longo 90] *Index of subfactors and statistics of quantum fields II.* Commun. Math. Phys. **130**, 285 (1990) IV.5
[Longo 92] *Minimal index and braided subfactors.* J. Funct. Anal. **101**, 98 (1992) IV.5
[Longo 95] – and Rehren, H. *Nets of subfactors.* Rev. Math. Phys. **7**, 567 (1995) IV.5
Lopuszanski, J.
See [Haag 75], bibliography
Loupias, G.
See [Kast 72]
Low, F.
See [Gell 51]
Ludwig, G.
See bibliography
Lüders, Ch.
[Lüd 90] – and Roberts, J. E. *Local quasiequivalence and adiabatic vasuum states.* Commun. Math. Phys. **134**, 29 (1990) VIII.3.2
Mack, G.
See [Buch 88]
[Mack 90] – and Schomerus, V. *Conformal field algebras with quantum symmetry from the theory of superselection sectors.* Commun. Math. Phys. **134**, 139 (1990) IV. 5
[Mack 91a] – and Schomerus, V. *Quasi Hopf symmetry in quantum theory.* Nucl. Phys. **B 370**, 185 (1992) IV. 5
[Mack 91b] – and Schomerus, V. *Quasi quantum group symmetry and local braid relations in conformal Ising model.* Phys. Lett. **267 B**, 207 (1991) IV. 5
Mackey, G. W.
[Mackey 57] *Borel structure in groups and their duals.* Trans. Amer. Math. Soc. **85**, 134 (1957) II.1.1
Majid, S.
[Maj 89] *Quasitriangular Hopf algebras and Yang Baxter equation.* Int. J. Mod. Phys. **A 5**, 1 (1990) IV.5

Maison, D.

[Maison 71] – and Reeh, H. *Properties of conserved local currents and symmetry transformations.* Nuovo Cimento **1A**, 78 (1971)

Mandula, J.

See [Col 67]

Marchetti, P. A.

See [Fröh 89]

Martin, A.

See [Ep 69]

Martin, P. C.

[Mart 59] – and Schwinger, J. *Theory of many particle systems: I.* Phys. Rev. **115**, 1342 (1959) V.1.1

Mayer, J. E.

[May 37] *The statistical dynamics of condensing systems I.* J. Chem. Phys. **5**, 67 (1937) II.2.2

Mebkhout, M.

See [Kast 72]

Messiah, A. M. L.

See [Greenb 64, 65]

Michel, L.

See [Kast 72]

[Mich 65] *Relations between internal symmetry and relativistic invariance.* Phys. Rev. **137B**, 405 (1965) VIII.2

Mielnik, B.

[Miel 69] *Theory of filters.* Commun. Math. Phys. **15**, 1 (1969) VII.2

[Miel 74] *Generalized quantum mechanics.* Commun. Math. Phys. **37**, 221 (1974) VII.2

Misra, B.

[Mis 65] *On representations of Haag fields.* Helv. Phys. Acta. **38**, 189 (1965)

Morchio, G.

See [Fröh 79a, 79b]

[Morch 85] – and Strocchi, F. *Spontaneous symmetry breaking and energy gap generated by variables at infinity.* Commun. Math. Phys. **99**, 153 (1985) III.3.2

[Morch 87] – and Strocchi, F. *Mathematical structure for long range dynamics and symmetry breaking.* J. Math. Phys. **28**, 622 (1987) III.3.2

Moser, J.

See bibliography

Mourre, E.

[Mour 79] *Link between the geometric and the spectral transformation approaches in scattering theory.* Commun. Math. Phys. **68**, 91 (1979) I.1

O'Raifeartaigh, L.
[O'Raif 65] *Mass differences and Lie algebras of finite order.* Phys. Rev. Lett. **14**, 575 (1965) VIII.2
Orzalesi, C.
[Orz 70] *Charges as generators of symmetry transformations in quantum field theory.* Rev. Mod. Phys. **42**, 381 (1970) VIII.2
Osterwalder, K.
See [Eck 73]
[Ost 73] – and Schrader, R. *Axioms for Euclidean Green's functions, I.* Commun. Math. Phys. **31**, 83 (1973) II.2.7
[Ost 75] – and Schrader, R. *Axioms for Euclidean Green's functions, II.* Commun. Math. Phys. **41**, 281 (1975) II.2.7
Parasiuk, O. S.
See [Bog 57]
Paunov, R. R.
See [Hadj 90]
Pedersen, G. K.
See bibliography
Peierls, R.
[Pei 52] *The commutation laws of relativistic field theory.* Proc. Roy. Soc. (London), **A 214**, 143 (1952) I.4
Penrose, R.
[Pen 67] *Twistor algebra.* J. Math. Phys. **8**, 345 (1967) I.2.1
Pimsner, M.
[Pim 86] – and Popa, S. *Entropy and index for subfactors.* Ann. Sci. Ecole Norm. Sup. **19**, 57 (1986) IV.5
Piron, C.
See [Jauch 69]
[Pir 64] *Axiomatique quantique.* Helv. Phys. Acta **37**, 439 (1964) I.1
Pohlmeyer, K.
[Pohl 72] *The equation curl $W = 0$ in quantum field theory.* Commun. Math. Phys. **25**, 73 (1972)
Polivanov, M. K.
[Poli 73] – Sushko, V. N. and Horuzhy, S. S. *Axioms of observable algebras and the field concept.* Theor. Math. Phys. **16**, 629 (1973)
Polley, L.
See [Kraus 77]
Pool, J. C. T.
[Pool 68a] *Baer *-semigroups and the logic of quantum mechanics.* Commun. Math. Phys. **9**, 118 (1968) VII.2
[Pool 68b] *Semimodularity and the logic of quantum mechanics.* Commun. Math. Phys. **9**, 212 (1968) VII.2

Popa, S.
See [Pim 86]
Porrmann, M.
See [Buch 90c, 91a]
Powers, R.
[Powers 67] *Representation of uniformly hyperfinite algebras and associated von Neumann rings.* Ann. Math. **86**, 138 (1967) V.2.4
Pusz, W.
[Pusz 78] and Woronowicz, S. L. *Passive states and KMS states for general quantum systems.* Commun. Math. Phys. **58**, 273 (1978) V.3.3
Radzikowski, M. J.
[Radz 92] *The Hadamard Condition and Kay's conjecture in axiomatic quantum field theory on curved spacetime.* PhD thesis, Princeton Univ. 1992 VIII.3.2
Reeh, H.
See [Maison 71], [Garb 78a, 78b]
[Reeh 61] – and Schlieder, S. *Bemerkungen zur Unitäräquivalenz von Lorentzinvarianten Feldern.* Nuovo Cimento **22**, 1051 (1961) II.5.2
Reents, G.
See [Kraus 77]
Regge, T.
See [Dopl 68]
Rehren, K. H.
See [Fred 89a, 92], [Longo 95]
[Rehr 88] *Locality of conformal fields in two dimensions: Exchange algebras on the light cone.* Commun. Math. Phys. **116**, 675 (1988) IV.5
[Rehr 89] – and Schroer, B. *Einstein causality and Artin braids.* Nucl. Phys. **B312**, 715 (1989) IV.5
[Rehr 91] *Field Operators for anyons and plektons.* Commun. Math. Phys. **145**, 123 (1992) IV.5
[Rehr 95] *On the range of the index of subfactors.* J. Funct. Anal. **134**, 183 (1995) IV.5
Rindler, W.
[Rind 66] *Kruskal space and the uniformly accelerated frame.* Am. J. Phys. **34**, 1174 (1966) IV.4.1
Ringrose, J. R.
See bibliography: Kadison
Roberts, J. E.
See [Buch 92], [Dopl 69a, 69b, 71, 74, 84b, 88, 89, 90, 95], [Drühl 70], [Lüd 90]

[Schroer 74] – and Swieca,J. A. *Conformal transformations for quantized fields.* Phys. Rev. D **10**, 480 (1974) V.4.2

Schwartz, L.

See bibliography

Schwinger, J.

See [Mart 59]

[Schwing 59] *Euclidean quantum electrodynamics.* Phys. Rev. **115**, 721 (1959) II.2.7

Segal, I. E.

[Seg 47] *Postulates for general quantum mechanics.* Ann. Math. **48**, 930 (1947) I.1, III.1, III.2.2

[Seg 57] *Caractérisation mathématique des observables en théorie quantique des champs et ses conséquences pour la structure des particules libres.* In: *Les problèmes mathématique de la théorie quantique des champs* (Lille 1957). Paris: Centre National de la Recherche Scientifique 1959 III.1

Sewell, G. L.

See [Ara 77], [Narn 81], bibliography

[Sew 80] *Relativity of Temperature and the Hawking-Effekt.* Phys. Rev. Lett. **79A**, 23 (1980) VIII.3.3

Sezgin, E.

See bibliography

Shilov, G. E.

See bibliography: Gelfand

Shirkov, D. V.

See bibliography: Bogolubov

Shultz, F. W.

See [Alf 78], [Alf 80], bibliography: Alfsen

Sigal, I. M.

See bibliography

Singer, I. M.

See [Dopl 68]

Sohnius, M.

See [Haag 75]

Speiser, D.

See [Fink 62]

Stapp, H. P.

See [Lands 67]

[Stapp 77] *Theory of Reality,* Found. of Phys. **7**, 313 (1977)

[Stapp 79] *Whiteheadian Approach to Quantum Theory and Generalized Bell's Theorem.* Found. of Phys. **9**, 1 (1979)

Stein, U.

See [Haag 84], [Buch 91a]

[Stein 89] Thesis, Hamburg Univ. 1989 VI.1.2

Steinmann, O.
See bibliography
[Steinm 91] *Asymptotic completeness in QED. 1. Quasilocal states.* Nucl. Phys. **B350**, 355 (1991) VI.2.3
Störmer, E.
[Störm 72] *Spectra of states and asyptotically abelian C*-algebras.* Commun. Math. Phys. **28**, 279 (1972) V.2.4
Stolt, R. H.
[Stolt 70] and Taylor, J. R. *Classification of paraparticles.* Phys. Rev. **D1**, 2226 (1970) IV.1
Stratila, S.
See bibliography
Streater, R. W.
See bibliography
[Streat 68] *On certain non-relativistic quantized fields.* Commun. Math. Phys. **7**, 93 (1968) V.1.4
[Streat 70] and Wilde, I. F. *Fermion states of a Boson field.* Nucl. Phys. **B24**, 561 (1970) IV.5
Strocchi, F.
See [Fröh 79a, 79b], [Morch 85, 87]
Summers, S. J.
See [Driess 86]
[Summ 87a] *From algebras of local observables to quantum fields: generalized H bounds.* Helv. Phys. Acta **60**, 1004 (1987) III.1
[Summ 87b] – and Werner, R. *Maximal violation of Bell's inequalities is generic in quantum field theory.* Commun. Math. Phys. **110**, 247 (1987)
Sushko, V. N.
See [Poli 73]
Sweeny, M.
See [Full 81a, 81b]
Swieca, J. A.
See [Haag 65], [Kast 66], [Ez 67], [Schroer 74]
[Swieca 67] *Range of forces and broken symmetries in many-body systems.* Commun. Math. Phys. **4**, 1 (1967) III.3.2
[Swieca 73] – and Voelkel, A. H. *Remarks on conformal invariance.* Commun. Math. Phys. **29**, 319 (1973) V.4.2
Symanzik, K.
See [Leh 55], [Leh 57]
[Sy 66a] *Euclidean quantum field theory I. Equation for a scalar model.* J. Math. Phys. **7**, 510 (1966) II.2.7
[Sy 66b] *Euclidean quantum field theory.* In: *Local quantum theory. Scuola Internaz. di Fisica "Enrico Fermi", Corso 45.* Jost, R. (ed.), New York: Acadamic Press 1969 II.2.7

[Sy 67] *Many particle structure of Green's functions.* In: *Symposia on Theoretical Physics, Vol. 3.* Ramakrishnan, A. (ed.), New York: Plenum Press 1967 — II.2.5
[Sy 70] *Small distance behaviour in field theory and power counting.* Commun. Math. Phys. **18**, 227 (1970) — II.2.4
[Sy 72] *Small distance behavior anlysis and Wilson expansion.* Commun. Math. Phys. **23**, 49 (1971) — II.2.4

Takesaki, M.
See [Ara 77c], [Kast 79], [Kast 82], bibliography

Taylor, J. R.
See [Stolt 70]

Thirring, W.
See [Hert 72], [Lieb 75a], [Narn 82, 91], bibliography

Thomas, L. H.
See [Baka 53]

Todorov, I. T.
See [Buch 88], [Hadj 90], bibliography: Bogolubov

Trych–Pohlmeyer, E. B.
See [Haag 74, 77]

Uhlenbeck, G. E.
See [Kahn 38], [Lap 31]

Uhlmann, A.
[Uhl 62] *Über die Definition der Quantenfelder nach Wightman und Haag.* Wiss. Zeits. Karl Marx Univ. **11**, 213 (1962) — VIII.3.2

Unruh, W. G.
[Unr 76] *Notes on black hole evaporation,* Phys. Rev. **D14**, 870 (1976) — V.4.1, VIII.3.3

Ursell, H. P.
[Urs 27] *The evaluation of Gibbs' phase-integral for imperfect gases.* Proc. Cambr. Phil. Soc. **23**, 685 (1927) — II.2.2

Varadarajan, V. S.
See bibliography

de Vega, H. J.
[de Vega 90] *Integrable theories, Yang-Baxter algebras and quantum groups.* Advanced Studies in Pure Mathematics **19**, 567 (1989) — IV.5

Verbeure, A.
See [Fann 77a, 77b]

Verch, R.
See [Buch 96]
[Verch 94] *Local definiteness, primarity and quasiequivalence of quasifree Hadamard quantum states in curved spacetime.* Commun. Math. Phys. **160**, 507 (1994) — VIII.3.2

Verlinde, E.
[Ver 88] *Fusion rules and modular transformations in 2D conformal field theory.* Nucl. Phys. **B300**, 360 (1988) IV.5
Vilenkin, N.
See bibliography: Gelfand
Vinberg, E. B.
[Vin 65] *The structure of the group of automorphisms of a homogeneous convex cone.* Trans. Mosc. Math. Soc. **13**, 63 (1965) I.1
Voelkel, A. H.
See [Swieca 73]
Volkov, D. V.
[Volk 65] *S-Matrix in the generalized quantisation method.* Sov. Phys. JETP **11**, 375 (1965) IV.1
Van der Waerden, B.
See bibliography
Wald, R. M.
See [Full 81a, 81b], [Kay 87], bibliography
[Wald 77] *The back reaction effect in particle creation in curved spacetime.* Commun. Math Phys. **54**, 1 (1977) VIII.3.2
[Wald 78] *Trace anomaly of a conformally invariant quantum field theory.* Rev. Phys. **D17**, 1477 (1978) VIII.3.2
Warner, F.
See bibliography
Weinberg, S.
See bibliography
v. Weizsäcker, C. F.
See bibliography
[Weiz 73] *Probability and quantum mechanics.* Br. J. Phil. Sci. **24**, 321 (1973) VII.3
Wentzel, G..
See bibliography
Wenzl, H.
[Wenzl 88] *Hecke algebras of type A_n and subfactors.* Invent. Math. **92**, 349 (1988) IV.5
Werner, R.
See [Summ 87b]
Wess, J.
See bibliography
[Wess 74] – and Zumino, B. *Supergauge transformations in four dimensions.* Nucl. Phys. B **70**, 39 (1974) VIII.2
Wheeler, J. A.
See bibliography
Wichmann, E. H.
See [Bis 75, 76], [Buch 86], [Driess 86]

Wick, J. C.

[Wick 50] *The evaluation of the collision matrix.* Phys. Rev. **80**, 268 (1950) I.5.2

[Wick 52] – Wightman, A. S. and Wigner, E. H. *The intrinsic parity of elementary particles.* Phys. Rev. **88**, 101 (1952) III.1

Wightman, A. S.

See [Wick 52], [Gard 54a, 54b], bibliography: Streater

[Wight 56] *Quantum field theories in terms of vacuum expectation values.* Phys. Rev. **101**, 860 (1956) II.2.1

[Wight 57] *Quelque problèmes mathématique de la théorie quantique relativiste.* In: Lecture notes, Faculté des Sciences, Univ. de Paris 1957 and in: *Les problèmes mathématique de la théorie quantique des champs* (Lille 1957). Paris: Centre National de la Recherche Scientifique 1959 II.1.2

[Wight 64] – and Garding,L. *Fields as operator valued distributions in relativistic quantum theory.* Ark. Fys. **23**, Nr. 13 (1964) II.1.2

Wigner, E. P.

See [Jord 34], [New 49], [Wick 52], bibliography

[Wig 39] *On unitary representations of the inhomogeneous Lorentz group.* Ann. Math. **40**, 149 (1939) I.3.1, I.3.2

[Wig 63] *Remarks on the mind-body question.* In: Symmetries and Reflections. Bloomington: Indiana Univ. Press 1967 VII.1

[Wig 78] *Interpretation of Quantum Mechanics.* In: Quantum Theory of Measurement. J. A. Wheeler and W. H. Zurek (eds.) VII.3

Wilczek, F.

[Wilc 82] *Quantum mechanics of fractional spin particles.* Phys. Rev. Lett **49**, 957(1982) IV.5

Wilde, I. F.

See [Streat 70]

Winnink, M.

See [Haag 67]

de Witt, B. S.

[de Witt 80] *Quantum gravity: the new synthesis.* In: *General Relativity.* Hawking, S. W. and Israel, W. (eds.) V.4.1

Wollenberg, M.

[Wol 85] *On the relation between quantum fields and local algebras of observables.* Rep. Math. Phys. **22**, 409 (1985)

Woods, E. J.

See [Ara 63b]

Woronowicz, S. I.

See [Sad 71], [Pusz 78]

[Wor 87] *Compact matrix pseudogroups.* Commun. Math. Phys. **111**, 613 (1987) IV.5

Yamagami, S.
See [Ara 82]
Yngvason, J.
See [Bor 91]
Zimmermann, W.
See [Leh 55], [Leh 57], [Glas 57], [Bor 64]
[Zi 58] *On the bound state problem in quantum field theory.* Nuovo Cimento **10**, 597 (1958) II.4.4
[Zi 70] In: Brandeis Univ. Summer Inst. in Theoret. Phys., Vol. 1, Deser, Grisam, Pendleton (eds.), MIT–Press 1970 II.2.4
Zsido, L.
See bibliography: Stratila
Zuber, J. – B.
See bibliography: Itzykson
Zumino, B.
See [Wess 74]
Zurek, W. H.
See bibliography

Subject Index